INTRODUCTION TO GENOMICS

SECOND EDITION

Arthur M. Lesk
The Pennsylvania State University

Immured the whole of Life
Within a magic Prison
– Emily Dickinson

OXFORD
UNIVERSITY PRESS

Great Clarendon Street, Oxford OX2 6DP

Oxford University Press is a department of the University of Oxford.
It furthers the University's objective of excellence in research, scholarship,
and education by publishing worldwide in

Oxford New York

Auckland Cape Town Dar es Salaam Hong Kong Karachi
Kuala Lumpur Madrid Melbourne Mexico City Nairobi
New Delhi Shanghai Taipei Toronto

With offices in

Argentina Austria Brazil Chile Czech Republic France Greece
Guatemala Hungary Italy Japan Poland Portugal Singapore
South Korea Switzerland Thailand Turkey Ukraine Vietnam

Oxford is a registered trade mark of Oxford University Press
in the UK and in certain other countries

Published in the United States
by Oxford University Press Inc., New York

First edition 2007

British Library Cataloguing in Publication Data
Data available

Library of Congress Cataloging in Publication Data
Data available

Typeset by Graphicraft Limited, Hong Kong
Printed in Italy on acid-free paper by L.E.G.O. S.p.A. – Lavis TN

ISBN 978–0–19–956435–4

10 9 8 7 6 5 4 3 2

For Victor and Valerie

PREFACE TO THE FIRST EDITION

Of all the claims on our curiosity, we want most to understand ourselves. What are we? What lies in our future? Many features of our lives depend on accidents of history. The time and place of our birth largely determine what language we first learn to speak and whether we are likely to be well-fed and well-educated and receive adequate medical care. Many aspects of our future depend on events outside ourselves and beyond our control.

Within ourselves, also, there are constraints on our lives that brook relatively little argument. In some respects, we are at the mercy of our genomes. Under normal circumstances, all of our basic anatomy and physiology, and eye colour, height, intelligence and basic personality traits, are ingrained in our DNA sequences. This is not to say that our genomes dictate our lives. Some of the constraints are tight – eye colour, for instance – but our genetic endowment also confers on us a remarkable robustness.

This robustness also is a product of evolution. When Shakespeare wrote of 'the thousand natural shocks that flesh is heir to', he coupled the challenges of life to heredity. Within the last century, lifestyles have changed with a rapidity hitherto unknown (except for the instants of asteroid impacts). Our talents have many opportunities to nurture themselves and develop in novel ways, and we can meet and survive brutal stresses. These are gifts of our genomic endowment: *What genes control is the response of an organism to its environment.*

The human genome is only one of the many complete genome sequences known. Taken together, genome sequences from organisms distributed widely among the branches of the tree of life give us a sense, only hinted at before, of the very great unity *in detail* of all life on Earth. This recognition has changed our perceptions, much as the first pictures of the Earth from space engendered a unified view of our planet.

Of course, superimposed on this basic unity is great variety. We ask: What is special about us? What do we share with our parents and siblings and how do we differ from them? What do we share with all other human beings and what makes us different from the other members of our species? What are the sources of our differences from our closest extant non-human relatives, the chimpanzees? What do we have in common and how do we diverge from other species of primates? of mammals? of vertebrates? of eukaryotes? of all other living things?

The complete sequences of human and other genomes give us complete information about the underlying text of this story. We are beginning to understand how our lives shape themselves under the influence of our genes plus our surroundings.

We are also beginning to intervene. Genetic engineering of microorganisms is an established technique. Genetically modified plants and animals exist and are the subjects of lively debate. To override the genes for hair colour is trivial. Changes in lifestyle or behaviour can – to some extent – avoid or postpone development of diseases to which we are genetically prone. Gene therapy offers the promise of rectifying some inborn defects.

In this book, we shall explore this new knowledge, what it tells us about ourselves and how we can apply it. With power derived from knowledge goes commitment to act wisely. We have responsibilities, to ourselves, to other people, to other species and to ecosystems ranging in size up to the entire biosphere.

Ethical, legal and social issues have been a prominent component of the human genome project. Most technical questions in genomics, as in other scientific subjects, have objectively correct answers. We do not know all the answers, but they are out there for us to discover. Ethical, legal and social issues are different. Many choices are possible. Their selection is not the privilege of scientists in individual laboratories, but of society as a whole. Scientists do have a responsibility to contribute to the informed public discussion that is essential for wise decisions.

One problem encountered in writing about genomics is the need to pick and choose from the many riches of the subject. The list of subjects that cannot be left out is too long and threatens to reduce the treatment of each to superficiality. There is also a serious organizational challenge: many phenomena must be approached from several different points of view. A reader may be relieved to conclude that a topic has been beaten thoroughly into submission in one chapter, only to encounter it again, alive and kicking, in a different context.

The speed at which the field is moving causes other problems. One is often pleased with a draft of a section only to find the carefully described conclusions modified in next week's journals. Yet, there is a great pleasure in seeing Nature's secrets emerging before one's eyes.

Another casualty of rapid progress is a loss of interest in history and biography. We are fantastically interested in the development of the sea urchin and the fruit fly, but not at all in the development of molecular genomics. Intellectual struggles that occupied entire careers leave behind only terse conclusions, often without any appreciation of the experiments that established the facts, much less of the alternative hypothesis tested and rejected. The force of the scientists' personalities, and their foibles, are forgotten. This is too bad: those who do not learn from the successes of history will find it harder to emulate them.

Genomics is an interdisciplinary subject. The phenomena we want to explain are biological. But many fields contribute to the methods and the intellectual approaches that we bring to bear on the data. Physicists, mathematicians, computer scientists, engineers, chemists, clinical practitioners and researchers, have all joined in the enterprise. This book will appeal, at least in part, to all of them. However, the central point of view remains focused on the biology.

More specifically, the focus is on human biology. In fact, on the biology of humans who are curious about other species, albeit primarily for what the other species tell us about ourselves. This choice naturally reflects the potential readership of this book. (If bacteria or fruit flies could read, genomics textbooks would look very different.)

This book assumes that the reader already has some acquaintance with modern molecular biology, and builds on and develops this background, as a self-contained presentation. It is suitable as a textbook for undergraduates or starting postgraduate students.

Exercises, problems and 'weblems' at ends of chapters test and consolidate understanding and provide opportunities to practise skills and explore additional subjects. Exercises are short and straightforward applications of material in the text. Answers to exercises appear on the web site associated with the book. Problems, also, make use of no information not contained in the text, but require lengthier answers or in some cases calculations. The third category, 'weblems', require access to the World Wide Web. Weblems are designed to give readers practice with the tools required for further study and research in the field.

PREFACE TO THE SECOND EDITION

Fast, inexpensive sequencing has transformed genomics. The landmark goal, the $US1000 human genome, will likely soon be achieved. At the time of writing, thousands of human individuals have had their full genomes sequenced, and many more are on the way. A very large number of people have had sequences determined for individual genes. For example, mutations in BRCA1 suggest an increased likelihood of developing breast or ovarian cancer.

Genetic testing for disease is one of many possible fields of application of genomics. Clinical medicine heads the list; no one doubts that it was the promise of improvements in health that motivated support for the original human genome project. Understanding the relationships between genes and disease will allow more precise diagnosis and warnings of increased risk of disease in patients and their offspring. It will allow design of treatment tailored to the biochemical characteristics of the patient, called pharmacogenomics. Genomes of other organisms also have implications for human health, especially those of pathogenic organisms that have developed, or are threatening to develop, antibiotic resistance. Other applications of genomics include improvement of crops and domesticated animals, enhancing food production, and support of conservation efforts dedicated to preserving endangered species. Development of alternative energy sources is a challenge to both physics and biology.

Underlying these applications, genomics offers us a profound understanding of fundamental principles of biology. On the personal level, genome exegesis will fundamentally alter our perception of ourselves: what does it mean to be human? The answer lies somewhere in the complex interplay of our genes and life histories. For some characteristics, there is a simple answer. Your eye colour, and whether or not you suffer from sickle-cell anaemia, depend exclusively on the sequences of particular genes. For most of your phenotypic traits, however, the assignment of contributions to their origin from genome, epigenetics, and life history is a severe challenge.

In this book, I have tried to present a balanced view of the background of the subject, the technical developments that have so greatly increased the data flow, the current state of our knowledge and understanding of the data, and applications to medicine and other fields. One aspect of the first edition that I liked was its concision. Unfortunately, this has had to be sacrificed to the stampeding progress of the field.

PLAN OF THE SECOND EDITION

Chapter 1, *Introduction to Genomics*, sets the stage, and introduces all of the major players: DNA and protein sequences and structures, genomes and proteomes, databases and information retrieval, and bioinformatics and the World Wide Web. Subsequent chapters develop these topics in detail. Chapter 1 briefly provides the framework of how they fit together and sets them in their context of biomedical, physical, and computational sciences.

Chapter 2 demonstrates that *Genomes are the Hub of Biology*. Whereas genome sequences are determined from individuals, to appreciate life as a whole requires extending our point of view spatially, to populations and interacting populations; and temporally, to consider life as a phenomenon with a history. We can study the characteristics of life in the present, we can determine what came before, and we can – at least to some extent – extrapolate to the future. The 'central dogma' and the genetic code underly the implementation of the genome, in terms of the synthesis of RNAs and proteins. Absent from Crick's original statement of the central dogma is the crucial role of regulation in making cells stable and robust, two characteristics essential for survival.

Chapter 3, *Mapping, Sequencing, Annotation, and Databases*, describes how genomics has emerged from classical genetics and molecular biology. The first nucleic acid sequencing, by groups led by F. Sanger and W. Gilbert, in the 1970s, were a breakthrough comparable to the discovery of the double helix of DNA. The challenges of sequencing stimulated spectacular improvements in technology. The first was the automation of the Sanger method. The original sequences of the human genome were accomplished by batteries of automated Sanger sequencers. Subsquently a series of 'new generations' of novel approaches have brought the landmark goal, the $US1000 human genome, within reach. Where do all the data go? Chapter 3 also introduces the databanks that archive, curate, and distribute the data, and some of the information-retrieval tools that make them accessible to scientific enquiry.

Chapter 4, *Comparative Genomics*, begins with a general survey of the different modes of genome organization with which living things have experimented. With complete sequences of genomes of many different species, we can confront and compare them. But, of course, different individuals of a species do not necessarily have identical genomes. What is the nature and extent of the variability? With intraspecies variability as a baseline, what are the similarities and differences between genomes of different species? Comparative genomics thereby allows us to address a question central not only to this book but to the field as a whole: what does it mean to be human?

Chapter 5, *Evolution and Genomic Change*, relates genomics to evolution, a major unifying principle of biology. (Arguably the laws of thermodynamics are another.) T. Dobzhansky famously said: 'Nothing in biology makes sense except in the light of evolution.' Description of some of the important ideas and tools – of taxonomy and phylogeny, on the classical species level and on the molecular level – will be useful in organizing the material in subsequent chapters.

Chapter 6, *Genomes of Prokaryotes*, surveys the genomes of bacteria and archaea in more detail. Taxonomy and phylogeny of prokaryotes present problems because of extensive horizontal gene transfer. This challenges the whole idea of a hierarchy of biological classification. Many bacteria have been cloned and studied in isolation, especially those responsible for disease. However, a new field, metagenomics, deals with the entire complement of living things in an environmental sample, allowing us to address questions about interspecies interaction in the 'real world'. Sources include ocean water, soil, and the human gut.

Chapter 7 surveys *Genomes of Eukaryotes*. It starts with yeast, which is about as simple as a eukaryote can get. Selected plant, invertebrate, and chordate genomes illuminate the many profound common features of eukaryotic genomes; and the very great variety of structures, biochemistry, and lifestyles that are compatible with the underlying similarities.

Chapter 8, *Genomics and Human Biology*, develops applications to the study of our own species. Although clinical applications are undoubtedly the most important, genomics has important contributions to make to human palaeontology, anthropology, and the law. The ability to extract DNA from extinct species, including Neanderthals, sheds light on our early evolution. Events in our history, including migrations and domestication of crops and animals, have left their traces in DNA sequences.

Chapter 9 deals with *Transcriptomics*, the measurement and application of protein expression patterns. These measurements have been carried out using microarrays. However, as sequencing technology grows in power, it may replace microarrays as the method of choice for these measurements. Applications treated include changes in different physiological states, such as the diauxic shift in yeast, or sleep and waking in rats; in plant and animal development; and in diagnosis and treatment of disease.

Chapter 10, *Proteomics*, describes the principles of protein structure and the high-throughput data streams that provide information about sets of proteins in cells. Proteomics is an essential complement to genomics. The interactions and relationships between the genome and proteome are intimate, both during cellular activity and in the longer term in evolution. (A colleague once entitled a keynote lecture: 'Genes are from Venus, proteins are from Mars.')

The last chapter emphasizes attempts to integrate our data and understanding, an area known as systems biology. Whereas classical biochemistry made great contributions to demonstrating the properties of proteins in isolation, our job now is to put things back together. Chapter 11, *Systems Biology*, presents a description of biological organization that is based on networks. Cells contain parallel sets of networks based on **physical** and **logical** interactions among molecules. Each network also has **static** and **dynamic** aspects. The ultimate, most profound, goal is a complete and integrated picture of all of life's activity, from the molecule to the biosphere.

NEW TO THIS EDITION

The most important change since publication of the first edition has been the spectacular progress in high-throughput sequencing. The resulting growth of the data produced – in quality, quantity, and type – have altered the entire landscape of genomics itself, and its influence has invaded surrounding fields. No area of biology has been left unscathed.

The new edition reflects this. Many more complete genomes are available. Instead of a few isolated snapshots of the evolution, we can trace its pathways through the phyla. Instead of sequencing single individuals, it is possible to measure directly the variation within populations. Palaeogenomics has opened a window onto extinct species.

High-throughput sequencing has created novel data streams, such as RNAseq to measure the transcriptome. The new techniques complement, and may for some purposes even supersede, some common experimental techniques such as microarrays.

Study of human disease through sequencing continues to be a major effort. Identification of genes responsible for particular diseases permits testing, genetic counselling, and risk assessment and avoidance. For many important diseases, a patient can expect more precise diagnosis and prognosis, and more precise recommendations for treatment. Cancer genomics – the comparison of sequences from normal and tumour cells from single patients – has become a major activity.

In the new edition, extended coverage is given both to the applications of genome sequences to working out of evolutionary relationships in microorganisms, plants and animals; and to clinical applications to humans. Non-clinical applications to human biology and history are sufficient to justify a chapter of their own. The genomics of crop domestication not only shed light on human – as well as plant – history, but emphasize the reciprocal interactions between humans and the rest of the biosphere.

Nevertheless, progress is happening so fast as to make unavoidable the feeling of frustration in aiming at a moving target. The hope is that the second edition has erected for the reader a sound framework, both intellectual and factual, that will make it possible, when encountering subsequent developments, to see where and how they fit in.

RECOMMENDED READING

Where else might the interested reader turn? This book is designed as a companion volume to three others: *Introduction to Protein Architecture: the Structural Biology of Proteins*; *Introduction to Protein Science: Architecture, Function, and Genomics*; and *Introduction to Bioinformatics* (all published by Oxford University Press). Of course there are many fine books by many authors, some of which are listed as recommended reading at the ends of the chapters. The goal is that each reader will come to recognize his or her own interests, and be equipped to follow them up.

Many applications of genomics to health care are discussed in the book. However, nothing here should be taken as offering medical advice to anyone about any condition.

INTRODUCTION TO GENOMICS ON THE WEB

Results and research in genomics make use of the web, both for storage and distribution of data, and methods of analysis. Readers will need to become familiar with web sites in genomics, and to develop skills in using them. Many useful sites are mentioned in the book. The author's *Introduction to Bioinformatics* offers a pedagogical approach to computational aspects of genomics. However, clearly the place to learn about the web is on the web itself.

To this end, an Online Resource Centre at www.oxfordtextbooks.co.uk/orc/leskgenomics2e/ accompanies this book. This contains material from the book – figures and 'movies' of the pictures of structures, answers to exercises, and hints for solving problems. In addition it contains a guided tour of web sites in genomics, coordinated with the printed book, with additional exercises and problems (of the 'weblem' variety). Some of these are suitable for use as practical or laboratory assignments.

ACKNOWLEDGEMENTS

I thank S. Ades, G.F. Anderson, M.M. Babu, S.L. Baldauf, P. Berman, B. de Bono, D.A. Bryant, C. Cirelli, A. Cornish-Bowden, N.V. Fedoroff, J.G. Ferry, R. Flegg, J.R. Fresco, D. Grove, R. Hardison, E. Holmes, H. Klein, E. Koc, A.S. Konagurthu, T. Kouzarides, H.A. Lawson, E.L. Lesk, M.E. Lesk, V.E. Lesk, V.I. Lesk, D.A. Lomas, B. Luisi, P. Maas, K. Makova, W.B. Miller, C. Mitchell, J. Moult, E. Nacheva, A. Nekrutenko, G. Otto, A. Pastore, D. Perry, C. Praul, K. Reed, G.D. Rose, J. Rossjohn, S. Schuster, B. Shapiro, J. Tamames, A. Tramontano, A.A. Travers, A. Valencia, G. Vriend, L. Waits, J.C. Whisstock, A.S. Wilkins and E.B. Ziff for helpful advice.

I thank the staff of Oxford University Press for their skills and patience in producing this book.

CONTENTS

Collect a sample of your cells by rinsing out your mouth with dilute salt water. Add some detergent to dissolve the cell and nuclear membranes, releasing the contents. Precipitate out the proteins by adding alcohol; stir, and let settle.

In a few minutes, fibres will rise out of the murk to the top of the vessel.

These fibres are your DNA (plus that of some bacteria that were living in your mouth). The sequence of your DNA is your lifelong endowment. It determined that you are human, that you are male or female, that you are brown- or blue-eyed, that you are right- or left-handed. But think of these fragile cables not as a leash that constrains you but as the cords that raise the curtains on the drama of your life.

–*The Slurry with the Fringe on the Top*

CHAPTER 1

Introduction to Genomics

LEARNING GOALS

- Knowing the basic facts about the human genome – how many base pairs it contains, estimates of how many genes it contains that code for proteins or RNAs.
- Recognizing the contributions to any individual's phenotype from the genome sequence itself, from life history, and from epigenetic signals within the fertilized egg.
- Appreciating that the human genome contains extensive repetitive regions of various kinds.
- Knowing the basic central dogma, that DNA is transcribed to RNA, which is translated to protein. Beyond this, knowing that many protein-coding genes show variable splicing, which adds another dimension of complexity to the scheme by which the genome specifies the proteome.
- Understanding the importance of comparative genome sequencing projects, to reveal processes of evolution, and to help interpret regions in the human genome.
- Appreciating the large number of genome projects treating different species, widely distributed among life forms, and including metagenomics, which produces very large amounts of data from environmental samples.
- Understanding how different types of human DNA sequencing projects are organized, including those carried out by large international organizations, the collection of data by law-enforcement organizations, those carried out for specific clinical tests, and direct-to-the-public sequencing usually motivated by questions of genealogy.
- Distinguishing the several types of potential applications of genome sequence data to medicine, including ways in which information about large numbers of people can support clinical research, and ways in which information about specific individuals can help treat disease more effectively, or – even better – prevent it.
- Understanding the importance of computer science and bioinformatics in producing the raw sequence data, in creating databases in molecular biology, in archiving and careful curation of the data, in distributing them via the web, and in creating information-retrieval tools to allow effective mining of the data for research and applications.
- Appreciating the ethical, legal, and social implications of collection of DNA sequence data and the conflicting demands of public safety and individual privacy.

The human genome

A human genome contains approximately 3.2×10^9 base pairs, distributed among 22 paired chromosomes, plus two X chromosomes in females and X and Y in males. The first human genomes were determined in 2001, the culmination of 10 years of pioneering work and dedication. Since then, advances in technology have made genomic sequencing cheaper and faster. Sequence data now flow copiously. This creates the challenges of understanding the information that our genomes contain, and applying the data and analysis to improve human welfare. Sequencing genomes of other species both facilitates these goals and extends them, by revealing general principles of biology.

How do the contents of our genomes determine who we are?

Phenotype = genotype + environment + life history + epigenetics

Each reader of this book is an individual, with physical, biochemical, and psychological characteristics. (Do not be surprised if these distinctions become more and more nebulous!) Each of you has a general form and metabolism that is common to all humans. At the molecular level, you have much in common with other species as well. But there is also substantial variation within our species, to give you your individual appearance and character. You are in a state of health somewhere within the spectrum between robust good health and morbid disease. You are currently in some psychological state, and in some mood, reflecting your personality and current activities.

- Your **genotype** is your DNA sequence, both nuclear and mitochondrial. (For plants, include also the sequence of the chloroplast DNA.)
- Your **phenotype** is the collection of your observable traits, other than your DNA sequence. These include macroscopic properties such as height, weight, eye and hair colour; and microscopic ones, such as possible sickle-cell anaemia, glucose-6-phosphate deficiency, or the retention beyond infancy of the ability to digest lactose.

  Of great importance to clinical applications of genomics are traits that govern susceptibility to disease and risk factors; and those that determine the effectiveness of different drugs in different individuals. These allow for personalized prevention and treatment of disease based on DNA sequences, or **pharmacogenomics**.
- Your **life history** includes the integrated total of your experiences, and the physical and psychological environment in which you developed. Your nutritional history has influenced your physical development. A nurturing environment and educational opportunities have influenced your mental development. Less obvious than most aspects of your life history is the growing recognition of the importance of your *in utero* environment in determining your development curve and even your adult characteristics.
- At the interface between the genome and life experience are **epigenetic factors**. It is largely true that all cells of your body, except sperm or egg cells and cells of the immune system, have almost the same DNA sequence (subject to usually modest amounts of accumulated mutations). Yet your tissues are differentiated, with different sets of genes expressed or silenced in liver, brain, etc.

  Now, some of these regulatory signals survive cell division. (When a liver cell divides, it divides into two liver cells.) Your parents' own life histories might have altered the epigenetic patterns in their cells, and the fertilized egg from which you were subsequently formed contained some of these 'pre-differentiation' signals. In this way inheritance of acquired characteristics has re-entered respectable mainstream biology.

The relative importance of these factors in determining your phenotype varies from trait to trait. Some are determined solely and irrevocably by your alleles for specific genes. Others depend on complex interactions between your genes and your life history, and epigenetic signals from your parents.

A genome is like a page of printed music. The page is a fixed physical object, but the notes are consistent with realizations, in time and space, in a variety of ways – a limited *variety of ways.*

Thus a genome constrains but does not dictate the features of an organism. Varied surroundings and experience lead organisms to explore different states consistent with their genomes. Even in microorganisms, expression of many genes is conditional. Synthesis of enzymes for lactose utilization in *E. coli* depends on the presence of substrate in the medium. The shift of yeast between anaerobic and aerobic metabolism is a *temporary and reversible* change in physiological state in response to changes in environmental conditions.

Most people would not regard simple and transient alterations in appearance that have no genetic component at all as true changes in phenotype. Trimming one's nails, getting a haircut, and wearing makeup are common examples.

Conversely, many environmental effects, thought of as lifestyles, have long-term effects on development. Exercise enhances musculature. Education enhances intellectual development. A phenylalanine-controlled diet can prevent the harmful effects of the metabolic disease phenylketonuria. Conversely, malnutrition, disease, injury, and cruel treatment can be physically and mentally debilitating.

Science has provided a variety of means of interposition between genotype and phenotype, widely applied and sometimes abused. Examples include simple changing of hair colour, tattooing, cosmetic surgery and 'cosmetic endocrinology' (including both the treatment of children deficient in human growth hormone and the use of performance-enhancing drugs by athletes), and even – at least arguably – psychoanalysis.

Some environmental effects have specific windows of responsiveness to give results that are subsequently irreversible. Growth hormone treatments and surgery to produce the famous castrati as opera singers are examples. In turtles and other reptiles, incubation temperature controls whether eggs develop into males or females. The effect depends on the temperature dependence of the expression of the enzyme aromatase, which converts androgens to oestrogens. (The corresponding enzyme in humans is the target of inhibitors used in treatment of cancers that are stimulated to grow by oestrogen, to reduce the exposure of the tumour to oestrogen.)

Given the variability of the contributions of heredity, environment, life history, and epigenetics in determining phenotype, how can we measure their relative importance for any particular trait? The classic method of distinguishing effects of genetics from those of surroundings and experience – 'nature' and 'nurture' – is controlled experiments with genetically identical organisms. For human beings, this means monozygotic twins. (Monozygotic human triplets are extremely rare.) Many studies have compared the similarities of identical twins reared apart, individuals who have the same genes but different environments. Less well-controlled are comparisons of identical with fraternal twins, or of non-twin siblings with adopted children reared in the same family. There have been suggestions that fraternal twins share greater similarity than non-twin siblings as a result of the shared uterine experience.

Human twin and sibling studies to distinguish hereditary components of traits have a long and contentious history, as they have formed the basis for important social decisions. Intelligence quotient, or IQ, is the ratio of mental age to chronological age. When properly measured, it remains constant in early childhood. Objective and quantitative measurement of intelligence in adults is extremely difficult. Attempts to measure inheritance of intelligence – or even of score on intelligence quotient tests; not the same thing if the tests are culturally biased – have produced controversial and unsatisfying results. It has proved very difficult to devise tests free of socioeconomic bias.*

- Phenotypic traits – both macroscopic and molecular – depend on a combination of influences from genome sequences themselves, the individual's life history, and epigenetic signals in the fertilized egg.

* See Gould, S.J. (1996). *The Mismeasurement of Man.* W.W. Norton, New York.

Contents of the human genome

A walk through the human genome is like a tour of a continent. One encounters centres of bustling activity, rich in genes and their regulatory elements. These are like villages and even cities. One passes also through large tracts of emptiness, or regions with unrelieved monotony of repeated elements.

An inventory of the human genome includes:

- The most prominent and familiar aspects of the genome, the regions that code for proteins. Protein-coding genes are transcribed into messenger RNA (mRNA). After processing, ribosomes translate mature mRNA to polypeptide chains.

> - Francis Crick encapsulated this scheme in the Central Dogma of Molecular Biology:
>
> DNA makes RNA makes Protein

Each triplet of nucleotides in the message corresponds to one amino acid, according to the genetic code (see Box 1.1). Box 1.2 lists the 20 canonical amino acids.

Despite their importance, protein-coding genes occupy a small fraction of the human genome – no more than about 2–3% of the overall sequence. They are distributed across the different chromosomes, but not evenly. Many protein-coding genes appear in multiple copies, either identical or diverged into families. For instance, humans have over 900 related olfactory-receptor genes, and some animals have many more.

- Some regions of the genome encode non-protein-coding RNA molecules (that is, RNAs exclusive of messenger RNAs), including but not limited to transfer RNAs, the RNA components of ribosomes, and microRNAs and small interfering RNAs that regulate translation (miRNAs and

Protein synthesis

Transcription of a protein-coding gene into RNA is followed in eukaryotes by splicing to form a mature messenger RNA (mRNA) molecule. The ribosome synthesizes a polypeptide chain according to the sequence of triplets of nucleotides, or **codons**, in the mRNA. The protein folds spontaneously to a native three-dimensional structure that accounts for its biological function.

The standard genetic code is shown here. The codons are those appearing in DNA rather than RNA; that is, the codons contain t rather than u. Both three-letter and one-letter abbreviations for the amino acids appear. Note that the code is redundant: except for Met and Trp, multiple codons specify the same amino acid. A mutation that changes a codon to another codon for the same amino acid is called a **synonymous mutation**. Three triplets are reserved as STOP signals, effecting termination of translation. Variations from this standard code occur in mitochondria and chloroplasts, and sporadically in individual species.

The standard genetic code

ttt	Phe	F	tct	Ser	S	tat	Tyr	Y	tgt	Cys	C
ttc	Phe	F	tcc	Ser	S	tac	Tyr	Y	tgc	Cys	C
tta	Leu	L	tca	Ser	S	taa	STOP		tga	STOP	
ttg	Leu	L	tcg	Ser	S	tag	STOP		tgg	Trp	W
ctt	Leu	L	cct	Pro	P	cat	His	H	cgt	Arg	R
ctc	Leu	L	ccc	Pro	P	cac	His	H	cgc	Arg	R
cta	Leu	L	cca	Pro	P	caa	Gln	Q	cga	Arg	R
ctg	Leu	L	ccg	Pro	P	cag	Gln	Q	cgg	Arg	R
att	Ile	I	act	Thr	T	aat	Asn	N	agt	Ser	S
atc	Ile	I	acc	Thr	T	aac	Asn	N	agc	Ser	S
ata	Ile	I	aca	Thr	T	aaa	Lys	K	aga	Arg	R
atg	Met	M	acg	Thr	T	aag	Lys	K	agg	Arg	R
gtt	Val	V	gct	Ala	A	gat	Asp	D	ggt	Gly	G
gtc	Val	V	gcc	Ala	A	gac	Asp	D	ggc	Gly	G
gta	Val	V	gca	Ala	A	gaa	Glu	E	gga	Gly	G
gtg	Val	V	gcg	Ala	A	gag	Glu	E	ggg	Gly	G

BOX 1.2

The twenty standard amino acids in proteins

Non-polar amino acids

G glycine A alanine P proline V valine
I isoleucine L leucine F phenylalanine M methionine

Polar amino acids

S serine C cysteine T threonine N asparagine
Q glutamine Y tyrosine W tryptophan

Charged amino acids

D aspartic acid E glutamic acid K lysine R arginine
H histidine

Amino acid names are frequently abbreviated to their first three letters – for instance Gly for glycine – except for isoleucine, asparagine, glutamine, and tryptophan, which are abbreviated to Ile, Asn, Gln, and Trp, respectively. The rare amino acid selenocysteine has the three-letter abbreviation Sec and the one-letter code U. The even rarer amino acid pyrrolysine has the three-letter abbreviation pyl and the one-letter code O.

Amino acid sequences are always stated in order from the N-terminal to the C-terminal. This is also the order in which ribosomes synthesize proteins: ribosomes add amino acids to the free carboxy terminus of the growing chain.

siRNAs). There are about 3000 genes coding for RNAs, exclusive of the mRNAs translated to proteins. It is becoming clear that the RNA-ome is much richer than had been suspected. Except for RNAs involved in the machinery of protein synthesis, such as transfer RNAs and the ribosome itself, most non-coding RNAs are involved in control of gene expression.

The regions that encode proteins and non-protein-coding RNAs correspond to molecules that form non-transient parts of the cell's contents. They do of course 'turn over', but at rates lower than messenger RNAs. mRNAs must have short lifetimes in order to turn off transcription, as part of the processes that regulate gene expression.

- Other regions contain binding sites for ligands responsible for regulation of transcription. In assessing the total amount of the genome dedicated to control, one would need to include both the regulatory sites themselves, and all the proteins and RNAs encoded that have regulatory functions, arguably including receptors.
- Repetitive elements of unknown function account for surprisingly large fractions of our genomes. Long and Short Interspersed Elements (LINES and SINES) account for 21% and 13% of the genome. Even more-highly repeated sequences – minisatellites and microsatellites – may appear as tens or even hundreds of thousands of copies, in aggregate amounting to 15% of the genome. (See Box 1.3).

- We know functions of some regions of the genome. That we cannot assign functions to others may merely reflect our ignorance. Some regions appear to have a life of their own, hitchhiking and reproducing within genomes, and contributing to evolution. In some cases they actively enhance rates of chromosomal rearrangements.

BOX 1.3

Repetitive elements in the human genome

Moderately repetitive DNA

- Functional
  - dispersed gene families, created by gene duplication followed by divergence
    - e.g. actin, globin
  - tandem gene family arrays
    - rRNA genes (250 copies)
    - tRNA genes (50 sites with 10–100 copies each in human)
    - histone genes in many species

→

- Without known function
  - short interspersed elements (SINEs)
    - Alu is an example
    - 200–300 bp long
    - 100 000's of copies (300 000 Alu)
    - scattered locations (not in tandem repeats)
  - long interspersed elements (LINEs)
    - 1–5 kb long
    - 10–10 000 copies per genome
  - pseudogenes

Highly repetitive DNA

- minisatellites
  - composed of repeats of 14–500 bp segments
  - 1–5 kb long
  - many different ones
  - scattered throughout the genome
- microsatellites
  - composed of repeats of up to 13 bp
  - ~100s of kb long
  - ~10^6 copies/genome
  - most of the heterochromatin around the centromere
- telomeres
  - 250–1000 repeats at the end of each chromosome
  - contain a short repeat unit (typically 6 bp: TTAGGG in human genome, TTGGGG in *Paramecium*, TAGGG in trypanosomes, TTTAGGG in *Arabidopsis*)

Genes that encode the proteome

It is now believed that the human genome contains about 23 000 protein-coding genes. Some regions of the genome are relatively poor in protein-coding genes. These include the subtelomeric regions, on all chromosomes, and chromosomes 18 and X. In contrast, chromosomes 19 and 22 are relatively rich in protein-coding genes.

Most human protein-coding genes contain exons (expressed regions) interrupted by introns (regions spliced out of mRNA and not translated to protein). The average exon size is about 200 bp. It is primarily the variability in intron size that causes the large size differences among protein-coding genes: the gene for insulin is 1.7 kb long, the LDL receptor gene is 5.45 kb, and the dystrophin gene is 2400 kb.

Genes appear on both strands. In many cases, unrelated genes are fairly well separated. However, there are examples of genes that partially overlap; and cases of an entire gene appearing, on the complementary strand, within an intron of another gene.

A typical protein-coding gene locus contains the exons and introns, with splice-signal sites at the intron–exon junctions. Transcription of the gene may be under the control of *cis*-regulatory elements near the gene, either upstream or down. Other regulatory elements may appear elsewhere in the genome, even on different chromosomes.

Often a neighbourhood of a gene contains a set of closely linked related genes. This is because a common mechanism of evolution is **gene duplication** followed by divergence. It is often possible to follow evolution through a set of successive duplications. However, in some cases a set of identical copies of a gene can appear on different chromosomes. The gene for ubiquitin is an example.

Ideally, it would be possible, following determination of a genome sequence, to infer the corresponding **proteome** – that is, the amino-acid sequences of the proteins expressed. However, several mechanisms introduce additional variety into the genome–proteome relationship:

- In eukaryotes, a mechanism of generating variety from a single gene sequence is **alternative splicing.** Alternative splicing involves forming a mature messenger RNA from different choices of exons from a gene, but always in the order in which they appear in the genome. It is estimated that ~ 95% of multi-exon protein-coding genes in the human genome produce splice variants. There are also some known cases in which multiple promotors lead to transcription of parts of the same region into different proteins. If the reading frames of the different transcripts are not in phase, there will be no relationship between the protein sequences.

- In both prokaryotes and eukaryotes, **RNA editing** can produce one or more proteins for which the amino-acid sequence may differ from that predicted from the genome sequence. For instance, in the wine grape *(Vinus vinifera)*, the mRNAs arising from every mitochondrial protein-coding gene are subject to multiple C→U editing events, most of which alter the encoded amino acid. In humans, nuclear protein-coding genes are subject to editing that changes adenine to inosine (inosine has the coding properties of guanine). The editing, and hence the final amino-acid sequence, can be tissue-specific.

Variable splicing and RNA editing describe the relationship between genome sequences and proteins *potentially* encoded in them. Of course, a large proportion of cellular activity is dedicated to the *regulation* of gene expression – the selection of *which* potentially encoded proteins are expressed, and in what amounts.

The immune system stands outside the general assertions about the protein-coding regions of the human genome. The number of antibodies that we produce dwarfs all the other proteins – it is estimated that each human synthesizes 10^8–10^{10} antibodies. The generation of such high diversity arises by special combinatorial splicing at the DNA, not the RNA, level.

Some regions of the genome contain **pseudogenes.** Pseudogenes are degenerate genes that have mutated so far from their original sequences that the polypeptide sequence they encode will not be functional. In some cases, pseudogenes have been picked up by viruses from mRNA, and reverse transcribed. This is recognizable from the fact that the introns have been lost.

- The human genome, and other eukaryotic genomes, contains genes that code for proteins and non-coding RNAs (that is, other than messenger RNAs), control regions, pseudogenes (non-functional sequences derived from genes by degeneration), and a wide variety of repetitive sequences. Genes that encode proteins in eukaryotes contain exons – regions that can be translated – and introns – regions that are spliced out before translation. The possibility of omitting one or more exons, called variable splicing, adds complexity to the relationship between base sequences of genes and amino acid sequences of proteins. In many cases, RNA editing creates additional differences between the DNA sequences in the genome and the amino acid sequences of the proteins.

The leap from the one-dimensional world of sequences to the three-dimensional world we inhabit

Gene sequences, mRNA sequences and amino acid sequences are all, from the logical point of view, one dimensional. To perform their proper catalytic, regulatory, or structural activities, proteins must adopt precise three-dimensional structures. The miracle is that the structure is inherent in the amino acid sequence. (See Figure 1.1.) For each natural amino acid sequence, there is a unique stable **native state** that under proper conditions is spontaneously taken up. The evidence is from the **reversible denaturation** of proteins: a native protein that is heated, or otherwise brought to conditions far from its normal physiological environment, will **denature** to a disordered, biologically inactive state. When normal conditions are restored, proteins **renature**, readopting the native structure, indistinguishable in structure and function from the original state. No information is available to the denatured protein, to direct its renaturation, other than the amino acid sequence. (See Box 1.4.)

We therefore have the paradigm:

- DNA sequence determines protein sequence;
- protein sequence determines protein structure;
- protein structure determines protein function.

Because amino acid sequence determines protein structure, we should be able to write computer programs to predict protein structures. This would be useful, because we know many more amino acid sequences than experimentally determined three-dimensional structures of proteins. Structure prediction methods have recently improved substantially (see Chapter 10). Reliable predictions will allow the creation of a library of the structures of the proteins encoded in any genome.

The principle that amino acid sequence dictates protein structure has been the most fundamental principle of structural molecular biology. Imagine, therefore, the shock produced by the observation of an effect of a synonymous mutation on a protein structure. In humans, the Multidrug Resistance 1 (*MDR1*) gene encodes a membrane pump, P-glycoprotein. In 2007, Kimchi-Sarfaty et al. observed that a synonymous mutation in *MDR1* produces a product with altered affinity for ligands. The hypothesis is that protein

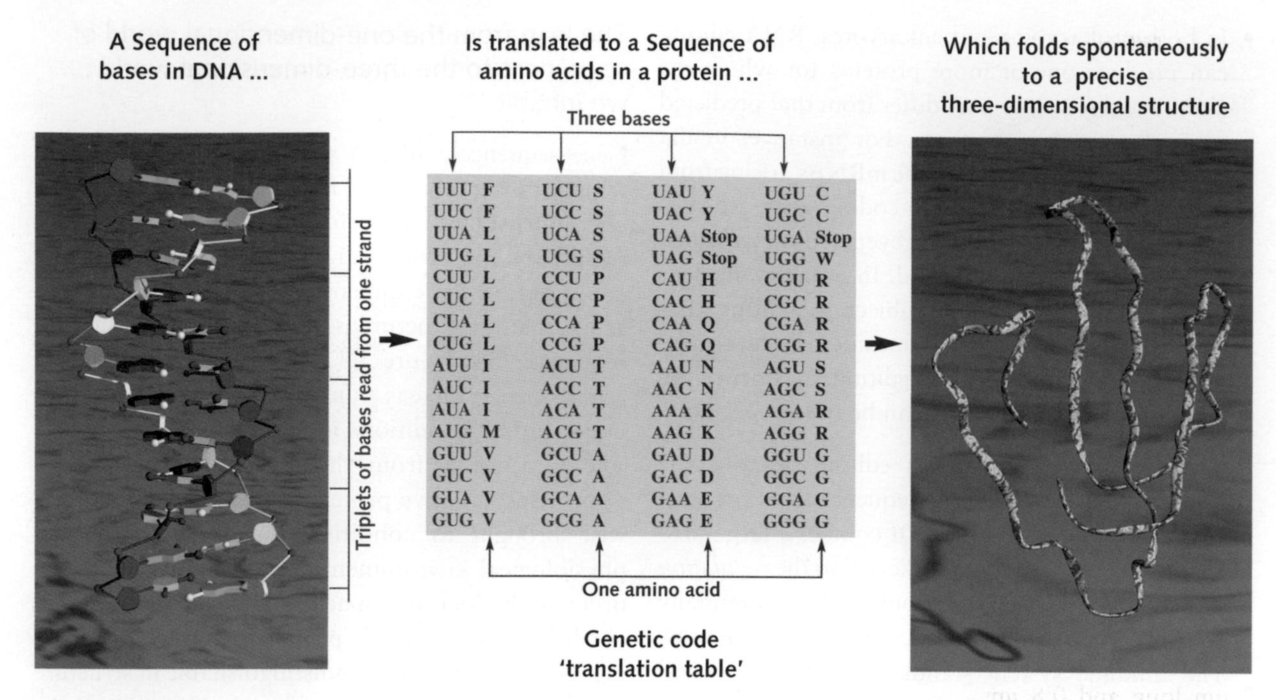
A Sequence of bases in DNA ...
Is translated to a Sequence of amino acids in a protein ...
Which folds spontaneously to a precise three-dimensional structure
Three bases
Triplets of bases read from one strand
One amino acid
Genetic code 'translation table'

UUU F	UCU S	UAU Y	UGU C
UUC F	UCC S	UAC Y	UGC C
UUA L	UCA S	UAA Stop	UGA Stop
UUG L	UCG S	UAG Stop	UGG W
CUU L	CCU P	CAU H	CGU R
CUC L	CCC P	CAC H	CGC R
CUA L	CCA P	CAA Q	CGA R
CUG L	CCG P	CAG Q	CGG R
AUU I	ACU T	AAU N	AGU S
AUC I	ACC T	AAC N	AGC S
AUA I	ACA T	AAA K	AGA R
AUG M	ACG T	AAG K	AGG R
GUU V	GCU A	GAU D	GGU G
GUC V	GCC A	GAC D	GGC G
GUA V	GCA A	GAA E	GGA G
GUG V	GCG A	GAG E	GGG G

Figure 1.1 A most ingenious paradox: the translation of DNA sequences to amino acid sequences is very simple to describe logically; it is specified by the genetic code. The folding of the polypeptide chain into a precise three-dimensional structure is very difficult to describe logically. However, translation requires the immensely complicated machinery of the ribosome, tRNAs, and associated molecules, but protein folding occurs spontaneously.

folding and membrane insertion are occurring during protein synthesis. A change in the rate of synthesis on the ribosome, perhaps because of differential concentrations of tRNAs, could produce the transient exposure of different partial sequences. This might bias the folding pathway. Ribosome pausing is like page turning in music: in principle it should not affect the result.

At this point, most people would agree that this observation represents an exception to the principle that amino acid sequence alone determines protein structure, rather than overturning it entirely.

- Proteins fold to native states based on information contained in the amino acid sequence. Supporting this principle are observations of reversible denaturation. However, those experiments were carried out on isolated proteins in dilute solution. The situation in the cell is much more crowded, and this makes a difference. For instance, proteins are in danger of aggregation, with fatal consequences for the cell. If you boil an egg and then cool it down, the proteins do not renature. What you have is an aggregate of denatured proteins.

Varieties of genome organization

Chromosomes, organelles, and plasmids

The biosphere as we know it includes living things based on cells and also viruses. The most general classification of cells, according to both their structure and molecular biology, divides **prokaryotes**, simple cells without a nucleus, from **eukaryotes**, cells with nuclei.

From the point of view of genomics, the most relevant difference between prokaryotic and eukaryotic cells is the form and organization of the genetic material.

Table 1.1 Differences between prokaryotic and eukaryotic cells

Feature	Typical feature of	
	Prokaryotic cell	Eukaryotic cell
Size	10 μm	~0.1 mm
Subcellular division	No nucleus	Nucleus
State of major component of genetic material	Circular loop, few proteins permanently attached	Complexed with histones to form chromosomes
Internal differentiation	No organized subcellular structure	Nuclei, mitochondria, chloroplasts, cytoskeleton, endoplasmic reticulum, Golgi apparatus
Cell division	Fission	Mitosis (or meiosis)

In structuring their DNA, cells have two problems to solve. The first is a packaging challenge. The DNA of *E. coli* is 1.6 mm long but must fit into a cell 2 μm long and 0.8 μm wide (1 μm = 0.001 mm). Eukaryotic cells have a harder version of the problem: the nucleus of a human cell has a diameter of 6 μm, but the total length of the DNA is 1 m. How can the DNA be deployed in cells in a form that is compact but accessible to proteins active in replication and transcription? The second problem is what to do during cell division, after DNA replication, to ensure that each daughter cell gets one copy of the DNA. Different types of cells, and viruses, solve these problems in different ways.

In the typical prokaryotic cell, most of the DNA has the form of a single closed, or circular, molecule. It is complexed with proteins to form a structure called a **nucleoid**. The DNA is attached to the inside of the plasma membrane but is accessible to molecules in the cytoplasm. Some bacteria have multiple circular DNA molecules; others have linear DNA. In addition, prokaryotic cells can contain **plasmids**, small pieces of circular DNA, neither complexed permanently with protein nor attached to the membrane. The development and spread of antibiotic resistance in pathogenic bacteria, an increasingly serious public health problem, are often associated with exchange of plasmids among strains. Bacterial plasmids are also used as vectors for genetic engineering.

In a eukaryotic cell, most of the DNA is sequestered in the nucleus. The nucleus is the site of DNA replication and RNA synthesis in gene transcription. Nuclear DNA is complexed with histones and other proteins to form chromosomes, large nucleoprotein complexes. Each chromosome contains a single linear molecule of DNA. The nuclei of different species contain different numbers of chromosomes and in each species the chromosomes vary in length. Humans contain 46 chromosomes – 22 pairs, plus two X chromosomes in females or one X and one Y chromosome in males. Deviations from the normal complement of chromosomes have clinical consequences; for example, the presence of three copies of chromosome 21 (trisomy 21) is associated with Down's syndrome.

The state – notably the accessibility to transcriptional machinery – of different regions of DNA in eukaryotic chromosomes is modulated by the local structure of the chromosome, notably the interaction with histones (Figure 1.2).

Subcellular organelles, including mitochondria and chloroplasts, are believed to have originated as intracellular parasites (see Box 1.4). These organelles contain additional DNA in the form of single closed or circular molecules, uncomplexed with histones, like the DNA of prokaryotes. Mitochondria and chloroplasts also contain their own protein-synthesizing machinery, using a slightly different dialect of the nearly universal genetic code.

Eukaryotic cells may also contain plasmids. Yeast artificial chromosomes (YACs) – plasmids in yeast cells – are of great utility in genome sequencing projects.

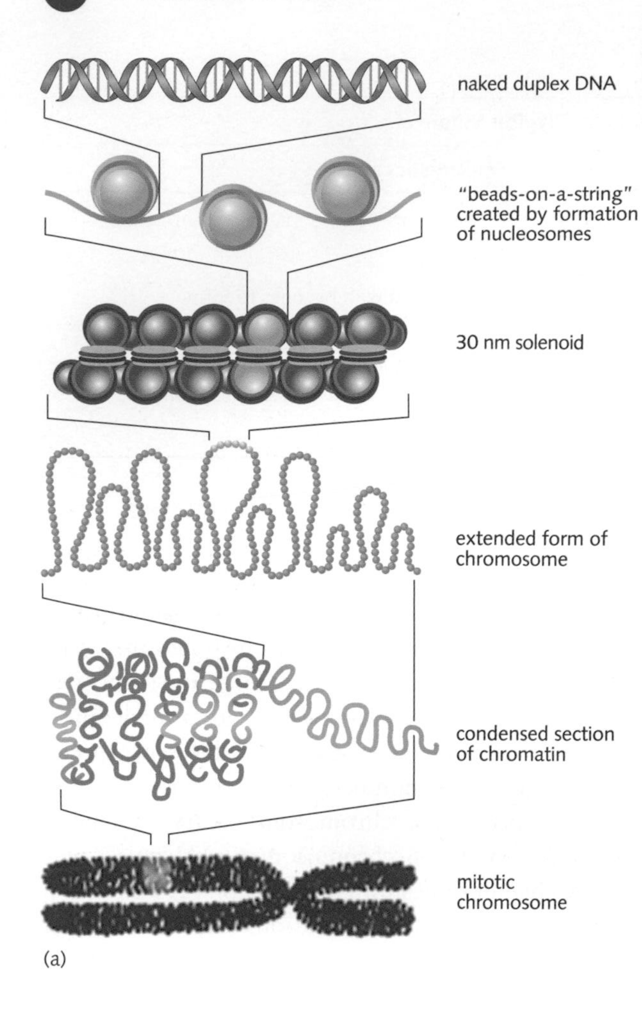

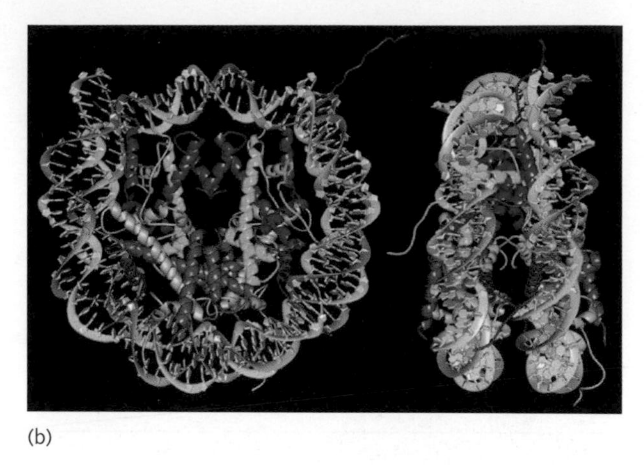

Figure 1.2 (a) Hierarchical organization of structure of DNA in eukaryotic chromosomes. From top: DNA double helix; if entirely straight, it would be much longer than dimensions of cell. DNA forms nucleosomes, containing a protein core of histones, around which nearly two turns of DNA (about 150 bp) are wrapped. The nucleosomes condense to form a 30 nm helical fibre called the solenoid. Further levels of compaction produce the bottom image, corresponding to the typical appearance of a chromosome in a metaphase karyotyping spread. (Reproduced with permission of themedicalbiochemistrypage.org) (b) The nucleosome core particle. The structure comprises 146 bp of DNA, the strands coloured brown and turquoise; and four pairs of histones, coloured blue, green, yellow, and red. (From: Luger, K., Mäder, A.W., Richmond, R.K., Sargent, D.F. & Richmond, T.J. (1997). Crystal structure of the nucleosome core particle at 2.8 Å resolution. Nature **389**, 251–260.)

BOX 1.4 Origin of intracellular organelles: the endosymbiont hypothesis

Mitochondria and chloroplasts are subcellular particles involved in energy transduction (Figure 1.3). Mitochondria carry out oxidative phosphorylation, the conversion into ATP of reducing power derived from metabolizing food. Chloroplasts carry out photosynthesis, the capture of light energy in the form of 'reducing power' – NADPH – and ATP.

Both types of particle lead a quasi-independent life within the cell. They are surrounded by membranes, they have their own genetic material and protein-synthesizing machinery, and they reproduce themselves within cells independently of cell division.

To perform their functions, it is essential for mitochondria and chloroplasts to be enclosed. This is because electrochemical and photochemical energy conversion require establishment of a pH gradient across the organelle membrane. The passage of protons through the membrane is coupled to generation first of mechanical energy and then of chemical bond energy through the action of the molecular motor ATP synthase: chemiosmotic energy stored in pH gradient → mechanical energy in ATP synthase → chemical bond energy of ATP.

Where might a cell look to find a small, self-enclosed, and largely self-sufficient object to serve as an organelle? Why, at another cell! There is a consensus that mitochondria and chloroplasts originated as prokaryotic endosymbionts that originally lived independently but took up residence inside other cells. Evidence for this includes: (1) the state of the DNA: organelle DNA is circular and uncomplexed with

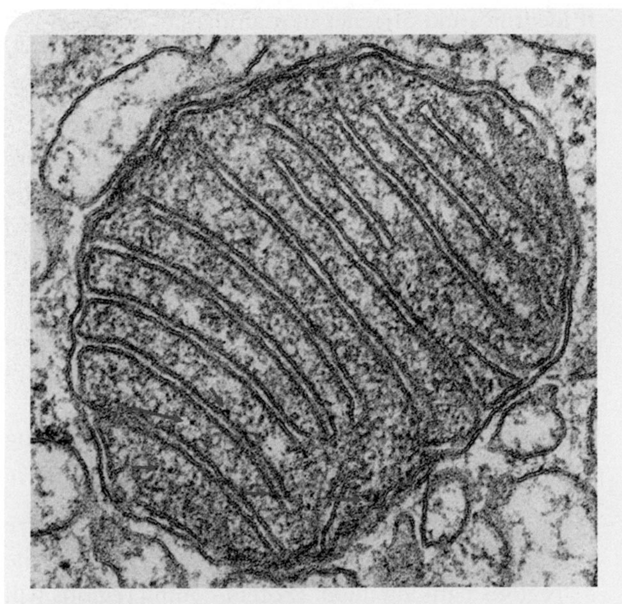

Figure 1.3 Structure of a mitochondrion. The convoluted inner membrane is easily visible. Red arrows indicate ribosomes. (Reproduced with permission of themedicalbiochemistrypage.org)

histones, like that of prokaryotes; and (2) the fact that organelle ribosomes resemble those of prokaryotes rather than those of eukaryotic cells.

We can identify the origins of mitochondria and chloroplasts from the similarity of their genome sequences to those of living species. Mitochondria are likely to have originated as a relative of *Rickettsia*, parasites that cause typhus and Rocky Mountain spotted fever. Chloroplasts evolved from cyanobacteria.

The genomes of mitochondria and chloroplasts have degenerated from their original forms. The human mitochondrial genome has 16 569 bp compared with the >1 Mb size of contemporary *Rickettsia*. Mitochondrial genome sizes range from 5966 bp in *Plasmodium* to 366 924 bp in *Arabidopsis*.

Chloroplast genomes contain 30 000–200 000 bp, compared with cyanobacteria, such as *Synechocystis*, with a 3.5 Mbp genome.

This shrinking of the genome is not entirely the effect of squeezing out the non-functional DNA in the interests of slimming down the organelle. Mitochondria and chloroplasts synthesize relatively few proteins, and certainly not all that they need. There has been a lot of transfer of genes from the mitochondria and chloroplast genomes to the nucleus. Proteins synthesized with a specific N-terminal targeting sequence are translocated into the appropriate organelle (see Chapter 4).

It is believed that chloroplasts of eukaryotic cells arose from cyanobacteria independently at least three times, in lineages leading to (1) green algae and higher plants; (2) red algae; and (3) a small group of unicellular algae called the glaucophytes.

Genes

As they come off the sequencing machines, genomes are long strings of As, Ts, Gs, and Cs, without captions or sign posts. Many people think that only a small fraction (possibly only ~5%) of the human genome is functional. The challenge is to identify the functional regions and figure out what they do.

- *Protein-coding regions*. Because three consecutive bases correspond to one amino acid, any nucleic acid sequence can be translated into an amino acid sequence in six ways. Beginning at nucleotide 1, 2, or 3 gives three possible phases of translation, and translating the reverse complement – the other strand – gives another three possible phases. A protein-coding region will contain **open reading frames** (ORFs) in one of the six reading phases. An ORF is a region of DNA sequence, of reasonable length, that begins with an initiation codon (ATG) and ends with a stop codon. An ORF is a potential protein-coding region.
- *Some regions are expressed as non-protein-coding RNA*. They will show regions of local self-complementarity corresponding to hairpin loops. For instance, genes for transfer RNA (tRNA) will contain the signature cloverleaf pattern (see Figure 1.4).
- *Other regions are targets of regulatory interactions*.

Gene identification in prokaryotes is easier than in eukaryotes. Prokaryotic genomes are smaller and contain fewer genes. Genes in bacteria are contiguous – they lack the introns characteristic of eukaryotic genomes. The intergene spaces are small. In *E. coli*, for instance, approximately 90% of the sequence is coding. Features such as ribosome-binding sites are conserved.

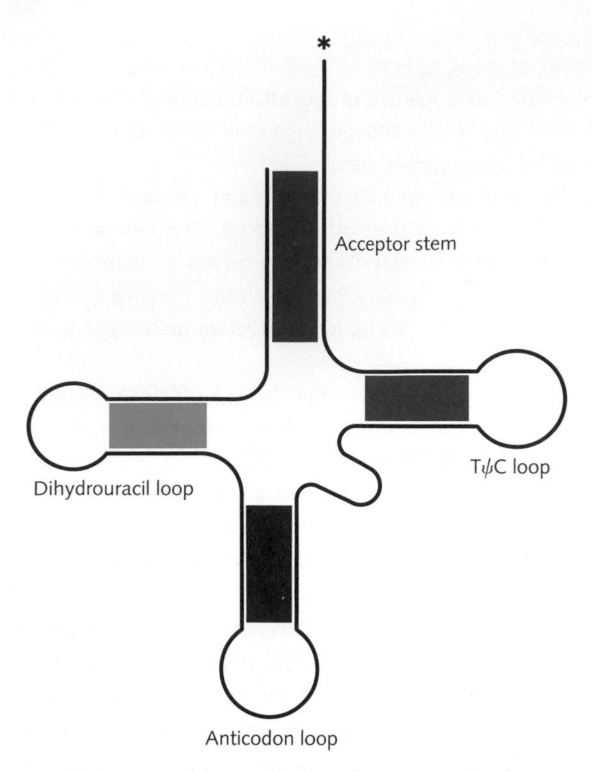

Figure 1.4 The framework of the structure of transfer RNA (tRNA). The * marks the 3′ end, the site of attachment of the amino acid. tRNA contains several regions of double-helical structure, indicated by coloured patches. These regions have almost perfect complementarity of base pairs. They therefore induce a pattern observable in the DNA sequence, by which it is possible to recognize genes for tRNAs in genomes.

BOX 1.5 Expressed sequence tags

An expressed sequence tag (EST) is a sequence that corresponds to at least a part of a transcribed gene.

In a cell, transcription of genes and – in eukaryotes – maturation of mRNAs by splicing out the introns, produces single-stranded RNA molecules that correspond directly to coding sequences. Collecting these RNA molecules, converting them to complementary DNA (cDNA) using the enzyme reverse transcriptase and sequencing the terminal fragments of the cDNA produces a set of ESTs. It is necessary to sequence only a few hundred initial bases of cDNA. This gives enough information to identify a known protein coded by the mRNA. Characterization of genes by ESTs is like indexing poems or songs by their first lines.

If identification of genes in mammalian genomes is – in a common simile – like hunting for needles in a haystack, identification of genes in prokaryotes is no worse than hunting for needles in a sewing kit.

Protein-coding genes in higher eukaryotes are sparsely distributed and most are interrupted by introns. Identification of exons is one problem and assembling them is another. Alternative splicing patterns present additional difficulty. Simpler eukaryotes, such as yeast, are not quite as bad: about 67% of the yeast genome codes for protein and fewer than 4% of the genes contain introns.

There are two basic approaches to identifying genes in genomes.

1. *A priori* methods, which seek to recognize sequence patterns within expressed genes and the regions flanking them. Protein-coding regions will have distinctive patterns of codon statistics, of course including – but not limited to – the absence of stop codons.
2. '*Been there, seen that*' methods, that recognize regions corresponding to previously known genes, from the similarity of their translated amino acid sequences to known proteins in another species, or by matching expressed sequence tags (ESTs) (see Box 1.5).

A priori methods are the blunter instrument. Combined approaches are also possible.

Characteristics of eukaryotic genes useful in identifying them include:

- The initial (5′) exon starts with a transcription start point, preceded by a core promotor site, such as the TATA box, typically ~30 bp upstream. It is free of in-frame stop codons and ends immediately before a GT splice signal. (Occasionally a non-coding exon precedes the exon that contains the initiator codon.)
- Internal exons, like initial exons, are free of in-frame stop codons. They begin immediately after an AG splice signal and end immediately before a GT splice signal.
- The final (3′) exon starts immediately after an AG splice signal and ends with a stop codon, followed by a polyadenylation signal sequence. (Occasionally a non-coding exon follows the exon that contains the stop codon.)

- All coding regions have non-random sequence characteristics, based partly on codon usage preferences. Empirically, it is found that statistics of hexanucleotides perform best in distinguishing coding from non-coding regions. Using a set of known genes from an organism as a training set, pattern-recognition programs can be tuned to particular genomes.

> - An early step in analysing a newly sequenced genome is to find the genes that code for proteins and RNAs, and to try to identify them.

Dynamic components of genomes

Transposable elements are skittish segments of DNA, found in all organisms, that move around the genome. They were discovered by B. McClintock in the 1940s in studies of maize. Transposable elements in Indian maize (or corn) create a genetic mosaic, giving the ears a mottled appearance (see Figure 1.5). In this case, transposition is fast enough to affect an individual organism. Other transposable elements move more slowly, on evolutionary timescales.

The focus of this section is transposable elements within nuclear genes of eukaryotes. We shall discuss the traffic of genes between organelle (mitochondria and chloroplasts) and nuclear genomes in Chapter 4.

Figure 1.5 The ears of Indian maize are mosaics. The dark pigments are anthocyanins. Yellow sectors arise when a jumping element or transposon interferes with expression or function of the genes for biosynthesis of anthocyanins, during development of individual kernels.

(Photograph courtesy of L.D. Graham, Rockingham, VA, USA)

Table 1.2 Transposable elements in the human genome

Element	Estimated number	% of total genome
SINE + LINE	2.4×10^6	33.9
LTR	0.3×10^6	8.3
Transposons	0.3×10^6	2.8
Total	3.0×10^6	~45

Data from: Bannert, N. & Kurth, R. (2004). Retroelements and the human genome: New perspectives on an old relation. Proc. Natl. Acad. Sci. USA **101**, 14572–14579.

Different types of element show alternative mechanisms of transposition.

Retrotransposons (class I) replicate via an RNA intermediate. Many, if not all, of them are degenerate retroviruses.

Transposons (class II) produce DNA copies without an intermediate RNA stage. They encode an enzyme called transposase, which recognizes sequences within the transposon itself, cuts it out, and inserts it elsewhere. Often the excision is sloppy, leaving a mutation at the original site. Sometimes, a bit of the surrounding sequence adheres to and accompanies the transposed material.

Because transposable elements can replicate, they are related to some of the types of repetitive sequence found in genomes (see Table 1.2). Mammalian genomes contain retrotransposons (RNA-mediated replication) called **long** and **short interspersed elements** (LINEs and SINEs). LINEs are typically 1–5 kb long, with tens to tens of thousands of copies. The most common LINE, L1, appears ~20 000 times in the genome. SINEs are typically 200–300 bp long, with hundreds of thousands of copies in the genome, at scattered locations. The human genome contains about 300 000 copies of the most common SINE, the Alu element, which is 280 kb long. The total amount of L1 + Alu is 7% of the human genome. LINEs encode a reverse transcriptase and can replicate autonomously. SINEs are too short to encode their own reverse transcriptase. SINEs depend on LINEs or other sources of the required activities for replication.

Transposons contain inverted repeats at their ends, which are the targets of the excision machinery (see Figure 1.6 and Box 1.6). Replication may occur in 'cut-and-paste' mode, moving the transposon from

Figure 1.6 Fragment of a chromosome containing a transposon (red). The ends of the transposon contain inverted repeat sequences, demarcating the region for excision. Within the transposon is a gene for the transposase enzyme, giving the region the capacity for autonomous replication.

one site to another. Alternatively, in 'copy-and-paste' mode, replication leaves the original copy behind, while creating another.

If two equivalent transposons are nearby, they can move a whole segment including all the material between them. Transfer of multiple genes to a plasmid is a common mechanism of generation of antibiotic resistance in bacteria. (The Tn3 transposon illustrated in Box 1.6 contains only one set of terminal repeats.)

Biological effects of transposable elements include:

- *Sequence broadcasting.* Multiple copies of elements of a sequence may be distributed to various locations in the genome.
- *Altering properties of genes.* The arrival of a fragment of sequence within or in the vicinity of a gene may, if inserted into a coding region, render the gene product non-functional, creating a 'knockout' effect. An inserted segment near a gene may affect its regulation or alter its splicing pattern. (Approximately 20% of human genes have transposable elements in flanking non-coding sequences.) Even an insertion in an intron can affect rates of transcription by slowing down the polymerase as it passes through.
- *Transposable elements as an important engine of evolution.* They provide a mechanism for gene evolution by gene fusion or exon shuffling. Transposable element insertion can cause species-specific alternative splicing patterns. This can produce new protein isoforms. It can also lead to disease; for example, the cause of a case of ornithine aminotransferase deficiency was a single base change that activated a cryptic 5′ splice site in an Alu element, introducing an in-frame stop codon leading to a truncated protein.
- *Causing chromosomal rearrangements.* This can include inversions, translocations, transpositions, and duplications, perhaps through mispairing of chromosomes during cell division. The deletions leading to Prader–Willi and Angelman syndromes (see Chapter 3) are associated with a mutation in the sequence of a nearby transposable element.
- *Leakage of epigenetic modification.* From the landlord's point of view, transposable elements are squatters. For example, they make up 70% of the maize genome. In one sense, it is bad enough that they clutter up the DNA, but, even worse, eukaryotes must defend themselves against the *expression* of transposable elements. The tactics of defence is to methylate transposable elements, or to use small interfering RNAs (siRNAs) (see pp. 6–7). Some cancers and other diseases that lead to hypomethylation of DNA can cause transcriptional reactivation of some transposable elements. Methylation also cuts down on the mobility of those transposable elements that require transcription for mobility. However, mechanisms of silencing transposable elements can also affect neighbouring genes.

The 5′ and 3′ ends of transposons contain inverted repeats

The beginning and end of the sequence of the Tn3 transposon of *E. coli* contain a terminal repeat. This is a plasmid encoding *β*-lactamase, conferring ampicillin resistance, as well as a transposase.*

```
   1 GGGGTCTGAC GCTCAGTGGA ACGAAAACTC
     ACGTTAAGCA ACGTTTTCTG CCTCTGACGC
  61 CTCTTTTAAT GGTCTCAGAT GACCTTTGGT
     CACCAGTTCT GCCAGCGTGA AGGAATAATG
                 . . .
4861 TTTTTAATTT AAAAGGATCT AGGTGAAGAT
     CCTTTTTGAT AATCTCATGA CCAAAATCCC
4921 TTAACGTGAG TTTTCGTTCC ACTGAGCGTA AGACCCC
```

* Heffron, F., McCarthy, B.J., Ohtsubo, H. & Ohtsubo, E. (1979). DNA sequence analysis of the transposon Tn3: three genes and three sites involved in transposition of Tn3. Cell **18**, 1153–1163.

- Transposable elements add a dynamic component to genetic change, at a higher level than point mutations. These changes may have significant biological effects.

Genome sequencing projects

The static contents of the human genome, and its dynamic aspects, are similar in general features to what other genomes contain. As genome sequencing techniques become easier, the field is progressing in the directions: (a) to determine more and more human genome sequences, especially those that may prove useful in research into disease anticipation and prevention, and (b) many different species have now had their genomes sequenced, for at least one individual.

The National Center for Biotechnology Information database currently reports:

Table 1.3 Genome sequencing projects. Figures refer to numbers of species and strains, not numbers of individuals[a]

Organism type	Number of genomes completed	Number of genomes in progress
Viruses and viroids[b]	3889	
Archaea	113	91
Bacteria	1588	4914
Eukaryotes	36	1175

[a] Source: http://www.ncbi.nlm.nih.gov/genome

[b] A viroid is a small single-stranded RNA which can replicate autonomously, but does not encode protein, nor is encapsulated within a coat.

There are many reasons for sequencing non-human genomes. The most important ones are that they reveal the processes of evolution, and that they help us to understand the functions of different regions of the human genome.

Other genomes are essential to illuminate ours. An important principle is that if evolution conserves something, it is essential. If evolution does not conserve something, it is not essential. In trying to understand the function of the ~98% of the human genome that does not recognizably code for proteins or non-coding RNAs, we gain important clues by comparing the human genome with other mammalian genomes. Regions that are conserved must be conserved for a reason. We can focus on these regions to try to determine their function. Regions that are not conserved can be reserved for later.

He who does not know foreign languages does not know anything about his own – Goethe, Kunst und Alterthum

Sequences of genomes of other species also have direct application to human welfare. The genomes of pathogens that have developed antibiotic resistance, or are threatening to, give clues that we can use to try to keep ahead of them. Other practical applications include improving crops and domesticated animals. Genomics can also support conservation efforts aimed at preserving endangered species.

Genomics, allied with anthropology and archaeology, helps recount the history of the human species. It can reveal patterns of migration. It can trace the domestication of plants and animals. These applications to many aspects of human biology are the subject of Chapter 8.

Many genome projects target individual species. In addition, a major component of public DNA sequence data repositories comes from **metagenomic data**. These are sequences determined from environmental samples, without isolating individual organisms. Sources include ocean water, soil samples, and the human gut.

- Some readers will be surprised to learn that within the volume of their bodies, there are more prokaryotic cells than human ones.

Nevertheless, the focus of genome sequencing efforts has been human subjects. Clinical applications have created a very large amount of sequence data for individual human genes. Many people have undergone genetic testing: for example, many prospective parents determine whether they are carriers of cystic fibrosis. Many women test for potentially dangerous mutations in genes that can predispose an individual to cancer, such as BRCA1 and BRCA2. Up to now, these tests have been carried out when a family history suggests increased risk. Reduced sequencing costs may lead to more widespread testing for these and other risk-alerting genes. Some people make use of publicly available genotyping services, outside medical supervision, to explore genealogy. Law-enforcement agencies determine DNA sequences from samples from crime scenes.

A separate area of comparative genomics involves comparing the genome sequences of normal cells and cancer cells from patients. These can differ in essential ways, which assist precise diagnosis and guide treatment.

Does this information appear in databases? For non-human sequences, the expectation is that the data will be publicly available. For human sequences, the situation is very different, as major questions of privacy arise (see section on ethical, legal, and social issues).

Genome projects and the development of our current information library

The first genome sequenced was that of the single-stranded DNA virus, bacteriophage ϕX-174. F. Sanger and co-workers published this result, 5386 bases, in 1977. Recognition of the importance of sequencing stimulated intensive efforts to improve and automate the techniques. A major breakthrough was the replacement of the autoradiography of gels, each nucleotide occupying a separate lane, with four fluorescent dyes, permitting a 'one-pot' reaction. Leroy Hood and colleagues automated this technique, developing a machine that supported a generation of sequencing projects. Recently, a number of novel approaches, or 'next-generation' sequencing techniques, have been developed. We shall discuss these in Chapter 3.

Vocabulary for high-throughput DNA sequencing

Fragment: a small piece of genomic DNA – typically several hundred bp in length – subject to an individual partial sequence determination, or *read*.

Single-end read: technique in which sequence is reported from only one end of a fragment (see Figure 1.7).

Paired-end read: technique in which sequence is reported from both ends of a fragment (with a number of undetermined bases between the reads that is known only approximately).

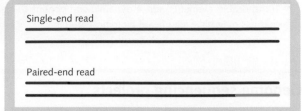

Figure 1.7 Sequence data reported from fragments using single-end sequencing and paired-end sequencing techniques. Data reported arise from red and green regions. The lengths of the black regions are known only approximately, from the fragment length. Typically, fragments used are about 200 bp long, varying in length by about 10%.

Read length: the number of bases reported from a single experiment on a single fragment.

Assembly: the inference of the complete sequence of a region from the data on individual fragments from the region, by piecing together overlaps.

Contig: a partial assembly of data from overlapping fragments into a contiguous region of sequence.

De novo *sequencing*: determination of a full-genome sequence without using a known reference sequence from an individual of the species to avoid the assembly step.

Resequencing: determination of the sequence of an individual of a species for which a reference genome sequence is known. The assembly process is replaced by mapping the fragments onto the reference genome.

Exome sequencing: targeted sequencing of regions in DNA that code for parts of expressed proteins (exons). There are approximately 180 000 exons in the human genome.

Variability, Single-nucleotide polymorphisms (SNPs): variability is the differences among genome sequences from different individuals of the same species. Much observed variability has the form of isolated substitutions at individual positions, or single-nucleotide polymorphisms.

RNAseq: sequencing the contents and composition of the RNAs in the cell, the transcriptome, by conversion of RNA to complementary DNA and sequencing the result.

ChIP-seq: sequencing of the fragments of DNA to which particular proteins are known to bind. ChIP-seq permits the identification of the targets, in the genome, of DNA-binding proteins.

Methylation-pattern determination: comparison of sequences of native DNA, and DNA after treatment with bisulphate to convert unmethylated C to U. Methylation of C residues in CpG dinucleotides at the 5′ position is the most common signal in animals for repression of transcription.

High-throughput sequencing

The past few years have seen truly astounding progress in the development of high-throughput sequencing techniques. Initial determination of a draft of the human genome took ten years, at an estimated cost of $US 3×10^9. At the time of writing, instruments exist that can produce 250 Gb per week. The largest dedicated institution in the field, the BGI – formerly the Beijing Genomics Institute, but currently in Shenzhen – has 128 such instruments. Each can produce 25×10^9 bp per day! This corresponds to one human genome at over 8X coverage. Running at full capacity, these resources could produce 10 000 human genomes per year.

Moreover, there is no reason to think that the technical progress will not continue to accelerate. Undoubtedly by the time this book is published, these specs will be outdated. (See Weblem 1.1.)

There are two aspects of a large-scale sequencing project. One is the generation of the raw data. Most methods sequence long DNA molecules by fragmenting them, and partially sequencing the pieces. To determine the first genome from a species, these short sequences must be assembled into the whole sequence, using overlaps between the individual fragments.

The typical length of the individual short sequences reported is called the **read length** of the method. The goals of contemporary technical development are to increase not only the number of bases sequenced per unit time and per unit cost, but the read length.

Both generation of raw data, and assembly, depend crucially on effective and efficient computer programs. Some contemporary genome centres have as many computational biologists on their staffs as 'wet-lab' scientists.

The very high throughput sequencing capacity of new instruments allows addressing several types of biological questions.

De novo sequencing

The most ambitious type of sequencing project is the determination of the complete sequence of the first genome from a species. The total DNA must be broken into fragments, typically about 200 bp long. High-throughput sequencing can produce partial sequence information for such fragments, either from one end only, or from paired ends (see Figure 1.7). Paired-end sequencing reports partial sequences from both ends of the same fragment. In either case, the number of bases reported is the read length. The number of unknown bases between the paired ends is limited (by the estimate of the overall fragment length) but the exact value is unknown.

Next it is necessary to assemble the genome, from the sequences of overlapping fragments. A **contig** is a partial assembly of fragments, into a contiguous stretch of sequence. As the assembly proceeds, the contigs grow in length, like the completed portion of a jigsaw puzzle.

Assembly requires a sufficient number of fragments to cover the entire genome, with enough replicates to be able to detect errors. The ratio of the total number of bases sequenced to the genome length is the **coverage** of the data set. There is now consensus that to achieve complete and accurate assembly of a novel genome requires collection of data with a coverage of 30 or 50 (30X or 50X).

For prokaryotic genomes, the process outlined, which corresponds to 'shotgun' sequencing of fragments, allows accurate assembly. For a large eukaryotic genome, achieving the highest quality result requires a genetic map, breaking the assembly problem into smaller pieces.

Eukaryotic genome assembly is a very computer-intensive problem. Development of algorithms and computer programs for effective assembly of fragmentary sequence data are a thriving field of research. The resources of many genome centres are such that assembly is the rate-limiting step in the process.

Resequencing

Once a reference genome for an individual of a species is available – for example, a published human genome – the sequences of genomes from other individuals of the species are considerably easier to determine. It is not necessary to assemble fragment sequences *de novo*, but merely to map them onto the reference genome. Except for highly repetitive regions, this is fairly straightforward. Coverage must be adequate so that the error rate of sequence determination is less than the frequency of natural variation. In the specific case of sequencing genomes from cancer cells, it is preferable to sequence normal cells from the same patient than to try to infer from the reference sequence the genome changes arising from the disease.

Exome sequencing

One goal of resequencing is to determine variation in the genome of an individual from the reference genome. By correlating these variations with phenotype – for example, the presence of an inherited disease – it is possible to identify the genetic origin of the lesion. Many inherited diseases result from loss of activity of particular proteins. The loss of activity frequently arises from a specific mutation in the sequence coding for the protein. To identify such a mutation, it is not necessary to sequence the entire genome but only the protein-coding regions; namely, the exons. There are approximately 180 000 exons in the human genome, amounting in total to approximately 30 Mb, or ~1% of the entire genome.

Variations within and between populations

Any two people – except for identical siblings – have genomic sequences that differ at approximately 0.1% of the positions. Measurements of multiple human genomes permit distinction between random components of this variation, and those that systematically characterize different populations. Much but not all of the variation takes the form of isolated base substitutions, or single-nucleotide polymorphisms.

Sequence variation in humans has applications in anthropology, to trace migration patterns, and in personal identification to prove paternity or for crime investigation. Sequence variability in other species gives clues to the history of the species, including but not limited to understanding the history of domestication of animals and crop plants.

Cancer genome sequencing

Healthy cells do accumulate mutations at a modest rate. Cancer cells that have lost checks on accuracy of DNA replication accumulate mutations copiously. To distinguish variations arising from the disease it is preferable to compare the sequences from tumour cells with those from normal cells from the same individual, rather than with a single reference genome.

- There are many different applications of high-throughput sequencing techniques, designed to address different types of questions.

Human genome sequencing

In October 2010, Nature published the estimate that over 2700 full human genomes will have been sequenced by the end of that month, and suggested that by the end of 2011 the total will be over 30 000. A few of the individuals are known. These include J. Craig Venter, the subject of the original Celera Corporation genome sequencing project, James D. Watson, Bishop Desmond Tutu, and Stanford University professor Stephen Quake. (Perhaps analysis might reveal a genetic locus associated with the Nobel Prize.) Actress Glenn Close had her genome sequenced, motivated by a family history that included several individuals who suffered from mental illness. The academic human DNA sequencing project used DNA primarily from a donor from Buffalo, New York. His identity has not been revealed.

Venter has also sequenced the genome of his poodle, Shadow. They are therefore the first human–pet combination with both sequences known.

Sequencing is now done under a variety of auspices:

- A number of international organizations with ambitious specific targets. These include the International HapMap Project http://hapmap.ncbi.nlm.nih.gov/ that focuses on the variations in sequences in populations distributed around the world. They are collecting in particular an atlas of **single-nucleotide polymorphisms** (SNPs) which are substitutions of individual bases. Clusters of SNPs that appear to be inherited in tandem are called **haplotypes.** (See Chapter 2.)

- The 1000-genome project is an extension of the HapMap project towards complete genome data, with an emphasis of discovering the conditions required to ensure appropriate data quality in projects of this type (http://www.1000genomes.org/). Specific goals include careful sequencing of family groups (mother + father + child), and detailed sequencing of 1000 protein-coding regions in 1000 individuals. Although one may wonder to what extent the stated goals will be overtaken by the growing 'background' of genome sequencing, the importance of the commitment of the international organizations to data quality control, curation, and free distribution should not be underestimated.
- Several companies now offer personal genome sequencing. Many provide sequencing of mitochondrial DNA or individual loci in nuclear DNA, for tracing of ancestry. The application of DNA to demonstration of legal relationships – most commonly, paternity testing – is well established.

Up to now, the cost of a full-genome sequence has been prohibitive for most people if the motivation is casual curiosity. However, it is already true that the cost of sequencing a person's DNA is comparable to the cost of a night's stay in hospital in the USA. As costs fall, and equipment becomes smaller, *and we can derive more and more clinically useful information from sequence data*, the conclusion seems inescapable that anyone entering a hospital will have at least a partial DNA sequence determination along with taking his or her pulse rate, temperature, and blood pressure. (See Problem 1.7.)

- It has taken 30 years for DNA sequencing to make the transition from Nobel Prize breakthrough research to secondary school science projects.

The teacher in a secondary school in New Jersey, USA, organized a project to analyse DNA samples of the students in her class. This was not full-genome sequencing but produced limited, genealogy-oriented data. The students compared the results with their own cultural backgrounds.

The following example is not human DNA sequencing, but usefully hints at the potential variety of applications, and the relative ease with which they can be carried out. In 2008, two teen-age students checked samples of fish from New York City sushi bars, using a genetic-fingerprinting technique called DNA Bar Coding (see p. 117). They discovered that, of the samples they could identify, 25% were mislabelled, in fact originating in less-expensive species than advertised. They identified some restaurants that were free of mislabelling.

The human genome and medicine

The development and delivery of health care challenge biomedical science, engineering, industry, and government – both separately and in their coordination. Not all components of these challenges are within the scope of this book. What is relevant is the increasing scientific sophistication of medical treatment and the shrinking of the distance and time between laboratory discovery and clinical practice: walls between 'pure' and 'applied' science have come tumblin' down.

Of course, immense improvements in human health could be achieved by quite low-tech measures. Obvious examples include educating people about the long-term dangers of obesity, smoking, and alcohol abuse; use of seatbelts in automobiles and aeroplanes; and the provision, to the many people who lack them, of basic medical care and hygienic living conditions – for instance, clean water. In all communities, preserving a mutually nurturing relationship with nature will improve people's mental and spiritual health, which – although in ways not entirely demonstrable yet in the laboratory – has important effects on physical well-being.

Nevertheless, genomics and proteomics have played a central role in the recent transformation of medicine and surgery, making contributions to prevention of disease, to detection and precise diagnosis, and to effective treatment.

Prevention of disease

- Vaccinations are pre-emptive strikes against infectious diseases. They prime the immune system to recognize pathogens. Some vaccines are (almost) completely effective. Vaccines have eliminated smallpox entirely, and polio from much of the world. Even a partially effective vaccine can tip the balance between a devastating pandemic – for example, of influenza – and a controllable set of localized outbreaks.

- Understanding individual genetic predispositions to disease can, for some conditions, help to preserve health through suggesting changes in lifestyle, and/or medical treatment. An example of a risk factor detectable at the genetic level involves α_1-antitrypsin, a protein that normally functions to inhibit elastase in the alveoli of the lung. People homozygous for the Z mutant of α_1-antitrypsin (342Glu→Lys) express only a dysfunctional protein. They are at risk of emphysema because of damage to the lungs from endogenous elastase unchecked by normal inhibitory activity, and also of liver disease because of accumulation of a polymeric form of α_1-antitrypsin in hepatocytes, where it is synthesized. Smoking makes the development of emphysema all but certain. Heavy smokers homozygous for Z-antitrypsin generally die from respiratory disease by the age of 50. In these cases, the disease is brought on by a *combination* of genetic and environmental factors.

> *'Genetics loads the gun and environment pulls the trigger' – J. Stern*

- In other cases, detection of genetic abnormalities will not prevent a disease but can dispel fear of the unknown (see Box 1.7). Genetic counselling is a potential preventative approach to avoiding abnormalities or diseases that arise from dangerous combinations of parental genes.

Detection and precise diagnosis

Early detection of many diseases permits simpler and more successful treatment. With a range of therapeutic strategies often available, precise classification may allow better prediction of the probable course of a disease, and dictate optimal treatment. For instance, oncologists classify leukaemias into seven subtypes. Determination of the subtype from gene expression patterns permits better prognosis and treatment.

Huntington's disease

Huntington's disease is an inherited neurodegenerative disorder affecting approximately 30 000 people in the USA. Its symptoms are quite severe, including uncontrollable dance-like (choreatic) movements, mental disturbance, personality changes, and intellectual impairment. Death usually follows within 10–15 years of the onset of symptoms. The gene arrived in New England, USA, during the colonial period in the 17th century. It may have been responsible for some accusations of witchcraft. The gene has not been eliminated from the population, because the age of onset – 30–50 years – is after the typical reproductive period.

Formerly, members of affected families had no alternative but to face the uncertainty and fear, during youth and early adulthood, of not knowing whether they had inherited the disease. The discovery of the gene for Huntington's disease in 1993 made it possible to identify affected individuals. The gene contains expanded repeats of the trinucleotide CAG, corresponding to polyglutamine blocks in the corresponding protein, huntingtin. (Huntington's disease is one of a family of neurodegenerative conditions resulting from trinucleotide repeats.) The larger the block of CAGs, the earlier the onset and more severe the symptoms. The normal gene contains 11–28 CAG repeats. People with 29–34 repeats are unlikely to develop the disease and those with 35–41 repeats may develop only relatively mild symptoms. However, people with >41 repeats are almost certain to suffer full Huntington's disease.

Discovery and implementation of effective treatment

Advances in basic science provide the background of understanding, the basis for applications. We study the biology of viruses and bacteria to take advantage of their vulnerabilities, and the biology of humans to ward off the consequences of ours.

A large component of the progress in effective treatments for diseases involves the development of drugs. It is a sobering experience to ask a classroom full of students how many would be alive today without at least one course of drug therapy during a serious illness (ignoring diseases escaped through vaccination) or to ask the students how many of their surviving grandparents would be leading lives of greatly reduced quality without regular treatment with drugs. The answers are eloquent. They engender fear of the new antibiotic-resistant strains of infectious microorganisms.

The traditional drug development process involved identifying a target – usually a protein – either from host or pathogen, the behaviour of which it is desired to affect. A drug is a molecule that intervenes in a living process by interacting with the target.

Recent scientific advances have accelerated the drug development process. Identifying metabolic features unique to a pathogen helps to identify targets for antibacterial and antiviral agents. Human proteins provide other drug targets to deal with molecular dysfunction or to adjust regulatory controls. Knowing the structure of a target permits computer-assisted drug design by molecular modelling.

Health care delivery

Many countries in which advanced treatments are available face severe economic impediments to the equitable delivery of medical and surgical care to their citizens. Research creates novel treatments, but many of them are extremely costly. In the USA, treatment for serious disease is already too expensive to be paid for out of most people's earnings, health insurance has not been universal and 'safety nets' to protect the poor and elderly have been cut back. Baby-boomers, born in the late 1940s, are now approaching elderly status. All of these factors will put further pressure on the system. The Health Care and Education Reconciliation Act, which became US law in March 2010, should ameliorate the situation.

Considerations of social and economic policy are outside the scope of this book. However, science can make a contribution by improving the efficiency of use of the resources available.

For example, many drugs vary in their effectiveness in different patients. A drug may be effective for some patients and useless for others. Some patients may tolerate a treatment easily; others may suffer side effects ranging from discomfort through disability to death.

Analysis of patients' genes and proteins permits selection of drugs and dosages optimal for individual patients, a field called **pharmacogenomics**. Physicians can thereby avoid experimenting with different therapies, a procedure that is dangerous in terms of side effects – sometimes even fatal – and in any case is wasteful and expensive. Treatment of patients for adverse reactions to prescribed drugs consumes billions of dollars in health care costs. Conversely, being able to predict individual patients' responses can make it possible to rescue drugs that are safe and effective in a minority of patients, but which have been rejected before or during clinical trials because of inefficacy or severe side effects in the majority of patients.

Some specific examples:

- Acute lymphoblastic leukaemia is a childhood cancer treated by thiopurines. In the patient, the enzyme thiopurine methyltransferase breaks down the drug. A genetic variant producing an inactive enzyme threatens build-up of toxic levels of the drug in patients. Screening for the deficiency allows monitoring to determine appropriate dosages.
- Abacavir is a drug used in treatment of AIDS. 4–8% of patients show a serious, potentially fatal hypersensitivity reaction. This is correlated with MHC allele HLA-B*5701. Genomic screening can thereby detect potential hypersensitivity, and guide treatment.
- Cytochromes P450 are a family of enzymes in the liver responsible for metabolizing a wide variety of drugs. Sequence variations affect the activities of these enzymes, to the point where lowered activity, or loss of activity, can cause drug toxicity. Genetic tests for variations in cytochrome P450 genes warn of potential overdose dangers. When J.D. Watson's genome sequence was determined, it emerged that he is homozygous for an unusual allele of the drug-metabolizing cytochrome gene CYP2D6. Individuals with this genotype metabolize some drugs relatively slowly. Watson had been taking β-blockers to reduce his blood pressure; however, the treatment made him unacceptably sleepy. Based on information from his genome he is now taking a lower dose.

Our ability to rationalize and anticipate individual differences in responses to drugs can improve the rate of clinical success. In improving the application of resources by avoiding treatments that are useless or even harmful, science can have a direct effect on the economics of health care delivery.

- Applications of genome sequencing to medicine already include: detection of diseases irrevocably implied by gene sequences, such as Huntington's disease, detection of diseases which an individual has an enhanced risk of developing but for which the risks can be lessened by changes in lifestyle or prophylactic surgery, genetic counselling of prospective parents, and the identification of optimal therapy depending on detailed diagnosis of the variety of the disease, and on the expected reaction of the patient to different drugs.

The evolution and development of databases

High-throughput sequencing methods are generating immense amounts of data. How can this information be archived and presented in useful form? This is the responsibility of databases.

Modern genomics combines biological data with computer science and statistics. From this union have emerged both an intellectual framework and a technological toolkit. These contribute essential components to research and applications of genomics and related fields. Access to data and software has become as necessary a part of the infrastructure of research as distilled water. Computer storage and software are essential for generating, collecting, archiving, curating, distributing, retrieval, and analysis of biological data. Many biologists specialize in computing; all find it an essential resource.

Sources of biological data include several high-throughput streams, including:

- *Systematic genome sequencing*
- *Protein expression patterns*
- *Metabolic pathways*
- *Protein interaction patterns and regulatory networks*
- *The scientific literature, including bibliographical databases. Now that much of the scientific literature is available online, it can itself be the subject of data mining.*

Bioinformatics started with clerical projects for archiving and distribution of data. Annotation and even curation of the data were, initially, minimal. Data entry structuring was rudimentary. Many databanks were distributed as a series of flat files – that is, plain text – to provide the lowest common denominator of intelligibility by different computer systems.

Recognition of the importance of access to the data led to policy decisions by journals and funding agencies that obliged scientists to make their sequence and structure data publicly available. This increased the harvest and permitted the combination of fragmentary and incoherent data into logically structured collections. Databanks began to adopt fixed formats and controlled vocabularies, and recognized the importance of careful curation and annotation of the data.

There followed a recognition of the power of information-retrieval tools, which require imposing a structure on the data. These made possible the selective search and retrieval of data needed to answer particular scientific questions. Other methods permit numerical and/or textual analysis. Sequence alignment is by far the most common example.

For the development and provision of these and many other tools, biology is indebted to computer science. Computer science is a young but nevertheless mature field, moving swiftly on the back of devices for high-capacity information storage and speed of calculation. Its background was mathematics and engineering. Its goal has been the development of methods for making the most effective use of the available technology. This involves understanding what determines the effectiveness of methods, devising efficient methods for the problems we want to solve, and production of computer programs to implement these methods.

The growth in the amount of information, together with the novel and unusual constellation of skills required for managing it, has led to the establishment of specialized institutions to organize the work.

The earliest databanks, the Protein Sequence Databank, started by Margaret O. Dayhoff at the National Biomedical Research Foundation, Georgetown University, Washington DC, USA; and the Protein Data Bank of macromolecular structures, started by Walter C. Hamilton of Brookhaven National Laboratory, New York, USA, were outgrowths of – and originally often found themselves uncomfortably competing for funding with – research activities.

With increasing recognition of the importance of databanks and the growing political attractiveness of their goals, several high-profile institutions have been established. Examples include the US National Center for Biotechnology Information (NCBI) at the US National Library of Medicine (NLM) in Bethesda, Maryland, USA, and the European Bioinformatics Institute outstation of the European Molecular Biology Laboratory, in Hinxton, Cambridgeshire, UK.

In addition to archiving and curating, databanks have been active in developing software for information retrieval and analysis. This is an integral part of making the resources in their care available on-line. However, the advantages of vesting responsibility for the archives in monolithic organizations do not preclude multiple routes of access to the data. Colloquially, anyone can design an individual 'front end'.

A databank without effective modes of access is merely a data graveyard.

Databank evo-devo

Archiving projects originally tended to specialize, matching the nature of the data with the skills of the scientists curating it. Typically, the Protein Data Bank employed crystallographers, whereas genomic databanks attracted sequence specialists. The databanks tended to develop along different lines, dictated in part by the nature of the data.

This independence and divergence has some drawbacks, notably the difficulty of answering the questions of greater subtlety that arise in studying relationships among information contained in separate databanks. For example: for which proteins of known structure involved in diseases of purine biosynthesis in humans are there related proteins in yeast? We are setting conditions on known structure, specified function, detection of relatedness, correlation with disease, and specified species. We need links that facilitate simultaneous access to several databanks.

In principle, the problems would go away if all the databanks merged into one. To some extent this is happening. For example, the umbrella database UniProt integrates the contents, features, and annotation of several individual databases of protein families, domains, and functional sites, and contains links to others.

Alternatively, databanks are taking advantage of the World Wide Web to include a dense network of links among different archives. Today, the quality of a database depends not only on the information it contains but also on the effectiveness of its links to other sources of information. The growing importance of simultaneous access to databanks has led to research in databank interactivity – how can databanks 'talk to one another' without sacrificing the freedom of each one to structure its own data in appropriate ways?

Specialized user communities may extract subsets of the data, or recombine data from different sources, to provide specialized avenues of access. Such 'boutique' databases depend on the primary archives as sources of information, but redesign the organization and presentation. Indeed, different derived databases can 'slice and dice' the same information in different ways. Similarly, an individual may improve a method for solving an important problem – for example, identification of genes within genome sequences – and make the method available via a web site.

A reasonable extrapolation suggests the idea of specialized 'virtual databases', grounded in the archives but providing individual scope and function, tailored to the needs and achievements of individual research groups or even individual scientists.

Genome browsers

There are many databases in the field of molecular biology. We shall survey them in Chapter 3. However, a particular species of database that deals with full-genome sequences and related information is a genome browser.

Genome browsers are projects designed to organize and annotate genome information, and to present it via web pages together with links to related data, such as evolutionary relationships or correlations with disease. Genome browsers are like encyclopaedias. There is a commitment to linking the actual sequence data to as many as possible of the resources of data about an organism. Two major genome browsers are Ensembl, a joint project of the Sanger Centre and the European Bioinformatics Institute in Hinxton, UK, and the University of California at Santa Cruz Genome Browser in the USA.

In addition to presenting the data, genome browsers provide tools for searching and analysis. You can scroll through chromosomes, zooming in on interesting regions. For any region, you can see its contents and annotated properties. For genes, you can see information about function, expression, and homologues.

Let us look briefly at a specific example. The globins form a family of proteins in humans and other species. The α-globin gene cluster provides an example both of the appearance of genome browser web pages, and of the contents of an interesting region of the human genome.

Definitely 'worth a detour' is the α-globin locus on chromosome 16. Our access is through a genome browser. Figure 1.8 shows a diagram of chromosome 16 (right) and the assignments of genes in the chromosome (left). The banding pattern on the chromosome is produced by differential uptake of Giemsa stain, a dye that reacts with DNA. The bands reflect the local structure of the chromosome, which depends on the DNA sequences. It is clear from this figure that the light bands are more gene-rich than the dark ones.

The α-globin gene cluster is near the upper tip of chromosome 16, in the band denoted p13.3. (For the nomenclature of chromosome bands see p. 85.) We can focus in on this region to see the distribution of individual genes (Figure 1.9). The α-globin locus contains five genes that are expressed, ζ, α_D, α_2, α_1, and θ_1, and two degenerate pseudogenes, $\psi\zeta$ and $\psi\alpha_1$ (see Figure 1.10).

These are not the only globins in the genome. Compare the α-globin locus with another multigene cluster, the β-globin locus, which appears on chromosome 11 (see Figures 1.10 and 1.11). In addition, there are single genes for myoglobin, cytoglobin, and neuroglobin (see Weblem 1.2). All of these genes arose through evolution by duplication and divergence (see Figure 1.11).

The order of the genes on the chromosome has another significance. Their transcription, leading to protein synthesis, follows a strict developmental pattern. A human embryo (up to six weeks after conception) primarily synthesizes two haemoglobin chains: ζ and ε. These molecules form a $\zeta_2\varepsilon_2$ tetramer. From six weeks after conception until about eight weeks after birth, the predominant species shifts to foetal haemoglobin, $\alpha_2\gamma_2$. This is succeeded by adult haemoglobin, $\alpha_2\beta_2$. As the organism develops,

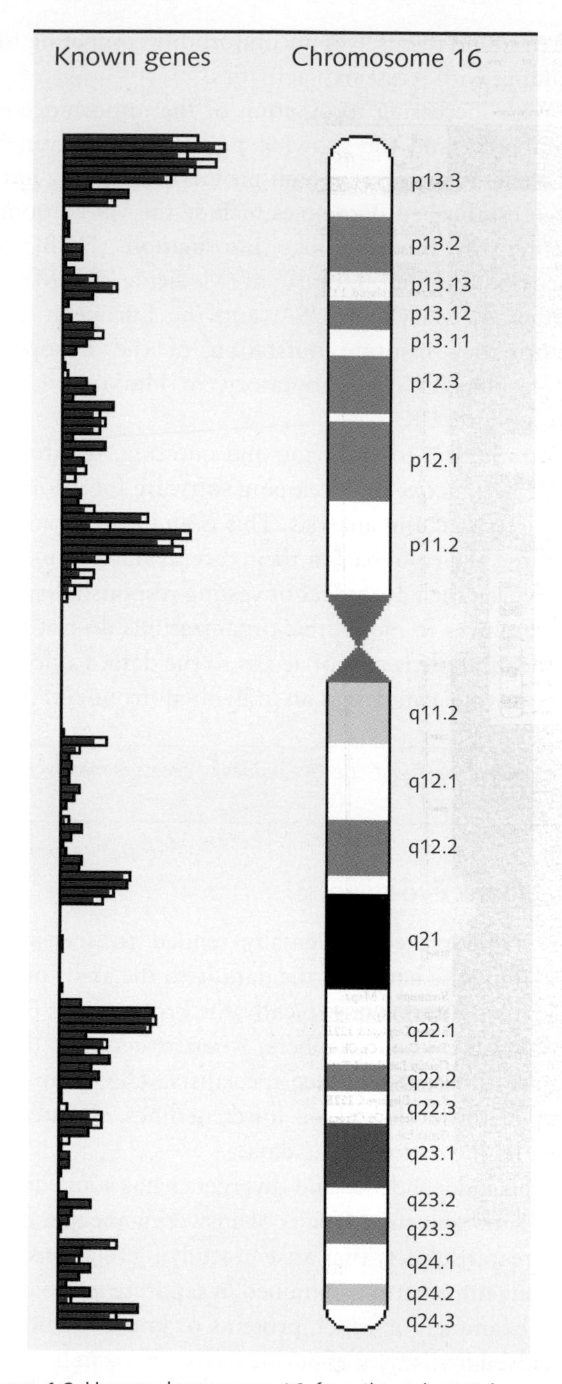

Figure 1.8 Human chromosome 16, from the web site of Ensembl, based at the Sanger Centre in Hinxton, Cambridgeshire. Right: a schematic diagram of the chromosome, showing the centromere, and the banding patterns. Left: assignments of genes to the DNA sequence along the chromosome. Genes of known function appear in red. Additional genes appear in white. There is a clear correlation between the absence of grey and black bands on the chromosome and gene-rich regions.

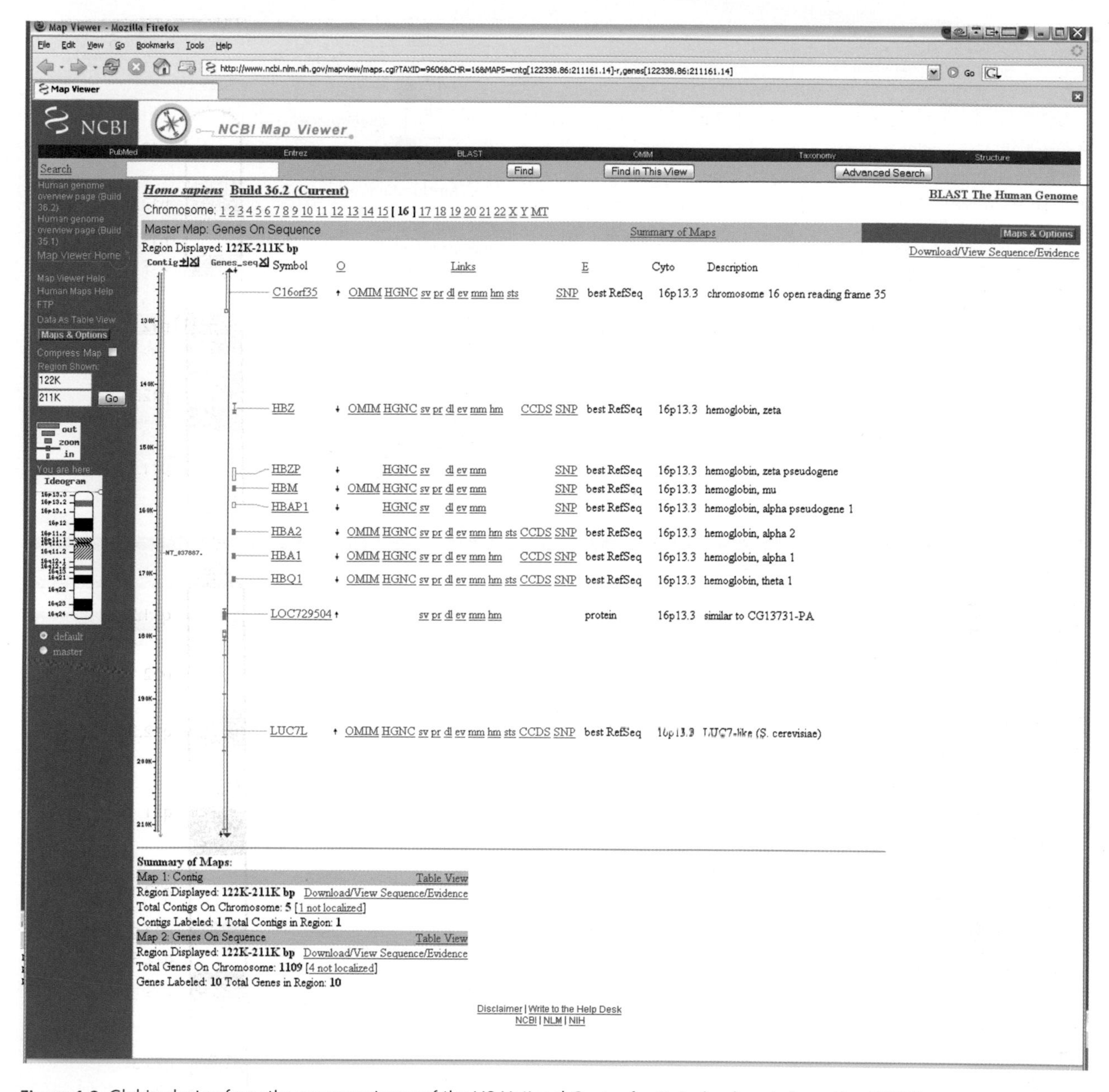

Figure 1.9 Globin cluster, from the genome viewer of the US National Center for Biotechnology Information (NCBI).

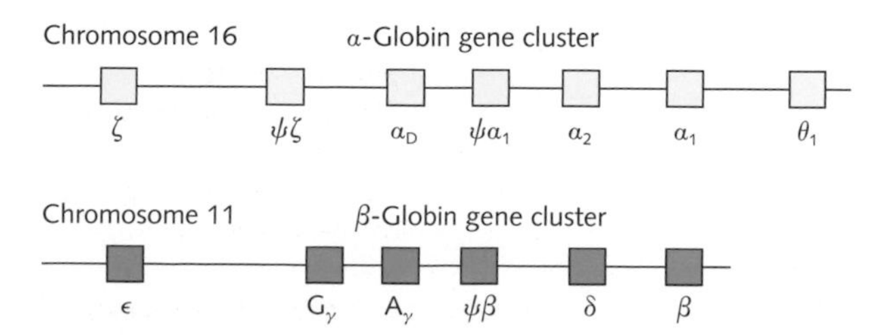

Figure 1.10 Distribution of protein-coding genes and pseudogenes in the α-globin cluster on chromosome 16, and the β-globin cluster on chromosome 11.

expression passes between genes in order of their position on the chromosome.

Focusing down still more finely within the α-gene cluster, we can see the structure of the gene for HBA, the α-subunit of adult haemoglobin (see Figure 1.12). Like most other eukaryotic genes, it is divided into exons (expressed segments of the gene) and introns (intervening regions) (Figure 1.13).

Now we have reached the level of the sequence itself (see Figures 1.14 and 1.15). The protein folds

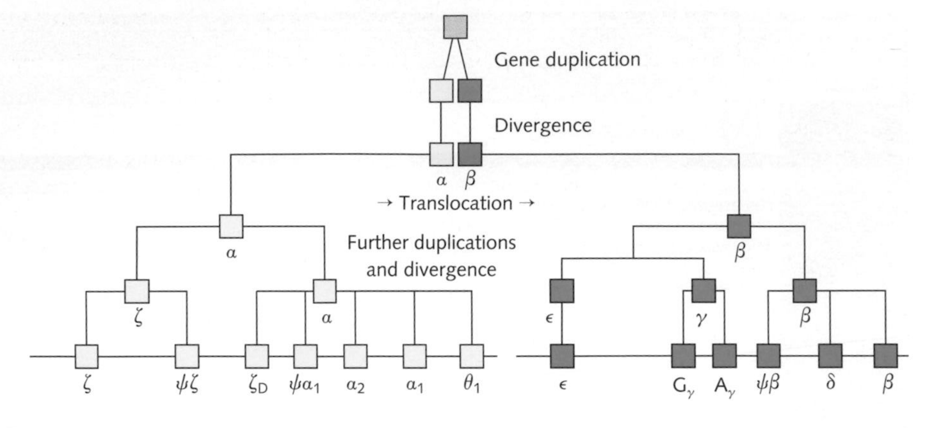

Figure 1.11 Haemoglobin genes and pseudogenes are distributed on their chromosomes in a way that appears to reflect their evolution via duplication and divergence. That is, adjacent genes are similar in sequence. The evolutionary tree can be drawn without any intersecting lines.

Figure 1.12 The structure of the gene for the α-subunit of human haemoglobin.

	Exon 1	Intron 1	Exon 2	Intron 2	Exon 3	

Figure 1.13 Structure of the human α_1-globin gene (HBA): 3′-untranslated region in red, exons in green, introns in black, and 5′-untranslated region in cyan. This exon/intron pattern is conserved in many expressed vertebrate globin genes, including haemoglobin α and β chains, and myoglobin. In contrast, the genes for plant globins have an additional intron, genes for *Paramecium* globins one fewer intron, and genes for insect globins contain none. The gene for human neuroglobin, a homologue expressed at low levels in the brain, contains three introns, like plant globin genes.

```
LOCUS       HSAGL1                  1138 bp    DNA     linear   PRI 24-APR-1993
DEFINITION  Human alpha-globin germ line gene.
ACCESSION   V00488
VERSION     V00488.1  GI:28546
KEYWORDS    alpha-globin; germ line; globin.
SOURCE      Homo sapiens (human)
  ORGANISM  Homo sapiens
            Eukaryota; Metazoa; Chordata; Craniata; Vertebrata; Euteleostomi;
            Mammalia; Eutheria; Euarchontoglires; Primates; Catarrhini;
            Hominidae; Homo.
REFERENCE   1  (bases 1 to 1138)
  AUTHORS   Liebhaber,S.A., Goossens,M.J. and Kan,Y.W.
  TITLE     Cloning and complete nucleotide sequence of human 5'-alpha-globin
            gene
  JOURNAL   Proc. Natl. Acad. Sci. U.S.A. 77 (12), 7054-7058 (1980)
   PUBMED   6452630
COMMENT     KST HSA.ALPGLOBIN.GL [1138].
FEATURES             Location/Qualifiers
     source          1..1138
                     /organism=''Homo sapiens''
                     /mol_type=''genomic DNA''
                     /db_xref=''taxon:9606''
     prim_transcript 98..929
     exon            98..230
                     /number=1
     CDS             join(135..230,348..551,692..820)
                     /codon_start=1
                     /product=''alpha globin''
                     /protein_id=''CAA23748.1''
                     /db_xref=''GI:28547''
                     /db_xref=''GOA:P01922''
                     /db_xref=''UniProtKB/Swiss-Prot:P01922''
                     /translation=''MVLSPADKTNVKAAWGKVGAHAGEYGAEALERMFLSFPTTKTYF
                     PHFDLSHGSAQVKGHGKKVADALTNAVAHVDDMPNALSALSDLHAHKLRVDPVNFKLL
                     SHCLLVTLAAHLPAEFTPAVHASLDKFLASVSTVLTSKYR''
     exon            348..551
                     /number=2
     exon            692..929
                     /number=3
ORIGIN
        1 aggccgcgcc ccgggctccg cgccagccaa tgagcgccgc ccggccgggc gtgcccccgc
       61 gccccaagca taaaccctgg cgcgctcgcg gcccggcact cttctggtcc ccacagactc
      121 agagagaacc caccatggtg ctgtctcctg ccgacaagac caacgtcaag gccgcctggg
      181 gtaaggtcgg cgcgcacgct ggcgagtatg gtgcggaggc cctggagagg tgaggctccc
      241 tcccctgctc cgacccgggc tcctcgcccg cccggaccca caggccaccc tcaaccgtcc
      301 tggccccgga cccaaacccc acccctcact ctgcttctcc ccgcaggatg ttcctgtcct
      361 tccccaccac caagacctac ttcccgcact tcgacctgag ccacggctct gcccaagtta
      421 agggccacgg caagaaggtg gccgacgcgc tgaccaacgc cgtggcgcac gtggacgaca
      481 tgcccaacgc gctgtccgcc ctgagcgacc tgcacgcgca caagcttcgg gtggacccgg
      541 tcaacttcaa ggtgagcggc gggccgggag cgatctgggt cgagggggcga gatggcgcct
      601 tcctctcagg gcagaggatc acgcgggttg cgggaggtgt agcgcaggcg gcggcgcggc
      661 ttgggccgca ctgaccctct tctctgcaca gctcctaagc cactgcctgc tggtgaccct
      721 ggccgcccac ctccccgccg agttcacccc tgcggtgcac gcttccctgg acaagttcct
      781 ggcttctgtg agcaccgtgc tgacctccaa ataccgttaa gctggagcct cggtagccgt
      841 tcctcctgcc cgctgggcct cccaacgggc cctcctcccc tccttgcacc ggcccttcct
      901 ggtctttgaa taaagtctga gtgggcggca gcctgtgtgt gcctgggttc tctctgtccc
      961 ggaatgtgcc aacaatggag gtgtttacct gtctcagacc aaggacctct ctgcagctgc
     1021 atggggctgg ggagggagaa ctgcagggag tatgggaggg gaagctgagg tgggcctgct
     1081 caagagaagg tgctgaacca tcccctgtcc tgagaggtgc cagcctgcag gcagtggc
//
```

Figure 1.14 EMBL Data Library entry for human haemoglobin, α chain.

Figure 1.15 (a) Nucleotide base sequence of mature messenger RNA produced from the gene sequence in previous figure. a = adenine, u = uracil, g = guanine, c = cytosine. (b) Amino acid sequence from translation of (a). Each letter stands for one of the twenty canonical amino acids. The convention is that bases appear in lower case, amino acids in upper case. For instance, a = adenine (a base) and A = alanine (an amino acid).

```
(a) Nucleotide sequence of mRNA:
auggugcugucuccugccgacaagaccaacgucaaggccgccugggguaaggucggcgcgcacgcuggcg
aguauggugcggaggcccuggagaggauguuccuguccuuccccaccaccaagaccuacuucccgcacuu
cgaccugagccacggcucugcccagguuaagggccacggcaagaagguggccgacgcgcugaccaacgcc
guggcgcacguggacgacaugcccaacgcgcuguccgcccugagcgaccugcacgcgcacaagcuucggg
uggacccggucaacuucaagcuccuaagccacugccugcuggugacccuggccgcccaccuccccgccga
guucaccccugcggugcacgccucccuggacaaguuccuggcuucugugagcaccgugcugaccuccaaa
uaccguuaa

(b) Translation to an amino acid sequence:
MVLSPADKTNVKAAWGKVGAHAGEYGAEALERMFLSFPTTKTYFPHFDLSHGSAQVKGHGKKVADALTNA
VAHVDDMPNALSALSDLHAHKLRVDPVNFKLLSHCLLVTLAAHLPAEFTPAVHASLDKFLASVSTVLTSK
YR
```

spontaneously to its proper three-dimensional structure. Two such haemoglobin α chains combine with two corresponding β chains to form the tetrameric structure (Figure 1.16).

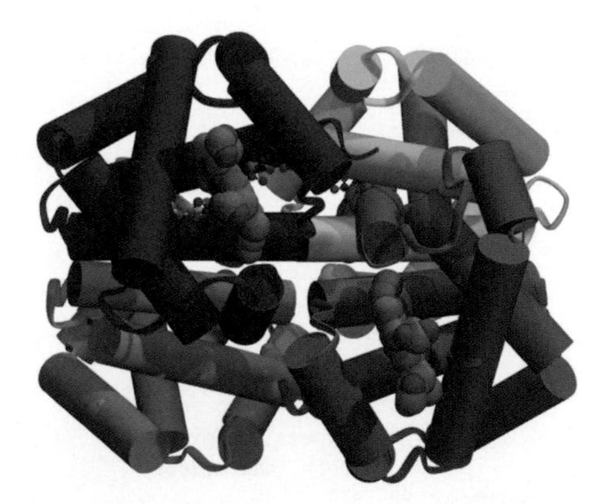

Figure 1.16 The structure of adult human haemoglobin. The protein contains four polypeptide chains – two α-subunits and two β-subunits – each binding a haem group. Cylinders represent α-helices. Spheres represent atoms of the haem groups. (Do not confuse the use of α to designate both a type of helix and a type of subunit of haemoglobin.)

- Database organizations archive data, curate and annotate them, make the data available over the World Wide Web, and provide information-retrieval tools facilitating research.
- The variety of specialities required of the database staff to provide all of these has resulted in large, often international, database institutions. A specific type of database aimed at presenting genomic sequences and related information is called a genome browser.

Protein evolution: divergence of sequences and structures within and between species

Different globins diverged from a common ancestor

Differences in gene sequences, differences in the corresponding amino acid sequences, and differences in three-dimensional structure reflect evolutionary divergence. The globins within the α and β clusters are more closely related than members of the α cluster are to members of the β cluster. Other globins in the human genome – myoglobin, neuroglobin, and cytoglobin – are more distant relatives. Corresponding globins in related species, such as human and horse, have also diverged. In general, the divergence at the molecular level parallels the divergence of the species according to classical taxonomic methods. But the power of comparative genomics and proteomics in tracing precise relationships both within and among species is immense.

The basic tool for investigating sequence divergence is the multiple sequence alignment (see Figure 1.17).

(a) Mammalian Globin Sequences

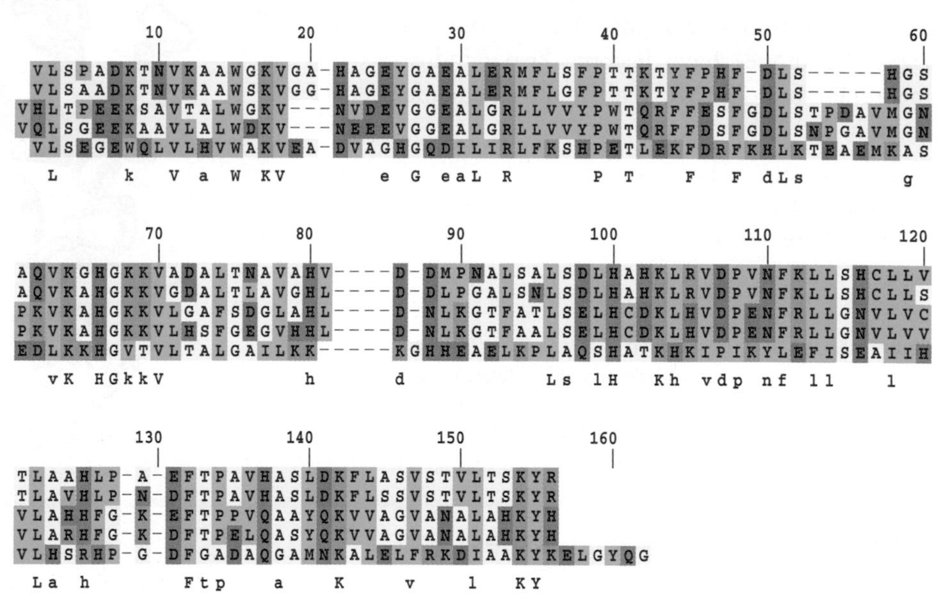

(b) Eukaryote and Prokaryote Full-length Globin Sequences

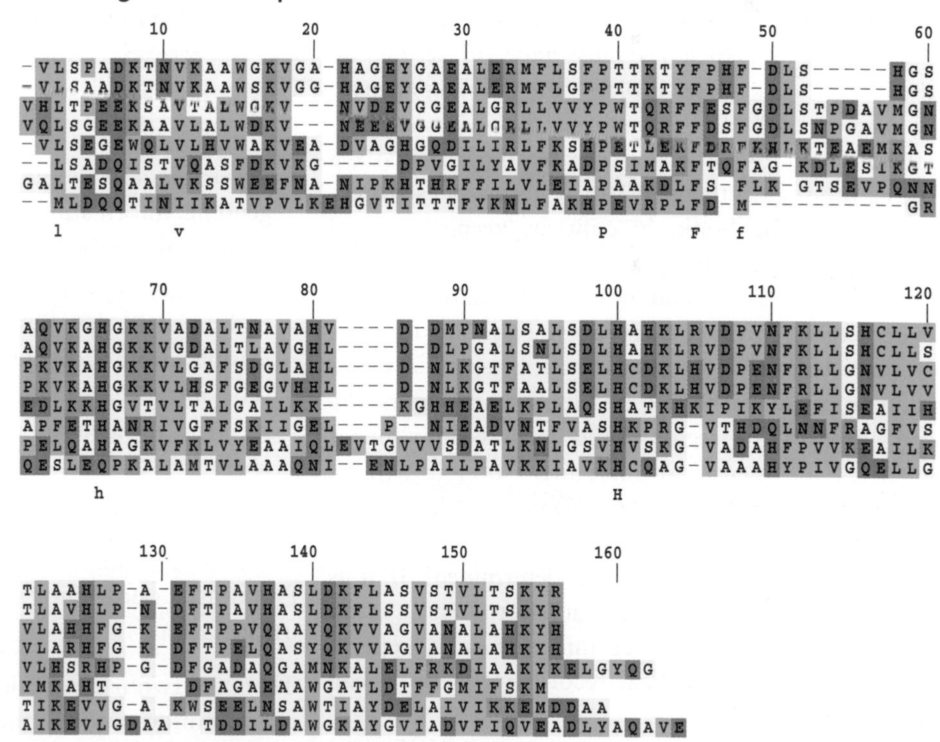

Figure 1.17 (a) Multiple sequence alignment of five mammalian globins: sperm whale myoglobin, and the α and β chains of human and horse haemoglobin. Each sequence contains approximately 150 residues. In the line below the tabulation, upper-case letters indicate residues that are conserved in all five sequences, and lower-case letters indicate residues that are conserved in all but sperm whale myoglobin. (b) Multiple sequence alignment of full-length globins from eukaryotes and prokaryotes. Many fewer positions are conserved than in the mammals-only case. In the line below this tabulation, upper-case letters indicate residues that are conserved in all eight sequences, and lower-case letters indicate residues that are conserved in all but the bacterial globin.

The basic tool for investigating structural divergence is superposition (see Figure 1.18). Applications of these techniques have become heavy industries in molecular biology. We shall make abundant use of them in the following chapters.

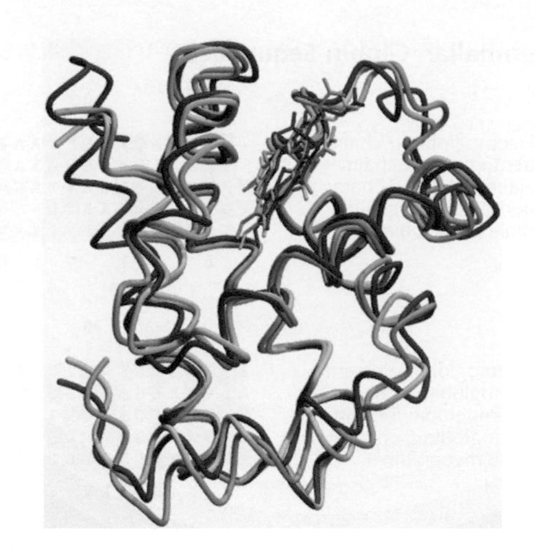

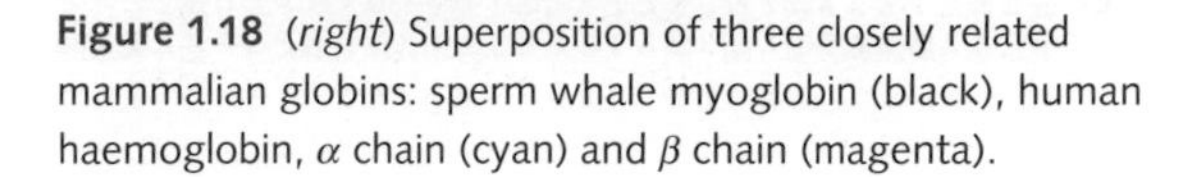

Figure 1.18 (*right*) Superposition of three closely related mammalian globins: sperm whale myoglobin (black), human haemoglobin, α chain (cyan) and β chain (magenta).

Ethical, legal, and social issues

Knowledge creates power. Power requires control. Control requires decisions.

Advances in genomics have created problems that individuals and societies must face. In setting up the Human Genome Project, the US Department of Energy and NIH recognized the importance of ethical, legal, and social issues by allocating 3–5% of the funding to them. We shall discuss these topics throughout the book, in context, keeping several general categories in mind.

DNA databases containing information about every citizen of a country are technically feasible. Routine testing of all newborns for genetic diseases is common, and it would be simple to determine the sequences of the regions used by law-enforcement agencies for identification. Fairly soon it will be relatively inexpensive to determine complete genome sequences of individuals.

Should this information be determined? If it is determined, who should have access to it?

Provided DNA sequence information is kept as private as normal medical records, sequencing can benefit individuals.

More controversial questions arise in allowing the information to be collected into a generally accessible databank. Most people would have no problem accepting an effort that would assist in the capturing of criminals, especially those criminals that are likely to repeat their offences. Most people would have no problem accepting an effort that would make it easier to identify victims of death on a battlefield or after a terrorist attack. Most people recognize that extensive data on genome sequences, in a form that can be correlated with clinical records, would be an extremely valuable source for research. Questions have been raised, however, over:

- *Privacy issues*: should inclusion in a databank of genomic information about individuals require the individuals' consent?
- *What data should be included?* Should the data be limited to the minimum required for standard identification procedures, or be more extensive? (For instance, should sufficient additional data be kept to identify physical features or ethnic characteristics?)
- *Access*: who should have access to the information?

Databases containing human DNA sequence information

There are two major national repositories of human genome information in the UK (see Box 1.8).

- The National DNA Databank (NDNAD) primarily supports law enforcement agencies.
- The UK BioBank has the goal of improving prevention, diagnosis, and treatment of illness. It has amassed a large collection of clinical data, and biological samples to provide sequence information.

Also based in the UK are the comprehensive nucleotide sequence databanks at the European Bioinformatics Institute and The Sanger Centre.

In England and Wales, police have been allowed to take samples, from any individual arrested on suspicion of all but the most minor crimes, and to retain the derived information *and the samples* even if the person arrested is never charged. Consent is not required. Access to the database is not strictly limited to police, but has been used to support research projects without consent of the individuals represented in the database.

In the current political climate, there is intense pressure to tip the scales towards giving governments powers that can be used to protect their citizens. The dangers are that such powers, although they appear to be innocuous to innocent people, may be compatible with abuse if safeguards are inadequate, or – in the worst case – deliberately violated. Even in the United Kingdom, a House of Commons committee in 2005* described as 'extremely regrettable' the fact 'that for most of the time that the NDNAD has been in existence there has been no formal ethical review of applications to use the database and the associated samples for research purposes'. If the situation in the UK could have been described as 'extremely regrettable', one need not be a Colonel Blimp to shudder to think what things are like elsewhere (see Box 1.8).

It has been suggested that the NDNAD be extended to the entire population. The current contents of the database show ethnic and gender biases, which a universal database would eliminate. A higher fraction of reported crimes might be solved, but there is no consensus about how significant this would be. Arguments against such an extension include what is at the moment still a very high cost. There is also intense debate over whether the loss of individual privacy would justify the benefits to society.

An important point about privacy of computer databanks is that, even if there is consensus that certain data *should* be kept confidential, computer security is simply not up to the task of ensuring that it will be.

DNA Sequence Databases, Law Enforcement, and the Courts

The UK National DNA Database (NDNAD) is one of the largest forensic DNA collections in the world. It stores both a database of sequence-derived information, and the biological samples from which the sequences were determined. In mid-2010 it contained profiles of an estimated 5.4 million individuals. Of these, 78.50% are male and 20.82% female; a small number are unassigned. The NDNAD contains personal identification 'DNA fingerprints', and gender. The probability that sequences from a sample collected at a crime scene will match an entry in the database is over 50%!

Entries in the NDNAD are of two types:

(a) Samples from known individuals. Police can collect samples, without consent, from anyone suspected of a 'recordable' offence, not only before trial and possible conviction, but even before being charged with a crime. In the UK, the list of recordable offences is steadily growing through incremental legislation. Roughly speaking, recordable offences include all but the most trivial antisocial actions. For example, under the Football (Offences) Act of 1991, 'unlawfully going onto the playing area' is a recordable offence.

For purposes of elimination, the NDNAD also contains samples from persons present at a crime scene, and from police officers.

(b) Samples collected at a crime scene. They may originate from the perpetrator of a crime, from a victim, or even from someone who had been at the scene at some time other than during the commission of the crime. In the event of a match, these become samples from known individuals. (In the case of a partial match, they may suggest that the crime scene sample came from a near relative of the individual matched.)

→

* House of Commons Science and Technology Committee, Forensic Science on Trial, 29 March 2005. http://www.publications.parliament.uk/pa/cm200405/cmselect/cmsctech/96/96i.pdf

In the UK, the governing legislation has been in an active state, particularly with respect to retention of samples; and, under that heading, particularly with respect to retention of samples from people not convicted of a crime. The Criminal Justice Act of 2003 authorized the widespread collection of samples for DNA profiles in England and Wales, without consent. (This was extended to Northern Ireland in the next year.) The act also allowed the indefinite retention of the information, even if the suspect were never charged with a crime. In 2008, the Counter-Terrorism Act extended the criteria that allow police to demand samples for DNA sequencing.

The law in Scotland, in this as in many other respects, varies from that of England and Wales. Scotland maintains a separate DNA database, and shares results with England. One salient legal difference is that Scotland does not permit automatic indefinite retention of samples from people not convicted of a crime.

The European Court of Human Rights has aligned itself more closely with the Scottish law.

A case against the UK government was brought to the European Court by two individuals who wanted their identifying information expunged from the NDNAD and their samples destroyed. One was tried for attempted robbery but acquitted, the other was never charged. The court held that there had been violation of Article 8 of the European Convention for the Protection of Human Rights and Fundamental Freedoms:

'In conclusion, the Court finds that the blanket and indiscriminate nature of the powers of retention of the fingerprints, cellular samples and DNA profiles of persons suspected but not convicted of offences, as applied in the case of the present applicants, fails to strike a fair balance between the competing public and private interests and that the respondent State has overstepped any acceptable margin of appreciation in this regard. Accordingly, the retention at issue constitutes a disproportionate interference with the applicants' right to respect for private life and cannot be regarded as necessary in a democratic society'.

In the US, in the absence of overriding Federal legislation, laws of individual states govern DNA collection. The variety of guidelines for collection and retention of samples expressed in state laws have had a patchy career in the (state) courts. Some have been declared unconstitutional.

In the US, the analogue of the UK NDNAD is the Combined DNA Index System and the National DNA Index System (CODIS/NDIS), maintained by the Federal Bureau of Investigation (FBI). In late 2010, NDIS contained almost 10 million DNA profiles. This is about twice the size of the corresponding UK NDNAD, but represents a smaller percentage of the national population.

The Genetic Information Nondiscrimination Act (GINA) of 2008 aimed at protecting individual privacy, with respect to genomic information. It prohibits:

(a) Health insurance companies from reducing coverage or increasing prices to individuals based on information from genetic tests.

(b) Employers from making hiring decisions based on DNA sequence information.

(c) Companies from demanding or even requesting a genetic test.

However, the law does not apply to people applying for life insurance or long-term care or disability insurance.

The Health Care and Education Reconciliation Act became US Federal law in March 2010. In theory, it should allay some of the fears associated with the potential consequences of improper release of genetic information.

In principle, use of DNA samples in research should require consent of the individuals donating the material: consent not only for its collection, but for the specific uses to which the samples will be put. There have been cases of 'research goal creep' in which permission was granted for analysis of restricted scope, but the samples were subsequently used for other studies.

A US case testing the propriety of use of samples collected voluntarily from subjects was settled in April 2010.

A Native American group, the Havasupai, who inhabit an inaccessible area of the Grand Canyon, has among the highest known incidences of Type II diabetes. In 1991, 55% of Havasupai women and 38% of Havasupai men were affected. Scientists from Arizona State University collected samples from the group, and carried out research projects, believed – by the subjects – to be focused on susceptibility to diabetes. However, using the same samples, the scientists also investigated genetic susceptibility for schizophrenia, and evidence for migration patterns. The subjects objected. Researchers pointed to the wording in the consent form; representatives of the Havasupai alleged (among other things) that the wording was too vague to constitute truly informed consent. The ensuing lawsuit was settled, with the Arizona state agency responsible for the university agreeing to pay the Havasupai US$7 000 000, and to return the samples.

The conclusion is that, in the UK and US at least, legislation is moving in the direction of greater protection of privacy of DNA sequence information. Belief in this protection, which may in some respects be illusory, may lead to increased genetic testing, both in regular medical practice, and by private companies. (a) Like mailing lists, testing companies may have the right to sell genetic information to outside parties. (b) Given the increased degree of international sharing of identification information, individuals need to be concerned not with the countries with the most secure databases, but those with the least secure ones. (c) Experience has shown that much private information in fact becomes disseminated, either through accident or design.

Ethical considerations for compiling DNA databases

Genomic databases are useful in identifying individuals from samples collected at crime scenes. In drafting the laws granting law-enforcement agencies authorization to maintain such databases, there is a tension between the desire to protect society against offenders and upholding individuals' rights of privacy.

RECOMMENDED READING

- The first two books are sources for the history of the development of molecular biology after the Second World War. Rosenfield, Ziff, & Van Loon's is a less formal but not less serious account of the founding people and events.

Judson, H.F. (1980). *The Eighth Day of Creation: Makers of the Revolution in Biology*. Jonathan Cape, London.

de Chadarevian, S. (2002). *Designs for Life/Molecular Biology after World War II*. Cambridge University Press, Cambridge.

Rosenfield, I., Ziff, E.B., & Van Loon, B. (1983). *DNA for Beginners*. Writers & Readers Publishing, London.

- The next five publications deal with genomics and the Human Genome Project, placing it in broad context. The book by Sulston & Ferry is a personal account by one of the major players.

Ridley, M. (1999). Genome: *The Autobiography of a Species in 23 Chapters.* HarperCollins Publishers, New York.

Lander, E.S. & Weinberg, R.A. (2000). Genomics: journey to the center of biology. Science **287**, 1777–1782.

Wolfsberg, T.G., Wetterstrand, K.A., Guyer, M.S., Collins, F.S., & Baxevanis, A.D. (2002). A user's guide to the human genome. Nat. Genet. **32** (Suppl.), 1–79.

Sulston, J. & Ferry, G. (2002). *The Common Thread: A Story of Science, Politics, Ethics and the Human Genome*. Bantam Press, London.

Choudhuri, S. (2003). The path from nuclein to human genome: a brief history of DNA with a note on human genome sequencing and its impact on future research in biology. Bull. Sci. Technol. Soc. **23**, 360–367.

The 18 February 2011 issue of *Science* magazine contains several articles recognizing the tenth anniversary of the sequencing of the human genome.

- The richness of non-protein coding RNAs in genomes:

Baker, M. (2011). Long noncoding RNAs: the search for function. Nat. Methods **8**, 379–383.

- Genomics and personalized medicine:

Cooper, D.N., Chen, J.M., Ball, E.V., Howells, K., Mort, M., Phillips, A.D., Chuzhanova, N., Krawczak, M., Kehrer-Sawatzki, H., & Stenson P.D. (2010). Genes, mutations, and human inherited disease at the dawn of the age of personalized genomics. Hum. Mutat. **31**, 631–655.

Marian, A.J. (2010). Editorial review: DNA sequence variants and the practice of medicine. Curr. Opin. Cardiol. **25**, 182–185.

Ma, Q. & Lu, A.Y.H. (2011). Pharmacogenetics, pharmacogenomics, and individualized medicine. Pharmacol. Rev. **63**, 437–459.

Brunicardi, F.C., Gibbs, R.A., Wheeler, D.A., Nemunaitis, J., Fisher, W., Goss, J., & Chen, C. (2011). Overview of the development of personalized genomic medicine and surgery. World J. Surg. **35**, 1693–1699.

Mahungu, T.W., Johnson, M.A., Owen, A., & Back, D.J. The impact of pharmacogenetics on HIV therapy. Int. J. STD AIDS **20**, 145–151.

- Finally, a collection of papers on ethical, legal, and social issues, describing some of the interactions, and collisions, between scientific advances and social policy; and a more recent paper, and a government report.

Gaskell, G. & Bauer, M.W. (eds) (2006). *Genomics & Society/Legal, Ethical & Social Dimensions*. Earthscan, London.

Levitt, Mairi. (2007). Forensic databases: benefits and ethical and social costs. Brit. Med, Bull. **83**, 235–248.

In March 2010, The Home Affairs Committee of the UK House of Commons published a report on The National DNA Database: http://www.publications.parliament.uk/pa/cm200910/cmselect/cmhaff/222/22202.htm.

EXERCISES, PROBLEMS, AND WEBLEMS

Exercises

Exercise 1.1 Make a very rough estimate of the average density of protein-coding genes in the human genome. (Total genome size ~3×10^9 bp; total number of genes ~3×10^4 genes.)

Exercise 1.2 Assume that most eukaryotes have approximately 25 000 protein-coding genes, and that the average eukarotic protein has a length of 300 amino acids. Assume that other functional regions, including RNA-coding genes and control regions, do not require more base pairs that the protein-coding genes themselves (an assumption for which there exists not the slightest justification other than ignorance). Estimate the minimum size that a eukaryotic genome could potentially have.

Exercise 1.3 For the standard genetic code (Box 1.1), give an example of a pair of codons related by a synonymous single-site substitution (a) at the third position and (b) at the first position. (c) Give an example of a pair of codons related by a non-synonymous single-site substitution at the third position. (d) Can a change in the second position of a codon ever produce a synonymous mutation?

Exercise 1.4 (a) Is it possible to convert Phe to Tyr by a single base change? If so, what would be possible wild-type and mutant codons? (b) Is it possible to convert Ser to Arg by a single base change? If so, what would be possible wild-type and mutant codons? (c) What is the minimum number of base substitutions that would convert Cys to Glu? (d) In the evolution of an essential protein encoded by a single gene, a Trp is converted to a Gln by two successive single-base changes. What is the intermediate codon?

Exercise 1.5 RNA editing in the mitochondria of higher plants often changes cytosine to uracil at the second position of codons. What amino acid changes could this effect? Note that higher plant mitochondria use the standard genetic code.

Exercise 1.6 A single base-pair deletion in an exon in a protein-coding gene would be very serious because it would throw off the reading frame. What would you expect to be the effect of a single base-pair deletion in a gene for a structural RNA molecule? (Consider transfer RNA, for example; see Figure 1.4.)

Exercise 1.7 Here is a fragment of length 200 from a mitochondrial genome. If this fragment were processed by a sequencer giving read length 35, what would be (a) the result of a single-end read?, (b) the result of a paired-end read? (See Figure 1.7.)

```
  1 5′-gttaatgtag cttaaactaa agcaaggcac tgaaaatgcc tagatgagtc tgcctactcc
 61    ataaacataa aggtttggtc ctagcctttc tattagttga cagtaaattt atacatgcaa
121    gtatctgcct cccagtgaaa tatgccctct aaatccttac cggattaaaa ggagccggta
181    tcaagctcac ctagagtagc tcatgacgcc ttgctaaacc acgcccccac gggatacagc-3′
```

Exercise 1.8 Which of the following bands of human chromosome 16 are gene-rich: p13.3, q22.1, q11.2? (See Figure 1.9.)

Exercise 1.9 On a photocopy of Figure 1.9, mark the approximate position of the α-globin gene cluster.

Exercise 1.10 From Figure 1.10, estimate the number of genes per base pair in the globin region.

Exercise 1.11 On a photocopy of Figure 1.14, mark with highlighters the regions shown in Figure 1.13, in the same colours as in Figure 1.13.

Exercise 1.12 On a photocopy of the amino acid sequence of the human haemoglobin α chain in Figure 1.15(b), indicate the regions arising from the three exons in the gene.

Exercise 1.13 What are the symmetries of the structure of the haemoglobin molecule? (See Figure 1.16.) If the α and β subunits were identical, what additional symmetries would there be?

Exercise 1.14 On a photocopy of the sequence of the Tn3 transposon of *E. coli* (Box 1.6), indicate the extents of the inverted terminal repeats.

Exercise 1.15 How might genome sequencing be applied to the conservation of endangered species?

Problems

Problem 1.1 Draw a rough sketch of the outline of a small eukaryotic plant cell (representing 10 000 nm diameter) to fill almost a whole page. Within this outline, draw in, at scale, the cellular components listed in the table. At the same scale, draw an *E. coli* cell (approximate diameter 2000 nm).

Globular protein diameter	4 nm
Cell membrane thickness	10 nm
Ribosome	11 nm
Large virus	100 nm
Mitochondrion	3000 nm
Length of chloroplast	5000 nm
Cell nucleus	6000 nm

Problem 1.2 From the following data, compute the number of genes per Mb for mitochondria, rickettsia, chloroplasts, and cyanobacteria. Compare the mitochondria with the rickettsia and the chloroplasts with the cyanobacteria. Is the smaller genome size of the organelles largely the result of eliminating non-functional DNA, or loss of genes, some of which were transferred to the nuclear genome?

Genome	Number of bp	Number of genes
Human mitochondrion	16 569	37
Rickettsia prowazekii	1 111 523	834
Arabidopsis thaliana chloroplast	154 478	128
Synechocystis sp. PCC 6803	3 573 470 (plus plastids)	~3500

Problem 1.3 It is estimated that the human immune system synthesizes 10^8–10^{10} antibodies. The portion of a typical antibody active in binding antigen consists of two variable domains each containing 100 amino acids. If every variable domain were separately encoded in the genome but could pair promiscuously, then the number of variable domains that needed to be encoded would be of order of magnitude the square root of the total number of antibodies. How many bases would be required to encode separately all the variable domains? Compare this with the size of the human genome.

Problem 1.4 Human haemoglobin α-subunits have helices A, B, C, E, F′, F, G, and H. Human haemoglobin β-subunits have helices A, B, C, D, E, F′, F, G, and H. The sites of interaction of the histidine residues with the iron in the haem group are in the F and G helices. In Figure 1.16, in what colours do the α-subunits appear and in what colours do the β-subunits appear?

Problem 1.5 Outline how you would search for tRNA genes in a genome sequence. Assume that the lengths of the double-helical regions and the lengths of the single-stranded regions between them vary within relatively tight limits and that there are only a limited number of deviations from perfect base pairing in the helical regions. How would your method have to be modified if tRNA genes contained introns?

Problem 1.6 In Figure 1.17(a), (a) find five positions in which the amino acid is the same in all haemoglobin chains but different in myoglobin. (b) Find four positions at which human and horse α chains have the same amino acid, human and horse β chains have the same amino acid, and myoglobin has a different amino acid. (c) Find two positions at which myoglobin and human and horse β chains have the same amino acid, but the α chains have a different one. (d) Classify the following sequence. Is it a haemoglobin α chain, a haemoglobin β chain, or a myoglobin?

```
MGLSDAEWQLVLNVWGKVEADIPGHGQDVLIRLFKGHPETLEKFDRFKHLKTEDEMKASE
DLKKHGTTVLTALGGILKKKGQHEAEIQPLAQSHATKHKIPVKYLEFISEAIIQVIQSKH
SGDFGADAQGAMSKALELFRNDIAAKYKELGFQG
```

Problem 1.7 The annual birth rate in the UK is approximately 700 000. Two 'back-of-the-envelope calculations': (a) If the goal of a $US1000 genome is attained, what would be the total cost (of) sequencing 700 000 babies? The current annual budget of the UK National Health Service is £120 000 000 000. What percentage of the current NHS budget would be required to sequence the genomes of all babies born in the UK? (Whether the money would be available, and, if it were, whether sequencing would be the best use of it, are separate questions.) (b) If the sequence data were stored completely and independently, by what percentage would they increase the current holdings of the Nucleotide Sequence Databanks?

Problem 1.8 Suppose that you knew your personal DNA fingerprint in the format stored in national DNA databases. Assume that the information did not reveal that you were unusually susceptible to any known disease. What are the arguments for and against your voluntarily depositing these data in the national DNA database of the country of your residence? What conditions would you want to impose on the use of the data?

Problem 1.9 Outline the arguments for and against the use of animals for testing in drug development. What restrictions might be imposed that would mitigate any of the arguments against use of animals in testing? How and to what extent would the effectiveness of drug development suffer if these restrictions came into force?

Weblems

Weblem 1.1 At the time of preparing the manuscript of this book, the sequencing capacity of the largest dedicated genomics institute was estimated as 3.2×10^{12} bp per day. This corresponds to 1000 human genome equivalents. What is the corresponding current maximal sequencing capacity?

Weblem 1.2 For each of the following traits of any human being, estimate the contributions of (1) sequences of particular genes, (2) life history and environment, (3) epigenetic factors: (a) blood type; (b) adult height; (c) native language; (d) whether or not a person will develop Huntington's disease; (e) whether or not a person will develop emphysema; (f) whether or not a person will develop Angelman syndrome.

Weblem 1.3 Which human chromosomes contain genes for ubiquitin?

Weblem 1.4 The gene for insulin is 1.7 kb long; the LDL receptor gene is 5.45 kb; and the dystrophin gene is 2400 kb. (a) What are the lengths of the amino acid sequences of these proteins? (b) What are the ratios of total exon length/total gene length and total intron length/total gene length for these three genes?

CHAPTER 2

Genomes are the Hub of Biology

LEARNING GOALS

- Recognizing that genomics has transformed our approaches to *all* the classical topics of biology and medicine.
- Appreciating that despite the individual variation among genomes within and among populations of humans, other animals, and plants, the idea that species are discrete entities is still valid. For prokaryotes, the situation is somewhat murkier.
- Adding to Crick's central dogma – DNA makes RNA makes protein – recognition of the necessity for regulating transcription, translation, and protein activity, for the health of cells and organisms.
- Distinguishing the static components of genomes – the full nucleotide sequences that appear in databases – from the dynamic aspects involved in the responsiveness of cells to internal and external signals, and in the programmes of development seen in higher organisms.
- Knowing the different types of control mechanisms in organisms, and their points of application, including but not limited to regulation of activities of proteins and control over gene expression.
- Realizing the importance of comparative genomics in revealing conserved elements that are likely to have interesting functions.
- Knowing the different mechanisms by which mutations can affect human health.
- Being familiar with a number of important diseases with genetic components, and knowing which ones can be treated, and for which ones lifestyle adjustments can reduce the danger inherent in genetic risk factors.
- Recognizing that the roster of species has been in flux. Mass extinctions have characterized the history of life. We are currently in another period of mass extinction.

Individuals, populations, the biosphere: past, present, and future

A genome sequence belongs to an individual organism. But, if the goal is to 'see life clearly and to see it whole' it is essential to relate genomes to one another. Genome comparisons are necessary across both space and time. We analyse differences in genomes within and among species that are currently extant, or that are accessible from sequencing preserved specimens of extinct species. Within species, we study genetic variation within and between populations. In humans, the intraspecific variation reveals the history of our origin and dispersal, and how we have adapted to different environments and lifestyles. An example is the development of adult lactose tolerance, a concomitant of the dietary change associated with cattle domestication. The original trait, loss of the ability to digest lactose after infancy – at the time of weaning – persists in many populations. For these populations, alternative sources of calcium include fermented dairy products such as cheese or yoghurt – bacteria hydrolyse the lactose – or soybean. (It is said that 'The soybean is the cow of Asia'.)

Comparing genomes between species can reveal their relationships and how they achieve their similarities and differences. We share 96% of our genomes with our nearest relatives, the chimpanzees, and many of our proteins are identical. This small amount of genomic change must account for the differences between humans and chimps. Indeed, genomics is essential to understanding and drawing species boundaries. As useful a concept as a species may be, there is consensus that its definition is a very tricky problem.

Genomes also contain records of their evolutionary history, which we can try to reconstruct. The organisms that populate the Earth are the product not only of the response to an imposed geological environment, but very largely show effects of interactions among individuals and species. There are many obvious examples of competition among members of the same species, and attack and defence between species. Our struggle against pathogenic viruses and bacteria are a salient example. Moreover, in the longer term the geological environment has in crucial respects been imposed by life. The early organisms that released oxygen through photosynthesis changed the composition of the atmosphere. This made possible the development of aerobic metabolism. The Earth has been the theatre of an enormous network of such interactions.

A feature of the history of life is the origin of new species and the extinction of many others. In some cases, extinctions arise from external events – for instance, an asteroid impact. In others, extinction is the result of deliberate activity – sailors killing the dodos of Mauritius, or twentieth-century scientists exterminating the smallpox virus in the wild. Today, habitat destruction threatens many other species, often an inadvertent result of human activity. To some threatened extinctions we can however plead not guilty: Examples include the molicutes infection ('elm yellow') which is devastating the elm trees of the eastern United States and southern Ontario (Figure 2.1), the fungal 'white-nose disease' threatening bats in eastern North America, and devil facial tumour disease, a transmissible cancer in Tasmanian devils (*Sarcophilus harrisii*), believed to have originated as a single mutation in a single individual.

We can study the present as thoroughly as we wish, and the past as extensively as we can. What of the future? One thing in which we can have confidence is that molecular biology will give us greater control over living things, including but not limited to ourselves. Clinical applications, of genomics and of other fields, have the potential to enhance the health of humans, and of animals and plants. It is already possible to intercede against some genetic diseases, by replacement of dysfunctional proteins through

Figure 2.1 Elm tree infected with elm yellows.

Pennsylvania Department of Conservation and Natural Resources – Forestry Archive, Bugwood.org

direct administration; for example, insulin therapy for diabetes and blood clotting factors for haemophilia. At the forefront of clinical developments are methods of rectification of mutant genes, through gene delivery by viruses (X-linked adrenoleukodystrophy), or by introduction of functional genes through stem cells (α_1-antitrypsin deficiency).

Genomes are now central to all of academic biology and clinical applications. In this chapter we shall emphasize the integrality of the relationships between genomes and other, formerly independent, areas of science. We shall begin with a cell, and then widen our point of view successively to organisms, populations, species, and the biosphere as a whole.

The central dogma, and peripheral ones

In 1958, F. Crick proposed the **'central dogma'** of molecular biology:

DNA makes RNA makes Protein.

Like many other dogmas, it provides important insights, even if it is not the whole story.

A less concise statement of the dogma would be: DNA sequences govern the synthesis of nucleotide sequences of RNA (transcription), which in turn govern the synthesis of amino acid sequences of proteins (translation). In his formulation, Crick emphasized the one-way flow of information: that amino acid sequences do *not* dictate the synthesis of nucleotide sequences of RNAs – this is still true – and that RNA sequences are *not* transcribed into DNA sequences. We now know that some viruses can 'reverse transcribe' RNA sequences into DNA.

Also sacrificed to concision in Crick's statement of the dogma is the question of regulation and control. In a healthy cell, the traffic through the dense network of metabolic pathways is coordinated, so that no intermediates build up to unwanted levels, and adequate amounts of products are created when and where they are needed. This involves regulation both of the activities of molecules already present in the cell, and control of synthesis of proteins and RNAs. An example of a mechanism of control over the activity of enzymes present in the cell would be **feedback inhibition**, in which the product of a concatenation of metabolic reactions inhibits an enzyme catalysing a step early in the sequence. Many other mechanisms exist to regulate the concentrations of enzymes and other proteins, by control over levels of transcription of the genes that encode them.

Thus cells contain two parallel networks: an enzymatic network manipulating metabolites, and a regulatory network manipulating genes, transcripts, and proteins.

- The central dogma – DNA makes RNA makes protein – describes not only a series of molecular events but a direction of *information* transfer. Information sensing and response are also essential for integration and coordination of the activities and expression patterns of many molecules. In this way, cells can achieve stability and robustness.

Expression patterns

The ~23 000 protein-coding genes in the human genome, exclusive of the immune system, can give rise to at least as many proteins – many more, if splice variants are taken into account. However, a healthy cell must synthesize only those proteins necessary for the requirements of its physiological state, and its differentiated type. Regulation of expression is common to all cellular life forms and even to some viruses. In prokaryotes, for instance, *Escherichia coli* will show robust synthesis of the genes in the *lac* operon only if growing on a medium containing lactose. Human cells in different organs will synthesize proteins appropriate for their cell types.

Not only must each cell choose which proteins to synthesize; it must control the rate of production of each. A variety of mechanisms achieve this control.

Some involve the binding of proteins to specific DNA sequences, to control protein synthesis at the level of transcription.

- The genome sets the parameters that circumscribe a potential life. Some constraints are tight, others relatively loose. Conversely, life implements the genome, in the form of synthesis of RNAs and proteins, and in the regulatory mechanisms that keep cellular activities stable and robust.

Regulation of gene expression

Living things must regulate the synthesis of proteins encoded in their genomes. Transcription and translation must be dynamic – to produce the right amount of the right protein at the right time at the right place. In this way, cells can respond to stimuli by altering their physiological state, or even their physical form. The driving force for these changes in profile of proteins produced may be changes in the environment, or internal signals directing different stages of the cell cycle or developmental programmes. We have mentioned that the appearance of lactose in the medium can trigger transcription of the lactose operon in *E. coli*. Transcription of another operon, encoding enzymes for the biosynthesis of the amino acid tryptophan, will be repressed if tryptophan is present in adequate concentrations. Similarly, a human cell may differentiate into a neuron, sprouting dendrites and an axon, and synthesizing tissue-specific or even cell-specific proteins.

The central dogma of DNA → RNA → protein suggests several possible leverage points for regulation of protein expression.

Most control takes place at the level of transcription

In prokaryotes, a specific focus of transcriptional regulation is at or near the binding site of RNA polymerases to DNA, just upstream of (5′ to) the beginning of the gene. Repressors can turn off transcription by occluding the binding site, blocking polymerase activity. In contrast, **promoters** can actively recruit polymerases through cooperative binding, along with polymerase, to a site on the DNA.

Gene regulation in eukaryotes is more complex. Transcription regulators bind to DNA at positions proximal to the gene as in prokaryotes, but also at remote sites. The control of human β-globin expression illustrates such a scheme (see Box 2.1). Regulatory interactions also govern the expression of other transcription factors. The resulting control networks show far greater complexity, in both their logic and their dynamics, than those of viruses or prokaryotes.

The gift of complexity is robustness. Eukaryotic control networks show an ability to reprogram themselves and to respond to stimuli by changing cell state. The source of robustness appears to be redundancy. Yeast (*Saccharomyces cerevisiae*) has about 6000 genes. Under 'normal' – non-stress – conditions, about 80% of them are being expressed. It is also true that yeast can survive approximately 80% of single-gene knockouts. (It would be interesting to know the overlap of these sets!) Many expressed genes must be redundant, and redundancy provides robustness. (Gene regulatory networks are the subject of Chapter 11.)

Other mechanisms of transcription regulation in eukaryotes involve changes in patterns of methylation of DNA associated with changes in the structure of chromatin. Eukaryotic chromosomes contain complexes of DNA with histones (see Figure 1.2). Chromatin remodelling is an important mechanism of transcriptional control. Reversible chemical modification of histones, by a mellifluous variety of reactions including deacetylation, methylation, decarboxylation, phosphorylation, ubiquitinylation, and sumoylation, leads to alterations of the DNA–histone interactions that render transcription-initiation sites more or less accessible.

In differentiation, DNA methylation is a regulatory mechanism that survives cell division (see Box 2.2). Methylation of cytosine in CpG islands silences the adjacent genes, possibly by stimulating chromatin remodelling. When a cell divides, enzymes copy the methylation patterns, preserving the settings of the regulatory switches.

- CpG islands are regions of high GC content, rich in the dinucleotide sequence GC, that appear at the 5′ ends of vertebrate genes. Methylation of C residues silences genes.

Control of β-globin gene expression

The protein-coding genes of the globin loci interact with many control regions. The β-globin region includes promoters proximal to individual genes; a locus control region occupying a region between 6 and 22 kb upstream of the most 5′ gene (ε); a 250 bp pyrimidine-rich region 5′ to the δ gene (YR); and enhancer regions, which may appear on the same chromosome, in some cases near to and in other cases distant from the gene they control, or on entirely different chromosomes.

Control of globin gene expression is asserted for both tissue specificity and developmental progression. In humans, transcription of the β-globin region switches from the embryonic ε-globin in the yolk sac to the foetal γ-globins in the liver (G_γ and A_γ) and finally to the adult β-globin produced in cells derived from bone marrow.

An essential mechanism of control is **modulation of local chromatin conformation**. Chromosomes contain nucleoprotein complexes called chromatin. DNA is associated with proteins called histones (Figure 1.2). Chromatin conformational changes can be induced by covalent modification of histones and by binding of chromatin-remodelling proteins.

Differential sensitivity of sites to DNAse I digestion measures differences in exposure. Some regions near actively transcribed genes are **hypersensitive** to DNAse I digestion (see Figure 2.2a). The locus control region upstream of the β-globin locus consists of five hypersensitive regions. The degree of their exposure is correlated with transcriptional activity.

Regulation of β-globin expression involves interaction of the locus control region with proximal promoters associated with specific genes. The interactions are mediated by a large complex of proteins recruited to the site. The alternative interactions of the locus control region, with foetal- and with adult-expressed genes, suggests that expression is determined by a competition between these interactions (see Figure 2.2b,c).

A well-known enhancer of globin expression is erythropoietin, a glycoprotein hormone encoded on chromosome 7. Erythropoietin does not interact with sequences in the vicinity of the globin locus but works indirectly by activating intracellular signalling pathways by binding to a receptor. Erythropoietin expression is sensitive to oxygen tension. Hypoxia increases erythropoietin production. This occurs naturally at high altitudes; as a result, people who live at 2500 m above sea level have about 12% more haemoglobin than people who live at sea level. Athletes take advantage of this. About 3 months of adaptation time is necessary to build up this differential.

A surprising player in the game of globin expression regulation is acetylcholinesterase, most famous for its physiological role in neural synapses and neuromuscular junctions. Acetylcholinesterase also regulates globin synthesis.

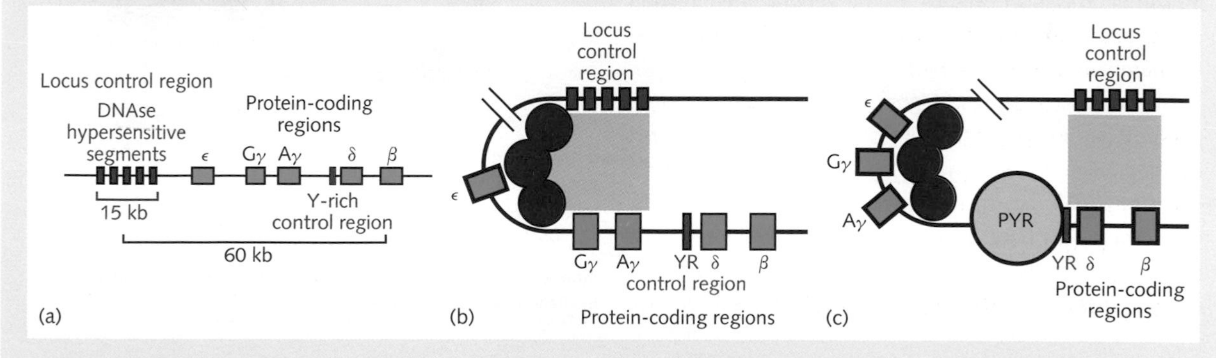

Figure 2.2 (a) β-Globin region showing protein-coding genes and locus control region, consisting of five DNAse hypersensitive segments. Pseudogene ψ_β not shown. (b,c) Schematic structural model for the control of globin gene expression. Part (b) shows the foetal structure, with interaction between the locus control region and the G_γ and A_γ genes, mediated by proteins (green). Blue circles indicate chromatin-remodelling complexes. In this configuration, the G_γ and A_γ genes will be expressed. In part (c), the cyan circle indicates the PYR complex of proteins, which binds to the pyrimidine-rich (YR; Y stands for pyrimidine) region just 5′ to the δ gene. This binding reconfigures the system: PYR blocks the foetal mode of interaction of the locus control region with the γ region (b). Instead, the locus control region interacts with, and promotes the expression of, the β gene.

Adapted from: Bank, A. (2005). Understanding globin regulation in β-thalassemia: it's as simple as α, β, γ, δ. J. Clin. Invest. **115**, 1470–1473.

BOX 2.2

Mammalian females are X-chromosome-silenced mosaics

An example of gene silencing by DNA methylation is the formation of Barr bodies in female mammals. Cells of mammalian females (except for oocytes) have two X chromosomes. The product of the *Xist* gene (an RNA molecule) on one of the X chromosomes inactivates that entire chromosome and causes it to form a compact, transcriptionally inert object, called a Barr body. The *Xist* gene on the other X chromosome is inactivated by cytosine methylation, leaving that chromosome normal in structure and activity. As a result, cells of both males and females have one active X chromosome. This is the mammalian solution of the 'dosage compensation' problem that arises because the genomes of males and females contain different numbers of copies of X chromosome genes.

In human females, and females of other placental mammals, each cell chooses at random which X chromosome to inactivate. Most mammalian females are, therefore, mosaics of cells expressing genes from alternative X chromosomes. A visible example of this mosaicity is a calico cat, necessarily a female (see Figure 2.3b). A calico cat has a yellow-coat allele on the X chromosome inherited from one parent and a black-coat allele on the other X chromosome. (In the white patches of the coat, neither allele is expressed.) The size of the coloured patches on the coat reveal when the genes were inactivated. In contrast, in female marsupials, all cells inactivate the paternally derived X chromosome. (This is an example of the general phenomenon of genetic **imprinting**, the dependence of phenotype on the parental origin of a gene.)

A difficulty in cloning of higher animals is how to restore pluripotency to the single cell from which the animal will develop. Figure 2.3 shows (a) a cloned cat, named Cc (for Copycat), and (b) its source organism (*very* loosely, its mother), Rainbow, a calico cat. Although the two cats are genetically identical as far as the nucleotide sequences of their DNA are concerned, Rainbow is a mosaic and Cc is not. The cell that produced Cc had an inactivated X chromosome and this inactivation was replicated in all of Cc's cells.

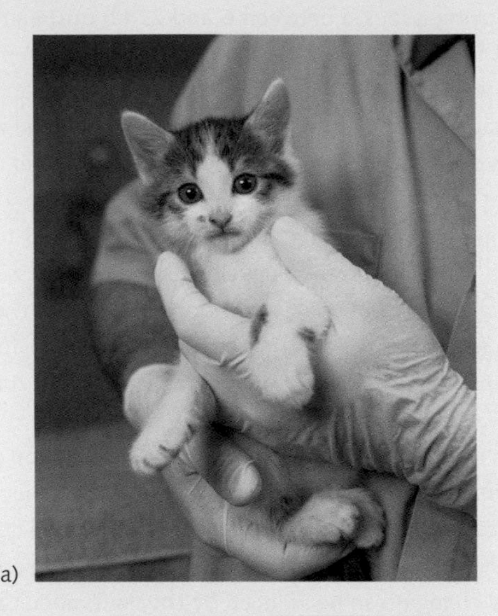

(a)

(b)

Figure 2.3 (a) Cc – or Copy Cat – a cloned cat. (b) Rainbow – a calico cat, the source organism for Cc. The varied-colour appearance is reminiscent of Indian maize (Figure 1.5), and indeed there are some similarities in the mechanisms that created them.

Photos reproduced courtesy of The College of Veterinary Medicine & Biomedical Sciences, Texas A&M University.

Some mechanisms of regulation act at the level of translation

Antisense RNA will form a double helix with mRNA, and block transcription. (Antisense RNA is single-stranded RNA complementary to mRNA.)

Introduction of genes for antisense RNA can silence genes. For example, ethylene is a plant hormone that stimulates ripening. Genetic modification – a controversial activity in the context of agriculture – has produced a tomato with longer shelf life. An artificial gene in the 'Flavr Savr' tomato is transcribed to an antisense RNA that greatly reduces translation of a gene involved in ethylene synthesis, delaying ripening.

In **RNA interference**, a short stretch of double-stranded RNA (~20 bp) elicits degradation, by a ribonucleoprotein complex, of mRNA complementary to either of the strands. RNA interference may have a natural function in defence against viruses. It has been applied in the laboratory to achieve effective gene knockouts in studies aimed at deducing gene functions.

Another mechanism of translation control is the attachment of ligands to the **Shine–Dalgarno sequence** in mRNA, preventing the RNA from binding to the ribosome. In *E. coli*, vitamins B1 (thiamin) and B12 (adenosylcobalamin) bind to an mRNA containing transcripts of genes encoding proteins involved in their biosynthesis. This is a kind of feedback inhibition – adequate amounts of a product inhibit the synthesis of more. However, unlike the more familiar product inhibition of an enzyme, this control is applied at the level of RNA, rather than protein.

Different modes of transcriptional control vary in their dependence on external conditions and all are reversible – some more readily than others. The lactose operon control in *E. coli* is a 'toggle' switch that can respond to both the appearance of lactose and its subsequent exhaustion. Other control processes are cyclic, such as cell-cycle control and diurnal rhythms. Cellular differentiation in higher eukaryotes is usually irreversible, except in certain forms of cancer.

Translation of eukaryotic genes may produce several different splice variants. Regulation of translation in eukaryotes may affect the distribution of splice variants produced. Mechanisms include (1) degradation of specific mRNA variants by microRNAs (miRNAs) and siRNAs, which may repress translation of specific splice variants; and (2) splicing factors, RNA-binding proteins that interact with the transcripts of specific exons (or even introns) and interact with the splicing machinery to direct maturation of the mRNA. At the transcriptional level, chromatin remodelling may render certain exons inaccessible and affect the splice variant expressed.

Some regulatory mechanisms affect protein activity

After expression of proteins, cells can modulate the activities of proteins by post-translational modifications, some of which are reversible. Binding of ligands can affect protein activities, through allosteric changes for example.

Cells regulate the activities of their proteins by mechanisms applied at the levels of transcription, translation, post-translational modification of proteins, and response of proteins to ligation (see Figure 2.4). Although these processes are biochemically distinct, cells apply them in coordinated ways. Chemical modifications of transcription factors can *quantitatively* regulate amounts of proteins in cells, rather than simply switching transcription on and off. Different control processes are effective over different time scales: cells must sometimes react quickly, to threat or stress; at other times rhythmically, over cell cycles; and sometimes in programmes unfolding over years or even decades during development of an organism. It is not possible to sort the different control mechanisms into fast- and slow-acting categories. There are layers of complexity here that we are only beginning to understand.

- There are many mechanisms, and types of targets, for regulation. These include control over expression patterns, and control over the activities of proteins and non-coding RNAs in the cell. For instance, allosteric changes are ligand-induced conformational changes in proteins that modify activity, often leading to cooperative binding curves, as in haemoglobin.

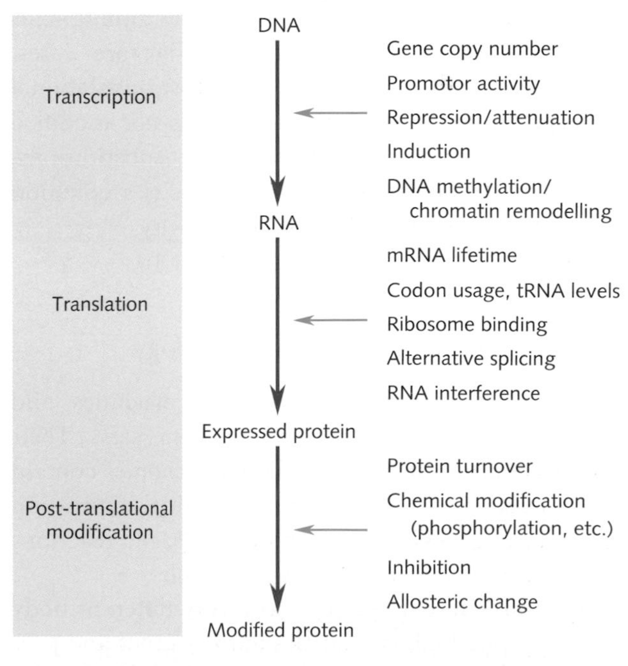

Figure 2.4 Steps in protein expression subject to control.

Proteomics

Proteins are the executive branch of the cell. Some proteins are structural, such as the keratins that form our hair and the outer horny layer of our skin. Some are catalytic: the enzymes that catalyse metabolic reactions. Others are involved in regulation, including but not limited to the proteins that bind DNA to control expression. Some are involved in signal transduction, receiving signals at the cell surface, and transmitting the signal to regulatory proteins.

- Think of intermediary metabolism, catalysed by enzymes, as the 'smokestack industries' of the cell, and the regulatory systems, including signal transduction and expression control, as the 'silicon valley'.

To carry out such a wide variety of functions, proteins show a great diversity of three-dimensional conformations. This diversity in structure and function is nevertheless compatible with many common structural features.

All proteins are polymers of amino acids. There is a common repetitive mainchain, with sidechains attached at regular positions. Each sidechain is one of a canonical set of 20, specified by the genetic code (see Box 1.1). At least two other amino acids naturally extend the genetic code in rare cases, and many proteins are subject to **post-translational modifications** whereby the sidechains are modified by a variety of rearrangements or substitutions. Phosphorylation of specific sidechains is a common mechanism for regulating protein activity.

Proteomics is the subject of Chapter 10.

Genomics and developmental biology

As they emerge from sequencing machines and assemblers, genomes are static data sets. Their implementation in cells is dynamic: genomes contain developmental programmes governing expression patterns of genes at different life stages, and reactions and responses to environmental stimuli.

Just as different species show very different body plans despite high similarities in their genes and proteins, different taxa can also show high similarities in their **developmental toolkits** – that is, sets of genes active in guiding development. For example, *HOX* genes are responsible for organization of anterior–posterior (head-to-tail) patterning in the body plans of flies and humans, and even *C. elegans*. The human Paired Box gene *PAX6* is required for proper eye development, but if expressed in *Drosophila* can transform embryonic wing tissue to an ectopic (= out-of-place) eye. The implication is that despite the very great differences in gross anatomy of the eyes in vertebrates, insects, and octopus, the visual systems arose from a common ancestor. Both the molecular structures of the initial light receptors – the rhodopsins – and the architecture of the neural pathways confirm this conclusion.

Full-genome sequences allow the tracking of similarities and differences in developmental processes during evolution. In particular, it is possible to identify homologues of genes from the developmental toolkit across different phyla. The results both make use of, and illuminate, phylogenetic relationships. This can be thought of as an extrapolation, to the molecular level, of the classical relationship between embryology and taxonomy.

- Correct taxonomic assignment of species with unusual features can reveal not only phylogenetic relationships, but developmental ones also. The phylum cnidaria contains almost 10 000 species of aquatic animals. Sea anemones and jellyfish are typical examples (Figure 2.5a,b). The worm *Buddenbrockia plumatellae* (Figure 2.5c) was until recently an enigma. Its body plan did not strikingly resemble the familiar radially symmetric cnidaria such as sea anemones or jellyfish; it might easily be thought to be related to nematodes. However, Holland and co-workers have recently shown, on the basis of sequence alignments of 129 proteins, that *B. plumatellae* is a cnidarian. The significance of this observation for developmental biology is that it extends the body plan observed for cnidaria, requiring investigation of unsuspected aspects of the developmental pathways in this species.
- *HOX* genes are a classic example of how genomes illuminate developmental biology, both within a species, and among species ('evo-devo'). Organisms with bilateral symmetry, including insects and

(a)

(b)

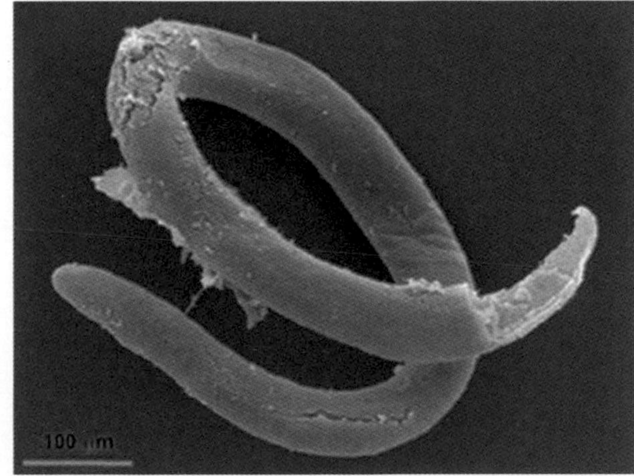

(c)

Figure 2.5 (a) Sea anemone, *Anthopleura sola*, showing the typical radially symmetric cnidarian body plan. (Photograph by Charles Halloran) (b) Jellyfish, photographed by David Burdick in the Mariana Islands. (from NOAA Photo Library, image reef1133, National Oceanic and Atmospheric Administration/US Department of Commerce). (c) Scanning electron microscopy image of a *Buddenbrockia plumatellae*. From: Jiménez-Guri, E., Philippe, H., Okamura, B., Holland, P.W. (2007). *Buddenbrockia* is a cnidarian worm. Science **317**, 116–118.

vertebrates, contain *HOX* genes, which encode a family of DNA-binding proteins. The expression of these genes varies along the anterior–posterior (head-to-tail) body axis, and controls the setting out of the body plan. *HOX* genes have overlapping domains of expression within the body. Different regions of the embryo develop into anatomically distinct regions of the adult, based on the subset of *HOX* genes expressed.

Indeed there is a fascinating mapping between (1) the order of the genes on the chromosome, (2) the relative times during development of the onset of their activity, and (3) the order of their action along the body. (The *β*-globin locus shares the first two of these but not the third (see p. 45).)

HOX genes reveal the duplications that have occurred during vertebrate evolution. Insects and amphioxus have a single *HOX* cluster. Humans have four *HOX* clusters. Zebrafish have seven *HOX* clusters, interpretable as a series of duplications: 1→2→4→8 followed by loss of one to reduce 8→7.

The *HOX* genes illustrate the conservation of the developmental toolkit in species with very different body plans.

- Conversely, the distribution of DNA methylases illustrates the diversification of developmental toolkit even in organisms with similar body plans.

DNA methylation patterns are important signals for transcription control in vertebrate development and tissue differentiation. In contrast, many invertebrates show little or no DNA methylation. Examples include *C. elegans* and *D. melanogaster*, the first two invertebrate genomes sequenced. These observations would suggest that DNA methylation arose in the vertebrate lineage. However, although *C. elegans* lacks genes for DNA methylases, and *D. melanogaster* has but an incomplete complement, the genome sequences of related nematodes and insects show that the genes were present in invertebrates, and some or all of them have been lost in the specific lineages leading to *C. elegans* and *D. melanogaster.* For instance, the honeybee contains a full, functional complement of DNA methylase genes, and its DNA is methylated. However, it appears that the invertebrates and vertebrates differ in the pattern and function of DNA methylation.

- The mapping from genome to proteins is a static correspondence. But in fact, implementation of the genome is a dynamic process, both (1) in the short term, as the response to internal and external stimuli, and (2) in the long term, as the unfolding of programmes of development and differentiation. Evolution illuminates both the static and dynamic aspects of the implementation of the genome.

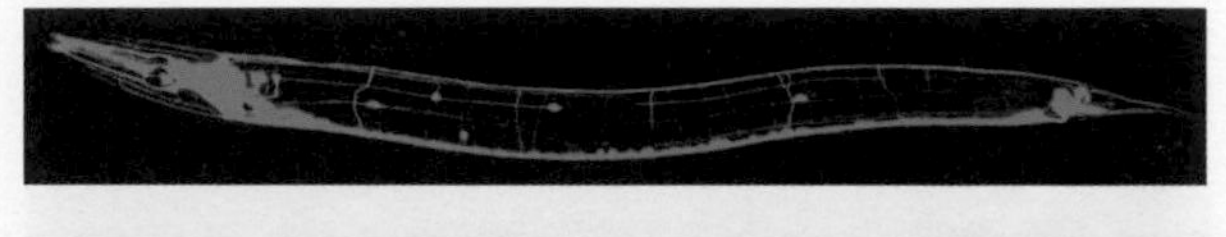

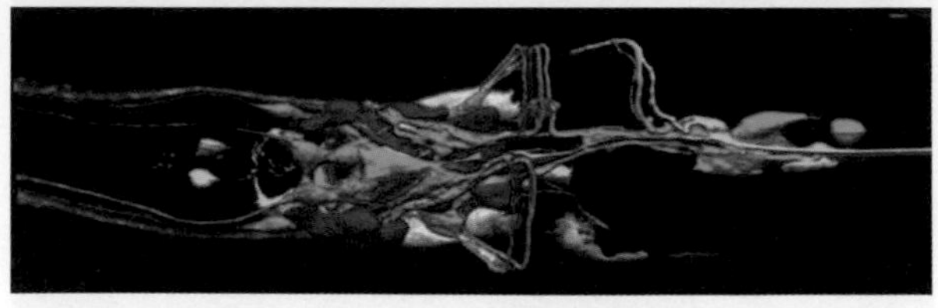

Figure 2.6 Top: the nervous system of *C. elegans*, head at left, labelled with green fluorescent protein. Bottom: the brain of *C. elegans*, with different groups of neurons labelled with different-coloured fluorescent proteins.

Reproduced by permission of Prof. H. Hutter.

Genes and minds: neurogenomics

Higher mental processes unique to humans are the last major unexplored territory in our understanding of life. Mental illness is a major cause of disability in Europe and the USA, but it has been difficult to integrate human mental achievements and diseases into mainstream biology and medicine. This is partly because the physical correlates of mental events are so complex and partly because of a clinical tradition of treating mental disease by conversation between therapist and patient, and by counselling. That tradition, with its origin in the pervasive influence of Freud, was based on the assumption that the causes of mental disease were either emotional trauma alone or emotional consequences of physical trauma. Even though we now know that the genetic component of neuropsychiatric disease is much more important than formerly suspected and that even environmentally caused neuropsychiatric diseases have biochemical effects, the fact remains that treatments alternative to counselling, based on a profound biological understanding, are not yet available. Drugs now play an important role in treating symptoms of neuropsychiatric disease. But Macbeth's anguished cry, 'Canst thou not minister to a *mind* diseased?', is still largely an unmet challenge.

One approach to mental processes in humans is to work our way up from simpler nervous systems. This was Sydney Brenner's motive for choosing *C. elegans* as a model organism. The adult hermaphrodite form has exactly 302 nerve cells and 7000 synapses (see Figure 2.6). J. White worked out the complete 'wiring diagram' by tracing individual cells through serial sections. We know not only the static structure and organization of the adult system, but the details of how it develops.

Some people espouse this approach. ('You have to walk before you can run.') Others retort that human mental achievements and diseases go far beyond those of *C. elegans*. It is undeniable that understanding the minds of worms – or even of chimpanzees – must have limited applicability to humans. Nevertheless, many surprising similarities have appeared, even in such distantly related organisms.

What analogues of human mental phenomena do model organisms show? The basic mechanisms of sensation, such as vision and olfaction, are similar in many animals. Learning and memory are widespread throughout the animal kingdom. Flies learn; even worms learn. Language is one uniquely human attribute and its genetic component has been traced through clinical disorders. One might assume that 'higher' emotions and sophisticated talents – love, despair, the ability to play chess, or talent for art or music or science – would also be absent from simple model organisms, but in many cases tantalizing analogies do exist. Fruit flies show courtship behaviour under the control of known genes. The objection that fruit flies are merely displaying unsophisticated instinctive behaviour raises the question of how rational and sophisticated is most human activity, including courtship (*especially* courtship).

Genomics provides the bridge between the minds of the different species. Two ways to use model organisms are the study of homologues to illuminate functions of human proteins and the insertion of human genes into the model organisms to study the effects of their expression.

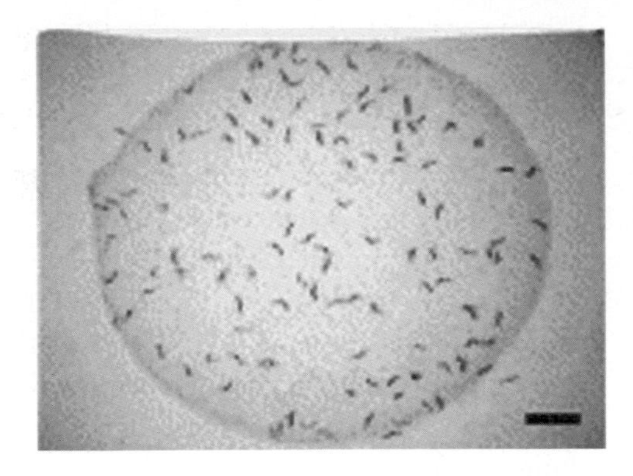
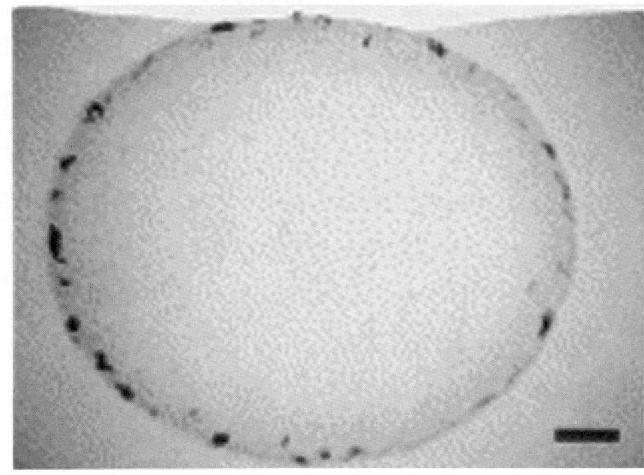

Figure 2.7 Alternative social behaviours of *C. elegans*, feeding on a lawn of agar in a Petri dish. Left: solitary feeding. Right: clumping. The agar is slightly thicker at the edges of the dish, which is why the worms congregate there.

From: de Bono, M. & Bargmann, C.I. (1998). Natural variation in a neuropeptide Y receptor homolog modifies social behavior and food response in *C. elegans*. Cell **94**, 679–689.

Models for neurological disease, in natural or transgenic *Drosophila* and *C. elegans*, include Parkinson's and Huntington's disease, Friedrich's ataxia (a degenerative disease of the nervous system), and early-onset dystonia (neuromuscular dysfunction producing sustained involuntary and repetitive muscle contractions or abnormal postures). Species more closely related to humans, including zebrafish and mice, provide models for most mental diseases.

Traditionally, there was a fairly rigid distinction between neurological/physiological/biochemical diseases affecting cognitive performance, and psychiatric diseases believed to have emotional causes and effects. However, we now recognize that even if an illness is caused by an unhealthy emotional environment (and leaving aside the genetic component of *susceptibility* to disease in response to such an environment), organic changes – down to the molecular level – underlie the psychological manifestations.

There is now good evidence for a substantial genetic component in numerous psychological conditions, including dyslexia, autism, schizophrenia, attention-deficit hyperactivity disorder, and others. A number of conditions appear in both **familial** forms, showing relatively simple patterns of inheritance, and **sporadic** forms, with more complex genetic components and greater influence of environment. Familial forms often show early onset.

For instance, familial Alzheimer's disease is a relatively early-onset form of the condition, with early onset defined in this case as appearing before the age of 65. It affects about 10% of those with Alzheimer's disease. The familial form is associated with mutations in genes on chromosomes 1, 14, and 21. In contrast, a mutation in the gene for ApoE on chromosome 19 causes increased risk of late-onset Alzheimer's disease (see pp. 61–62).

Genetics of behaviour

- *Different strains of* C. elegans *show different social behaviour in dining. C. elegans* can be grown on an agar lawn in a Petri dish. The wild type collected from Australia will congregate in groups to feed. The wild type collected from Britain will eat separately (see Figure 2.7). The difference has been traced to a single amino acid change in a seven-transmembrane helix protein, NPR-1.

> *The unity of life is a theme of this book, but perhaps it would be wrong to read too much into this example.*

- *In fruit flies, T. Tully and co-workers have identified a gene associated with memory. CREB* (cyclic AMP *r*esponse *e*lement-*b*inding protein) encodes a transcription factor, part of a large family of paralogues in mammals and distributed widely in eukaryotes and prokaryotes. Some engineered changes in *CREB* produced flies that could learn but not store memories; other changes produced flies that learned substantially faster than normal. Alterations in mouse *CREB* have produced memory impairment.

- *The Lesch–Nyhan syndrome was the first correlation discovered between a specific human genetic defect and behavioural anomalies*. It is an X-linked deficiency of a single protein, the enzyme hypoxanthine–guanine phosphoribosyltransferase. The consequent inability to metabolize uric acid properly leads to physical symptoms including gout and kidney stones. But patients also show poor muscle control, mental retardation, facial grimacing, writhing and repetitive limb movements, and uncontrollable lip and finger biting to the point of severe self-mutilation.
- *Seasonal affective disorder, a mood change in response to prolonged darkness, is related to the regulation of circadian rhythms*. Jet lag is a related condition. The system of genes involved in circadian rhythms was originally worked out in fruit flies and has homologues in many animals, including humans and *C. elegans*, and in plants.
- *Mutations in an X-linked human gene,* DLG3, *cause severe learning disability*. Knockout of the mouse homologue *PSD95* produces learning-impaired mice. The protein encoded by *PSD95* binds to the NMDA receptor, a protein involved in synaptic plasticity. Overexpression of NMDA receptor gives mice superior learning and memory abilities.
- *Several genes are implicated in schizophrenia*. A common effect, hypersensitivity to the neurotransmitter dopamine in the brain, plays a role in schizophrenia. Some antipsychotic drugs block dopamine receptors on neuronal surfaces; conversely, amphetamines, which release dopamine, aggravate the symptoms.
- *Depression is a common response to stress*. Although we all suffer stressful episodes in our lives, the development of debilitating mental disease depends on our alleles in the promoter region of the gene for the serotonin transporter (*5-HTT*). Individuals with one or two copies of the shorter allele of this gene are more likely to exhibit depression and suicidal tendencies than individuals homozygous for the longer allele.
- *Maltreatment in childhood is a cause of antisocial behaviour*. Obviously. However, the likelihood of development of antisocial behaviour depends on the allele for monoamine oxidase A (MAOA). Maltreated children with a genotype producing high expression levels of MAOA are less likely to become antisocial or violent offenders. A. Caspi and colleagues found that, although only 12% of a sample of people had the combination of low-activity MAOA genotype and childhood maltreatment, these individuals accounted for 44% of subsequent convictions for violent offences.

- Cognitive abilities are our species' proudest achievement. Neuropsychiatric illness is a correspondingly difficult problem to understand and treat. Both have genetic components that are beginning to be understood, in part through study of analogues in other species.

Populations

An interacting group of individuals of the same species is a population. How do genomes vary in a population? Study of the variation reveals the population structure and its history. A population that has passed through a bottleneck; or that has developed in isolation, from a small 'founder' group, will show very narrow variation. All cheetahs, for example, are as closely related as human siblings. A population with relatively high variation is likely to have a longer evolutionary history. It is such observations that suggest that humans originated in Africa.

Single-nucleotide polymorphisms (SNPs) and haplotypes

All people, except for identical siblings, have unique DNA sequences.

Comparisons between unrelated individuals reveal overall differences between whole-genome sequences of ~0.1%. Any change in DNA sequence is a mutation, including substitutions, insertions and deletions, and translocations. Many of the differences between individuals have the form of individual isolated base

BOX 2.3 Guide to databases of SNPs and related databases

General SNP links	http://www.snpforid.org/snpdata.html
NCBI dbSNP	http://www.ncbi.nlm.nih.gov/entrez/query.fcgi?db=snp
The SNP Consortium	http://snp.cshl.org/
HapMap	http://www.hapmap.org/
Applied Biosystems Assays-on-Demand	http://myscience.appliedbiosystems.com/cdsEntry/Form/assay_search_basic.jsp
Ensembl	http://www.ensembl.org/Homo_sapiens/
HGVBase	http://hgvbase.cgb.ki.se/
SeattleSNPs	https://gvs.gs.washington.edu/GVS/
dbSNP database at NCBI	http://www.ncbi.nlm.nih.gov/snp
Human Gene Mutation Database	http://www.hgmd.cf.ac.uk
OMIM (Online Mendelian Inheritance in Man)	http://www.ncbi.nlm.nih.gov/omim

substitutions, or **single-nucleotide polymorphisms** (SNPs). There are also many short deletions.

- Single-nucleotide polymorphisms (SNPs) are single-base variations between genomes.

Databases now contain over 100 000 mutations, in 3700 genes (see Box 2.3). This is 6.2% of the total ~23 000 genes. Of course this number is growing rapidly; approximately 10 000 new mutations are discovered each year.

Each of us bears an accumulated collection of SNPs, reflecting mutations that occurred in our ancestors. Some constellations of SNPs are co-inherited as blocks. Others are not. Mutations in *different* DNA molecules of diploid chromosomes become separated within a single generation, by assortment. Mutations on the *same* chromosome become separated more slowly, by recombination. Haploid sequences, such as most of the human Y chromosome, or mitochondrial DNA, are not subject to recombination. Mutations in these sequences remain together.

Mutations in the same DNA molecule in diploid chromosomes will become unlinked by recombination events that occur between their loci. The greater the separation between two sites, the greater the frequency of recombination. However, recombination rates vary widely along the genome, by several orders of magnitude. SNPs on opposite sides of recombinational 'hot spots' are more likely to be separated in any generation. SNPs lying within recombination-poor ('cold') regions will tend to stay together.

In humans, many 100 kb regions tend to remain intact. They show the expected number of SNPs, but relatively few of the possible combinations. An average SNP density of 0.1%, or 1 SNP/kb, suggests ~100 SNPs per 100 kb. The genome of any individual may possess, or may lack, each of them, giving a very large number (2^{100}) of possible combinations. However, many 100 kb regions show fewer than five combinations of SNPs. These discrete combinations of SNPs in recombination-poor regions define an individual's **haplotype**, or '*haplo*id geno*type*'.

Haplotypes provide a very economical characterization of entire genomes. They simplify the search for genes responsible for diseases, or any other phenotype–genotype correlations. For field biologists, including anthropologists, haplotypes permit detection of migratory and interbreeding patterns in populations.

- Haplotypes are local combinations of genetic polymorphisms that tend to be co-inherited.

In looking for genes responsible for diseases, or other phenotypic traits, haplotypes provide a magnifying glass. The goal is to correlate phenotype with genome sequence. The target may be to identify one base out of 3×10^9. By correlating phenotype with haplotype, only enough sequence must be collected to localize the site to within the typical length of a haplotype block, perhaps ~100 kb, containing only a few genes.

- Think of boundaries between haplotype blocks as being like the grooves in a bar of chocolate that permit it to be broken easily into bite-size fragments.

Variations in human genomes are the subject of several large-scale projects.

The SNP Consortium (http://snp.cshl.org) collects human SNPs. Its database currently contains nearly 4.2 million SNPs.

The International HapMap Project collects and curates haplotype distributions from several human populations. SNPs are its raw material, from which it identifies the correlations among them. Phase I of the project, published in October 2005, had the goal of measuring the distributions of at least one SNP every 5 kb across the whole human genome. Blood samples were provided by 269 individuals from four continents (see Box 2.4). Over 1 million SNPs of significant frequency (>5%) were documented. In addition, ten selected 500 kb regions were fully sequenced from 48 of the samples. Phase II will extend the analysis of the samples to determine an additional 4.6 million SNPs from the same individuals.

The work of the International HapMap Consortium, together with other studies, show that:

- Most of the variations appear in all populations sampled. Some of the inter-population differences reflect different relative amounts of the same SNPs.
- A very few SNPs are unique to particular populations. For example, out of over 1 million SNPs, only 11 are consistently different – in the sample studied – between all individuals of European origin and all individuals of Chinese or Japanese origin.
- The genomes of individuals from Japan and China are very similar, suggesting more recent common ancestry than other population pairs in the study.
- The X chromosome varies more between different populations than other chromosomes. This may arise from the fact that males contain only one X chromosome, the genes on which are, therefore, more subject to selective pressure. Recombinations of X chromosomes can occur, but only in females.
- Lengths of haplotype blocks vary among the different sources of samples. They tend to be shorter among populations from Africa, consistent with the idea of an African origin of the human species. The idea is that the older the population – more accurately, the larger the number of generations – the greater the chance of recombination.

BOX 2.4

Origin of samples for the International HapMap Project*

Population origin	Location	Number of individuals	Relationships
Yoruba	Ibadan, Nigeria	90	30 parent–offspring trios
Northern and western European descent	Utah, USA	90	30 parent–offspring trios
Han Chinese	Beijing, China	45	–
Japanese	Tokyo, Japan	44	–

Why the choice of parent–offspring combinations? A difficulty in determining haplotypes in heterozygous regions of diploid chromosomes is how to determine which SNPs lie in the *same* DNA molecule. Comparison of parental and child sequences can sort the observed SNPs into haploid contributions.

* The International HapMap Consortium (2005). A haplotype map of the human genome. Nature **437**, 1299–1320.

Ethical issues faced in the HapMap project

The International HapMap Consortium paid due attention to ethical, legal, and social issues. Informed consent of the donors preceded collection of samples. The procedure for informed consent involved not only individual agreement, but also community engagement, including interactive public explanation of the project. Samples were labelled anonymously. In fact, more samples were collected than used (similar in some ways to the principle of issuing blank cartridges to a firing squad). Nevertheless, characteristics of a population constitute personal information, the release of which may affect all individuals in the population, including those who were never asked to contribute a sample, and even those who refused. For this reason the HapMap Consortium did not collect medical information about the sample contributors, even under the protection of consent and anonymity.

Application of haplotypes to infer relationships between populations: the Barbary macaques

Comparison of genomes between populations can reveal histories of migrations. Here we shall consider the Barbary macaques; Chapter 8 will treat human history.

The macaques (*Macaca sylvanus*) on the island of Gibraltar are the only wild primates on the continent of Europe – except for certain football fans. The island has been host to these animals for several hundreds of years. Other populations of the same species exist across the Mediterranean, in northwest Africa.

Although there is fossil evidence for the ancient presence of Barbary apes in Europe, Gibraltar is not a refuge for survivors of that population.

Instead, almost all the population arrived in Gibraltar during the Second World War, deliberately imported to enhance morale during those dark days. Politically, Gibraltar is the last remnant of the British Empire in continental Europe: It is said that Gibraltar will remain British as long as the macaque population survives. With just four individuals left on the island in 1943, Winston Churchill ordered that the population be restocked. He minuted the Colonial Secretary: 'The establishment of the apes should be 24. Action should be taken to bring them up to this number at once and maintain it thereafter'.

Where did they come from? Modolo, Salzburger, & Martin analysed a 428 bp segment of mitochondrial DNA from individuals from the Gibraltar colony and from seven natural populations in Algeria and Morocco.* Among the individuals sampled, this region contained 24 different haplotypes. The haplotypes differed by between 1 and 26 mutations, all but one mutation being an SNP.

Figure 2.8 shows the clustering of the sequences. Each haplotype is represented by a circle with a radius proportional to the number of individuals bearing the haplotype. Colour coding indicates the location of sample collection. For instance, individuals of haplotype M16 are found mostly in Gibraltar, sometimes in Algeria, and infrequently in Morocco.

The topology of the relationship has several interesting implications.

- There are three major clusters. One cluster (upper right) contains most of the individuals from Morocco. The individuals from Algeria show two populations, from different collection areas, a major one (upper left) and a minor one (bottom centre). One location, *Pic des Singes*, provided individuals with haplotypes linked to each of the well-separated Algerian clusters.
- Dating of the divergence suggests that the Moroccan and Algerian populations separated over 1.2 million years ago.
- It is likely that the female line in the current Barbary macaques originated from *both* the Moroccan and Algerian population.

A clinically important haplotype: the major histocompatibility complex

In the human genome, proteins of the major histocompatibility complex (MHC; known in humans as the human leucocyte antigen (HLA) system) are

* Modolo, L., Salzburger, W., & Martin, R.D. (2005). Phylogeography of Barbary macaques (*Macaca sylvanus*) and the origin of the Gibraltar colony. Proc. Natl. Acad. Sci. USA **102**, 7392–7397.

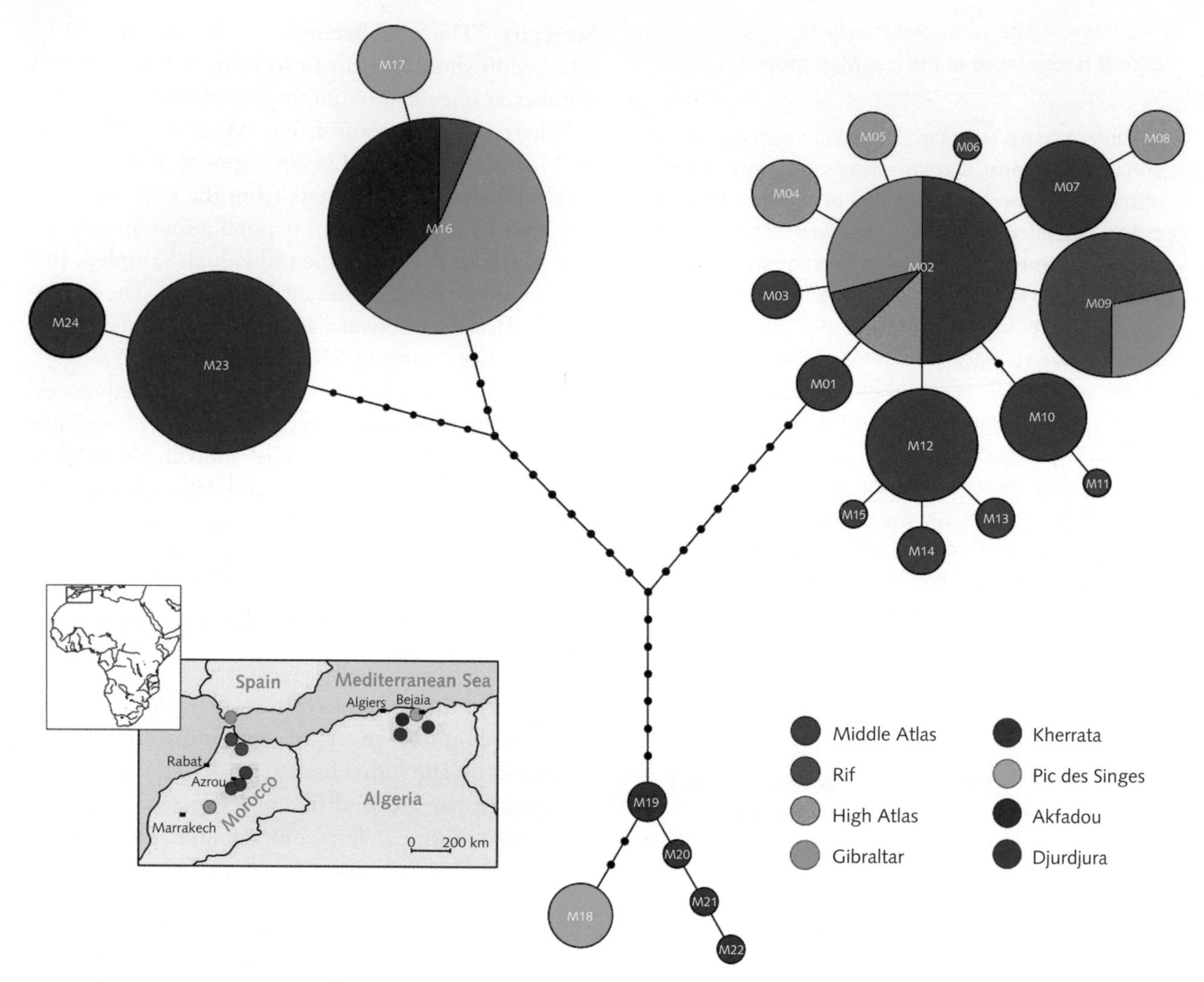

Figure 2.8 Haplotype network from 428-bp segments of mitochondrial DNA collected from free-living Barbary macaques (*Macaca sylvanus*) inhabiting Gibraltar, Algeria, and Morocco. The size of each coloured circle is proportional to the number of individuals bearing each haplotype. Each line segment represents a single mutation. Thus, the sequences represented by M18 and M19, just to the right of the inset map, differ by three mutations. The colours of the circles in the graph indicate locations of sample collection (see inset map).

encoded in a ~4 Mb region on chromosome 6 (6p21.31). In vertebrate species, each individual expresses a set of MHC proteins selected from a diverse genetic repertoire of the species. The system is highly polymorphic, with 50–150 alleles per locus, higher sequence variation than found in most polymorphic proteins. The set of MHC proteins expressed defines a partial haplotype of an individual. Compared with other haplotype blocks, the MHC region shows unusually wide individual variation.

The MHC region contains over 120 expressed genes, coding for proteins that:

- provide the mechanism by which the immune system distinguishes 'self' molecules – those to be tolerated – from 'non-self' molecules – those recognized as foreign invaders that must be repelled;
- determine individual profiles of competence for resistance to diseases;
- are useful markers for determining relationships among populations of humans and animals, and for tracing large-scale migrations and population interactions.

MHC haplotypes control donor–recipient compatibility in transplants

Surgical patients, if not immunosuppressed by drugs, will reject transplanted organs – unless the donor is an identical sibling – because the transplant is recognized as foreign. MHC proteins bind peptides and present them on cell surfaces. The triggering event in alerting the immune system to the presence of a foreign protein is the recognition by a T-cell receptor of a complex between an MHC protein and a peptide derived from the foreign protein (see Figure 2.9).

MHC haplotype influences autoimmune diseases – breakdowns in self-/non-self-distinguishability that result in a person's immune system attacking his or her own tissues. Examples of autoimmune diseases include rheumatoid arthritis, multiple sclerosis, type I diabetes, and systemic lupus erythematosus.

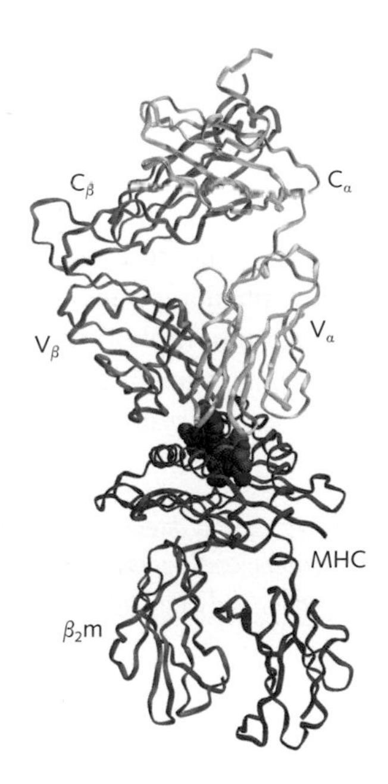

Figure 2.9 The immunological distinction between 'self' and 'non-self' resides in the proteins of the major histocompatibility complex (MHC) and their interaction with T-cell receptors. This picture shows a human T-cell receptor in complex with a class I MHC protein and a viral peptide. MHC proteins have broad specificity, each binding many peptides including those of self and non-self origins. Cell surfaces contain large numbers of MHC–peptide complexes, among which those binding foreign peptides are a small minority. T-cell receptors, in contrast, have narrow specificity, and pick out the complexes containing foreign peptides, like a professional antiques dealer spotting a valuable item in a rummage sale.

In addition to triggering immune responses in mature individuals, MHC–peptide complexes are also involved in the removal of self-complementary T cells in the thymus during development, at the stage when the distinction between self and non-self is 'learnt'.

MHC haplotypes determine patterns of disease resistance

Different MHC molecules have different binding specificities and can present different sets of peptides. People whose MHC molecules do not effectively present epitopes from a particular pathogen will be more susceptible to infection. For instance, MHC haplotype is a predictor of survival horizon in people infected by human immunodeficiency virus.

MHC haplotypes influence mate selection

Opposites attract: one person will tend to find another romantically attractive if they have *different* MHC haplotypes. The mechanism is apparently through linkage of MHC haplotype and body scent. This effect will tend to produce offspring that have MHC molecules that can present a broader repertoire of peptides, producing broader resistance to infection.

- A person's MHC haplotype is the set of alleles for over 100 highly polymorphic sites. MHC haplotype is an explicit signature of individuality, governing transplant rejection, resistance to infection, and even mate selection.

Mutations and disease

Many mutations, even if they are not synonymous mutations, are consistent with a healthy life, and typical life-span, provided that the individual practises a reasonable lifestyle. Such mutations contribute to our ethnic and individual variety. Loss of some proteins is surprisingly innocuous. Mice lacking myoglobin thrive, and even show athletic performance comparable to normal mice.

In some cases, species-wide loss of biosynthetic enzymes is not generally considered a disease, but contributes to the list of essential nutrients. For instance, whereas most animals can synthesize vitamin C, we must provide it in our diet. If not, the result is the disease scurvy.

Nevertheless, evolution has largely optimized proteins for their roles in healthy organisms. Therefore, most mutations causing amino acid sequence changes are deleterious, impairing protein function and threatening to produce disease.

Mutations associated with human disease are collected by the organization Online Mendelian Inheritance in Man (OMIM)™ (http://www.ncbi.nlm.nih.gov/omim). A corresponding site, Online Mendelian Inheritance in Animals (OMIA) (http://www.ncbi.nlm.nih.gov/omia) collects mutations associated with diseases in animals, other than human and mouse.

By what mechanisms can mutations affect human health?

Mutations causing defects in some proteins can be accommodated by adjustments in lifestyle. Other mutations cause disease only in combination with unusual features of lifestyle or specific triggering events. We have already mentioned the Z-mutation of α_1-antitrypsin, smoking enhancing its tendency to cause emphysema.

Loss-of-function mutations are often recessive, so that the homozygosity for the mutant allele typically has more severe consequences than heterozygosity. Every individual is heterozygous for some deleterious mutations that, if homozygous, would be lethal.

Many diseases are associated with the formation of insoluble aggregates, usually of misfolded proteins. These include classical amyloidoses, Alzheimer's and Huntington's disease, aggregates of misfolded serpins, and prion diseases. Polymerization of insulin creates problems in production, storage, and delivery in diabetes therapy. Mutations that destabilize proteins can increase the proportion of misfolded proteins, which show a greater tendency to form aggregates.

Some mutations that produce defective proteins show complex interactions with other traits, with the result that some deleterious mutations have not been eliminated from populations by selection, because they carry some compensating advantages. For example, the genes for sickle-cell anaemia and for glucose-6-phosphate deficiency confer resistance to malaria (see Box 2.5).

Dysfunction of a regulatory protein or receptor can disorganize the operation of a pathway even if all components of the pathway regulated are normal. Some abnormal regulatory proteins cannot be activated at all, whereas others are constitutively activated and cannot be shut off. The effects include:

Physiological defects: a number of diseases are associated with mutations in G protein-coupled receptors. Some mutations in opsins are associated with colour blindness. Certain mutations in the common G protein target of olfactory receptors lead to loss of sense of smell.

Developmental defects: several types are traceable to mutations in hormone receptors. For instance, Laron syndrome, a phenotype including diminished stature, arises from a mutation in the human growth hormone receptor. Administration of exogenous growth hormone does not restore normal growth.

- Many diseases are caused directly by mutations; many others have a genetic component and arise in the context of an interaction between genetics and lifestyle.

Glucose-6-phosphate deficiency, food taboos, folk medicine, pharmacogenomics, and mosquito breeding seasons

Glucose-6-phosphate dehydrogenase (G6PDH) is the most common enzyme deficiency, affecting over 400 million people worldwide. It is a recessive X-linked genetic defect, affecting up to 10% of populations in which mutations are common.

Glucose-6-phosphate dehydrogenase (G6PDH) catalyses the reaction:

glucose-6-phosphate + NADP $\rightarrow$
6-phosphogluconate + NADPH,

the first step in the pentose phosphate shunt. This reaction produces reduced glutathione needed to dispose of hydrogen peroxide (H_2O_2). It is particularly important in red blood cells, which, lacking nuclei and mitochondria, are metabolically impoverished, and have no alternative mechanism for detoxifying H_2O_2. Without active G6PDH, build-up of H_2O_2 will oxidize and denature haemoglobin, leading to destruction of red blood cells, producing a condition called haemolytic anaemia.

Eating fava beans, especially if uncooked, can induce anaemic episodes in people deficient in G6PDH. The danger of eating fava beans has been recognized since antiquity, and has been associated with food taboos, and preparation techniques designed to reduce toxicity. Pythagoras, for example, banned eating of fava beans in his school. We now know that fava beans contain the compounds vicine and convicine, metabolized in the intestine to isouramil and divicine, which react with oxygen to produce hydrogen peroxide, subjecting cells to oxidative stress.

Other chemicals, including certain drugs, present the same danger to G6PDH-deficient people. During World War I, some patients were observed to suffer dangerous side effects from the antimalarial drug primaquine. Many drugs, including sulphonamides, are now contraindicated for G6PDH-deficient patients, as is the taking of large doses of vitamin C. The observation of variations in effectiveness and toxicity of different drugs in different people has developed into the new field of **pharmacogenomics**, the tailoring of drug treatments to the genotype of the individual patient.

Why have dysfunctional G6PDH genes remained at such a high level in the population? Why does primaquine produce haemolytic anaemia in G6PDH-deficient patients, and does this have a relationship to its antimalarial activity? And why have fava beans continued to be grown if non-toxic alternatives are available?

The malarial parasite invades the red blood cell of its host, and competes metabolically with normal activity. Primaquine and related drugs, such as chloroquine, subject the red blood cells to oxidative stress. Cells stressed by *both* parasite and drug are the most vulnerable, and if they die they take the parasite down with them. Because consumption of fava beans subjects cells to oxidative stress, they also provide an antimalarial effect, recognized in folk medicine. Indeed, fava beans have some effect against malaria even for people with normal G6PDH activity; those with abnormal G6PDH have a greater advantage, until the maturing *Plasmodium* produces its own G6PDH.

The link with malaria is the likely explanation of the persistence of the gene in the population and the fava bean in agriculture. A final clue appears in the calendar: there is good overlap between the fava bean harvest period and the peak *Anopheles* breeding season.

Genetic diseases – some examples of their causes and treatment

Haemoglobinopathies – molecular diseases caused by abnormal haemoglobins

Sickle-cell anaemia: Pauling and co-workers showed in 1949 that haemoglobin isolated from patients with sickle-cell anaemia differed in electric charge from normal haemoglobin. Patterns of inheritance showed that sickle-cell anaemia is a genetic disease. Pauling's discovery was therefore the first evidence that genes precisely control the structures of proteins. This preceded the first determination of the amino acid sequence of a protein.

Recall that haemoglobin contains four polypeptide chains, two α chains and two β chains (Figure 1.16). The sickle-cell mutation changes residue 6 of the β chain from a charged sidechain, glutamic acid, to a nonpolar one, valine (β6Glu$\rightarrow$Val).

This creates a 'sticky patch' on the surface of the molecule. As a result, the mutant haemoglobin forms polymers within the erythrocyte, in the unligated or deoxy state (in the deoxy form, typical of venous blood, haemoglobin is not binding oxygen). To flow through small capillaries, erythrocytes must be deformable, as their typical size, 7.8 μm (in humans), is larger than the diameter of small capillaries. The formation of the polymers has a rigidifying effect on the erythrocyte, impeding its flow and blocking capillaries (Figure 2.10). In the traffic jam building up behind a plugged capillary, arriving red cells release their oxygen to surrounding hypoxic tissues, become deoxygenated, and thereby aggravate the problem. This produces pain – because of reduced oxygen supply resulting from capillary congestion – and anaemia and jaundice, consequences of the rapid breakdown of red blood cells.

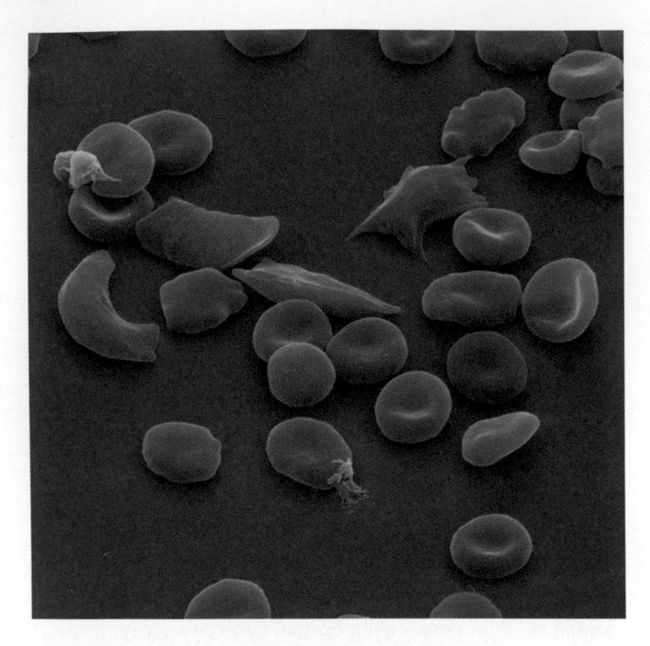

Figure 2.10 Coloured scanning electron micrograph of normal (round) and sickle-shaped erythrocytes. The shape of the cells is caused by aggregation of mutant haemoglobin in its deoxy state. The symptoms of the disease are caused not so much by the shape of the cells but by their inability to distort their shape in order to pass through the narrow capillaries.

(Science Photo Library).

The most common SNP associated with sickle-cell anaemia changes the codon gag to gtg. It is possible to test specifically for this mutation. Alternatively, sequencing the entire region would pick up other possible mutations, that might lead to thalassaemias.

Thalassaemias are genetic diseases associated with defective or deleted haemoglobin genes: most Caucasians have four genes for the α chain of normal adult haemoglobin, two alleles of each of the two tandem genes α_1 and α_2. Therefore α-thalassaemias can present clinically in different degrees of severity, depending on how many genes encode normal α chains. Only deletions leaving fewer than two active genes present as symptomatic under normal conditions. Observed genetic defects include deletion of both genes (a process made more likely by the tandem gene arrangement and sequence repetition, which make crossing over more likely) and loss of chain termination leading to transcriptional 'read through', creating extended polypeptide chains, which are unstable.

β-thalassaemias are usually point mutations. These may be:

- missense mutations (amino acid substitutions): the sickle-cell mutation is an example of a missense mutation;
- nonsense mutations (changes from a triplet coding for an amino acid to a stop codon) leading to premature termination and a truncated protein;
- mutations in splice sites;
- mutations in regulatory regions;
- certain deletions, including the normal termination codon and the intergenic region between δ and β genes, creating δ–β fusion proteins.

Phenylketonuria

Phenylketonuria (PKU) is a genetic disease caused by deficiency in a metabolic enzyme, phenylalanine hydroxylase, the enzyme that converts phenylalanine to tyrosine (see Figure 2.11). If untreated, phenylalanine accumulates in the blood, to toxic levels. If untreated, the high levels of phenylalanine cause a variety of developmental defects, including mental retardation, microcephaly, and seizures. The disease cannot be cured, but symptoms can be avoided by lifestyle control: a phenylalanine-free diet. Screening of newborns for PKU is legally required in the United States and many other countries.

PKU is an example of how understanding the molecular mechanism of a disease can, for some conditions, help restore and preserve health through suggested changes in lifestyle and/or medical treatment.

PKU is an autosomal recessive trait, associated with mutations in the phenylalanine hydroxylase gene on the long arm of chromosome 12 (12q22–12q24.1). In the UK and USA the prevalence is about 1 in 10 000 individuals; 1 out of 50 are carriers. PKU is the subject of neonatal screening in many countries.

A large number of known mutations are associated with PKU, appearing in all 13 exons of the gene and in flanking sequences (http://www.pahdb.mcgill.ca/). Many but not all of them are SNPs. These include many that affect catalytic activity, somewhat fewer that affect regulation, and even fewer that affect the assembly of the tetrameric enzyme.

The classical test for PKU depended on the build-up of phenylalanine and its degradation products such as phenylpyruvate in neonatal blood and urine.

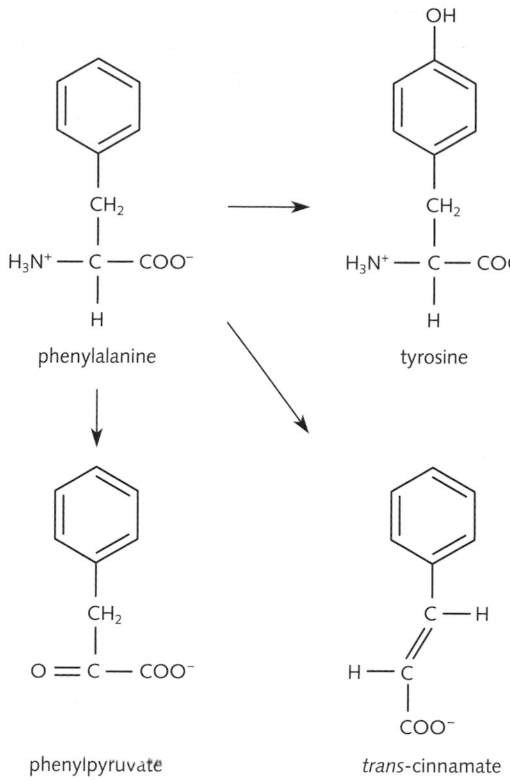

Figure 2.11 Phenylalanine (top left) and three metabolites. Top right: tyrosine, produced by the normal phenylalanine hydroxylase enzyme. Bottom left: phenylpyruvate, a degradation product formed in large quantities by people suffering from untreated phenylketonuria. Bottom right: *trans*-cinnamate, produced from phenylalanine by the enzyme phenylalanine ammonia-lyase. A derivative of this enzyme is now in phase II clinical trials for treatment of phenylketonuria.

(Phenylpyruvate is a ketone, hence the name of the disease. See Figure 2.11.) From a blood sample taken from a neonate, mass spectrometry can measure abnormal concentrations of phenylalanine or tyrosine. It is also possible to detect mutations in the gene by sequencing, although should a novel mutation appear it might not be possible to conclude with confidence that the mutant protein is dysfunctional. Genomic sequencing could also detect carriers, allowing counselling of potential parents.

Management of PKU depends on enforcing a low-phenylalanine diet. This is not entirely satisfactory: compliance is a common problem, as low-phenylalanine foods are relatively unpalatable and do not provide complete nutrition (see Weblem 2.6). Artificial mixtures of amino acids – phenylalanine-free formulas – replace high-protein foods.

It is particularly tricky to manage PKU women in pregnancy. Remember that PKU is an autosomal *recessive* trait. A woman with PKU must be homozygous for defective phenylalanine hydroxylase (although the two alleles need not necessarily bear the same mutation). If such a woman becomes pregnant, it is likely that the foetus is only a carrier (unless the father is also a carrier). The problem is to control phenylalanine levels in the mother so as to provide adequate nutrition to the foetus, without subjecting the mother to toxic levels of phenylalanine.

A current topic of PKU research is enzyme replacement therapy. It is not satisfactory to administer functional phenylalanine hydroxylase itself, as is done with insulin for diabetes. Phenylalanine hydroxylase is a tetramer, requires a cofactor, and is subject to complex regulatory controls. An alternative is to use phenylalanine ammonia-lyase, an enzyme found in plants, fungi, and bacteria, which converts phenylalanine to *trans*-cinnamic acid (see Figure 2.11).

This enzyme is more stable, and does not require cofactors. The product, *trans*-cinnamic acid, degrades to hippuric acid, excreted in the urine. The problem of antigenicity is addressed by attachment of polyethylene glycol (PEG). As a drug, pegylated phenylalanine ammonia-lyase is currently in phase II clinical trials.

Alzheimer's disease

The symptoms of Alzheimer's disease are loss of cognitive functions, characterized by: loss of train of thought, progressive memory problems, missing important appointments, etc. The most common form is late-onset Alzheimer's, appearing in people over the age 65. Approximately 50% of people over age 85 suffer from it. Alzheimer's disease is a very severe public-health problem, especially in view of increased life-spans. Early-onset Alzheimer's, defined as first appearing at age <65, is rarer. Even rarer is familial Alzheimer's, involving $<1\%$ of cases, appearing at age 40–60.

The risk of late-onset Alzheimer's disease is correlated with SNPs in apolipoprotein E (ApoE). The basic function of this 317-residue protein is to remove cholesterol from the blood. The gene for ApoE is on chromosome 19. There are four common alleles, which differ by SNPs:

ApoE1 = rs429358(C) + rs7412(T) [minor variant]
ApoE2 = rs429358(T) + rs7412(T)
ApoE3 = rs429358(T) + rs7412(C) [~55%]
ApoE4 = rs429358(C) + rs7412(C)

The correlations with risk of Alzheimer's disease are:

At least one E4 allele → increased risk of Alzheimer's
At least one E2 allele → decreased risk of Alzheimer's

SNPs and cancer

SNPs are relevant to cancer research and treatment in several ways:

(1) Mutations detectable in the genome indicate propensity for development of cancers. Mutations in BRCA1 and BRCA2, as indicators for likelihood of breast and ovarian cancer development, are probably the best known.
(2) Sequence analysis can predict disease progression and outcome.
(3) Sequence analysis can help choose optimal treatment.
(4) Tumour progression often involves mutations and divergence of cell lines.

The onset of cancer is associated with loss of genome integrity. Cancer results from accumulated mutations that break down the controls on cell growth. The source can be in three classes of genes: genes that regulate cell proliferation, genes required for repair of DNA damage, and genes that control apoptosis.

The 'two-hit hypothesis' interprets the relation between sporadic and familial forms of disease to the need to mutate both copies of such genes.

Consider, as an example, retinoblastoma, a rare childhood tumour of the eye. Approximately 30–40% of cases are familial; the rest sporadic. The familial form shows an autosomal dominant inheritance pattern. Clinical characteristics of familial retinoblastoma that distinguish it from the sporadic picture are early onset, and the appearance of multiple tumours, affecting both eyes.

The two-hit hypothesis offers an explanation for the differential age of onset, and severity, of familial and sporadic retinoblastoma. The idea is that non-familial cases require inactivation of both copies of retinoblastoma gene, each of which was originally functional. Separate and independent mutations are necessary. In contrast, familial retinoblastoma affects a person who has inherited one defective and one functional copy of the gene. That is, the first hit is inherited; all that is needed is the second hit (see Figure 2.12).

Tumour suppressor genes protect cells against development of cancer. They encode proteins that inhibit tumour formation. Their normal function can

Sporadic retinoblastoma: two hits required

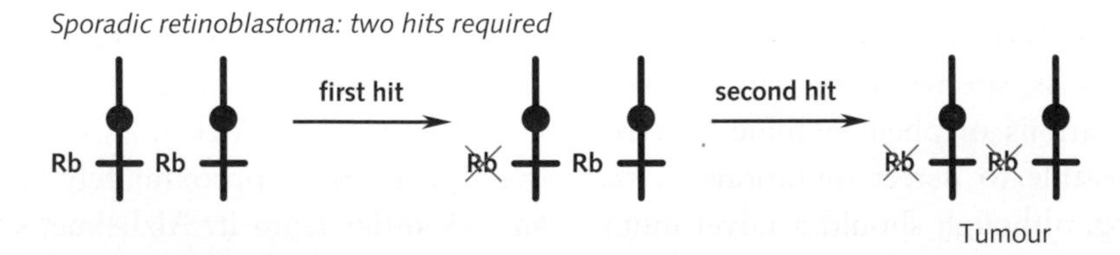

Familial retinoblastoma: first hit inherited, one more required

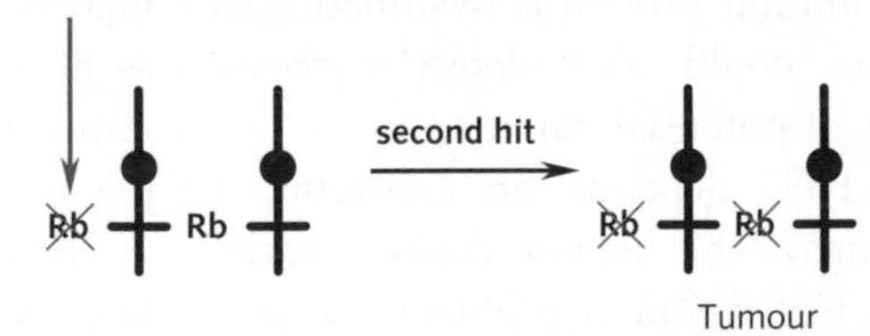

Figure 2.12 Explanation of the difference between sporadic and familial retinoblastoma, a rare cancer of the retina. According to the two-hit theory, two copies of a gene must be inactivated. In sporadic retinoblastoma, both copies are originally functional, and two separate, independent mutations are required to inactivate them. In familial retinoblastoma, one defective copy of the gene is inherited. Only a single mutation, in the other allele, is required to produce the disease.

BOX 2.6

The genes *BRCA1* and *BRCA2*

Gene	Chromosome band	Gene length	Protein length (amino acids)	Number of exons
BRCA1	17q21	>100 kb	1863	24 (22 coding, exon 11 very large)
BRCA2	13q12–q13	>200 kb	3418	27

be to inhibit cell growth; mutations 'take the foot off the cell-growth brake'. Mutants in these genes raise the risk of developing cancer.

Well-known examples of tumour suppressor genes are: *BRCA1* and *BRCA2*. In the general population:

~12% of women will develop breast cancer;

~1.4% of women will develop ovarian cancer.

Of women with a harmful mutation in *BRCA1* or *BRCA2*:

~60% will develop breast cancer;

~15–40% will develop ovarian cancer.

BRCA1 and *BRCA2* encode long proteins unrelated in sequence and structure (see Box 2.6). Both proteins are required for chromosome stability, participating in mechanisms of repair of DNA double-strand breaks.

Many *BRCA1* mutations are known. (Many are not SNPs.) Their prevalence varies among populations, showing strong founder effects (see Table 2.1). Testing for mutations in these genes is now quite common (see Box 2.7).

Table 2.1 Common *BRCA1* and *BRCA2* mutations

Population	Common *BRCA1* mutations	Common *BRCA2* mutations
Ashkenazi Jews	185delAG, 5382insC	6174delT
Iceland		999del5
Denmark	2594delC, 5208T→C	
Lithuania	4153delA, 5382insC, 61G→C	
China	589delCT, IVS7–27del10, 1081delG, 2371–2372delTG	3337C→T

BOX 2.7

Genetic testing for mutations in breast cancer genes *BRCA1 and BRCA2*

Breast cancer is a leading killer of women in western Europe and the USA, affecting over 13% of the population (100 times as many women as men) and causing death in about 3% of all women. Among the known genetic factors that raise the risk of breast cancer are mutations in the genes *BRCA1* and *BRCA2*.

Screening for mutations in these genes can provide a risk-alerting system for breast and ovarian cancer. However, planning of population-wide screening programmes must take into consideration cost/benefit analysis. In many countries, medical care delivery policy is set by a national health service. (The USA is a notable exception.) Governments must decide how to apportion resources, taking into account the cost of any procedure and the utility of the information it produces. Mutations in *BRCA1* and *BRCA2* are associated with only about 5–10% of breast cancers. Conversely, not all women with a mutation in either of these genes develop cancer, although over 50% do. The *BRCA1* and *BRCA2* genes are long, multiexon sequences each >100 kb long, making total resequencing of the genes a complicated procedure. Many mutations are known, spaced widely within the exons. Given that even complete resequencing of the genes would not provide a helpful prognosis in many cases, it is not deemed useful,

→

with current technology, to screen the entire population fully for *BRCA1* and *BRCA2* mutations.

The logic of the decision changes for individuals suspected to be at high risk. These include women who:

- have a close relative known to have a *BRCA1* or *BRCA2* mutation;
- have close relatives who have been diagnosed with early-onset (age <50 years) breast or ovarian cancer; or
- have themselves been diagnosed with breast cancer and want to know their likelihood of developing ovarian cancer.

For these individuals, the likelihood of finding useful information certainly justifies genetic testing.

In some genetically relatively homogenous populations, practical advantage can be taken of the observation that specific mutations in *BRCA1* and/or *BRCA2* are common in that population. For instance, Ashkenazi Jews are about 20 times more likely to bear a mutant in one of the two genes than the general population. Three mutations – 185delAG and 5382insC in *BRCA1* and 6174delT in *BRCA2* – account for 90% of *inherited* breast and ovarian cancers in this group. Testing for these three mutations or for any specific mutation known in a relative – whether one of these three

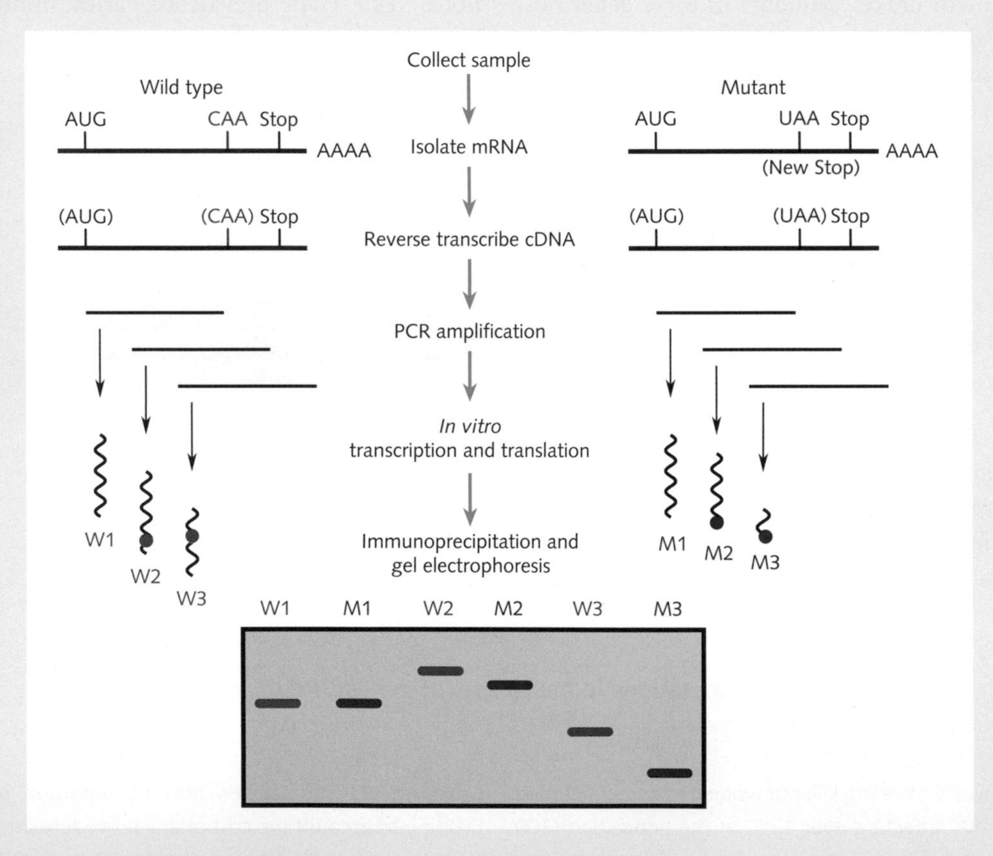

Figure 2.13 Schematic explanation of the protein truncation test. Starting from mRNA, or in some cases from genomic DNA, a series of fragments covering the gene or exon of interest is amplified. *In vitro* transcription and translation produces the corresponding polypeptide chains. A mutation that introduces an internal stop codon or a deletion leading to a frame shift produces some shortened fragments that can be distinguished on a gel.

Straight lines indicate nucleic acids; wavy lines indicate polypeptides. W1, W2, and W3 are peptides derived from the wild-type gene. M1, M2, and M3 are derived from the mutant gene, *using the same primers*. The red and blue dots in the peptides correspond to the position of the CAA in the wild-type mRNA and the UAA in the mutant mRNA. In this example, the peptides W1 and M1 have the same size, but M2 is shorter than W2 and M3 is shorter than W3. Note that the mobilities of the peptides are not strictly proportional to their molecular weights.

The triplets in parentheses refer to the original mRNA sequence. Different bases appear in the molecules that are created during the procedure. (See Exercises 2.6–2.9.)

or any other – is much simpler and more cost-effective than a full resequencing of the genes. Resequencing all of the exons of the gene is at present about ten times more expensive and can always be considered as a subsequent step if none of the specific mutations appear. This is likely to change, as the cost of sequencing diminishes.

What techniques are used to screen for common mutations?

- *The protein truncation test* detects premature stop codons by amplifying the coding region of exon 11 of *BRCA1* or exons 11 and 12 of *BRCA2* (see Figure 2.13).
- *The single-stranded conformational polymorphism test* is applied to analysis of exons 2–10 and 12–24 of *BRCA1* and exons 2–10 and 12–27 of *BRCA2* (see Figure 2.14).
- *Full resequencing of the gene.*

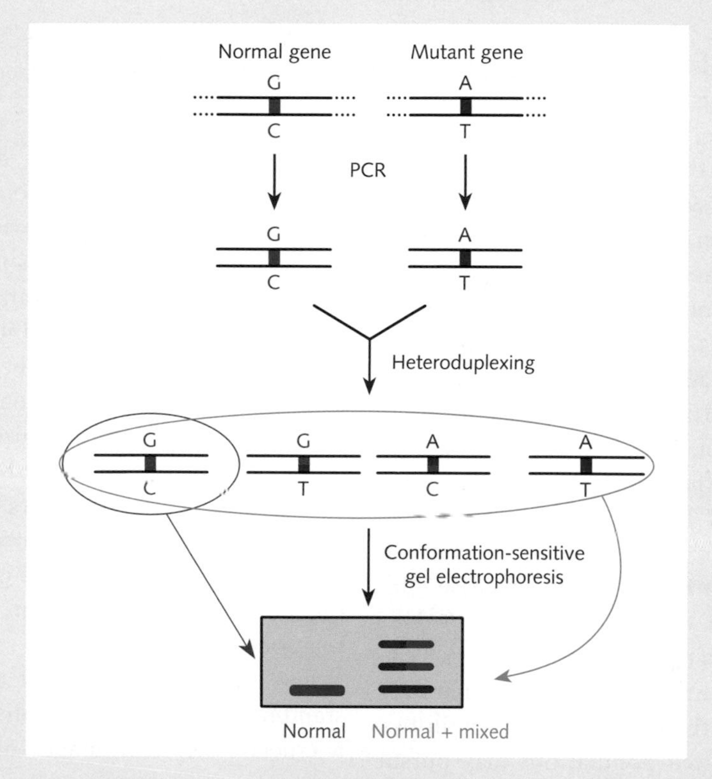

Figure 2.14 Conformation-sensitive gel electrophoresis, a method for detecting localized mutations. This figure shows the procedure for analysing a sample from a heterozygote, containing normal DNA from one chromosome (red) and a mutation on the other (blue). The region surrounding the site of suspected mutation is amplified by PCR. Melting and annealing produces the two original double-stranded regions and also two mismatched pairs. The regions are sufficiently similar to form a hybrid pair despite the single-base mismatch, but the mismatch causes sufficient conformational deformation to alter mobility on the gel, especially under partially denaturing conditions. A normal sample would produce only a single band (left lane on the gel), but the mixture of normal and mutant would produce additional bands, the mismatched pairs showing altered mobility.

- Applications of genomics in cancer research include:
  (a) understanding the cellular events required to transform normal to cancer cells
  (b) improving the precision of diagnosis and prediction of outcome
  (c) guidance in choosing the most effective therapy
  (d) surveying populations for high-risk individuals, who carry mutations in tumour-suppressor genes.

Species

The next step up from populations is species.

Species are a fundamental unit of evolution. Species represent nature's experiments in structures and lifestyles. It is species that were the elements of Linnaeus's taxonomy, and it was the emergence of novel species that gave Darwin his life's work and the title of his major book.

It is extremely rare in science that a concept so fundamental to a field as species is in biology would be so difficult to define. Genomics has greatly illuminated but not solved the problem.

To understand what we mean by species we need not only an explanation of the concept, but a criterion for deciding whether two organisms – or, more appropriately, two populations – belong to the same species or different ones. Ernst Mayr enunciated the classic biological approach, focusing on the idea that different species are reproductively isolated in nature. Mayr's definition applies to sexually reproducing organisms: individuals that can mate with each other to produce viable and fertile offspring when they encounter each other in nature are members of the same species. When reproductive barriers arise – for instance, when groups of animals are trapped on different islands by rising sea levels – the species divides into two or more populations that, separately, interbreed only within themselves and maintain two gene pools. The separated populations may pursue different evolutionary paths and ultimately diverge to form separate species. Of course, geographic isolation is only one possible cause of speciation.

Many other definitions have been proposed, including those based on comparisons of features of phenotypes, and divergence of genomic sequences. For higher organisms at least, the different definitions are usually consistent, but they are not entirely equivalent.

- Many modern biologists define a species as a group of similar organisms that interbreed naturally to produce fertile offspring. An alternative approach is to base species definitions on genomic sequences. This approach is extremely powerful even for higher organisms, and essential for prokaryotes.

Any valid definition must take into account the fundamental observation that species of higher organisms are *discrete* entities. The taxonomic hierarchy is in principle a classification of species. We now recognize that the static classification is largely congruent with the evolutionary tree of ancestor–descendant relationships.

- Why living things should be 'quantized' into discrete species is a very subtle question.

A fundamental paradox about the biology of higher organisms is that although species are discrete, new species can evolve. Darwin's finches are a classic example of reproductive isolation of populations that have diverged into separate species. Resolution of this paradox is quite a difficult challenge, but perhaps a basic component of the answer would be that it is the discreteness of ecological niches – to which individual species are optimized – that accounts for the discreteness of species of higher organisms.

For prokaryotes, in contrast, the species concept is in serious trouble. In particular the presence of large amounts of horizontal gene transfer overturns the idea that there is a hierarchy of ancestor-descent relationships.

Few people doubt that if the problem of understanding species can be resolved, genomics will play a crucial role. Advantages of genomic approaches over others are that they allow:

(1) more accurate assignment of relationships (as we saw in the case of *Buddenbrokia plumatellae* (see p. 48)),

(2) quantitative measurements of the divergences between species, and

(3) estimation of the time elapsed since the last common ancestor. (Now, classical palaeontology also allows measurements of dates of origins of species that have left fossil records. In fact from fossil records we can date extinctions, which is not possible from genomics!)

- Here we have described both the importance of the species concept and some of the problems associated with it. Both the concept and the problems will be themes of much of the rest of this book.

A census of species?

We simply do not know the number of species that have ever existed. Even for the species alive today, only a fraction have been formally described. About 1.5 million species are known to science (see below), but estimates suggest that at least twice as many exist, and probably even an order of magnitude more than that.

Most palaeontologists agree that living species amount to less than 1% of the number that have ever lived, but it is impossible to infer precise estimates.

Group of organisms	Number of species described
Insects	751 000
Other animals	281 000
Higher plants	248 400
Fungi	69 000
Protozoa	30 800
Algae	26 900
Prokaryotes	4 800
Viruses	1 000

Over half of the described species are insects. Almost 40% of the insects are beetles (order *Coleoptera*). When J.B.S. Haldane was asked what he had learned about God from his study of biology, he replied that God must have '. . . an inordinate fondness for beetles.'

Given the large numbers of even fairly closely related species, one might be tempted to think that the species that lack scientific description are only minor variations on familiar themes.

This is not true.

The Burgess Shale, in the Canadian Rockies near Banff, British Columbia, Canada, contains some of the world's best-preserved fossils from the Cambrian era, ~530 million years ago. The locality contained a rich, predominantly invertebrate, community that left fossils in an unusually high-quality state of preservation. Some of the forms are similar to known animals, both extant and extinct, but others show profound structural differences, including completely different body architectures. The loss of these species has impoverished the natural world in general and the science of biology in particular.

These fossils strongly suggest that the living forms we see are, to a large extent, the result of historical accident and that the course of evolution might easily have been very different.

The biosphere

Palaeontologist G.E. Hutchinson wrote a book entitled *The Ecological Theater and the Evolutionary Play*. Life on Earth has been a 3.5 billion year epic of exploration and adventure. It has included comedy and tragedy. Hutchinson emphasized the interdependence of environment and living things, which reciprocally affect each other. (We have already mentioned the example of the photosynthetic origin of atmospheric oxygen.)

We shall never know the full extent of the species that Nature has generated in its explorations (see Box 2.8). Organisms alive today represent only a fraction of all that have existed.

Extinctions

We now recognize that the roster of extant species is in flux. Although Darwin was the first to argue convincingly that new species arise, the idea of extinction was propounded by Cuvier in 1796. Nevertheless, the idea of the immutability of species retained strong adherents, a problem Darwin faced later. When US President Thomas Jefferson sent Meriwether Lewis and William Clark to explore the North American continent, he urged them to be on the lookout for mammoths. Anyone could have *hoped* that mammoths, extinct in Eurasia, might have survived in the New World; the idea of a fixed roster of species shaded hope into confidence. Cuvier believed that extinction of species was the result of major natural catastrophes. Indeed, sometimes, natural catastrophes do cause large-scale simultaneous extinctions (see Figure 2.15 and Box 2.9). An example is the asteroid that landed in what is now the Yucatan Peninsula of Mexico, approximately 65 million years ago, at the boundary between the Cretaceous and Tertiary ages. However, in other cases, species reach a more peaceful end of their existence – 'dying

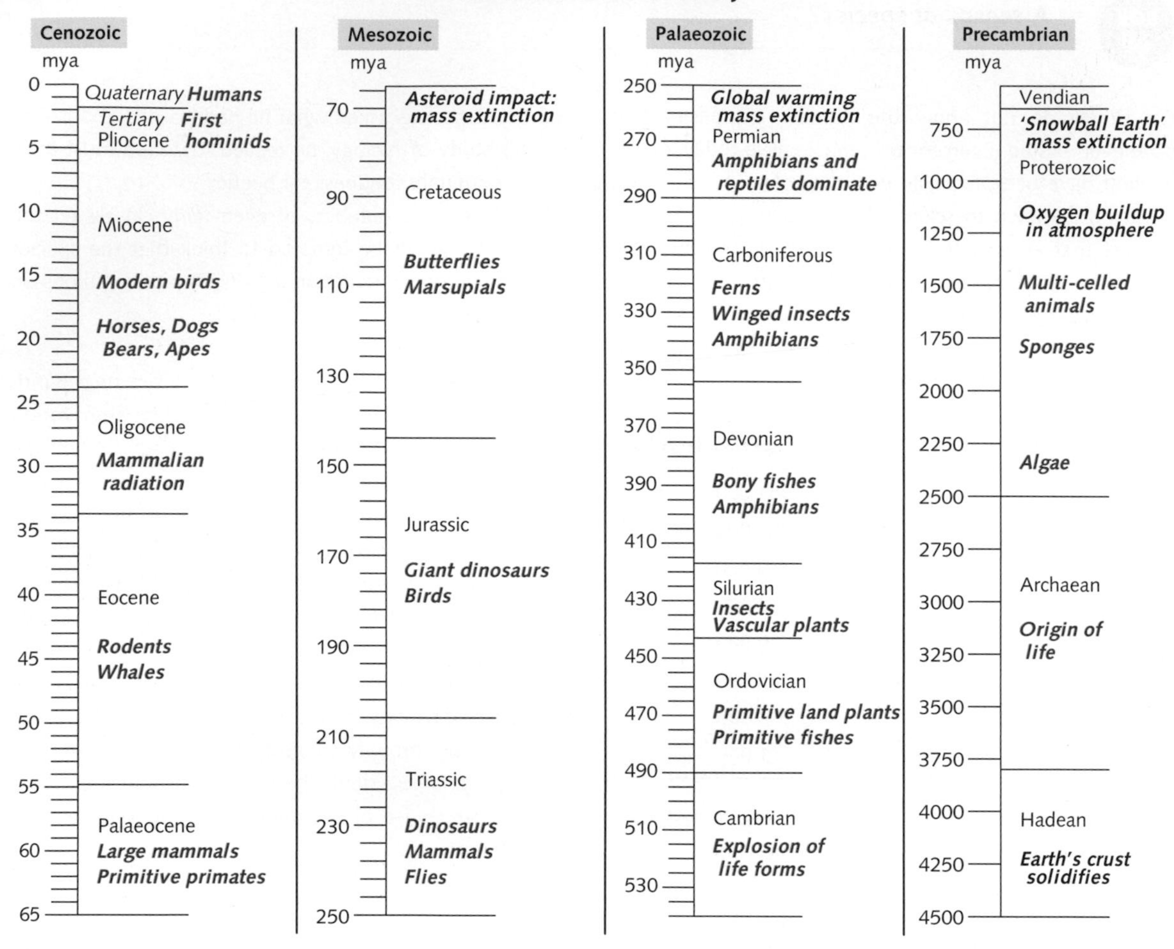

Figure 2.15 Geological ages (e.g. Cenozoic), epochs (e.g. Quaternary), and cataclysmic events (e.g. asteroid impact: mass extinction) in black. First appearance of, or prevalence of, different life forms in red. mya = millions of years ago.

peacefully in their beds' so to speak – as a result of gradual environmental change.

A major mass extinction is going on right now and we are largely, although not entirely, to blame. Human activity is responsible for the disappearance of species through excessive hunting, habitat destruction, transport of species that replace native ones, and the introduction of toxic chemicals into the environment. Moreover, extinction of one species may cause the loss of others. For instance, some plants are dependent on a particular insect for pollination. Loss of the insect is fatal to the plant, and the network of dependence may be quite far ranging. North American bees are being killed in great numbers, by the *Varroa* mite, a parasite that has developed resistance to the pesticides that formerly controlled it, and by the mysterious colony collapse disorder, which may be at least partly viral in origin. Many Californian almond growers must now import honeybees to ensure the pollination of their trees. (The problem is not restricted to almonds: about one-third of our food depends on honeybee pollination.)

*Extinction of the thylacine (*Thylacinus cynocephalus*)*

It is a sad event when the last individual of a species dies. The last thylacine, or Tasmanian wolf, named Benjamin, died in the Hobart Zoo on 7 September 1936 (see Figure 2.16). (On the same day, Boulder Dam, now Hoover Dam, started operation. Straddling the Colorado River, it provides water and electrical power to the south-western USA, including the rich agricultural lands of Southern California and the city of Los Angeles. The opening of the Hoover Dam

BOX 2.9 Mass extinctions

Period	Approximate date (mya)*	Likely cause	Estimated loss of taxa
Ordovician-Silurian	439	Formation and melting of glaciers	60% of marine genera 25% of marine families
Late Devonian	364	?	57% of marine genera 22% of marine families
Permian–Triassic	251	Asteroid impact (?)	95% of *all species!* 53% of marine genera 53% of marine families 70% of land species
End-Triassic	199–214	Volcanic eruptions, possibly leading to global warming	52% of marine genera 22% of marine families ? vertebrate species
Cretaceous–Tertiary	65	Asteroid impact	47% of marine genera 16% of marine families 18% of land vertebrate families (including dinosaurs)
Present	0	Human activity, in part	???

* mya, millions of years ago.

Figure 2.16 The Tasmanian wolf, or thylacine (*Thylacinus cynocephalus*), was a marsupial showing striking convergent evolution in overall body form to the placental wolf/dog family. It had stripes on its back and tail. It lived on 'mainland' Australia and New Guinea several thousand years ago, but its recent distribution was confined to south-western Tasmania. The thylacine was challenged by the introduction of dogs, and hunted to extinction in the early 20th century. Despite claims of sightings since the 1930s, no convincing evidence for survival has emerged.

The preservation of specimens in museums has prompted suggestions that the thylacine might be cloned, but it appears that degradation of the DNA has been too extensive.

and the extinction of the thylacine bear simultaneous witness to human control over nature.)

The thylacine appears prominently on the label of a well-known Tasmanian beer.

Conversely, it is a happy event when a threatened species is rescued.

*Survival of Père David's deer (*Elaphurus davidianus*)*

In 1865, Père Armand David (1826–1900), a French Basque priest and biologist, learned of an unusual animal living in the Imperial Hunting Park, near Beijing, China, within precincts reserved to the Emperor. He was able to observe the animals and to take an antler and two hides back to Europe. Recognized as a novel species, the deer were named for the missionary (see Figure 2.17). (Père David also discovered the giant panda.)

The animal's range had extended over a large area of western and northern China, but it became extinct in the wild. (A picture of this deer illustrates a

Figure 2.17 Père David's Deer.

(Photograph by Tony Pearse, taken at Woburn Abbey in 1993.)

17th century map of Xensi province.) A herd in the hunting park survived. The emperor gave about a dozen deer as gifts to zoos in Europe and Japan. Herbrand Arthur Russell, 11th Duke of Bedford, installed a small herd on his Woburn Abbey estate.

It was well that he did. The Beijing deer staggered under several blows. There were severe floods in 1894–1895, reducing the Chinese herd to 20–30 animals. The survivors were lost in 1900 during the military campaign to lift the siege of European legations during the Boxer Rebellion.

The remaining Père David's deer were collected from European zoos and pooled in the Woburn Abbey herd. This contained 18 individuals, of which 11 were fertile. Some – probably not all – of these were the founder individuals of the entire current population. The deer bred, with the size of the Woburn Abbey herd reaching 250 in 1945.

In 1956, some deer were sent to Beijing. In 1957, a calf was born in China for the first time in over 50 years. They are now maintained in dedicated nature reserves and have also been reintroduced into the wild. The site of the Emperor's hunting garden is now a park devoted to Père David's deer. The deer have come home!

RECOMMENDED READING

- A standard textbook in this field:

  Strachan, T. & Read, A. (2010). *Human Molecular Genetics*, 4th ed. Garland Science, New York.

- Genomics and developmental biology:

  Cañestro, C., Yokoi, H., & Postlethwait, J.H. (2007). Evolutionary developmental biology and genomics. Nat. Rev. Genet. **8**, 932–942.

- Genomics and neurobiology:

  Boguski, M.S. & Jones, A.R. (2004). Neurogenomics: at the intersection of neurobiology and genome sciences. Nat. Neurosci. **7**, 429–433.

  Berger, M.S., Couldwell, W.T., Rutka, J.T., & Selden, N.R. (2010). Introduction: neurogenomics and neuroproteomics. **28**, E1.

  Sforza, D.M. & Smith, D.J. (2003). Genetic and genomic strategies in learning and memory. Curr. Genomics **4**, 475–485.

  Nestler, E.J. & Hyman, S.E. (2010). Animal models of neuropsychiatric disorders. **13**, 1161–1169.

- The general consequences of mutation:

  Hill, W.G. & Loewe, L. (2010). The population genetics of mutations: good, bad and indifferent. Philos. Trans. Roy. Soc. Lond. B: Biol. Sci. **365**, 1153–1167.

- A description of the large number of haemoglobin mutants then known, set in a public-health context:

  Weatherall, D.J. & Clegg, J.B. (2001). Inherited haemoglobin disorders: an increasing global health problem. Bull. World Health Organ. **79**, 704–712.

- Cancer genomics:

  Ding, L., Wendl, M.C., Koboldt, D.C., & Mardis, E.R. (2010). Analysis of next-generation genomic data in cancer: accomplishments and challenges. Hum. Mol. Genet. **19**, R188–R196.

  Chin, L., Hahn, W.C., Getz, G., & Meyerson, M. (2011). Making sense of cancer genomic data. Genes & Dev. **25**, 534–555.

  Majewski, I.J. & Bernards, R. (2011). Taming the dragon: genomic biomarkers to individualize the treatment of cancer. Nat. Med. **17**, 304–312.

  Stratton, M.R. (2011). Exploring the genomes of cancer cells: progress and promise. Science **331**, 1553–1558.

- A fascinating general essay about evolutionary biology:

  Hutchinson, G.E. (1965). *The Ecological Theater and the Evolutionary Play*. Yale University Press, New Haven, CT, USA.

- Discussion of the problem of biological species. Papers from a colloquium in honour of 100-year old Ernst Mayr.

  Hey, J., Fitch, W.M., & Ayala, F.J. (2005). Systematics and the origin of species: An introduction. Proc. Natl. Acad. Sci. USA **102**, 6515–6519.

  de Queiroz, K. (2005). Ernst Mayr and the modern concept of species. Proc. Natl. Acad. Sci. USA **102** (Suppl 1), 6600–6607.

- Current and past extinctions:

  May, R.M. (2010). Ecological science and tomorrow's world. Philos. Trans. Roy. Soc. Lond. B: Biol. Sci. **365**, 41–47.

  Barnosky, A.D., Matzke, N., Tomiya, S., et al. (2011). Has the Earth's sixth mass extinction already arrived? Nature **471**, 51–57.

- Two celebrated palaeontologists interpret the Burgess Shale fossils:

  Gould, S.J. (1990). *Wonderful life: the Burgess Shale and the Nature of History*. W.W. Norton, New York.

  Conway Morris, S. (1998) *The Crucible of Creation: The Burgess Shale and the Rise of Animals*. Oxford University Press, Oxford.

  See also the exchange between Conway Morris and Gould (1998): Showdown on the Burgess Shale. Nat. Hist. **107**, 48–55.

EXERCISES, PROBLEMS, AND WEBLEMS

Exercises

Exercise 2.1 A randomly chosen pair of humans will show an average nucleotide diversity that lies between 1 base pair in 1000 and 1 in 1500. Any human and any chimpanzee differ in approximately 1 base pair in 100. (Note that the chimpanzee genome is somewhat larger than the human.) (a) Estimate the number of differences in the total sequences between two randomly chosen humans. (b) Estimate roughly the number of differences in the total sequences between a human and a chimpanzee.

Exercise 2.2 Why is it possible to have calico cats but not calico kangaroos (barring abnormal cell division at an early embryonic stage or a skin disease)?

Exercise 2.3 From material presented in this chapter, give examples of mutations with clinical consequences that involve: (a) dysfunction of a metabolic enzyme, (b) 'read-through' producing extended proteins, (c) proteins with lower stability than the normal version, (d) increase in the risk factor for a disease not known to be associated with the primary (?) function of a protein, (e) enhanced probability of developing cancer. In each case state, if possible, the nature of the mutation, and the mechanism by which it produces clinical consequences.

Exercise 2.4 Most newborn mammals can digest lactose in infancy, consistent with their dependence on maternal milk for feeding. The enzyme lactase hydrolyses the disaccharide lactose, the major carbohydrate component in milk, to glucose + galactose. In most mammalian species, expression of lactase ceases at the time of weaning. Many human populations follow the mammalian paradigm: lactase expression is permanently turned off when a child is about 4 years old. Loss of lactase expression produces subsequent intolerance to dairy products. A single-site mutation that causes lifetime lactase expression – or lactase persistence – was selected for in populations that domesticated cattle and depended on dairy products in their adult diet. In Europe, lactase persistence is more common in the north, where less exposure to sunlight makes northern Europeans more dependent on dairy products as sources of calcium and vitamin D precursors. Ninety-six per cent of Swedes and Finns are lactose persistent.

(a) Would you expect the mutation to be found within the coding regions of the lactase gene itself?

(b) Heterozygotes produce approximately half the amount of lactase. (This is sufficient to avoid the symptoms of lactose intolerance.) From this observation, what additional inference can you make about the location of the mutation?

Exercise 2.5 Suppose that there are ten SNPs in a 10 kb region. If the region is on the Y chromosome, how many possible haplotypes are there? If the region is on a diploid chromosome, how many possible haplotypes are there?

Exercise 2.6 In the diagram of the protein truncation test (Figure 2.13), what actual triplets appear in place of the triplets shown in parentheses?

Exercise 2.7 In the diagram of the protein truncation test (Figure 2.13), which pair of peptides has the larger difference in length: W2 and M2 or W3 and M3?

Exercise 2.8 In the protein truncation test (see Figure 2.13), in cases in which the novel stop codon arises from a single-site substitution, as shown in the figure, why would it not work to subject the amplified cDNA fragments directly to electrophoresis, i.e. to skip the *in vitro* transcription and translation step?

Exercise 2.9 What would the gel in Figure 2.13 look like if the person from whom the sample was taken had the mutation C→G at the same site?

Exercise 2.10 In the conformation-sensitive gel electrophoresis technique for detecting mutations (Figure 2.14), the gel is run under mildly denaturing conditions to accentuate conformational differences, thereby increasing the differences in mobility. Why would a gel fail to give useful information if it were run under fully denaturing conditions?

Exercise 2.11 Why might the prescription of monoamine oxidase inhibitors to a child presenting with symptoms of anxiety, in the case of suspected maltreatment, be contraindicated?

Exercise 2.12 In Figure 2.8 showing the data on Barbary macaques, where did the animal corresponding to the isolated orange point come from?

Problems

Problem 2.1 Survival of a SNP. Suppose a person is heterozygous for a novel, selectively neutral mutation. Suppose the person has two children that survive to reproductive age. The probability of loss of the mutation in that one generation is 25%. If each descendant has two children that survive to reproductive age, what is the probability of complete disappearance of the mutation in 200 years. Assume 25 years per generation.

Problem 2.2 Replication of RNA viruses is error-prone. It is estimated that the replication of HIV-1 introduces one mistake per replication of its $\sim 10^5$ bp genome. If the estimated generation time of HIV-1 in a human body is ~ 1 day and 10^{10} progeny viruses are produced per patient per day, and assuming that an AIDS patient is initially infected by viruses with identical genomes, estimate whether the patient will (a) generate a mutation at every possible site in the genome every day; and (b) generate mutations at every possible *pair* of sites in the genome every day.

Problem 2.3 Assume the model of expression switching in the human β-globin region shown in Figure 2.2(b,c). (a) What developmental progression of globin expression would you expect if the region containing the $G\gamma$ and $A\gamma$ genes were interchanged with the region containing the δ and β genes? (b) What developmental progression would you expect in patients with deletions in the δ- and β-regions, including the pyrimidine-rich control region YR?

Problem 2.4 (a) In the experiment that produced Cc (Copycat) as a clone of Rainbow (see Figure 2.3), the X-linked inactivation of the cell selected was not reversed. If a number of other cats were cloned by the same procedure, from other cells taken from Rainbow, what coat colours would the cats produced show? (b) Suppose a way were discovered to reverse the X-linked inactivation in the cell from Rainbow from which another cat were cloned. Would the cloned cat have a coat indistinguishable in appearance from Rainbow's? (c) Would monozygotic twin (= 'identical' twins, in genome sequence at least) natural daughters of Rainbow show identical coat colour patterns? (d) If your answer to (c) suggests that the twin daughters might not look the same, how then could you be sure that two daughter cats are *monozygotic* twins, without sequencing their entire genomes?

Problem 2.5 A *transitive relationship* is one such that if A is related to B, and B is related to C, then A is related to C. (In arithmetic, equality is a transitive relation – if A = B and B = C, then A = C.) Consider whether 'belong to the same species' is a transitive relationship. There are examples of 'ring species', sets of geographically connected populations such that members of neighbouring populations can interbreed, but at least two populations, not necessarily the most geographically distant, cannot. A classic example is the *Larus* gulls, which inhabit regions surrounding the North Pole (Figure 2.18).

For which definitions of species (see p. 66) is 'belong to the same species' a transitive relationship?

Problem 2.6 We know that viruses can deliver genes to mammals, including humans. Examples include curing X-linked adrenoleukodystrophy in humans, and increasing the apparent intelligence of mice. (a) What kinds of problems would you expect to arise in setting guidelines for what genetic modifications of humans should be allowed? (b) How far could you go towards drafting such a set of guidelines?

Problem 2.7 In a study of brown bear (Figure 2.19) populations in northwest North America, samples of mitochondrial DNA were collected and sequenced from 317 free-ranging brown bears (*Ursus arctos*) from 22 localities. Forty-six variable sites corresponded to 29 haplotypes, which clustered into four major clades (see Table 2.2). Table 2.3 identifies the location(s) at which bears with the corresponding haplotypes were found.

(a) On a copy of a map showing Alaska, northwestern Canada, and the lower 48 states as far south as northern Wyoming, mark in different colours the sites of appearance of bears with mitochondrial DNA in the different classes. Use the data in Table 2.3. Describe the geographical distribution of the different classes. Do they overlap substantially?

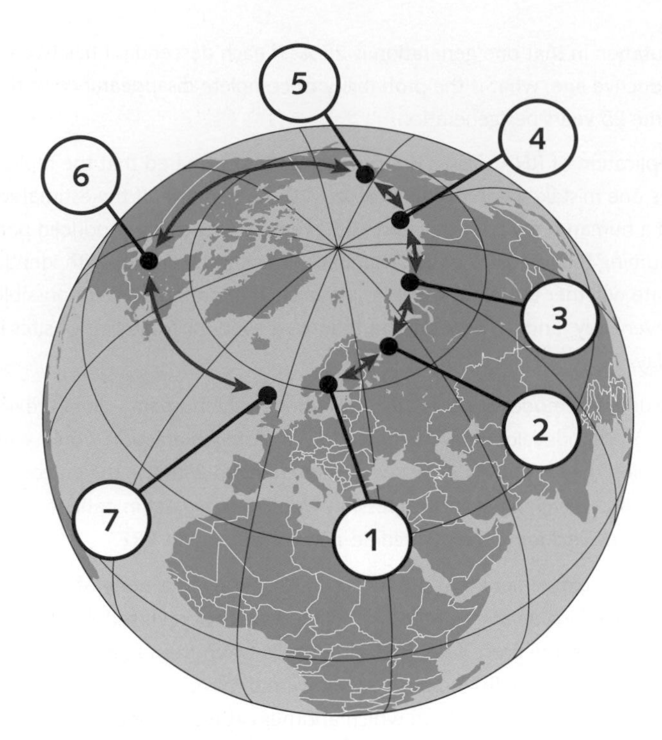

Figure 2.18 North circumpolar distribution of *Larus* gull species.

1. Lesser Black-backed Gull (*Larus fuscus*)
2. Lesser Black-backed Gull (Siberian population of *Larus fuscus*)
3. Heuglin's Gull (*Larus fuscus heuglini* or *Larus heuglini*)
4. Birula's Gull (*Larus argentatus birulai*)
5. East Siberian Herring Gull (*Larus argentatus vegae* or *Larus vegae*)
6. American Herring Gull (*Larus argentatus smithsonianus* or *Larus smithsonianus*)
7. Herring Gull (*Larus argentatus*)

Species linked by arrows can interbreed. Notably, however, Lesser Black-backed Gulls (1) and Herring Gulls (7) cannot.

By Global_European_Union.svg: S. Solberg J. this file: Own work: Frédéric Michel (Global_European_Union.svg) [CC-BY-3.0 (www.creativecommons.org/licenses/by/3.0)], via Wikimedia Commons.

Figure 2.19 North American brown bear cubs playing.

(Photograph by S. Hillebrand, from US Fish and Wildlife digital library.)

Table 2.2 Mitochondrial sequence data from brown bears in northwest North America. These data do not represent continuous sequences, but only the variable positions. A dot indicates that the base at this position is the same as in the reference sequence in the first line

	Sequence data	Location of collection
	CCCTCCCAACGTTAACATTACGTAATCGAACAGCGGGTTAGGGAAC	O
	..T.T.T.......................................	Q
Clade I	..T.TTT.......................................	PU
	....T...	Q
	....T.T.......................................	MOPRSV
	....TTT....................................G..	U
	....TTT.......................................	T
	......TGG......T.......G.........T.......AA...	HN
	......TGG.....GT.........................AA.G.	G
	......TGG......T............G..........G.AA.G.	LM
Clade II	......TGG......T............G............AA.G.	IJK
	......TGG......TG................T.......AA.G.	K
	......TGG.....GT...G.................C...AA.G.	H
	......TGG......T.................T...C...AA.G.	H
	......TGG......T.................T.......AA.G.	H
	......TGG.....GT.......G.............C...AA.G.	H
	TT..T.T..T..CG...................T......A.A..T	BE
	TT..T.T..T..CG........C..........T........A..T	ABCDE
	TT..T.T.....CG........C........ATA........A..T	B
Clade III	TT..T.T.....CG........C..........T...........T	BC
	TT..T.T..T..CG........C..........T...........T	A
	TT..T.T..T.CCG........C........ATA........A..T	B
	TT..T.T..T..CG........C..........T..A.....A..T	A
	TT..T.T..T..CG...............G...T..A.....A..T	C
	.T.C......A......C..TA.GGCT...T....A..C...A..T	F
Clade IV	.T.C....G.A......C.GTA.GGCT...T....A..C...A..T	F
	.T.C......A......C.GTA.GGCT...T....A..C...AG.T	F

(b) You are lost somewhere in northwest North America. You collect a tissue sample from a brown bear in the vicinity and determine the following mitochondrial DNA haplotype (relative to the reference sequence in Table 2.2):

```
.T.C......A......C.GTA. GGCT...T....A..C...A..T
```

Where are you?

(c) Additional sequences were collected from four polar bears (*Ursus maritimus*) from zoos:

```
Reference sequence   CCCTCCCAACGTTAACATTACGTAATCGAACAGCGGGTTAGGGAAC
Polar bear 1         .T........A......CCGTA.GG.....T....A..C...A..T
Polar bear 2         .T........A......CCGTA.GG..A..T....A..C...A..T
Polar bear 3         .T........A......CCGTA.GG..A...GA..A..C...A..T
Polar bear 4         .T........A......CCGTA.GG..A....A..A......A..T
```

Table 2.3 Sample collection locations of bears with sequences in Table 2.2. For instance, bears with the first haplotype of clade III were found at location B (latitude 48.0°N, longitude 113.0°W) and location E (latitude 51.0°N, longitude 118.1°W)

Location	Latitude, °N	Longitude, °W
A	44.4	110.3
B	48.0	113.0
C	48.5	116.5
D	51.0	114.2
E	51.0	118.1
F	57.2	134.6
G	58.7	133.5
H	60.8	139.5
I	67.8	115.3
J	69.2	124.0
K	69.4	129.0
L	69.0	138.0
M	69.2	143.8
N	60.1	142.5
O	63.0	145.5
P	60.0	154.8
Q	63.1	151.0
R	68.5	158.0
S	65.5	165.0
T	55.2	162.7
U	58.3	155.0
V	60.9	161.2

(The reference sequence, from a brown bear, is the same as the reference sequence in Table 2.2.) To which class of brown bears are these polar bears most closely related? Where are the brown bears most closely related to polar bears found?

(d) For the brown bear sequences, compute the average number of sequence differences between classes. Assuming a divergence rate of ~0.125 sequence changes per 10 000 years, estimate the times since divergence of the classes. Note that the total length of the region sequenced was 294 bases.

Brown bears immigrated to North America from Asia over a temporary land bridge over the Bering Strait. The first fossil evidence for brown bears in the New World is from 50 000–70 000 years ago. Is it likely that the classes diverged in North America or that they had already diverged in Asia?

(e) Assume that (1) the original North American brown bear population contained all of the currently observed haplotypes; and (2) there is now a continuous brown bear habitat covering all areas listed in Table 2.3. What then accounts for the current geographical distribution of the different haplotype classes? There are two questions to address:

- what accounted for their initial separation?
- what is continuing to keep them separated?

Consider the following possible scenarios:

- Climate changes in the past, associated with glaciation, fragmented the population and left pockets that reflect 'founder effects'.
- Climate changes in the past, associated with glaciation, fragmented the population, and mutations within each group of bears accumulated to form separate haplotype groups.
- Female bears tend to be more philopatric than males, i.e. male bears have larger home ranges and disperse greater distances than females. Daughters often set up home ranges within their mother's home range. (Recall that mitochondrial DNA sequences tell us *only* about patterns of maternal inheritance.)

To what extent do these scenarios account for the observations? What additional experiments would you suggest to illuminate the situation further?

(f) Ethical, legal, and social issues. The division of North American brown bears into subspecies affects conservation issues. The brown bear populations of the lower 48 states of the USA are endangered. If these populations represent an **evolutionary significant unit** (ESU), they could be protected as a **distinct population segment** under the US Endangered Species Act. To qualify as an ESU, a population must (1) be substantially reproductively isolated from other populations of that species; and (2) contain an important component in the evolutionary legacy of a species (69 Fed. Reg. at 31355). Outline a petition to the US Secretary of the Interior or the US Fish and Wildlife Service arguing that the US brown bear population of the lower 48 states should be declared an ESU and listed as a distinct population segment. What, if any, additional data would you recommend collecting that might strengthen the case? (In fact, such a petition has been successful, and the brown bears in the lower 48 states are currently protected under the Endangered Species Act.)

Weblems

Weblem 2.1 Using the Online Mendelian Inheritance in Man™ site (http://www.ncbi.nlm.nih.gov/omim), what are the clinical consequences of the mutations associated with (a) haemoglobin Adana? (b) haemoglobin Malhacen?

Weblem 2.2 What animal models are available for the human mental diseases schizophrenia, depression, and bipolar disorder?

Weblem 2.3 Mammalian females achieve dosage compensation by silencing one of the X chromosomes in each cell. In birds, however, males are homogametic (ZZ) and females are heterogametic (ZW), with the W chromosome being gene deficient like the mammalian Y chromosome. Do male birds achieve dosage compensation by silencing one of their Z chromosomes?

Weblem 2.4 Determine a feature of a human MHC haplotype that is associated with a relative slow progression to AIDS after HIV infection.

Weblem 2.5 Which MHC haplotypes give indications for effectiveness of use of the following drugs: (a) carbamazapine (epilepsy, bipolar disorder) (b) ximelabatran (an anticoagulant)?

Weblem 2.6 Suggest menus, as close as possible to normal diets, for breakfast, lunch, and dinner, appropriate for a university student with PKU. Include recommended amounts.

Weblem 2.7 The three most frequent mutations in *BRCA1* and *BRCA2* genes in Ashkenazi Jews are 185delAG and 5382insC in *BRCA1* and 6174delT in *BRCA2*. In what exons of these genes do these mutations appear?

Weblem 2.8 What are the most common mutations in *BRCA1* among Swedish women?

CHAPTER 3

Mapping, Sequencing, Annotation, and Databases

LEARNING GOALS

- To know some of the important landmarks in the historical background, including the classical work of Darwin and Mendel, through Morgan and Sturtevant, to the more recent research leading to the discovery of the double-helical structure of DNA and the development of the human genome project.
- To distinguish different types of map – genetic linkage maps, chromosome banding patterns, restriction maps, and DNA sequences – and the relationships among them.
- To understand the relationships among linkage, linkage disequilibrium, haplotypes and the collection of single-nucleotide polymorphism (SNP) data.
- To understand how the basic principles of DNA sequencing developed from the initial breakthroughs to the current automated high-throughput systems.
- To understand the primer extension reaction catalysed by DNA polymerase and its termination by dideoxynucleoside triphosphates.
- To understand the basis and importance of the polymerase chain reaction (PCR) as a method for amplification of selected DNA sequences within a mixture.
- To grasp the significance and relationships of reads, overlaps, contigs, and assemblies as parts of an overall strategy and organization of a sequencing project.
- To be familiar with sequencing gels using autoradiography and to appreciate the advantages of using fluorescent chain-terminating dideoxynucleoside triphosphates.
- To understand new developments in high-throughput sequencing, and the goals set for the next few years of development.
- To be able to contrast hierarchical strategies of whole-genome sequencing based on maps and BAC clones, with the whole-genome shotgun approach.
- To understand techniques of genetic testing based on sequence determination.

Classical genetics as background

Similarities between parents and offspring, within human families, and in animals and plants, have always been obvious. An understanding of *how* heredity works is more recent.

The story begins in a 10-year period starting in the late 1850s. Threads then cast on would require a further century to ramify and intertwine.

- On 1 July 1858, Charles Darwin and Alfred Russel Wallace presented to the Linnaean Society of London their ideas on the development of species through natural selection of inheritable traits.* Darwin's book, *The Origin of Species*, appeared on 22 November 1859.
- In 1860, Louis Pasteur published the observation that the mould *Penicillium glaucum* preferentially metabolized the L-form of tartaric acid, one of two mirror-image molecules that have identical structures except that one is left-handed and the other right-handed. This work, based on Pasteur's earlier separation of racemic tartaric acid by manual selection of crystals of different shape, brought the idea of three-dimensional molecular structure into biology: *to understand how a biological process works, we must know the detailed spatial structure of the relevant molecules*. Thus began a long courtship between biology and molecular structure that has flowered into an intimate and fecund marriage.
- On 8 February and 8 March 1865, the monk Gregor Mendel read his paper, *Experiments on Plant Hybridization*, to the Natural History Society of Brünn in Moravia. (US President Abraham Lincoln was assassinated on 15 April.) His paper was published the following year in the Society's Proceedings. Mendel had entered the monastery, instead of becoming a teacher, because he had failed the botany exam. Twice! The price of insisting on original ideas.

A lack of understanding of the mechanism of heredity was a barrier to the development of Darwin's insights. Mendel's work supplied the crucial missing ideas: the discreteness and persistence of the elements of hereditary transmission, later called **genes**. Nevertheless, although copies of the *Proceedings of the Natural History Society of Brünn* were distributed around the scientific community, Mendel's work went largely unnoticed until it was rediscovered early in the 20th century.

There are several ironic aspects to this situation.

1. Among the recipients in Britain of the *Brünn Society Proceedings* were the Royal Society of London and the Linnaean Society. Darwin was a member of both and had access to their libraries. Even more, Darwin personally owned a book by W.O. Focke, *Plant Hybridisation*, published in 1880, which included a section describing Mendel's work and its implications. When J.G. Romanes, preparing an article on hybrids for the *Encyclopaedia Britannica*, appealed to Darwin for help in making his review complete, Darwin sent his copy of Focke's book. But, in one of the nearest near-misses in scientific history, neither Darwin nor Romanes read the section on Mendel's work. (How do we know this? The relevant pages were never cut open! The book is now in the Cambridge University Library, with the pages *still* intact.)
2. Darwin came close to an independent statement of Mendel's conclusion, that traits that differ between parents persist in the offspring, rather than blend. In a letter to Thomas Huxley in 1857, Darwin wrote:

   'I have lately been inclined to speculate, very crudely and indistinctly, that propagation by true fertilisation will turn out to be a sort of mixture, and not true fusion, of two distinct individuals, or rather of innumerable individuals, as each parent has its parents and ancestors. I can understand no other view of the way in which crossed forms go back to so large an extent to ancestral forms. But all this, of course, is infinitely crude.'

* Several earlier writers had suggested the idea of natural selection, including William Charles Wells and Patrick Mayhew. It appears, for example, in the appendix to Mayhew's 1831 book, *On Naval Timber and Arboriculture*, explicitly alluding to the possibility of creating novel species. (Mayhew's interest was in optimizing the growth of trees for building warships for the Royal Navy.) Mayhew complained when *The Origin of Species* first appeared, and Darwin gave credit to Wells and Mayhew in subsequent editions.

Indeed, Darwin himself hybridized pea plants and observed segregation of traits! In 1866 he wrote to Wallace:

'I crossed the Painted Lady and Purple sweetpeas, which are very differently coloured varieties, and got, even out of the same pod, both varieties perfect but none intermediate.'

In the same letter, Darwin pointed out, as an obvious example of discrete rather than blending inheritance, that male and female parents give rise to male and female offspring.

The legacy of the 1860s – Darwin, Pasteur, Mendel – was completed by the discovery of DNA by Friedrich Miescher in 1869. The structure and function of DNA were equally unknown. However, microscopic observations of the role of chromatin in fertilization led quite early on to suggestions that Miescher's substance was 'responsible . . . for the transmission of hereditary characteristics'.* The cell biologists got there first, and got it right.

The idea of DNA as the hereditary material then vanished for many years. It encountered considerable resistance when it was subsequently proposed again.

What is a gene?

In 1931, Frederick Griffith studied virulent and non-virulent strains of *Streptococcus pneumoniae*, showing that the virulent strain, even if killed, contained a substance that could transform a non-virulent strain into a virulent one and that the induced virulence is heritable. In 1944, Oswald Avery, Colin MacLeod, and Maclyn McCarty tested different chemical components of the cell for transforming activity. They identified the DNA from the virulent strain as the molecule that induced the transformation. (We now interpret bacterial transformation as 'horizontal gene transfer'.) As controls, they showed that transformation was inhibited by enzymes that destroy DNA but not by enzymes that destroy proteins. However, their contemporaries were not receptive to their conclusions. General acceptance of the idea that DNA is the hereditary material awaited the 1952 experiments of Alfred Hershey and Martha Chase, who showed that when bacteriophage T2 replicates itself in *Escherichia coli*, it is the viral DNA and not the viral protein that enters the host cell and carries the inherited characteristics of the virus.

- DNA was discovered in 1869. Avery, McLeod, and McCarty showed in 1944 that DNA was the active substance in bacterial transformation. Hershey and Chase showed that during bacteriophage infection, DNA but not protein entered the host cell.

Maps and tour guides

Maps tell us where things are. More specifically, they tell us where things are in relation to other things. In genomics, maps have been essential in revealing the organization of the hereditary material.

Different types of map describe different types of observation:

1. Linkage maps of genes
2. Banding patterns of chromosomes
3. Restriction maps – DNA cleavage fragment patterns
4. DNA sequences.

Genes, as discovered by Mendel, were entirely abstract entities. Chromosomes are physical objects, with banding patterns as their visible landmarks. Only with DNA sequences are we dealing directly with stored hereditary information in its physical form. Restriction maps are in effect partial DNA sequences – they give the positions of particular oligonucleotides within DNA molecules.

It was the very great achievement of the last century of biology to forge connections between these maps.

A crucial idea that emerged from mapping is that the organization of hereditary information is **linear**.

* Hertwig, W.A.O. (1885). Das Problem der Befruchtung und der Isotropie des Eies, eine Theorie der Vererbung. Jenaische Zeitsch. f. Medizin u. Naturwiss. **18**, 276–318. For an article about Miescher's scientific career, see Dahm, R. (2005). Friedrich Miescher and the discovery of DNA. Dev. Biol. **278**, 274–288.

The first steps – and giant strides they were indeed – proved that, within any chromosome, linkage maps are one-dimensional arrays. In fact, all of these types of map are one-dimensional, and indeed they are co-linear. Any school child now knows that genes are strung out along chromosomes and that each gene corresponds to a DNA sequence. But the proofs of these statements earned a large number of Nobel Prizes.

The complete sequences of genomes are the culmination of the entire mapping enterprise. But genome sequences describe the hereditary information of organisms in *only* a one-dimensional and static form. What they don't tell us is (a) how this information is implemented in space and time; (b) how gene expression is choreographed by orderly developmental programmes; and (c) the influence of surroundings and experience on the structure and activities of the organism.

Genetic maps

Gene maps were classically determined from patterns of inheritance of phenotypic traits (see Box 3.1).

Mendel discovered that elements of heredity are discrete and persist through a lineage. He observed recessive characteristics that disappear in one generation but re-emerge in a later one. Expressed in modern terminology, he observed traits that depend on a single locus, with a dominant and a recessive allele – denote them D and d. Mendel further observed that the distribution of phenotypes followed simple statistical rules. The offspring of Dd and Dd parents showed the dominant phenotype three times as often as the recessive. He inferred that the genotypes of the offspring, DD, Dd, dD (all three producing the dominant phenotype) and dd (producing the recessive phenotype) occur at the frequencies expected from segregation in the gametes, and independent transmission to the offspring, of the elements of heredity.

Studying the simultaneous inheritance of several traits, Mendel found further statistical regularities consistent with the existence of stable, independent, and persistent elements of heredity that distribute themselves randomly. The cards are shuffled and a new hand is redealt to each offspring.

Mendel had no concept of the physical nature of the elements of heredity that he was studying and was content to describe their behaviour in abstract terms. It is interesting to compare Mendel's contemporary, the physicist James Clerk Maxwell. Despite the success of Maxwell's kinetic theory – based on an abstract model of particles in random motion – the idea that matter is actually composed of tiny particles gained widespread acceptance only with Perrin's studies of Brownian motion, over half a century later. Indeed, it is questionable whether a general belief in the physical reality of atoms came before or after a general belief in the physical reality of genes!

BOX 3.1 Vocabulary inherited from classical genetics

Here are traditional meanings of these terms. We must reconsider how they should be defined in the light of recent understanding.

Term	Meaning
Gene	A bearer of hereditary information.*
Trait	An observable property or feature of an individual organism.
Phenotype	The collection of observable traits of any individual, other than genomic sequence.
Genotype	The sequence of any individual's genome.
Allele	One of the set of possible genes that govern a particular trait.
Homozygote	An individual that has two identical alleles at some locus.
Heterozygote	An individual that has two different alleles at some locus.
Segregation	The separation of corresponding alleles during the reproductive process.
Independent assortment	The uncorrelated choices of genes for different characters that each parent transmits to children.
Linkage	Absence or reduction of independent assortment of parental genes, which are usually transmitted together because they lie on the same chromosome.

* With apologies to A.S. Eddington.

Linkage

Mendel did not report that in some cases genes for different traits do *not* show independent assortment but are **linked**, i.e. their alleles are co-inherited.

Linked traits are governed by genes on the same chromosome. However, in many cases linkage is incomplete. During gamete formation, alleles on different chromosomes of a homologous pair can recombine. This occurs as a result of **crossing over**, the exchange of material between homologous chromosomes during copying in meiosis (see Figure 3.1).

Thomas Hunt Morgan, at Columbia University in New York City, USA, observed varying degrees of linkage in different pairs of genes. He suggested that the extent of recombination could be a measure of the distance between the genes on the chromosome.

Morgan's student Alfred Sturtevant, then an undergraduate, made a crucial observation: the data were consistent with a *linear* distribution of genes. What he found was that genetic distance, as measured by crossing-over frequency, was *additive*. Consider three genes *A*, *B*, and *C*. Suppose that the distance from *A* to *B* is 5 and the distance from *B* to *C* is 3. Then, if the distance from *A* to *C* is 8 = 3 + 5, the observations are consistent with a linear and additive structure with gene order *A*–*B*–*C*. (Alternatively, the distances would also be additive if the distance from *A* to *C* were 2, implying gene order *A*–*C*–*B*.) Note that additivity of distances does not hold for points at the vertices of a triangle rather than on a line.

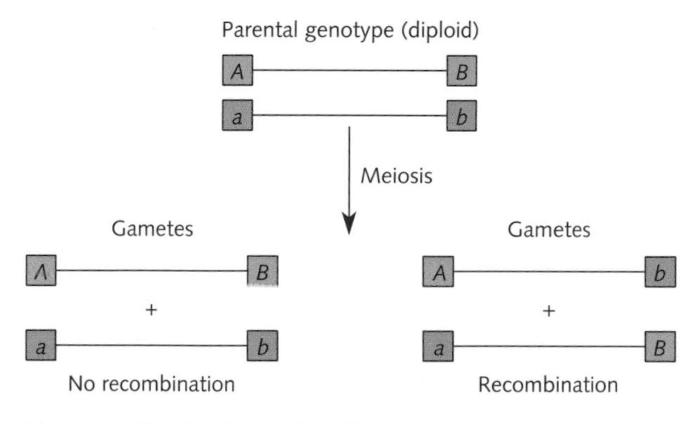

Figure 3.1 Consider two loci on the same chromosome. One locus has alleles *A* and *a*, the other has alleles *B* and *b*. One individual has alleles *A* and *B* on one chromosome (pink) and alleles *a* and *b* (blue) on the other (top). Gametes from this individual may form without recombination to give a haploid gamete containing alleles *A* and *B* and another haploid gamete containing alleles *a* and *b* (lower left). Each gamete contains a chromosome identical (at least as far as these loci are concerned) to one of the parental chromosomes. Alternatively, crossover between the two loci may produce recombinant gametes: one haploid gamete containing alleles *A* and *b* and another haploid gamete containing alleles *a* and *B* (lower right). Neither gamete contains a chromosome identical to a parental one. The fraction of gametes showing recombination depends on several factors, notably the distance between the loci. (The same fractions of recombinants and non-recombinants would be produced even if the parent were homozygous, although they would be likely to be indistinguishable.)

In the absence of selection, the fraction of viable recombinant and non-recombinant gametes produced depends only on the genotype of the individual that produced them. *It has nothing to do with the allele distribution in the population.* If a remote descendant had the parental genotype shown here, its gametes would show the same fractions of recombinants and non-recombinants.

Sturtevant's analysis made it possible to determine the order of genes along each chromosome and to plot them along a line at positions consistent with the distances between them. The unit of length in a gene map is the Morgan, defined by the relation that 1 cM corresponds to a 1% recombination frequency. We now know that 1 cM is ~10^6 bp in humans, but it varies with the location in the genome, the distance between genes, and the gender of the parent: for males 1 cM is ~1.05 Mb; for females, 1 cM is ~0.88 Mb. Crossing over is reduced in pericentromeric regions. Other regions are 'hot spots' for crossing over. It is estimated that ~80% of genetic recombination takes place in no more than ~25% of our genome.

Linkage guides the search for genes. To identify the gene responsible for a disease, look for a marker of known location that tends to be co-inherited with the disease phenotype. The target gene is then likely to be on the same chromosome, at a position near to the marker.

Linkage disequilibrium

Figure 3.1 showed the gametes arising from one parent, heterozygous for two traits. The recombination rate depends only on the structure of the chromosomes of this individual – whether this genotype is rare or common in a population.

Now suppose that we have a large interbreeding population and that every individual in the population has the *same* parental genotype shown in Figure 3.1. How will the genotype distribution in the population develop? (Assume that no combination of alleles for these two traits has any selective advantage or produces any preferential mating pattern.) Recombination at meiosis, followed by zygote formation, can in principle produce individuals of three genotypes: *AB/ab*, *Ab/aB* = *aB/Ab* and *ab/ab* (where, for instance, *AB/ab* signifies an individual with alleles *AB* on one chromosome and *ab* on the other; this is the parental genotype in Figure 3.1). Starting from the original completely *AB/ab* population, *eventually* recombination will randomize the allelic correlation, producing a population in which the ratio of genotypes is: *AB/ab*:*Ab/aB* = *aB/Ab* : *ab/ab* is 1:2:1. (This assumes that the overall gene frequency of the population is $A = a$ and $B = b$; see Problem 3.1.)

How fast the randomization occurs depends on the recombination rate, which depends on the genetic distance between the loci. *In the short term*, if recombination is infrequent, the parental genotype *AB/ab* will continue to predominate. The deviation of the genotype distribution in the population from the ultimate 1:2:1 ratio is called **linkage disequilibrium**.

- Two markers are in **linkage disequilibrium** if the observed distribution of different combinations of alleles differs from that expected on the basis of independent hereditary transmission of the individual alleles.

In the absence of linkage disequilibrium, the frequencies of allelic combinations will be proportional to the products of the frequencies of the individual alleles. Linkage disequilibrium measures the deviation from this equilibrium distribution.

If the overall fractions of the alleles at two loci are p_A, $p_a = 1 - p_A$, p_B, $p_b = 1 - p_B$ then:

equilibrium value of $p_{AB} = p_A \times p_B$
equilibrium value of $p_{Ab} = p_A \times p_b$
equilibrium value of $p_{aB} = p_a \times p_B$
equilibrium value of $p_{ab} = p_a \times p_b$.

Note that:

- there is no necessary relationship between p_A and p_B;
- at equilibrium:

$$p_{AB} \times p_{ab} = p_{Ab} \times p_{aB} = p_A \times p_B \times p_a \times p_b;$$

- a measure of linkage disequilibrium is:

$$D = p_{AB} \times p_{ab} - p_{Ab} \times p_{aB}$$

where $D = 0$ implies that the system is at equilibrium, $D > 0$ implies that chromosomes with *AB* and *ab* are more common than expected, $D < 0$ implies that chromosomes with *Ab* and *aB* are more common than expected.

Suppose there is no selection or gene import into the population, i.e. the overall frequencies of individual alleles in the population remains constant. *Then linkage equilibrium will decay as a result of recombination.* With each successive generation, the value of D will become closer to 0.

Linkage and linkage disequilibrium are closely related but distinct concepts

Linkage is about the distribution of loci among chromosomes. Linkage disequilibrium is about the distribution of allelic patterns in populations. Close linkage of two loci on a chromosome is a common source of long-term persistence of linkage disequilibrium. Two genes at opposite ends of the same chromosome, although formally linked, may not show significant linkage disequilibrium, because crossing over is frequent. Conversely, it is possible – although rare – to observe linkage disequilibrium between two genes on *different* chromosomes. (This can happen in two ways: (1) a community of immigrants imports a particular set of single-nucleotide polymorphisms (SNP) into a larger population and they preferentially inter-marry for many generations, or (2) theoretically by interactions between gene products that permit only certain combinations of alleles to be viable.)

Classical linkage maps typically involved markers no less than 1 cM apart (~1 Mb in humans) (this is the situation shown in Figure 3.2). Linkage disequilibrium is detectable between markers ~0.01–0.02 cM apart (~10–20 kb). Therefore, linkage disequilibrium is a much finer tool for localizing a target gene.

Chromosome banding pattern maps

Banding patterns are visible features on chromosomes (see Box 3.2). The most commonly used pattern is G-banding, produced by Giemsa stain. The bands reflect base composition and chromosome loop structure. The darker regions tend to contain highly condensed heterochromatin of relatively low GC/AT ratio and sparse in gene content.

The **karyotype** of an individual comprises the structures of the individual chromosomes. The karyotype

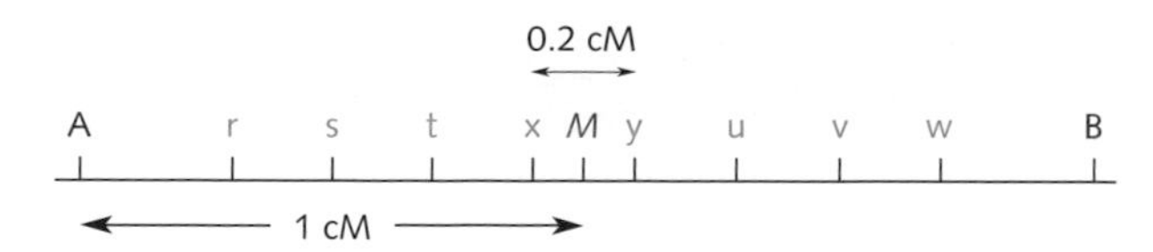

Figure 3.2 Suppose a mutation to a disease gene, *M*, occurred in a human population 50 generations ago. Consider a portion of the genome that includes the site of the disease mutation, *M*, plus genes for two known phenotypic traits, A and B, and two closely spaced markers, *x* and *y*. The genes for traits A and B are 1 cM away from the mutated locus. The markers *x* and *y* are 0.1 cM from *M*.

It is highly probable that markers *x* and *y* will be co-inherited with *M* in any pedigrees for which records exist, as the probability of recombination between *x* and *M* or between *y* and *M* in any generation is small: 0.1% = 0.001. The probability of recombination in 50 generations is approximately $0.001 \times 50 = 0.05$. On the other hand, the probability that markers *A* and *M* or *B* and *M* have been separated by recombination is very high. The probability of recombination between *A* and *M* in any generation is 1%. The probability of recombination in 50 generations is approximately 0.4 (see Problem 3.2).

In the history of transmission of this disease gene over 50 generations, markers *x* and *y* are likely to be co-inherited with the disease, but genes for traits A and B will not be reliably coupled with *M*. Therefore, genes *A* and *B*, separated by 1 cM (~1 Mb in the human), will not be a reliable guide to localizing the target gene *M* by looking in family pedigrees for genes co-inherited with *M*. However, a distribution of markers such as *x* and *y*, separated by 0.2 cM (~200 kb) or less, *is* likely to provide a reliable guide to localizing the target gene *M* through correlation of disease occurrence and genetic markers in family pedigrees.

For humans, we do not have access to 50 generations of records and DNA samples (which would amount to about 1000 years). However, the effects of recombination during the 50 generations since the mutation filter out all but the most closely linked genes from the co-inheritance pedigree.

Later we shall see that haplotype groupings simplify the identification of gene–marker correspondences.

is largely constant for all individuals within a species, but varies between species. This is the result of chromosome rearrangement during evolution. The inability of cells with incongruent karyotypes to pair properly is one barrier to fertility that contributes to species divergence.

Although most individuals of a species have the same karyotype, occasionally aberrant chromosomes appear. Some of them are lethal and others are correlated with disease. (For example, Prader–Willi or Angelman syndromes; see Box 3.2.) Studies of chromosome banding patterns support several types of investigations.

BOX 3.2

Nomenclature of chromosome bands

In many organisms, chromosomes are numbered in order of size, 1 being the largest. The two arms of human chromosomes, separated by the centromere, are called the p (petite = short) arm and q (queue) arm. Regions within the chromosome are numbered p1, p2 . . . and q1, q2 . . . outward from the centromere. Additional digits indicate band subdivisions. For example, certain bands on the q arm of human chromosome 15 are labelled 15q11.1, 15q11.2, and 15q12. Originally, bands 15q11 and 15q12 were defined; subsequently, 15q11 was divided into 15q11.1 and 15q11.2.

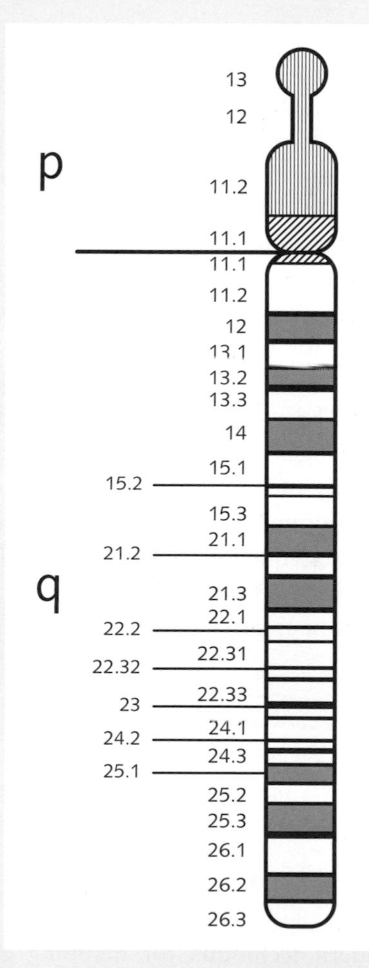

Deletions in the region 15q11–13 are associated with Prader–Willi and Angelman syndromes. These syndromes have the interesting feature that alternative clinical consequences depend on whether the affected chromosome is paternal (leading to Prader–Willi syndrome) or

maternal (leading to Angelman syndrome). This observation of **genomic imprinting** shows that the genetic information in a fertilized egg is not simply the bare DNA sequences contributed by the parents. Chromosomes of paternal and maternal origin have different states of methylation, signals for differential expression of their genes. The process of modifying the DNA, which takes place during differentiation in development, is already present in the zygote.

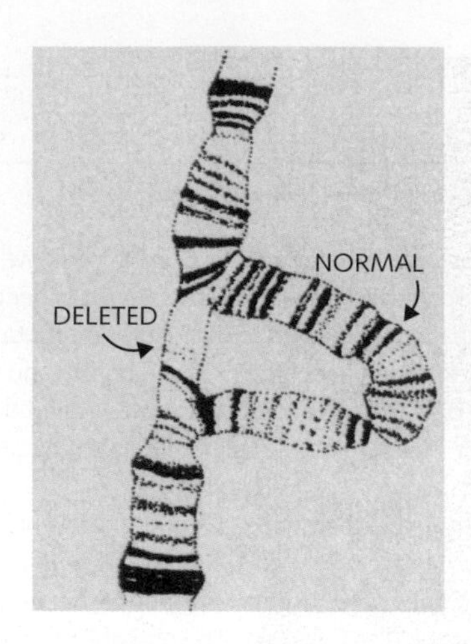

Figure 3.3 Loss of part of one of a pair of homologous chromosomes leads to a structure during chromosome replication called a 'deletion loop'. This drawing, from original work carried out shortly after the discovery in the 1930s of the large chromosomes in the salivary gland of *Drosophila*, shows two chromosomes that are paired except in a region in which one chromosome has suffered a deletion. Genes that map to this region can, in affected individuals, show 'pseudodominance' – the appearance of a recessive phenotype as a result of deletion of a masking dominant allele.

From: Painter, T.S. (1934). Salivary chromosomes and the attack on the gene. J. Hered. **XXV**, 465–476.

Correlation between genetic linkage maps and chromosome structure

Chromosome aberrations include deletions, translocations (of material from one chromosome to another), and inversions. The genetic consequences of a short deletion in only one of a pair of homologous chromosomes – only one allele instead of two for traits that map to the deleted region – allowed for direct mapping of genes to positions on chromosomes. In this way, the abstract genetic linkage maps could be superimposed onto the chromosome. This was first done in the 1930s after the discovery of the very large chromosomes in the *Drosophila melanogaster* salivary glands. The correlation of chromosome aberrations with changes in genetic linkage patterns proved the co-linearity of the two maps. Together with the mapping of sex-linked traits to the X chromosome, the genetic consequences of chromosomal deletions confirmed what had previously been a hypothesis – that chromosomes carry hereditary information (see Figure 3.3).

- Complete sequencing of yeast chromosome III in 1992 gave the first opportunity for direct comparison of a genetic linkage map and positions in the DNA sequence.

The modern technique for mapping genes onto chromosomes is **fluorescent *in situ* hybridization** (FISH). A probe oligonucleotide sequence labelled with fluorescent dye is hybridized to a chromosome. The location where the probe is bound shows up directly in a photograph of the chromosome. Typical resolution is ~10^5 bp, but specialized new techniques can achieve high resolution, down to 1 kb. Simultaneous FISH with two probes can detect linkage and even estimate genetic distances. FISH can also detect chromosomal abnormalities (see Figures 3.4 and 3.5, and Box 3.3).

Studies of evolutionary changes in karyotype

If we compare our chromosomes with those of a chimpanzee, we see that some large-scale rearrangements have taken place. Human chromosome 2 is split into two separate chromosomes in the chimpanzee (see Figure 3.6). However, most regions in the corresponding chromosomes of the two species show conservation of banding patterns. Such regions are called **syntenic blocks**. For human and chimpanzee, full genome sequences are available. They confirm that the relationships suggested by the comparisons of banding patterns reflect conservation at the level of DNA sequences.

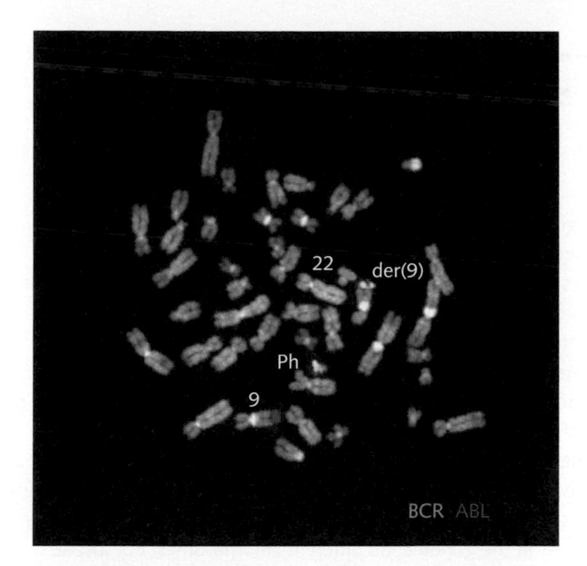

Figure 3.4 Detection by FISH probes of the Philadelphia (Ph) chromosome, associated with chronic myeloid leukaemia. It arises from a reciprocal exchange between chromosomes 9 and 22. The figure shows a representative image of a metaphase cell from a patient with chronic myeloid leukaemia analysed by FISH with fusion (red/green or yellow) signals marking the Ph and der(9) chromosomes (der stands for derivative: a **derivative chromosome** combines segments from two or more normal chromosomes). The normal 9 and 22 homologues are shown by a red and a green signal, respectively.

Figure courtesy of Dr E. Nacheva, Royal Free and University College Medical School, UK.

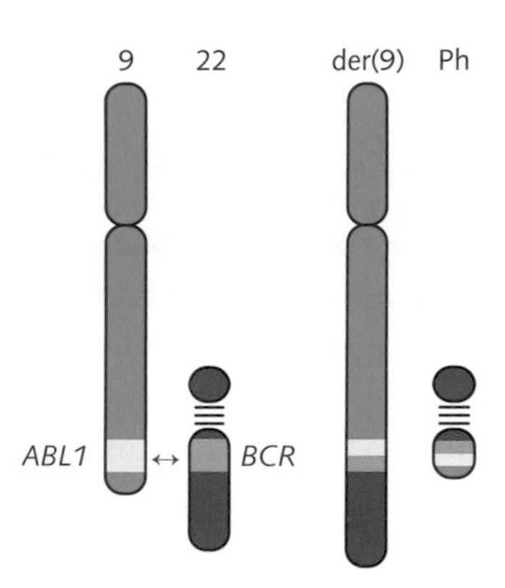

Figure 3.5 Transposition of material between chromosomes 9 and 22 forms the Philadelphia chromosome (Ph), associated with chronic myeloid leukaemia. Left: normal chromosome 9 (green), containing the *ABL1* gene (yellow) and normal chromosome 22 (magenta) containing the *BCR* gene (cyan). The arrow shows the positions of the breakpoints for the transposition. Right: the der(9) and Ph chromosomes, showing exchange of material at the bottom. Note that the site of translocation on each chromosome is within each gene.

Chromosome abnormalities are frequently associated with cancer

The Philadelphia chromosome is an abnormal, shortened chromosome 22, arising from a translocation – an exchange of chromosomal segments – between chromosome 22 and chromosome 9. The breakpoints are at bands 9q34 and 22q11, and this translocation is denoted t(9;22)(q34;q11) (see Figures 3.4 and 3.5).

The disastrous effects of the Philadelphia translocation arise because both breakpoint sites are within genes. This results in genes for two chimaeric, or fusion, proteins, combining *ABL1* (Abelson murine leukaemia viral oncogene homologue 1), a tyrosine kinase, from chromosome 9, and *BCR*, the breakpoint cluster region gene, from chromosome 22. In normal cells, *ABL1* and *BCR* are separate. As a result of the translocation, chromosome 22 encodes a *BCR–ABL1* fusion and chromosome 9 encodes an *ABL1–BCR* fusion.

The *BCR–ABL1* gene on chromosome 22 encodes a fusion protein with tyrosine kinase activity insensitive to the normal regulation of *ABL1*. (The *ABL1–BCR* gene on chromosome 9 is apparently silent.) There is good evidence for the association with cancer of the product of this abnormal gene: (1) the fusion protein is expressed in the proliferating cells; and (2) a drug – imatinib mesylate (Gleevec®) – that inhibits the kinase activity is therapeutically effective.

Because the translocation produces unique DNA sequences at the site of the join, it is possible to design FISH probes that span the common breakpoints of both the *ABL1* and *BCR* genes. A red signal detects sequences on the *ABL1* side of the breakpoint; a green signal detects sequences on the *BCR* side of the breakpoint. Normal cells would show two green and two red signals, on different chromosomes. Diseased cells show overlapping red and green signals (appearing yellow) on the aberrant chromosomes, and single red and green signals for the normal homologues.

Applications to diagnosis of disease

Many diseases are caused by – or at least correlated with – chromosomal abnormalities, including many forms of cancer (see Figures 3.4 and 3.5). It seems likely that partial or complete DNA sequencing of

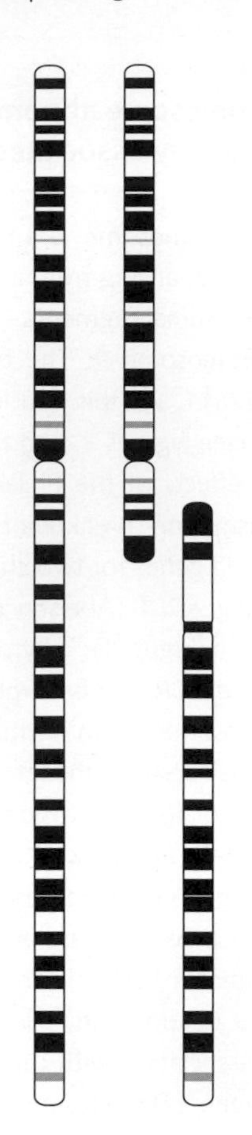

Figure 3.6 Left: human chromosome 2. Right: matching chromosomes from a chimpanzee.

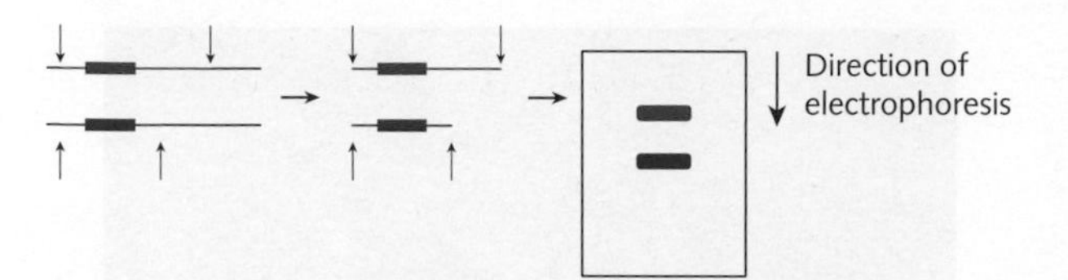

Figure 3.7 Restriction enzymes cleave DNA at the sites of specific sequences to produce fragments separable, according to size, by gel electrophoresis. Left: Regions of two homologous chromosomes showing two alleles that differ in the length of a restriction fragment. The fragment contains a marker (red or blue rectangle), a region that has *the same sequence* on both chromosomes. Vertical arrows show the cut sites for one restriction enzyme. The common fragment is 'labelled' red on one chromosome and blue on the other. After cleavage (middle), the fragments are loaded on a gel (right) and separated according to size. Cut sites for the restriction enzyme may appear in many places in the genome, on different chromosomes, not just at the positions of the allele illustrated. In order to see only those fragments that originate at the locus of interest, a radioactive or fluorescent probe complementary to the common marker (the rectangle) is used via Southern blotting to make visible only the selected fragments.

individuals will supersede the cytogenetic detection of large-scale mutations.

High-resolution maps, based directly on DNA sequences

Formerly, we could see genomes only by the reflected light of phenotypes. Now, markers are no longer limited to genes with phenotypically observable effects, which are anyway too sparse for an adequately high-resolution map of the human genome. Now that we can interrogate DNA sequences directly, any features of DNA that vary among individuals can serve directly as markers.

The first genetic markers based directly on DNA sequences, rather than on phenotypic traits, were restriction fragment length polymorphisms (RFLPs). The genetic marker is the size of the restriction fragments that contain a particular sequence within them (see Figure 3.7).

A restriction endonuclease is an enzyme that cuts DNA at a specific sequence, typically 4, 6, or 8 bp long (see Box 3.4). Many specificity sites for restriction enzymes are palindromic, in the sense that the sequence is equal to its reverse complement. For instance:

```
EcoRI recognition site:  GAATTC
                         CTTAAG
```

Other useful types of marker include:

- *Variable number tandem repeats (VNTRs)*, also called *minisatellites*. VNTRs contain regions 10–100 bp long, repeated a variable number of times – same sequence, different number of repeats. In any individual, VNTRs based on the same repeat motif may appear only once in the genome or several times, with different lengths on different chromosomes. The distribution of the sizes of the repeats is the marker. Inheritance of VNTRs can be followed in a family and correlated with a disease phenotype like any other trait. VNTRs were the first genetic sequence data used for personal identification – genetic fingerprints – in paternity and in criminal cases (see Chapter 8).
- *Short tandem repeat polymorphisms (STRPs), also called microsatellites*. STRPs are regions of only

*Eco*RI illustrates the nomenclature of restriction enzymes

Eco abbreviates the name of the species of origin (*Escherichia coli*) as the first letter of the genus name and the first two of the species name. The letter R specifies the strain. (Not all restriction enzyme names contain a strain identifier.) The Roman numeral, in this case I, distinguishes different restriction enzymes from the same strain of the same organism. Another example is *Taq*I, from *Thermus aquaticus*.

approximately 2–5 bp but repeated many times, typically 10–30 consecutive copies. They have several advantages as markers over VNTRs, one of which is a more even distribution over the human genome.

There is no reason why these markers need lie within expressed genes and usually they do not. The CAG repeats in the gene for Huntington's disease and certain other disease genes are exceptions (see p. 22). Markers have applications in several different fields. Medical applications of markers include their use in tissue typing to identify compatible donors for transplants, detection of disease susceptibility, and prediction of individual drug-response variation (pharmacogenomics). Anthropologists use them to trace migrations and relationships among populations.

The DNA sequence itself can also be used as a marker. Physically, it is a sequence of nucleotides in the molecule, computationally a string of characters A, T, G, and C. Collection and application of sequence data directly often makes use of mitochondrial DNA or haplotypes (see below).

Restriction maps

Splitting a long molecule of DNA – for example, the DNA in an entire chromosome – into fragments of convenient size for cloning and sequencing requires additional maps to report the order of the fragments, so that the entire sequence can be reconstructed from the sequences of the fragments.

Cutting DNA with a restriction enzyme produces a set of fragments (as in Figure 3.7). Cutting the same DNA with other restriction enzymes, with different specificities, produces overlapping fragments. From the sizes of the fragments produced by individual enzymes and in combination, it is possible to construct a **restriction map**, stating the order and distance between the restriction enzyme cleavage sites (see Figure 3.8 and Box 3.5).

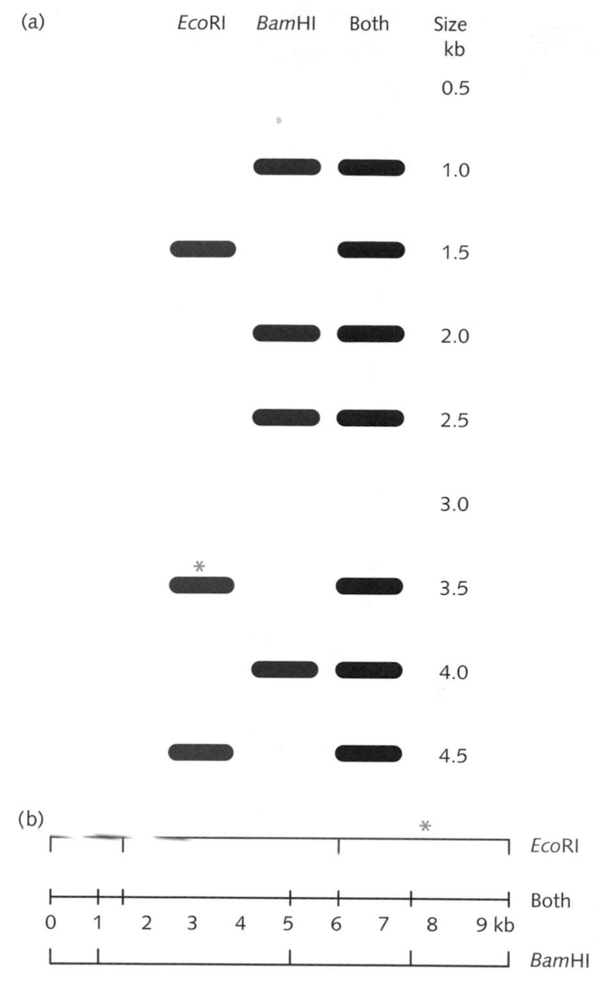

Figure 3.8 (a) Pattern of a gel showing sizes of restriction fragments produced by *Eco*RI (red), *Bam*HI (blue), or both (black). (Direction of migration: up the page.) (b) Positions of the cutting sites, deduced from the measurements in (a). Each lane of (a) corresponds to a strip of the same colour in (b) in the following way: each band on the gel in (a) corresponds to the length of a fragment between two successive cutting sites in (b). For instance, the green* indicates a band on the gel and the fragment that gave rise to it. In contrast to Figure 3.7, in this case we do *not* want to limit our observation of the fragments to those at a single locus.

In this case, it is relatively easy to determine the unique order of fragments that accounts for all of the data. In more complicated cases, it is possible to simplify the problem by placing radioactive tags on either end of the segment being mapped, or by carrying out partial digests to produce additional fragments (see Exercise 3.6).

BOX 3.5

Give my regards to restriction maps

As an analogue of a restriction map: walk the entire length of Broadway in New York (see http://www.marktaw.com/local/MarksWalkingTour.html). Mark on the map the location of every Starbucks coffee shop and calculate the number of blocks between successive sites; then mark the location of every CitiBank office and calculate the number of blocks between successive sites; then calculate the number of blocks between every occurrence of *either* Starbucks *or* Citibank. A mutation in one of these 'sites' will change the sizes of the fragments, allowing the mutation to be located in the map.

(Thanks to Professor B. Misra, New York University)

Restriction enzymes can produce fairly large pieces of DNA. Cutting the DNA into smaller pieces, which are cloned and ordered by sequence overlaps, produces a finer dissection of the DNA called a **contig map.**

In the past, the connections between chromosomes, genes, and DNA sequences have been essential for identifying the molecular deficits underlying inherited diseases, such as Huntington's disease or cystic fibrosis. Sequencing of the human genome has changed the situation radically.

Discovery of the structure of DNA

Life involves the controlled manipulation of matter, energy, and information. Biochemists were familiar with the structures and mechanisms of enzymes, living molecules catalyzing conversions of *matter* and *energy*. What kind of molecule could store and manipulate *information*?

Chemical analysis of DNA during the early part of the 20th century characterized the constituents – the bases and the sugars – and the nature of their linkage. DNA is a polynucleotide chain, containing a repetitive backbone of sugar–phosphate units, with small organic bases – adenine, thymine, guanine, and cytosine – attached to each sugar (see Figure 3.9).

Not only was the distribution of the bases along the chain unknown, its significance was entirely unsuspected (except for a prescient comment by physicist Erwin Schrödinger – based on ideas of Max Delbrück – in his influential 1944 book, *What is Life?*: 'We believe a gene – or perhaps the whole chromosome fibre – to be an aperiodic solid.' Do not be misled by the reference to a solid. Schrödinger himself included a footnote: 'That it [the chromosome] is highly flexible is no objection, so is a thin copper wire.') Indeed, many people accepted P.A. Levene's idea that DNA contained a regular repetition of a constant four-nucleotide unit. It was, therefore, considered that DNA simply lacked the versatility required to convey hereditary information, compared, for example, with proteins. We now know that eukaryotic DNA *does* contain repetitive sequences, but is not limited to them.

- This attitude contributed to the general lack of acceptance of the experiments of Avery, MacLeod, and McCarty as proof that DNA was the genetic material.

An understanding of how DNA worked in biological processes required a detailed three-dimensional structure. In the 1950s, the method of choice for determination of molecular structure was X-ray crystallography. The pioneer of X-ray structure determination, Sir Lawrence Bragg, Cavendish Professor of Physics at Cambridge, was an enthusiastic supporter of the efforts of Max Perutz and his group to extend methods of X-ray crystallography to biological molecules as large as proteins.

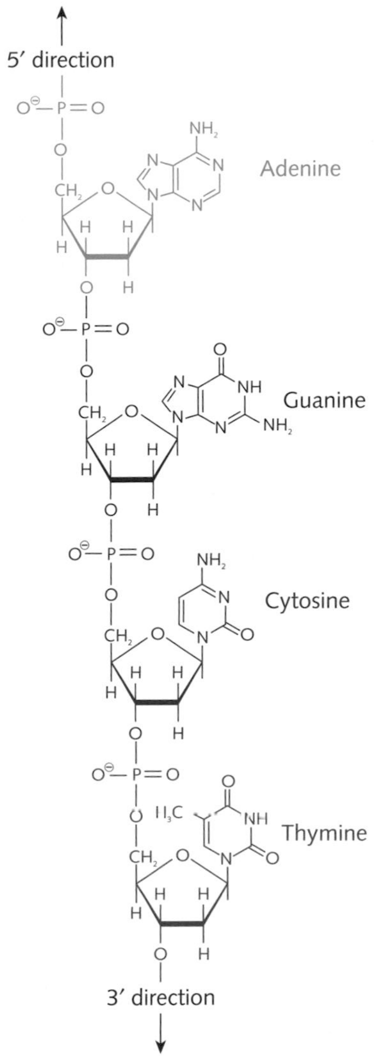

Figure 3.9 The chemical structure of one strand of DNA. The backbone consists of deoxyribose sugars linked by phosphodiester bonds. In its chemical bonding pattern, the backbone is a regular repetitive structure, independent of the bases. Each base, one attached to each sugar, can be one of four choices. In the sequence of bases, DNA is logically equivalent to a linear message written in a four-letter alphabet.

However, DNA molecules are flexible and do not form classic crystals. They can be drawn into fibres, which can be thought of as crystalline in two dimensions, and disordered around the fibre axis. This disorder severely impoverishes the available data. X-ray diffraction patterns of three-dimensional crystals permit an objective determination of the individual atomic positions in a structure. In contrast, fibre diffraction data present a trickier puzzle. Guided by all available clues, the biochemist must imagine a model that is consistent with the known covalent structure and with the X-ray diffraction pattern.

X-ray diffraction patterns of DNA fibres were measured at King's College, London, by a group that included Rosalind Franklin and Maurice Wilkins. In May 1951, Wilkins spoke at a conference in Naples. In the audience was a young American postdoctoral scientist named James Watson. Watson was converted. He took his newly kindled interest in X-ray structure determination to Cambridge, where he met a graduate student named Francis Crick, and together they sought the structure of DNA.

For DNA, the clues included: (a) titration curves that showed the bases to be involved in internal hydrogen-bonding interactions; and (b) E. Chargaff's observations that, although the amounts of the four bases differ among samples of DNA from different organisms, the amounts of adenine and thymine are always equal and the amounts of guanine and cytosine are always equal. Chargaff published his results in 1949 and described them to Watson and Crick during a visit to Cambridge in July 1952.

It was recognized as a race. Lined up against Watson and Crick were Franklin and Wilkins at King's College and Linus Pauling in California. Pauling was a frightening contender, with many major discoveries already to his credit. He had recently beaten the Cambridge group to the structure of the α-helix in proteins.

Watson and Crick did no experimental work themselves, but confined their efforts to rationalizing all of the data and clues that they knew, including fibre diffraction data. Watson attended talks in which the King's group's fibre diffraction photographs were shown. Franklin continued to improve the data, producing in 1952 the photograph in Figure 3.10. This photograph appeared in a Medical Research Council report, which was shown to Watson and Crick without Franklin's knowledge. The fibre diffraction data implied that the chains formed helices, with a repeat distance of 34 Å, containing 10 residues per turn (i.e. 3.4 Å between successive bases). Although manuscripts left by Franklin suggest that she was close to solving the structure herself, Watson and Crick beat her to the goal.

What animated Watson and Crick's model was the idea of the complementarity of specific pairs of bases,

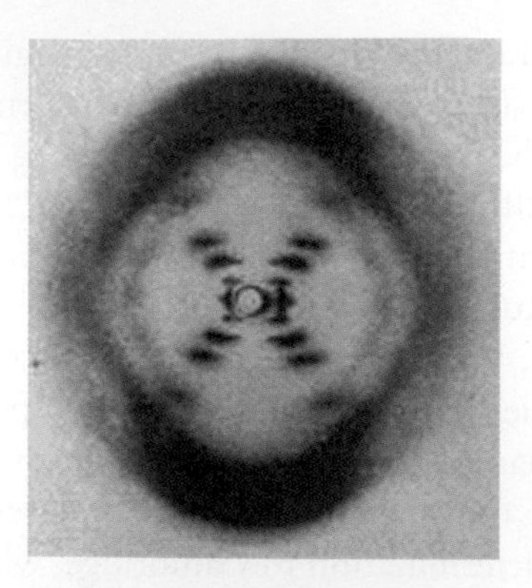

Figure 3.10 X-ray diffraction pattern of DNA fibre, by R. Franklin. The X-shaped pattern at the centre of the picture is diagnostic of a helical structure. From the distribution of intensity, it is possible to deduce the number of residues per turn and the symmetry of the structure.

From: Franklin, R.E., & Gosling, R.G. (1953). The structure of sodium thymonucleate fibres. II. The cylindrically symmetrical Patterson function, Acta Cryst. **6**, 678–685.

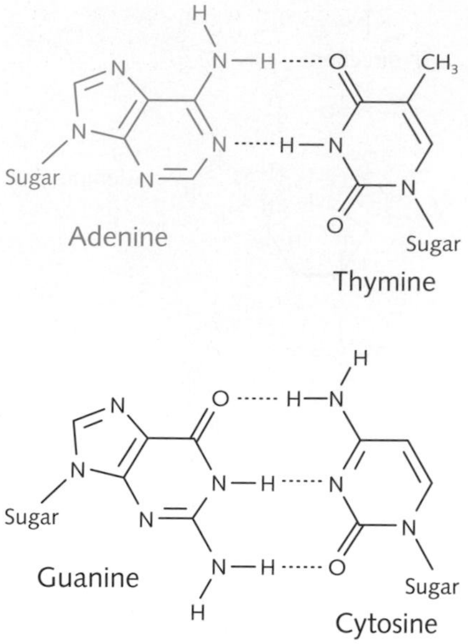

Figure 3.11 The complementary base pairings: adenine–thymine and guanine–cytosine.

together with the recognition that AT and GC pairs had compatible stereochemistry. There had been several premonitory hints at this crucial idea. A Cambridge mathematician, John Griffith (nephew of the Frederick Griffith who discovered bacterial transformation), told Crick of a preferential attraction between A and T, and between G and C, that he had deduced from theoretical calculations. This, together with Chargaff's rules, immediately suggested the idea of complementarity.* (A colleague has commented: in biology, theory suggests and experiment proves; in physics experiment suggests, theory proves. A.S. Eddington, an astrophysicist, once warned against putting 'too much confidence in observational results until they have been confirmed by theory'.)

What in retrospect screams hydrogen-bonded base pairing (see Figure 3.11) only whispered, before the model was imagined. The leap from energetic and compositional complementarity to the correct structural complementarity was *not* (as it may perhaps now appear) an easy step. Even the fact that the model rationalized the hints from Griffith and Chargaff and the titration curves, was more confirming evidence than proof. The famous sentence from Watson and Crick's paper, 'It has not escaped our notice that the specific pairing we have postulated immediately suggests a possible copying mechanism for the genetic material' distracts by its coyness from the underlying logic: it was the clear structural basis for the biological activity that made the model immediately and utterly convincing (see Figure 3.12).

Once the base pairing was recognized, the pieces of the puzzle quickly fell into place. On 28 February 1953, Crick strode into the Eagle, a pub in downtown Cambridge near the laboratory, and announced to anyone listening that he and Watson had discovered the secret of life. The 25 April issue of *Nature* contained three papers on the structure of DNA, one by Watson and Crick, and two from the King's group. A second paper by Watson and Crick, on the genetic implications of the structure, appeared in the 30 May issue of *Nature*.

It was an eventful spring: Stalin died on 5 March; Edmund Hillary and Tenzing Norgay reached the summit of Mount Everest on 29 May; Queen Elizabeth II was crowned on 2 June.

Rosalind Franklin died in 1958. Only the living can win a Nobel Prize. The 1962 Prize for Physiology and Medicine was awarded to Watson, Crick, and Wilkins.

* Lagnado, J. (2005). From pabulun to prions (via DNA): tale of two Griffiths. The Biochemist **27**, 33–35.

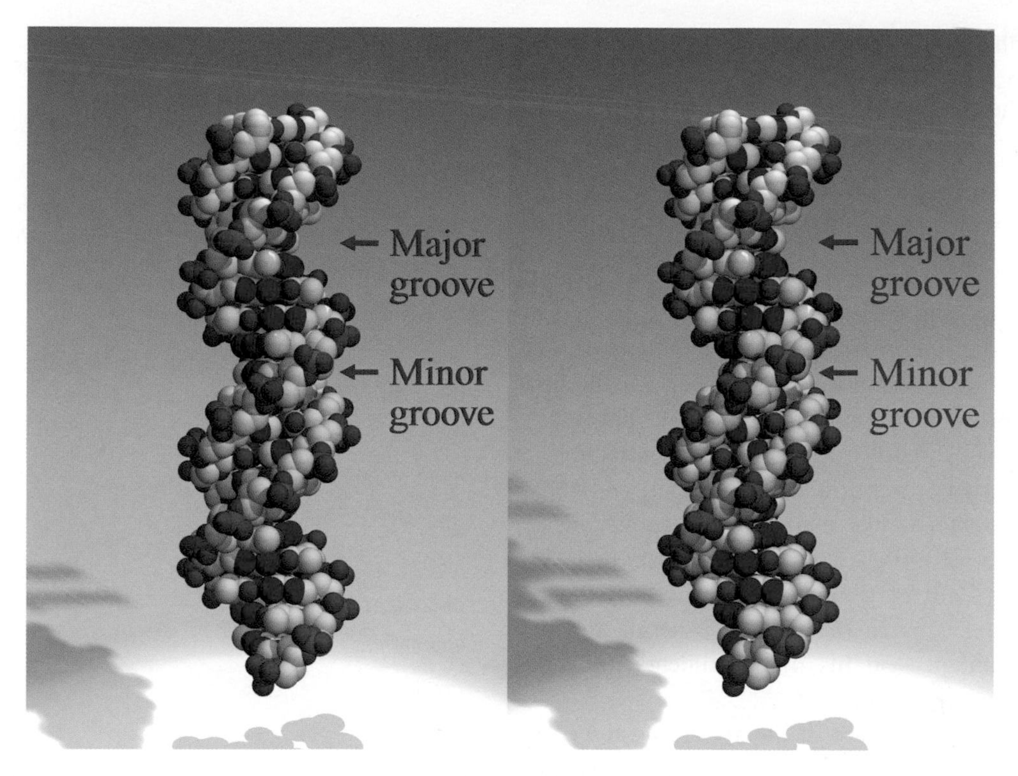

Figure 3.12 The structure of DNA. The two helical strands wind around the outside of the structure. The bases are inside, stacked like the treads of a staircase with their planes perpendicular to the axis of the double helix. The bases are visible through the major and minor grooves and are thereby accessible for interaction with proteins. This picture shows a stereo pair, most easily viewed with a standard stereo viewer or lorgnette.

The story has been told on many occasions, in print and on film. Notable for its unusually sensational treatment of scientific history is Watson's 1968 autobiography, *The Double Helix*. A collection of the reviews it elicited still makes interesting reading.† There is now consensus that Franklin's contributions to the discovery of the structure of DNA were underestimated, not least by Watson, who in *The Double Helix* disparages her mercilessly, both professionally and personally. Of numerous recent attempts to redress the balance, Sir Aaron Klug's lecture is the most authoritative.‡

DNA sequencing

In 1953, after the Hershey–Chase experiment and the announcement of the double helix, molecular biologists knew that DNA contained the hereditary information. They saw how cellular machinery could achieve access to it, using base-pair complementarity. But if the sequence of the bases was like a text everyone wanted to read, not only was it a text in an unknown language, but there were not even any examples of the language, because the sequences were unknown. The importance of the problem of

† Stent, G. (ed.) (1980). *The Double Helix: Text, Commentary, Reviews, Original Papers*. W.W. Norton, New York & London.

‡ Klug, A. (2003). The discovery of the DNA double helix. In *Changing Science and Society*. T. Krude (ed.) Cambridge University Press, Cambridge, pp. 5–43.

determining DNA sequences was obvious, but the difficulties frightened most people away from confronting the challenge.

Frederick Sanger and the development of DNA sequencing

Cambridge biochemist Frederick Sanger devoted his career to sequence determination of biological macromolecules. First, he determined the amino acid sequence of the protein insulin, proving for the first time that proteins had definite sequences. This work won him the Nobel Prize in Chemistry in 1958. He next turned his attention to sequencing of RNA, and then to DNA. His methods for sequencing of DNA won a second chemistry Nobel Prize, in 1980, shared with Paul Berg and Walter Gilbert. Sanger's professional legacy to the field in general, and his personal influence on his many students and colleagues, is unsurpassed.

To appreciate the state of the art during the early 1970s, it is important to recognize that it was very difficult even to prepare pure samples of single-stranded DNA for attempts at sequencing. Restriction enzymes were a new discovery. They did not provide the mature and versatile technology that we take for granted today. Bacteriophage ϕX-174, providing 5386 bp of purifiable, single-stranded DNA, was a natural source of material for Sanger's early work, yielding the first whole genome sequence.

Even ϕX-174 is too long to sequence 'in one go'. It depended, as has every subsequent genome sequencing project, on solving the 'Humpty Dumpty problem': putting fragments together again. (Box 3.6 introduces the sequence assembly problem.) Sanger's achievement was a reliable and convenient method for sequencing relatively long fragments, making unambiguous assembly possible.

Sanger's method used DNA polymerase to synthesize a new strand of DNA, embodying, as part of the synthesis, reactions that reveal the sequence.

DNA polymerase is a replication enzyme that synthesizes the strand complementary to a piece of single-stranded DNA. It requires a primer: a short stretch of complementary strand to be extended by successive addition of nucleotides (see Figure 3.13). The polymerase requires a supply of nucleoside triphosphates. In successive steps, the enzyme adds to the growing primer strand the nucleotide complementary to the next unpaired base in the template. This reaction also forms the basis of the **polymerase chain reaction** (PCR) (see Figure 3.14).

BOX 3.6 The problem of sequence assembly from short fragments

To illustrate what is involved in assembling a long sequence from overlapping short fragments, consider the first two verses of Shakespeare's *Richard III*:

Now is the winter of our discontent
Made glorious summer by this sun of York

Breaking this up into overlapping 10-letter fragments (treat the line-break as a space), and presenting them in random order, gives:

ntent Made

this sun o

f Yor

discontent

ious summe

r by this

er of our

glorious

summer by

our disco

winter of

Made glor

Now is the

s the wint

It is easy to reassemble the pieces, and would be easy even if you didn't know the answer. Observe that the longer the fragments, the easier the challenge. (See Exercise 3.1.) Recognize also, that repetitive sequences are harder. In Act 2 of *Hamlet*, Polonius says:

'tis true, 'tis true 'tis pity,
And pity 'tis 'tis true

It would be more difficult to reassemble this from fragments, because the repetitions create ambiguities. Indeed, the very large amounts of repetitive sequence in eukaryotic genomes do create problems for assembly algorithms.

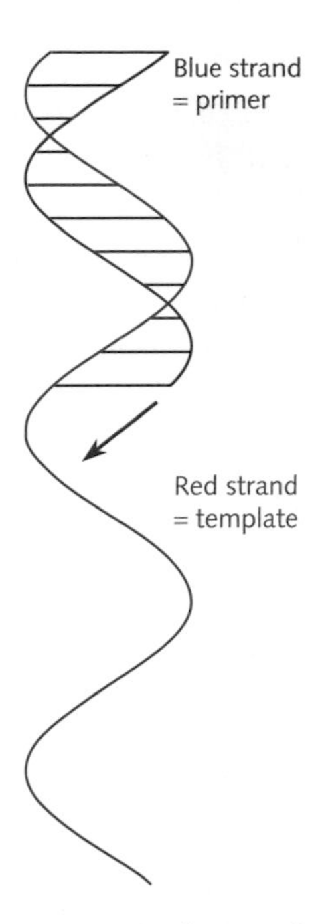

Figure 3.13 Primer extension. Duplication of a piece of single-stranded DNA, the template (red), occurs by extension of a short primer (blue) by addition of complementary nucleotides to the primer. The arrow shows the direction of synthesis. The basis of the Sanger method for DNA sequencing is the termination of the extension by poisoning the reaction with a modified base – a dideoxynucleoside triphosphate – lacking the reactive group necessary for continuing the extension.

DNA polymerase links the 5′-hydroxyl of the newly added nucleotide to the 3′ position of the end of the primer. The activating triphosphate provides the energy. Sanger's idea was to include in the mixture of nucleoside triphosphates a fraction of molecules containing no reactive 3′ position, a dideoxynucleoside triphosphate (see Figure 3.15). One reaction mixture contains normal deoxyATP, deoxyGTP, deoxyTTP, and a mixture of normal deoxyCTP plus dideoxyCTP (Figure 3.15). As primer extension proceeds, when the next unpaired base in the template is a guanine, the enzyme will randomly incorporate either a normal deoxycytidine, after which the extension will continue, or a dideoxycytidine, after which no further extension will occur. (Three other reactions are run in parallel containing all four normal triphosphates plus dideoxyATP, dideoxyGTP, or dideoxyTTP.)

This reaction produces a mixture of fragments of different lengths, each of which ends in a (modified) deoxycytidine:

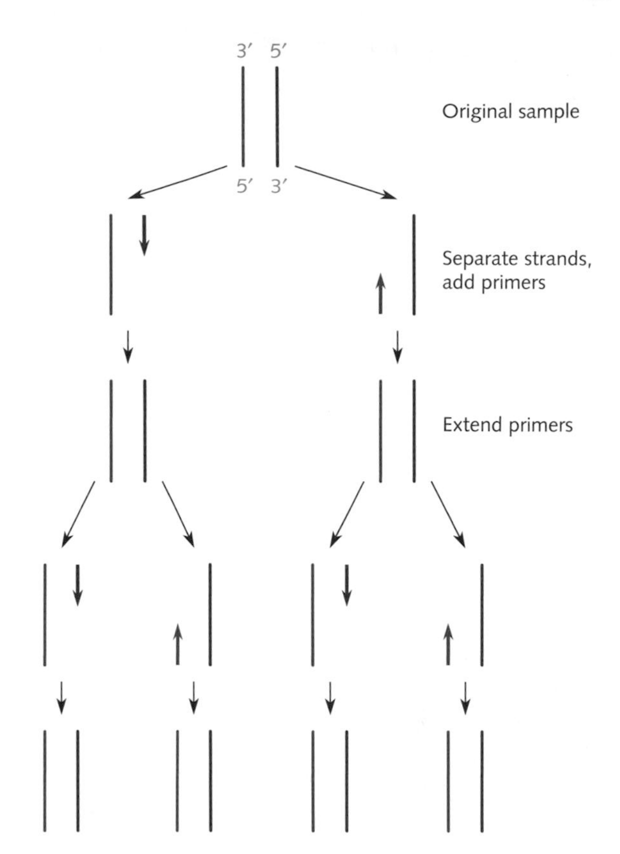

Figure 3.14 Schematic diagram of the polymerase chain reaction (PCR), a molecular copying technique that can *selectively* amplify very small quantities of DNA in a sample. In this figure, red strands are complementary to blue strands.

The target sequence is called the **amplicon**. One cycle doubles the amount. First, separate the strands of the original sample and add primers complementary to both ends of the target region. (The short red and blue arrows represent the primers.) Primer extension by DNA polymerase reconstitutes two copies of the original sample. Repetition of these operations doubles the amount of amplicon at each step. Only the first and second cycles are shown in this diagram. Exponential growth leads quickly to the production of large quantities of the desired sequence.

Specificity is achieved by tailoring the primers to the target sequence. Amplification is effective even if the target sequence appears in a small amount within a large background of other DNA.

A development called real-time PCR allows the quantitative measurement of levels of mRNA in a sample by first converting it to cDNA and then following the rate of the amplification reaction by measurement of incorporation of a fluorescent label. The basic idea is that the more abundant a sequence in a mixture, the faster the build-up of the product of its PCR amplification.

Extremely useful in understanding PCR are animations available on the web. Searching will turn up a number of sites, for example. http://users.ugent.be/~avierstr/principles/pcrani.html.

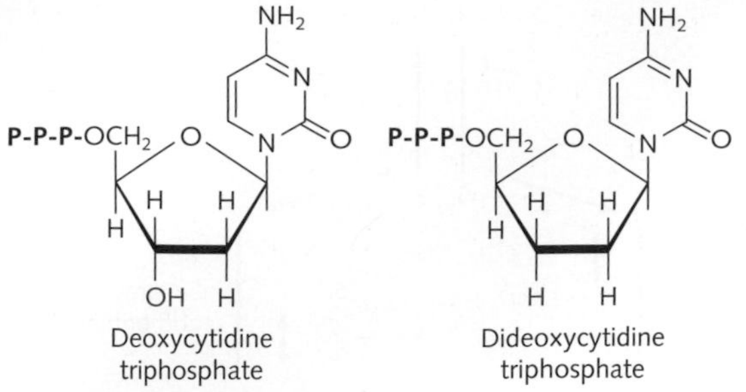

Deoxycytidine triphosphate

Dideoxycytidine triphosphate

Figure 3.15 Left: normal deoxycytidine triphosphate, containing both a reactive 5′-hydroxyl, activated as the triphosphate derivative permitting it to be added to the growing primer strand, and a 3′-hydroxyl (red), to which the *next* nucleotide will be attached. Right: dideoxycytidine triphosphate, in which the 3′ position is unreactive. Dideoxycytidine can be incorporated into the growing strand, but no subsequent extension is possible. **P-P-P** is a simplified representation of the triphosphate group.

```
3'-atacagagaatctagatacagagttgttcgag  Template
5'-tatgtctcttagat →                  Primer extension

5'-tatgtctcttagatctatgtctcaacaagctc
5'-tatgtctcttagatctatgtctcaacaagc    Fragments
5'-tatgtctcttagatctatgtctcaac         produced
5'-tatgtctcttagatctatgtctc
5'-tatgtctcttagatctatgtc
5'-tatgtctcttagatc
5'-tatgtctc
5'-tatgtc
```

The crucial point is that the fragments are *nested*. To determine the positions of the cytosine residues in the extended primer, it is sufficient to determine the lengths of the fragments. Polyacrylamide gel electrophoresis can separate oligonucleotides according to their lengths accurately enough for sequence determination. It is possible to separate polynucleotides differing in length by a single base, up to molecules about 1000 bases long.

- Almost all DNA sequencing methods, from the earliest to those in most common use today, depend on breaking a long DNA molecule into fragments, sequencing the fragments, and then assembling them, via overlaps, into the complete sequence. The methods of Sanger and of Maxam and Gilbert worked by creating a set of nested fragments, each terminating in a known nucleotide, that could be separated on a gel. Knowing the length of the fragment and the terminal nucleotide allowed the sequence to be 'read off' the gel.

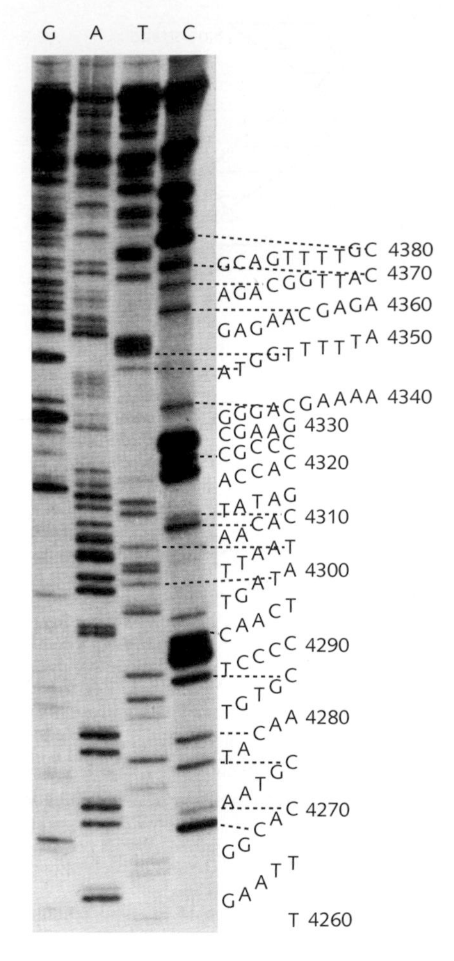

Figure 3.16 A gel from early in the history of DNA sequencing, showing part of the genome of bacteriophage ϕX-174.

From: Sanger, F., Nicklen, S., & Coulson, A.R. (1977). DNA sequencing with chain-terminating inhibitors. Proc. Natl. Acad. Sci. USA **74**, 5463–5467.

In Sanger's original procedure, the dideoxycytidine carried a radioactive label and the gel was developed by autoradiography. To determine the positions of the other three nucleotides, it was necessary to run four reactions, in parallel, each containing radioactive dideoxy analogues of one of the four nucleoside triphosphates. Four separate reactions are necessary because each nucleotide gives the *same* signal – the darkening of the film from the radioactive spot. Running the products of the four reactions in parallel lanes on the same gel allows the sequence to be read from the autoradiograph (see Figure 3.16).

- Labelling the four dideoxynucleoside triphosphates with different fluorescent dyes, with distinguishable colours, allows 'one-pot-one-lane' sequencing reactions (see pp. 18 and 97).

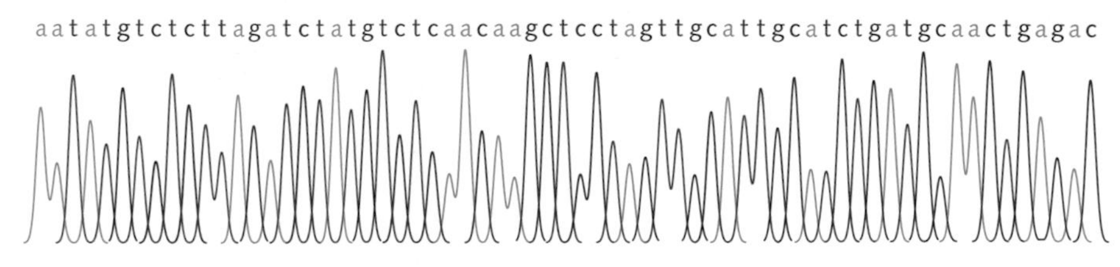

Figure 3.17 A tracing of a sequencing fluorescent chromatogram (simulated). Different-coloured peaks correspond to different bases.

The Maxam–Gilbert chemical cleavage method

The Maxam–Gilbert method for DNA sequencing, also developed in the mid-1970s, also worked by comparing nested fragments. A sample of single-stranded DNA, labelled at its 5′-end, is cleaved by base-specific reagents. As in Sanger's method, polyacrylamide gel electrophoresis separates the fragments by size. The separate cleavage reactions produce fragments that share the 5′-labelled end, but differ in the bases at which the specific 3′-cleavage occurs. The sequence can be read from an autoradiograph.

The Maxam–Gilbert method had early successes, for which Gilbert shared the Nobel Prize. It was Sanger's method that spawned subsequent developments, however. Because the Maxam–Gilbert method does not use primed DNA synthesis, its applicability is inherently limited to sequences adjacent to restriction sites or other fixed termini. Another disadvantage of the Maxam–Gilbert method was reagent toxicity, notably of hydrazine, a neurotoxin.

Automated DNA sequencing

Using fluorescent dyes as reporters, rather than radioactivity, was an important technical advance in sequencing. Radioisotopes present health hazards in both use and disposal, and are expensive. (One of Sanger's 'postdocs', who wore his hair long in the fashion of the 1970s, was obliged to have a haircut because of contamination from frequently pushing his hair out of his eyes.) By attaching different fluorescent dyes to the four dideoxynucleoside triphosphates, each fragment produced gives a *different* signal, depending on which dideoxynucleotide terminated the extension. All four reactions can be done 'in the same pot', and electrophoresis separates them in a single lane. A laser focused at a fixed point identifies the fragments as they pass. The result can be displayed as a four-colour chromatogram in which successive peaks correspond to successive bases in the sequence (see Figure 3.17).

Quality is important too: The phred score q measures sequencing accuracy. $q = 20$ implies a probability of $\leq 1\%$ error per base (see Box 3.7). A common unit of sequencing costs is US$ per 1000 bases determined to $q = 20$ accuracy. *A sequence of 1000 $q = 20$ bases would be expected to contain no more than ten errors.*

In 1986, L. Hood, L. Smith, and co-workers described an instrument, based on detection of base-specific fluorescent tags, that led to the automation of DNA sequencing. Instruments produced by Applied Biosystems implemented the work of Hood, Smith, and co-workers, with improvements by J.M. Prober and co-workers at DuPont. Capillary systems for fragment separation to replace flat-sheet gels were another essential advance. A population explosion in

Phred scores: a measure of quality of sequence determination

The phred score of a sequence determination is a measure of sequence quality. It specifies the probability that the base reported is correct.

If p = the probability that a base is in error, then the corresponding phred score $q = -10 \log_{10}(p)$.

Here is a short table:

Quality score q	Probability of error	Error rate
10	0.1	1 base in 10 wrong
20	0.01	1 base in 100 wrong
30	0.001	1 base in 1000 wrong
40	0.0001	1 base in 10 000 wrong

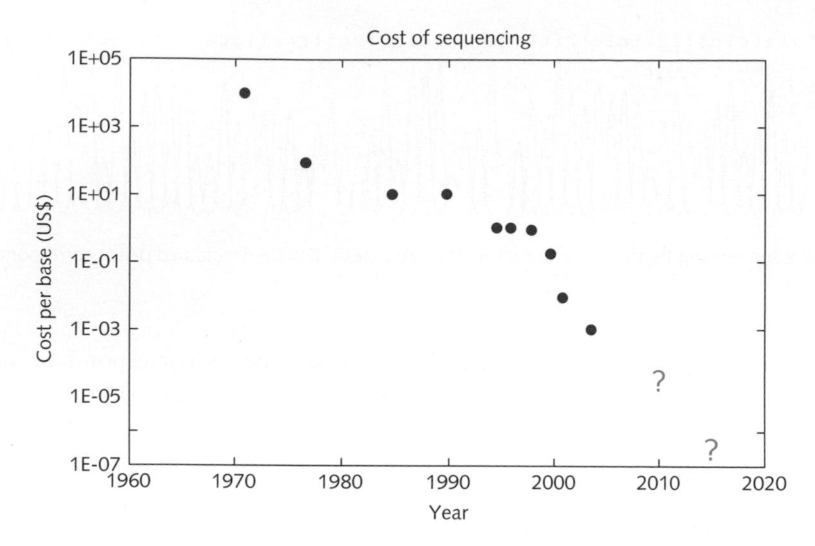

Figure 3.18 Fall in the cost of sequencing over time. Note logarithmic scale on the cost axis. The sea change in 1998 from the introduction of next-generation sequencing platforms is striking. The US National Institutes of Health set goals for US$100 000 human genome in 2010, and a US$1000 human genome in 2016. Has the 2010 target been met? (See Weblem 3.4.)

(Data from: http://www.genome.gov/sequencingcosts/)

automatic DNA sequencing machines filled the high-throughput installations that produced the human and other complete genomes. Instruments available in 1998 were able to produce ~1 Mb of sequence per day.

Current goals are to improve the technology by additional orders of magnitude. The US National Institute of Health (NIH) has set goals for a US$100 000 human genome by 2009 and a US$1000 genome by 2014 (see Figure 3.18).

Organizing a large-scale sequencing project

Two general approaches to genome sequencing projects are:

1. The hierarchical method, in which the whole genome is first fragmented and cloned into bacterial artificial chromosomes (BACs) (see Box 3.8), and the order of the fragments is established *before* sequencing them.
2. The whole-genome shotgun method, which works directly with large numbers of smaller fragments, with a concomitantly more challenging assembly problem.

Bring on the clones: hierarchical – or 'BAC-to-BAC' – genome sequencing

One approach to organizing the sequencing of a large DNA molecule involves dividing the sample into pieces *of known relative position*.

- First, cut the DNA into fragments of about 150 kb. Clone them into BACs. For example, *Arabidopsis thaliana* has a haploid genome size of about 10^8 bp. A 3948 clone BAC library for *A. thaliana* contains ~100 kb inserts per clone, giving approximately fourfold coverage.
- Identify a series of clones in the library that contains overlapping fragments. Although referred to as 'fingerprinting', this process depends on *shared* (rather than unique) features of overlapping clones, including:
  1. overlap of restriction fragment size patterns
  2. amplification of single-copy DNA between interspersed repeat elements and checking for similar size patterns of fragments
  3. mapping sequence-tagged sites (STSs) and looking for fragments sharing STSs.

BACs: bacterial artificial chromosomes

A plasmid is a small piece of double-stranded DNA in a bacterial cell, in addition to the main genome. A bacterial artificial chromosome, or BAC, is a plasmid containing foreign DNA – for instance, a fragment of the human genome. A typical BAC, in an *Escherichia coli* cell, can carry about 250 000 bp.

The idea of a BAC is to provide storage for relatively small DNA molecules. The storage should be stable and, by growing up the cells, replicable.

The number of copies per cell varies from plasmid to plasmid. A high copy number would be advantageous if we wanted to use a bacterial culture as a factory to produce large amounts of protein. However, multiple copies of plasmids in cells can undergo recombination and are therefore not suitable for *stable* storage of DNA fragments. The BACs in common use in genome sequencing projects are based on choosing an *E. coli* plasmid that maintains a low copy number and shows low recombination rates.

- Using the overlaps, order the clones according to their position along the original large target DNA molecule.
- Subfragment each clone, sequence the fragments and assemble them. The idea is that the clones are small enough that the ~1500 bp sequenced subfragments can be assembled to give the complete sequence of the ~150 kb BAC clones. Then the clones can be assembled using their known order in the original sequence.

Whole-genome shotgun sequencing

The idea of the whole-genome shotgun approach is to sequence random pieces of the DNA and then put them together in the right order. If this can be done, one can skip the laborious stage of creating a map as the basis for assembling partial sequences.

In the whole-genome shotgun sequencing of the *Drosophila melanogaster* genome, the DNA was sheared into random pieces of approximately 2, 10, and 150 kb. For each piece, the sequences of approximately 500 bp from each end were determined;

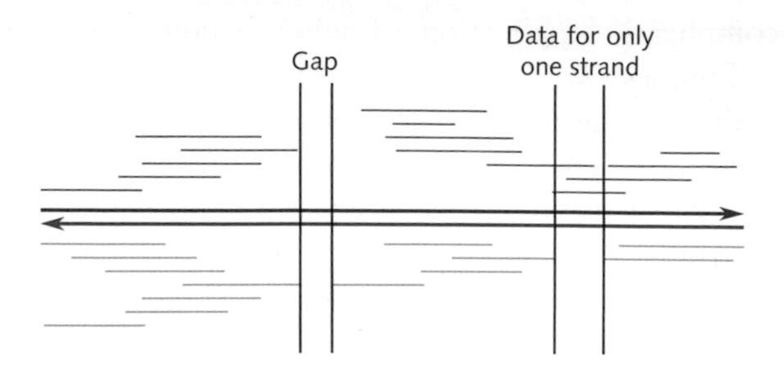

Figure 3.19 Diagram of an assembly of a region of DNA from shotgun fragments. Strands of DNA are shown in red and blue. 'Reads' of fragments from the red strand are shown in cyan; those from the blue strand are shown in magenta. In practice, sequences are determined only for about 500 bp at both ends of longer fragments. This assembly, based on overlaps, leaves one gap. Although for one region all data arise from fragments of only one strand, this does *not* leave a gap.

these are called **reads.** A computer program then assembled the results into a maximal set of contiguous sequence, or a **contig** (see Figure 3.19). The fully assembled genome sequence, built by coalescence of contigs, is known as the 'Golden Path'.

The **coverage** is the average number of times each base appears in the fragments. If G = genome length, N = number of reads and L = length of a read, the coverage = NL/G.

E. Lander and M. Waterman derived formulas for the number of gaps expected as a function of coverage and genome size. For instance, for a 1 Mbp genome, eight- to tenfold coverage should permit assembly of the reads into five contigs. Completion of the process, called **finishing**, involves synthesis and sequencing of specific fragments to close the gaps.

Whole-genome shotgun sequencing worked smoothly for prokaryotes, which contain relatively less internal repetitive sequence. Repeats create problems in assembly, and this led to scepticism about the feasibility of the shotgun approach for a complex eukaryotic genome. In fact, the *Drosophila* genome has fewer repeats than mammalian genomes and this contributed to its successful sequencing by shotgun methods. In the event, the publication of the *Drosophila* genome in 2000 contained 120 Mb of finished sequence, with about 1600 gaps. In the latest release of the *Drosophila* genome (April 2004), the number of gaps has been reduced to 23.

Highly skewed base composition, as in *Plasmodium falciparum* – which contains ~80 mol% AT – also complicates application of whole-genome shotgun techniques.

Even if the ultimate goal is a complete, finished, genome sequence, it may be possible to identify genes in a partly assembled genome with many gaps, provided that the genes are contained within contigs.

Celera took the success of whole-genome shotgun sequencing of the fruit fly as 'proof of principle', justifying its use in their human genome project. Despite the simultaneous announcement of academic and commercial human genomes, on 26 June 2000 (see Chapter 1), the academic group made its results publicly available (largely to preclude patenting) and it has been alleged that Celera made use of these data in their assembly.*

Comparison of BAC-to-BAC and whole-genome shotgun approaches

Suppose the sample of DNA for sequencing comes from a diploid organism. Fragments arising from homologous regions of two chromosomes of a pair may have sequence differences. Correct assembly must place them at the same location, noting the discrepancies, and must *not* split these reads into different contigs because of the imperfect matches.

BAC-to-BAC methods are more robust than whole-genome shotgun methods with respect to this problem. Each BAC is a clone and, therefore, has a unique sequence. Successive clones may come from different chromosomes and may, therefore, show mismatches, but this will not affect the order of assembly of the clones.

Note that an unambiguous success of whole-genome shotgun methods, the *Drosophila* genome, was based on a highly inbred laboratory strain. Sequencing the DNA of an individual from a natural, outbred population – or, even worse, DNA pooled from several individuals from such a population – would present a more severe challenge.

Box 3.9 contrasts the alternative approaches.

Common and different steps in 'BAC-to-BAC' and whole-genome shotgun methods

'BAC-to-BAC' method	Whole-genome shotgun method
1. Make random cuts to produce fragments of:	
~150 kb	~2000 kb and 10 000 kb
2. Make plasmid library in BACs.	
3. Fingerprint, overlap, and order BAC clones.	3. Skip this step.
4. Partially sequence 1500 bp subfragments of individual clones.	
5. Assemble overlaps by computer.	

High-throughput sequencing

Several rewards await the developer of a breakthrough improvement in DNA sequencing technology. Sequencing has shown that it can enhance human welfare. Some clinical applications have already entered medical practice; sequencing supports research that will provide additional ones. Other applications include more effective and safer production of food, and biotechnological approaches to generating safe and abundant consumable energy.

Intellectually, all of modern biology rides on the back of DNA sequencing. High quantities of data allow addressing more subtle questions. Think of higher coverage, or being able to treat a larger sample cohort, as a magnifying glass that brings details into sharper focus.

There are also significant financial rewards. The market for sequencing machines was over US$1 billion in 2010. The X Prize Foundation, which sponsored a US$10 million prize for space flight, has launched another US$10 million competition, for sequencing 100 human genomes in 10 days, for less than $10 000 per genome.

General approaches to improving the throughput/cost ratio are: (a) Miniaturization, and (b) Parallelization, or multiplexing. Common to many but not all of the new methods are preparation steps in which the target DNA is fragmented, common adaptors are attached to one or both ends, and – in most methods

* See Sulston, J. & Ferry, G. (2002). *The Common Thread: a Story of Science, Politics, Ethics and the Human Genome.* Bantam Press, London; and Ashburner, M. in Jobling, M.A., Hurles, M.E., & Tyler-Smith, C. (2003). *Human Evolutionary Genetics: Origins, Peoples and Disease*, Garland Science, New York, p. 27.

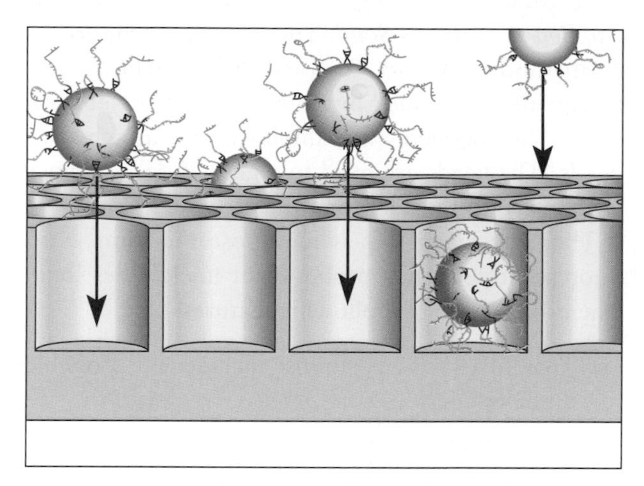

Figure 3.20 Each bead, attached to a clone of amplified fragments, occupies a separate well in the Roche 454 Life Sciences high-throughput sequencing platform.
(454 Sequencing © Roche Diagnostics)

– the results are amplified. The results are distributed spatially, either in an array of wells, or fixed to a ground.

Major players in the field, and the features of their methods, include:

Roche 454 Life Sciences. The 454 system achieves multiplexing by forming 'polonies' – single-molecule replicates. Digests of DNA are ligated to a common PCR primer, and amplified by replication in individual wells (Figure 3.20). The samples in the wells generate individual signals, detected in parallel.

As in the Sanger method, DNA polymerase extends the primer. The fragments are exposed, in four successive operations, to each of the four bases. The base added is identified 'on the fly' through detection of the pyrophosphate released (see Box 3.10). In this case, each addition generates the same signal. Each nucleotide is presented separately, so we know which base is added to each well because we know which nucleotide triggered the signal. Figure 3.21 shows an integrated view of the whole system.

Helicos. The playing field for helicos sequencing is a flow-cell surface to which billions of oligo-T molecules are fixed. The subject DNA is fragmented into 100–200 bp fragments, and an oligo-A primer is added to the 3′-end of each fragment. These primers bind to the oligo-Ts fixed to the surface.

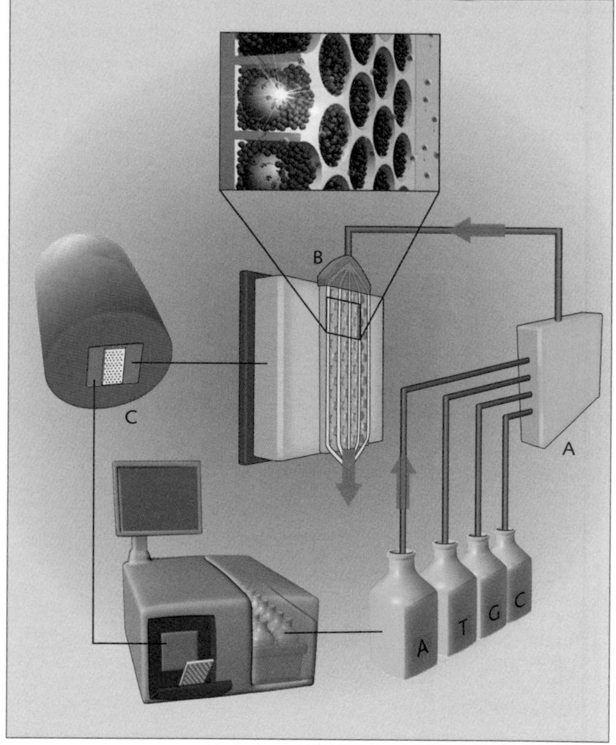

Figure 3.21 System overview of the Roche 454 Life Sciences high-throughput sequencing platform. The bead that is flashing at you has just added a nucleotide, signalled by the luciferase detector (see Figure 3.22).
(454 Sequencing © Roche Diagnostics)

BOX 3.10

Pyrosequencing

Pyrosequencing detects incorporation of a specific nucleotide into a growing strand by detecting the pyrophosphate (PP_i) released in a step of the reaction catalysed by polymerase:

Template: 3′-ATACAGAGAATCTAGAT . . . + TTP → ATACAGAGAATCTAGAT + PP_i
Primer: 5′-TATGTCTCT → TATGTCTCTT

In each cycle, the reaction is separately exposed to each of the four triphosphates.* Incorporation of the complementary nucleotide releases pyrophosphate. ATP sulphurylase converts the pyrophosphate to ATP:

adenosine 5′-phosphosulphate + PP_i → ATP

→

which is detected by a light-producing luciferase reaction (see Figure 3.22). Apyrase destroys unincorporated nucleoside triphosphates and ATP produced upon successful incorporation.

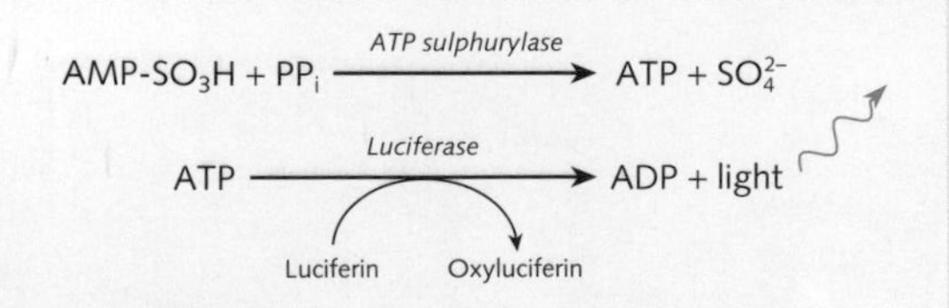

$$AMP\text{-}SO_3H + PP_i \xrightarrow{\text{ATP sulphurylase}} ATP + SO_4^{2-}$$

$$ATP \xrightarrow[\text{Luciferin} \rightarrow \text{Oxyluciferin}]{\text{Luciferase}} ADP + \text{light}$$

Figure 3.22 Detection of matching nucleotides by the luciferase reaction to signal the appearance of PP_i released when a nucleotide is incorporated.

Each cycle of successive exposure to the four nucleoside triphosphates produces one sequenced base.

* As in many plays, including The Merchant of Venice, Turandot, etc., in which an eligible woman is offered several suitors in turn.

The fragments are sequenced by synthesis. Flooding the system with a polymerase and one fluorescently labelled nucleotide results in incorporation of the nucleotide onto each fragment that has the complementary base adjacent to the growing primer. As in the Roche 454 Life Sciences system, an image of the system shows fluorescent spots at the positions at which the nucleotide was incorporated. After washing and removal of the fluorescent tag, the process repeats with the other three nucleotides, in succession. The four images reveal the distribution of the incorporated nucleotides. Then the process repeats, for the next position.

Note that in this system there is no amplification step.

Illumina Solexa. In the preparation step, fragments bound to a surface are amplified *in situ*. They form distinct clusters on the surface. At each step, the polymerase adds a base. The four bases have four different fluorescent tags. Therefore the distribution of colours, in an image of the field, identifies which base was added to each cluster. Remove the fluorescent tags, and repeat. The result is a kaleidoscopic movie of shifting colours, one frame per position.

Applied Biosystems Solid. This is perhaps the trickiest method to explain. To prepare the system, a set of fragments is extended by standard adaptors at both ends. The fragments are amplified and attached to beads. The beads are fixed to a glass slide. A primer is annealed to the adaptor sequence.

The sequencing step is to expose the samples to fluorescently labelled probes eight nucleotides long. The probes have the following features:

(a) The first two positions include all possible dinucleotides.

(b) The remaining six positions are 'wild cards' – molecules that will pair with all four bases.

(c) The 5′ end of the probe bears one of four fluorescent tags. Because there are four tags and 16 dinucleotides, the system is degenerate, each tag representing four of the sixteen dinucleotides.

Figure 3.23 shows how the system works. The first probe bound to one of the fragments starts with TC, implying that the first two bases from the original fragment (after the adaptor) are AG. However, because the fluorescent tag, yellow in this diagram, represents AG, CT, GA, *or* TC, we only know that the first dinucleotide must be one of these four. We don't yet know the identity of any one base.

Remove the last three nucleotides, and repeat the process. After the next step, we know that the dinucleotide offset by three positions from the first must be one of four possibilities corresponding to the red tag. In summary, what we know about the target sequence at this point is that it must have the following form:

```
position 1 2 3 4 5 6 7
         A G ? ? ? A T
      or C T      or C G
      or G A      or G C
      or T C      or T A
```

This process continues for a total seven cycles, giving us additional partial information about 35 bases in the sequence.

How are the ambiguities resolved? By generating overlapping information. The newly synthesized strand is removed. A new primer is added, which binds at one position offset from the first. Now the first dinucleotide probe that binds has a blue tag (Figure 3.23)

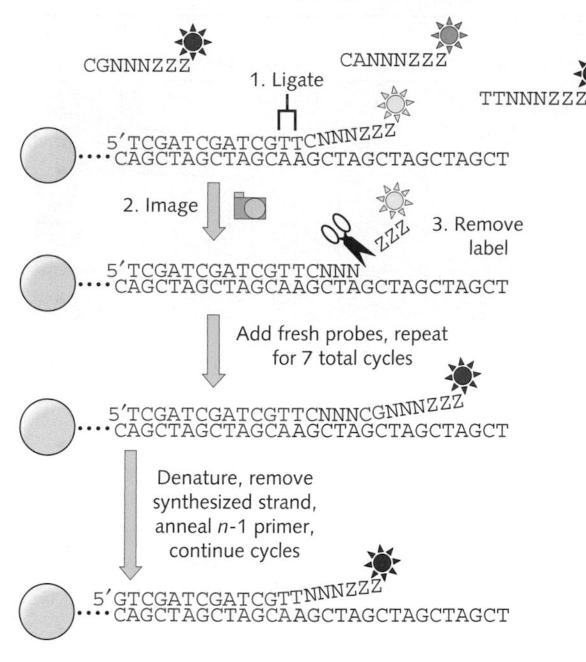

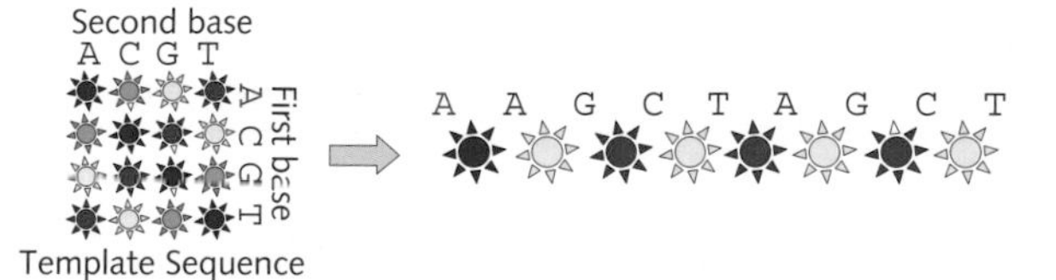

Figure 3.23 DNA sequencing with Applied Biosystems' Solid technology. Fragment (black) lengthened by adaptor CAGCTAGCTAGCA (pink), and fixed to beads. From a mixture of probes containing 5′ dinucleotides plus 6 'wild cards' (shown as grey N and Z) the probe starting with TC binds, and is ligated. The yellow fluorescence shows that the dinucleotide must be TC, GA, CT, or AG; at this point we don't know that it is TC Seven such cycles of extension provide partial information about separated dinucleotides in the fragment.

After removing the synthesized strand, the process is repeated with a displacement of one residue. Now the blue fluorescence shows that the new dinucleotide sequence must be TT, GG, CC, or AA. Knowing the initial A from the adaptor shows that the dinucleotide, complementary to the adaptor, must therefore be TT, and that the adjacent dinucleotide, known from the previous round to be TC, GA, CT, or AG, must be TC and hence that the second base in the fragment must be G. We can walk along the sequence, resolving successive ambiguities (see Figure 3.24).

(From: Anderson, M.W., & Schrijver, I. (2010). Next generation DNA sequencing and the future of genomic medicine. Genes **1**, 38–69.)

and therefore must be AA, CC, GG, or TT. But the adaptor base is an A; therefore the dinucleotide must be TT, and the target sequence must contain AA (of which the first A is from the adaptor).

Lining these possibilities up with the earlier information means that the sequence must have the form:

```
position 1 2 3 4 5 6 7
         A G ? ? ? A T
      or C T    or C G
      or G A    or G C
      or T C    or T A
         A A ? ? ? ? ?
```

Therefore the base incorporated at the first position of the original run must be A (which we knew from the adaptor). Therefore the second base must be a G. Now we know that the sequence begins AG. Knowing the base at position 2 resolves the ambiguity at position 3, etc., and we can step along the sequence (see Figure 3.24).

Life in the fast lanes*

The turkey genome offers an example of how high-throughput technology facilitated the recent sequencing of the first genome of a species. As is common, an international consortium carried out the project. The lead investigators were from Viginia Tech, the University of Minnesota, the University of Maryland, and the US Department of Agriculture. The 1.1 Gb turkey genome was completed in Autumn 2010.

The success of the project depended on:

1. Two high-throughput platforms for 'brute force' sequencing power.
   - A Roche 454 GS-FLX Titanium platform generated ~5X coverage
   - An Illumina Genome Analyzer II platform generated ~25X coverage
2. A genetic map and a BAC library, to assist in the assembly. In order to support the assembly, Sanger sequencing of elements of the BAC library added about another 6X coverage. (Sanger sequencing was used only because the BAC map construction

* Section title is a malaprop: lanes are older-generation technology.

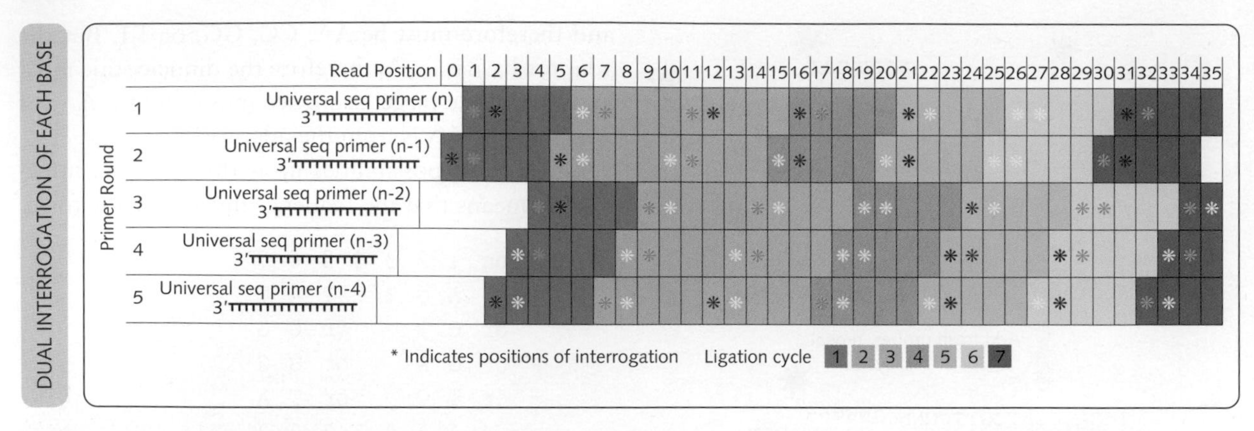

Figure 3.24 In the Applied Biosystems Solid high-throughput sequencing platform, five rounds of primer reset are completed for each primer. By reprocessing the same fragment with a shifted primer, overlapping signals provide enough information to resolve the ambiguities arising from the fact that sets of four of the sixteen possible dinucleotide probes have the same fluorescent tag. Through the sliding-primer process, almost every base is interrogated in two independent binding and ligation reactions by two different primers. For example, the base at read position 20 is tested by primer *n-1* in its fifth cycle, and by primer *n-2* in its fourth cycle. This double testing is not only necessary to resolve ambiguities in labelling, but provides a sensitive mechanism of error detection.

(Photo courtesy of Life Technologies.)

Table 3.1 Turkey genome data produced by high-throughput sequencing platforms

Platform	Type of read	Number of reads	Average read length
Roche 454 GS-FLX Titanium:	shotgun	13 million	366 bp
	3 kb fragment paired end	3 million	180 bp
	20 kb fragment paired end	1 million	195 bp
Illumina Genome analyser II:	shotgun	200 million	74 bp
	180 bp paired end	200 million	74 bp

was carried out before the newer equipment became available.)

The high-throughput sequencing platforms produced the amounts of raw data shown in Table 3.1 (see Exercise 3.14).

Sequencing of BACs produced an integrated physical and genetic map of the turkey genome. 725 contigs assembled from the BAC sequence data comprised an average of 76 clones. The average length of the contigs was 2300 kb.

The genome was assembled using a genetic map, maps based on BAC clones, and comparison with the chicken genome, which was already available.

Only for the Z chromosome was the chicken relied on for assembly. It would be appropriate to regard the turkey genome as an independenet, *de novo*, assembly.

Databanks in molecular biology

After completion of sequence and annotation, a genome enters the databanks of molecular biology – 'to take its place in society'.

The many databanks form interlocking networks. Release of a genome into any of the major archival projects is like casting a stone into a lake, sending ripples through the whole system. The genome itself is a nucleic acid sequence, but the protein-coding genes it contains will, after translation into amino acid sequences, contribute to protein sequence databanks.

Databases in molecular biology have grown with astonishing rapidity recently. Their development follows rules of their own. Here we can only describe general principles and guide the reader to his or her own exploration of this world. (See the Online Resource Centre associated with this book.)

The original databases were small, specialized, and – by post-web standards – isolated. It has always been true that different types of information need to be curated by people with the appropriate expertise. Specialists in different areas of biology organized the archiving of data related to their interests. One problem was that even where data overlapped – for instance, amino acid sequences are common to protein sequence and structure databases – there was relatively little effort to use controlled vocabularies and to make storage formats compatible. The International Scientific Unions and CODATA (the Committee on Data of the International Council of Scientific Unions) made important contributions, but problems remained.

With (1) the growing recognition of the importance of bioinformatics to research in biology, (2) the spectacular increases in the quantity of data, and – above all – (3) the emergence of the internet and World Wide Web, came pressure towards growth and integration of databanks. Requirements for large-scale funding, and the need for combining biological and computational expertise, led to the creation of national and international institutions responsible for archiving and curating the data.

The term 'databank' suggests a metaphor that perhaps is outdated: a bank as a safe place for something valuable, from which you can make a withdrawal and then go off shopping up and down the high street. This description emphasizes the archiving and curation activities of the databanks. Of course, these activities remain absolutely essential. However, the databanks also provide facilities – or at least links – for the computational analysis of the information recovered. (More like banks within large shopping malls.)

The realization that different data types were not intellectual islands, but that researchers needed coordinated access to them, led to the forging of links among the databases. The World Wide Web made this possible. The results were (1) systematization of formats and vocabularies, so that data in different collections became compatible; (2) pointers or links from each databank to others, facilitating access to related material; and (3) development of information-retrieval software that would streamline access to different databanks and smooth the passage from retrieving data to subjecting the results to computational analysis. Gathering all sequences in birds homologous to a given human protein is a problem in information retrieval; forming a multiple sequence alignment of these sequences is computational analysis. Smooth passage means a simple pipeline from the sequences returned by the database search into a multiple sequence alignment program.

In summary, the requirements of a major database project include the following:

1. Harvest the data plus annotations, curate them – that is, check both for accuracy and format – and distribute them.
2. Track and back up the data so that it does not get lost. Should any question arise, it should be possible to trace the data back to their origin and review all subsequent actions performed on them.
3. Provide links from the data to relevant items in other databanks, including bibliographical libraries such as PubMed.
4. Provide information retrieval and analysis software to support a research pipeline that includes both recovery of selected data and calculations with them.
5. Provide ample documentation and tutorial information, so that users can make effective use of the facilities.
6. Keep up with scientific advances in both biology and informatics. These may suggest improvements in the presentation and facilities.
7. Be responsive to users' needs.

Primary data collections related to biological macromolecules include:

- nucleic acid sequences, including whole-genome projects;
- amino acid sequences of proteins;

- protein and nucleic acid structures;
- small-molecule crystal structures;
- protein functions;
- expression patterns of genes;
- metabolic pathways and networks of interaction and control; and
- publications.

Nucleic acid sequence databases

The worldwide nucleic acid sequence archive, The International Nucleotide Sequence Database Collaboration, is a partnership of EMBL-Bank at the European Bioinformatics Institute (EBI), the DNA Data Bank of Japan (DDBJ) at the Center for Information Biology (CIB), and GenBank at the National Center for Biotechnology Information (NCBI).

The groups exchange data daily. As a result, the raw data are identical, although the format in which they are stored and the nature of the annotation vary among them. These databases curate, archive, and distribute DNA and RNA sequences collected from genome projects, scientific publications, and patent applications.

Entries have a life cycle in the database. Because of the desire on the part of the user community for rapid access to data, new entries are made available before annotation is complete and checks are performed. Entries mature through the classes:

unannotated → preliminary →
unreviewed → standard.

Rarely, an entry 'dies' – a few have been removed when they are determined to be erroneous.

- In addition to the International Nucleotide Sequence Database Collaboration, we have discussed several other DNA sequence databanks:

  Genome browsers – databanks organized around one or more genome sequences, with links to other information sources about the organism.

  The International HapMap Consortium – database of single-nucleotide polymorphisms.

  Forensic DNA databases – for which law enforcement agencies in various countries are responsible.

Protein sequence databases

In 2002, three protein sequence databases – the Protein Information Resource (PIR, at the National Biomedical Research Foundation of the Georgetown University Medical Center in Washington DC, USA), and SWISS-PROT and TrEMBL (from the Swiss Institute of Bioinformatics in Geneva and the European Bioinformatics Institute in Hinxton, UK) – coordinated their efforts, to form the UniProt consortium. The partners in this enterprise share the database but continue to offer separate information-retrieval tools for access.

The PIR grew out of the very first sequence database, developed by Margaret O. Dayhoff – the pioneer of the field of bioinformatics. SWISS-PROT was developed at the Swiss Institute of Bioinformatics. TrEMBL contains the translations of genes identified within DNA sequences in the EMBL Data Library. TrEMBL entries are regarded as preliminary. They mature – after curation and extended annotation – into full-Fledged UniProt entries.

Today, almost all amino acid sequence information arises from translation of nucleic acid sequences. Information about ligands, disulphide bridges, subunit associations, post-translational modifications, glycosylation, splice variants, effects of mRNA editing, etc. are not available from gene sequences. For instance, from genetic information alone, one would not know that human insulin is a dimer linked by disulphide bridges. Protein sequence databanks collect this additional information from the literature and provide suitable annotations.

Databases of genetic diseases – OMIM and OMIA

Online Mendelian Inheritance in Man™ (OMIM™) is a database of human genes and genetic disorders. Its original compilation, by V.A. McKusick, M. Smith, and colleagues, was published on paper. The NCBI of the US National Library of Medicine has developed it into a database accessible from the web and introduced links to other archives of related information, including sequence databanks and the medical literature. OMIM is now well integrated with the NCBI information-retrieval system ENTREZ. A related database, the OMIM Morbid Map, deals with genetic diseases and their chromosomal locations. OMIA (Online Mendelian Inheritance in Animals) is a corresponding database for disease and other inherited traits in animals – excluding human and mouse.

Databases of structures

Structure databases archive, annotate, and distribute sets of atomic coordinates.

Approximately 80 000 protein structures are now known. Most were determined by X-ray crystallography or nuclear magnetic resonance (NMR). The Worldwide Protein Data Bank (wwPDB) now comprises four collaborating primary archival projects to integrate the archiving and distribution of experimentally determined biological macromolecular structures:

- The Research Collaboratory for Structural Bioinformatics (RCSB), in the USA
- The Protein Data Bank in Europe (PDBe), at EBI, UK
- The Protein Data Bank Japan (Osaka, Japan)
- The Biological Magnetic Resonance Data Bank (BMRB), in the USA.

The wwPDB sites accept depositions, process new entries, and maintain the archives.

These and many other web sites organize and provide access to these data, including but not limited to pictorial displays. Naturally, there is considerable overlap among them. Each has its own strengths, based in many cases on the research interests of the contributing scientists: the PDBe has recently embarked on an ambitious software development program for structural analysis. Many sites offer search facilities to identify structures of interest, based on the presence of keywords (or a logical combination of keywords), or numerical values such as the year of deposition. Different sites differ also in their 'look and feel', and users will discover their own preferences.

The wwPDB overlaps in scope with several other databases. The Cambridge Crystallographic Data Centre (CCDC) archives the structures of small molecules. This information is extremely useful in studies of conformations of the component units of biological macromolecules, and for investigations of macromolecule–ligand interactions, including but not limited to applications to drug design. The Nucleic Acid Structure Databank (NDB) at Rutgers University, New Brunswick, New Jersey, USA, complements the wwPDB.

Classifications of protein structures

Several web sites offer hierarchical classifications of the entire PDB according to the folding patterns of the proteins. These include:

- SCOP: structural classification of proteins
- CATH: class/architecture/topology/homologous superfamily
- DALI: based on extraction of similar structures from distance matrices
- CE: a database of structural alignments.

These sites are useful general entry points to protein structural data. For instance, SCOP offers facilities for searching on keywords to identify structures, navigation up and down the hierarchy, generation of pictures, access to the annotation records in the PDB entries, and links to related databases.

Specialized or 'boutique' databases

Many individuals or groups select, annotate, and recombine data focused on particular topics, and include links affording streamlined access to information about subjects of interest. For instance, the Protein Kinase Resource is a specialized compilation that includes sequences, structures, functional information, laboratory procedures, lists of interested scientists, tools for analysis, a bulletin board, and links. It has recently been redesigned, with a view to integrating expanded information content with a workbench equipped with embedded tools for launching analyses of the data within the user interface.

Expression and proteomics databases

Recall the central dogma: DNA makes RNA makes protein. Genomic databases contain *DNA sequences.*

Expression databases record measurements of *mRNA* levels, usually via ESTs (expressed sequence tags: short terminal sequences of cDNA synthesized from mRNA) describing patterns of gene transcription. Proteomics databases record measurements on *proteins*, describing patterns of gene translation.

Comparisons of expression patterns give clues to (1) the function and mechanism of action of gene products; (2) how organisms coordinate their control over metabolic processes in different conditions – for instance, yeast under aerobic or anaerobic conditions; (3) the variations in mobilization of genes in different tissues, or at different stages of the cell cycle, or of the development of an organism; (4) mechanisms of antibiotic resistance in bacteria and consequent suggestion of targets for drug development; (5) the response to challenge by a parasite; (6) the response to medications of different types and dosages, to guide effective therapy.

There are many databases of ESTs. In most, the entries contain fields indicating tissue of origin and/or subcellular location, stage of development, conditions of growth, and quantification of expression level. Within GenBank, the dbEST collection currently contains almost 70 million entries, from 2281 species. The species with the largest numbers of entries in dbEST are shown in Table 3.2.

Some EST collections are specialized to particular tissues (e.g. muscle, teeth) or to species. In many cases, there is an effort to link expression patterns to other knowledge of the organism. For instance, the Jackson Laboratory Gene Expression Information Resource Project for Mouse Development coordinates data on gene expression and developmental anatomy.

Many databases provide connections between ESTs in different species, for instance, linking human and mouse homologues, or relationships between human disease genes and yeast proteins. Other EST collections are specialized to a type of protein, for instance cytokines. A large effort is focused on cancer, integrating information on mutations, chromosomal rearrangements, and changes in expression patterns to identify genetic changes during tumour formation and progression.

Although of course there is a close relationship between patterns of transcription and patterns of translation, direct measurements of protein contents

Table 3.2 Species with the largest numbers of entries in dbEST

Species	Number of entries
Homo sapiens (human)	8 314 483
Mus musculus + *domesticus* (mouse)	4 853 533
Zea mays (maize)	2 019 105
Sus scrofa (pig)	1 620 479
Bos taurus (cattle)	1 559 494
Arabidopsis thaliana (thale cress)	1 529 700
Danio rerio (zebra fish)	1 481 937
Glycine max (soybean)	1 461 624
Xenopus (Silurana) tropicalis (western clawed frog)	1 271 375
Oryza sativa (rice)	1 251 304
Ciona intestinalis	1 205 674
Rattus norvegicus + sp. (rat)	1 162 136
Triticum aestivum (wheat)	1 071 367
Drosophila melanogaster (fruit fly)	821 005
Xenopus laevis (African clawed frog)	677 806
Oryzias latipes (Japanese medaka)	666 891
Brassica napus (oilseed rape)	643 874
Gallus gallus (chicken)	600 423
Panicum virgatum (switchgrass)	546 245
Hordeum vulgare + subsp. *vulgare* (barley)	501 620
Salmo salar (Atlantic salmon)	498 212
Caenorhabditis elegans (nematode)	393 714
Phaseolus coccineus	391 150
Porphyridium cruentum	386 903
Canis lupus familiaris (dog)	382 629
Vitis vinifera (wine grape)	362 392
Physcomitrella patens subsp. *patens*	362 131
Ictalurus punctatus (channel catfish)	354 466
Ovis aries (sheep)	338 364
Branchiostoma floridae (Florida lancelet)	334 502
Nicotiana tabacum (tobacco)	332 667
Pinus taeda (loblolly pine)	328 662
Malus domestica (apple tree)	324 512
Picea glauca (white spruce)	313 110
Bombyx mori (domestic silkworm)	309 472
Aedes aegypti (yellow fever mosquito)	301 596
Solanum lycopersicum (tomato)	297 104
Oncorhynchus mykiss (rainbow trout)	287 967
Linum usitatissimum	286 852
Neurospora crassa	277 147
Gasterosteus aculeatus (three-spined stickleback)	276 992
Medicago truncatula (barrel medic)	269 238
Gossypium hirsutum (upland cotton)	268 797
Pimephales promelas	258 504
Aplysia californica (California sea hare)	255 605

of cells and tissues – proteomics – provides additional valuable information. Because of differential rates of translation of different mRNAs, measurements of proteins directly give a more accurate description of patterns of gene expression than measurements of transcription. Post-translational modifications can be detected *only* by examining the proteins.

Databases of metabolic pathways

The Kyoto Encyclopedia of Genes and Genomes (KEGG) collects individual genomes, gene products, and their functions, but its special strength lies in its integration of biochemical and genetic information. KEGG focuses on interactions: molecular assemblies, and metabolic and regulatory networks. It has been developed under the direction of M. Kanehisa.

KEGG organizes five data types into a comprehensive system:

1. catalogues of chemical compounds in living cells
2. gene catalogues
3. genome maps
4. pathway maps
5. orthologue tables.

The catalogues of chemical compounds and genes contain information about particular molecules or sequences. Genome maps integrate the genes themselves according to their chromosomal location. In some cases, knowing that a gene appears in an operon can provide clues to its function.

Pathway maps describe potential networks of molecular activities, both metabolic and regulatory. A metabolic pathway in KEGG is an idealization corresponding to a large number of possible metabolic cascades, combining reactions occurring in different organisms. It can generate a real metabolic pathway of a particular organism, by matching the proteins of that organism to enzymes within the reference pathways.

One enzyme in one organism would be referred to in KEGG in its orthologue tables, which link the enzyme to related ones in other organisms. This permits analysis of relationships between the metabolic pathways of different organisms.

Bibliographic databases

MEDLINE (based at the US National Library of Medicine) integrates the medical literature, including very many papers dealing with subjects in molecular biology not overtly clinical in content. It is included in PubMed, a bibliographical database offering abstracts of scientific articles, integrated with other information retrieval tools of the NCBI within the National Library of Medicine (http://www.ncbi.nlm.nih.gov/PubMed/).

One very effective feature of PubMed is the option to retrieve *related articles*. This is a very quick way to 'get into' the literature of a topic. Here's a tip: if you are trying to start to learn about an unfamiliar subject, try adding the keyword *tutorial* to your search in a general search engine, or the keyword *review* to your search in PubMed.

Surveys of molecular biology databases and servers

It is difficult to explore any topic in molecular biology on the web without quickly bumping into a list of this nature. Lists of web resources in molecular biology are very common. They contain, to a large extent, the same information but vary widely in their 'look and feel'. The real problem is that, unless they are curated, they tend to degenerate into lists of dead links.

Each year the January issue of the journal Nucleic Acids Research *contains a set of articles on databases in molecular biology. This is an invaluable reference.*

This book does not contain a long annotated list of relevant and recommended sites, for the following reasons. First, you do not want a long list; you need a short one. Second, the web is too volatile for such a list to stay useful for very long. *It is much more effective to use a general search engine to find what you want at the moment you want it.*

My advice is spend some time browsing; it will not take you long to find a site that appears reasonably stable and has a style compatible with your methods of work. Alternatively, here is a site that is comprehensive and shows signs of a commitment to keeping it up to date: http://www.expasy.org. It is a suitable site for starting a browsing session.

RECOMMENDED READING

- Discussions of haplotypes in general, and the important MHC complex in particular:

Neale, B.M. (2010). Introduction to linkage disequilibrium, the HapMap, and imputation. Cold Spring Harb Protoc; 2010; doi:10.1101/pdb.top74.

Vandiedonck, C., & Knight, J.C. (2009). The human Major Histocompatibility Complex as a paradigm in genomics research. Brief. Funct. Genomic Proteomic. **8**, 379–394.

- The following describe the recent advances that have produced the high-throughput sequencing platforms on which contemporary sequencing depends:

Davies, K. (2010). *The $1000 Genome: The Revolution in DNA Sequencing and the New Era of Personalized Medicine.* Free Press, New York.

Mardis, E. (2008). The impact of next-generation sequencing technology on genetics. Trends Genet. **24**, 133–141.

Ng, P.C. & Kirkness, E.F. (2010). Whole genome sequencing. Methods Mol. Biol. **628**, 215–226.

- A collection of papers on new techniques and their applications in medicine:

Janitz, M., ed. (2008) *Next-Generation Genome Sequencing: Towards Personalized Medicine.* Wiley-VCH, Weinheim.

- The next two articles describe the development of databases of sequences and structures.

Smith, T.F. (1990). The history of the genetic sequence databases. Genomics **6**, 701–707.

Bernstein, H. & Bernstein, F. (2005). Databanks of macromolecular structure. In *Database Annotation in Molecular Biology: Principles and Practice*, Lesk, A.M. (ed.) J. Wiley & Sons, Chichester, pp. 63–79.

EXERCISES, PROBLEMS, AND WEBLEMS

Exercises

Exercise 3.1 Two loci with alternative alleles *A/a* and *B/b*, respectively, are 1 cM apart. A cross between parents of genotype *AB/AB* and *ab/ab* produces a large number of offspring. Assuming no selective difference between genotypes, estimate the fraction of the next generation that has genotype *Ab/aB*.

Exercise 3.2 Gene *A* has two alleles, A_1 and A_2. Gene *B* has two alleles, B_1 and B_2. In a population, the following haplotype frequencies are observed: $A_1B_1 = 0.2$, $A_2B_2 = 0.45$, $A_1B_2 = 0.15$, $A_2B_1 = 0.2$. Calculate *D*, the extent of linkage disequilibrium.

Exercise 3.3 A fruitfly with a chromosome deletion shows pseudodominance for a trait. On a photocopy of Figure 3.3, indicate with an 'X' where the locus for this trait might be.

Exercise 3.4 The Philadelphia translocation occurs in a bone marrow cell, resulting in the development of chronic myeloid leukaemia. Would the patient transmit this leukaemia to his or her offspring?

Exercise 3.5 On a photocopy of Figure 3.6, indicate two positions in the human chromosome where one would look for genes that are linked in humans but unlinked in chimpanzees?

Exercise 3.6 (a) On a photocopy of Figure 3.8(a), indicate with an 'A' the band on the gel that corresponds to the *Bam*HI fragment in Figure 3.8(b) which begins at 1 kb and ends at 5 kb. (b) On a photocopy of Figure 3.8(b), indicate with a 'B' the fragment that gives rise to the band in the *Eco*RI lane of the gel which corresponds to the lowest molecular mass fragment.

Exercise 3.7 The lengths of blocks that define human haplotypes vary, partly because of the variation of recombination rates along the genome. Would you expect the haplotype blocks to vary more if their sizes are measured in terms of number of base pairs or in terms of genetic distance in centimorgans?

Exercise 3.8 Suppose that there are ten SNPs in a 10 kb region. If the region is on the Y chromosome, how many possible haplotypes are there? If the region is on a diploid chromosome, how many possible haplotypes are there?

Exercise 3.9 On a photocopy of Figure 3.9, indicate, by crossing atoms out and writing atoms in, how the structure would have to be changed to illustrate an RNA molecule with the equivalent base sequence.

Exercise 3.10 From a photocopy of Figure 3.11, cut out the individual bases and show that guanine and thymine could form a non-canonical base pair containing two hydrogen bonds. By comparing with a copy of the standard base pairs, show the extent to which a guanine–thymine pair would not match the correct relative position and orientation of the sugars to be stereochemically compatible with standard base pairs in a double helix of standard structure. Uracil has the same hydrogen-bonding specificity as thymine. Guanine–uracil 'wobble' base pairs are implicated in codon–anticodon interactions between tRNA and mRNA.

Exercise 3.11 The tetranucleotide illustrated in Figure 3.9 is self-complementary. (a) What does this mean? (b) Make two photocopies of Figure 3.9. From one of them trim off the names of the bases. From the other, cut out the individual bases and mount them adjacent to the first in position to form Watson–Crick base pairs. Draw in the hydrogen bonds between bases. (It will not work simply by turning one copy upside down and mounting it next to the other. In a double helix, the two copies are symmetrically disposed in three dimensions but not in two dimensions.)

Exercise 3.12 If all of the DNA in all of the cells of your body were laid end to end, would you be surprised if it were longer than the diameter of the solar system? Calculate the result and compare. The semi-major axis of Pluto's orbit is 5 906 376 272 km. The number of cells in an adult human body has been estimated as 10^{13}.

Exercise 3.13 In Figure 3.13, the region of the template strand not complexed with the primer is shown as continuing the helical structure. Although justifiable pedagogically for clarity, why is this not a structurally correct representation?

Exercise 3.14 On a photocopy of Figure 3.19, indicate the longest contig available from the data given.

Exercise 3.15 In Figure 3.19, (a) What is the minimal coverage of any position (this is obvious)? (b) What is the maximal coverage of any position (i.e. the largest number of fragments in which the same position appears)? (c) Estimate the average coverage of the entire region? (Hint: measure the total lengths of the fragments and divide by the length of the region.)

Exercise 3.16 The International Human Genome Mapping Consortium fingerprinted 300 000 BAC clones. Assuming an average insert size of 150 kb and a 3.2 Gb genome size, what coverage would be expected?

Exercise 3.17 One difficulty in extracting reads that correspond to mitochondrial DNA from sequencing mixed fragments of nuclear and mitochondrial DNA is that the nuclear genome contains segments homologous to regions of the mitochondrial genome, called **numts**. Mammalian genomes contain 50–450 kb of numts. (The human genome contains 1005 such segments, of average length 446 bp.) Estimate the fraction of reads from fragments of mammoth DNA that are likely to be numts. The mammoth genome is 4.7 Gb long.

Exercise 3.18 Referring to Figure 3.24, by what primers, in which cycle, is the base at position 10 tested?

Exercise 3.19 Referring to Figure 3.23, suppose that the dinucleotide that bound to positions 23456789 gave a green fluorescence. What is the base at position 3?

Exercise 3.20 How much raw sequence data was generated for the turkey genome project? How many human genome equivalents does this amount to?

Problems

Problem 3.1 Consider two linked traits in a population in which half of the individuals are double heterozygotes with genotype *AB*/*ab* and the other half are double homozygotes (*AB*/*AB*). Assuming no selective advantage of any combination of alleles for these traits and no preferential mating, after recombination brings the population to equilibrium, what will be the ratio of *AB*/*AB*, *Ab*/*aB* = *aB*/*Ab* and *ab*/*ab* individuals?

Problem 3.2 Consider two markers 1 cM apart. (a) What is the probability that there will be recombination between them in one generation? (b) What is the probability that there will *not* be recombination between them in one generation? (c) What is the formula for the probability that there will *not* be recombination between them in *n* generations? (d) Evaluating this formula, what is the probability that there will not be recombination between them in $n = 10$, 20, 30, 40, and 50 generations?

Problem 3.3 As a simplified but illustrative example of sequence assembly, we saw the first two verses of *Richard III* chopped into overlapping 10-character fragments. (a) Chop these lines into consecutive overlapping 5-character fragments, and scramble these fragments into random order. Is it still possible to reconstruct the lines without ambiguity? Why is it more difficult to do so than to reconstruct the lines from 10-character fragments? (b) Try generating, and then trying to reassemble, 10- and 5-character fragments, presented in random order, of the lines of Polonius (ignore punctuation marks):

... 'tis true, 'tis true 'tis pity,
And pity 'tis 'tis true

In each case, is it still possible to reconstruct the lines without ambiguity?

(c) Try the same with Richard II's speech:

Your cares set up do not pluck my cares down.
My care is loss of care, by old care done;
Your care is gain of care, by new care won:
The cares I give I have, though given away;
They tend the crown, yet still with me they stay.

Problem 3.4 Extend Problem 3.3 to simulate the effect of paired-end reads. Take any text of about 100 words in length (a sonnet is about the right length) and write a program to create fragments with a distribution of lengths distributed roughly normally around 30 ± 5 characters. Print reads of 8 characters from each end. Tabulate the data from different fragments in random order. Try to reassemble the text. Study how the difficulty of the assembly depends on the read length, fragment length, and coverage.

Problem 3.5 Lander & Waterman* derived formulas for the expected completeness of an assembly as a function of coverage (G = genome length, N = number of reads, L = read length, $c = NL/G$ = coverage):

probability that a base is *not* sequenced = e^{-c}
total expected gap length = $G \times e^{-c}$
total number of gaps = Ne^{-c}

* Lander, E.S. & Waterman, M.S. (1988). Genomic mapping by fingerprinting random clones: a mathematical analysis. Genomics 2, 231–239.

(a) What fraction of a genome could you expect to assemble from eightfold coverage? (b) What total gap length would you expect in an assembly of a 2 Mb target genome size from eightfold coverage? (c) How many gaps would you expect in an assembly of a 2 Mb target genome size from an eightfold coverage of fragments with a read length of 500? (d) You want to sequence a 4 Mb genome by the shotgun method, by assembling random fragments with read length 500. What coverage would you require, to expect no more than four gaps, assuming no complications arising from repetitive sequences or far-from-equimolar base composition?

Problem 3.6 Figure 3.25 shows a sequencing gel. (a) What is the sequence of this fragment? (b) Can you see any self-complementary regions in this fragment that might form hairpin loops? (c) On the basis of your answer to part (b), would you guess that this region encodes RNA or protein?

Problem 3.7 Figure 3.26(a) shows a series of measurements from the Solid technology (see p. 102). Figure 3.26(b) shows the colour coding of the dinucleotides. The known template sequence implies that the first base is an A, therefore the dinucleotide at positions 0-1 is A-? (a) What is the sequence of the fragment? (b) Suppose another fragment differs by a SNP at position 15, and suppose that this SNP is a transition mutation (that is, a purine to the other purine, or a pyrimidine to the other pyrimidine). How would a figure, corresponding to Figure 3.26(a), that presents SOLID results from the mutated fragment, differ from Figure 3.26(a)?

Problem 3.8 From Figure 3.18, (a) determine the rate of change of the cost of sequencing over the years 2005–2007. (b) determine the rate of change of the cost of sequencing over the years 2008–2010. (c) According to these figures, what would be the cost in 2010 of determining a human-sized genome at 10X coverage?

Problem 3.9 Many people have asked whether the author is related to Filippo Brunelleschi, who was born in 1377, died in 1446 and is buried beneath the cathedral in Florence (the dome of which he famously created). Assume that you were granted permission to exhume his body and collect a tissue sample. (a) Estimate the number of generations between Brunelleschi and the author. (b) Assuming that the author is a direct descendant of Brunelleschi, would you expect to be able to prove it by DNA sequencing? Explain your answer.

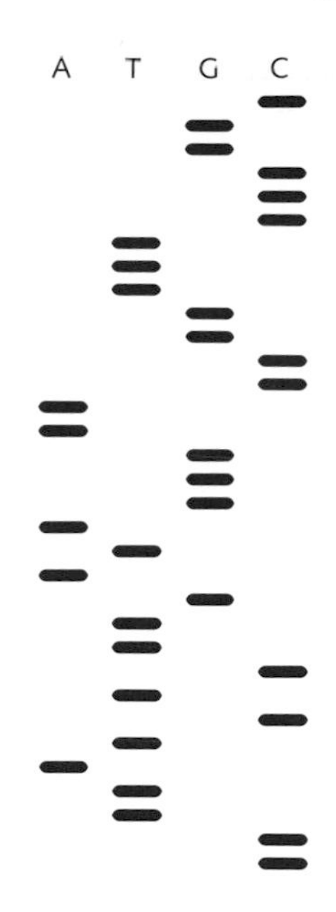

Figure 3.25 Autoradiograph of a sequencing gel (simulated). The shortest fragment travels the farthest. In this diagram, the direction of travel is down the page.

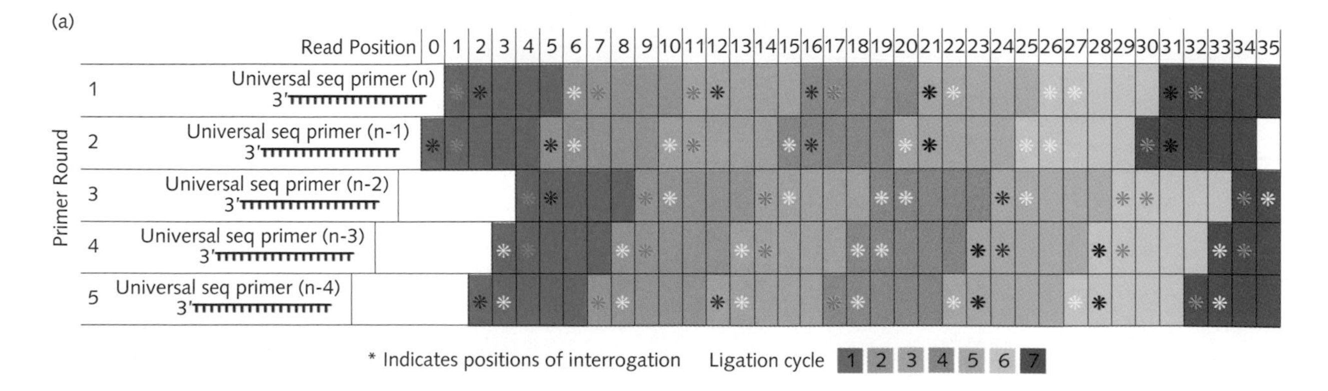

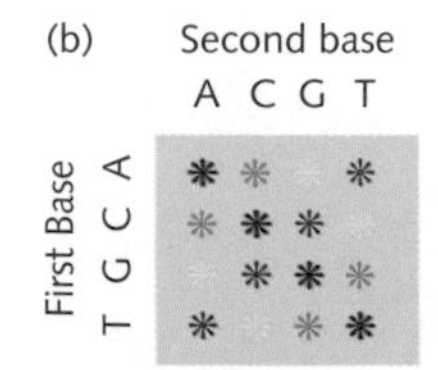

Figure 3.26 (a) a series of measurements from the Solid technology. (b) The colour coding of the dinucleotides.

Problem 3.10 You are asked to devise a Master's degree programme to train annotators for databases in molecular biology. (a) What background would you require for entry to the programme? (b) What courses would you require students to take during the programme?

Weblems

Weblem 3.1 Which of the following families have genes appearing in tandem arrays in the human genome and which have genes dispersed among several chromosomes? (a) Actin, (b) tRNA, (c) all globins, (d) *HOX* genes, (e) the major histocompatibility complex.

Weblem 3.2 What institution currently has the highest sequencing throughput power? How many Gb per week can this institute produce?

Weblem 3.3 For the major companies providing sequencing equipment. What is the current throughput rate and typical read length of their state-of-the-art instrument?

Weblem 3.4 On a photocopy of Figure 3.18, add points to bring the figure up-to-date.

CHAPTER 4

Comparative Genomics

LEARNING GOALS

- Knowing the three major divisions of living things – archaea, bacteria, and eukaryotes – based on analysis of the sequences of 16S rRNA genes.
- Recognizing the prevalence of horizontal gene transfer, especially among prokaryotes, and to understand that horizontal gene transfer is inconsistent with the hierarchical 'tree of life' picture that the Linnaean classification scheme suggests.
- Being familiar with major events in the history of life.
- Appreciating the general distribution of genome sizes and numbers of genes.
- Distinguishing the characteristics of different types of genome organization in viruses, prokaryotes, and eukaryotes.
- Recognizing the effects of gene duplication on genome evolution.
- Being able to distinguish the meanings of homologue, orthologue, and paralogue.
- Understanding the mechanism of genome change at the levels of individual bases, genes, chromosome segments, and whole genomes.
- Understanding the limits of what genomes determine and what they do not determine, and the limits of what we can currently explain on the basis of genetics and what we cannot.
- Appreciating, as far as possible, what makes us human.
- Understanding the idea of a model organism in the study of human disease.
- Appreciating the goals and plans of the Encyclopedia of DNA Elements (ENCODE) project, and the related project, modENCODE.

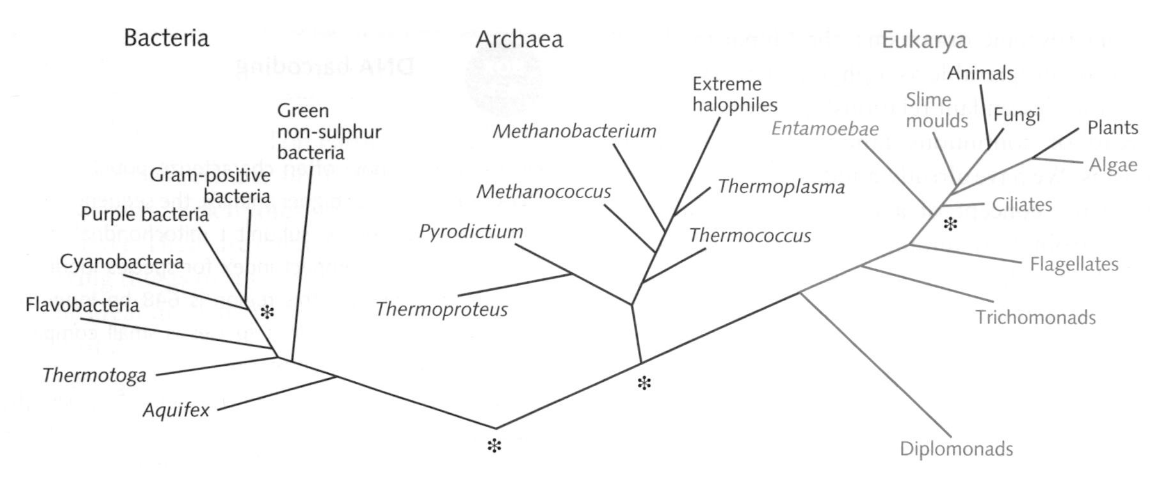

Figure 4.1 Major divisions of the tree of life. **Bacteria** (blue) and **archaea** (magenta) are prokaryotes; their cells do not contain nuclei. Bacteria include the typical microorganisms responsible for many infectious diseases and, of course, *Escherichia coli*, the mainstay of molecular biology. Archaea include, but are not limited to, extreme thermophiles and halophiles, sulphate reducers, and methanogens. We ourselves are **eukarya** – organisms containing cells with nuclei (green and red). Asterisks mark crucial splitting points (see Exercise 4.4). This phylogenetic tree was derived by C. Woese from comparisons of ribosomal RNAs. These RNAs are present in all organisms, and show the right degree of divergence. (Too much or too little divergence and relationships become invisible.) Figure 4.2 shows in more detail the group that includes us – animals, fungi, and plants (red).

- *All life on Earth has enough general similarity to show that all life forms had a common origin.* Evidence includes the universality of the basic chemical structures of DNA, RNA, and proteins, the universality of their general biological roles, and the near-universality of the genetic code.
- *On the basis of 16S rRNAs, C. Woese divided living things most fundamentally into three domains: bacteria, archaea, and eukarya.* A domain occupies a level in the hierarchy *above* kingdom.

Figure 4.1 shows the major divisions of the tree of life. At the ends of the eukaryote branch are the metazoa, including yeast and all multicellular organisms – fungi, plants, and animals (see Figure 4.2). We and our closest relatives are in the vertebrate branch of the deuterostomes (see Figure 4.3).

Although archaea and bacteria are both unicellular organisms that lack a nucleus, at the molecular level archaea are in some ways more closely related to eukarya than to bacteria. It is also likely that the archaea are the closest living organisms to the root of the tree of life.

- *Dating of historical events from sequence differences.* As species diverge, their sequences diverge. L. Pauling and E. Zuckerkandl suggested that if sequence divergence occurred at a constant rate, it would provide a 'molecular clock' that would allow dating of the splits in lineage between species.

Although the clock is not universal, judicious calibration of rates of sequence change with palaeontological data permits dating of events in the history of life (see Box 4.2 and Figure 2.15).

Molecular phylogeny and chronology

Molecular approaches to phylogeny developed against a background of traditional taxonomy, based on a variety of morphological characters, embryology, geographical distribution and, for fossils, information about the geological context (stratigraphy). The classical methods have some advantages. Traditional taxonomists have much greater access to extinct organisms via the fossil record. They can **date** the appearance and extinction of species by geological methods (see Figure 2.15).

Molecular biologists, in contrast, have very limited access to extinct species. Some subfossil remains of species that became extinct as recently as within the last two centuries have legible DNA, including specimens of the quagga (a relative of the zebra), the thylacine (Tasmanian 'wolf', a marsupial), the mammoth from the permafrost in Russia, the dodo of Mauritius, the 'elephant bird' of Madagascar, and some New Zealand birds, for instance, moas. It has been possible to sequence mitochondrial DNA from ~10 000-year-old remains of an 'Irish elk'. DNA sequences from Neanderthal man have been recovered from individuals who died approximately 40 000 years ago. But *Jurassic Park* remains fiction!

A crucial event in the acceptance of molecular methods occurred in 1967 when V.M. Sarich and A.C. Wilson dated the time of divergence of humans from chimpanzees at 5 million years ago, based on immunological data. At that time, traditional palaeontologists dated this split at 15 million years ago and were reluctant to accept the molecular approach. Reinterpretation of the fossil record led to acceptance of a more recent split and broke the barrier to general acceptance of molecular methods. It is now generally accepted that human and chimpanzee lineages diverged between ~6 and 8 million years ago.

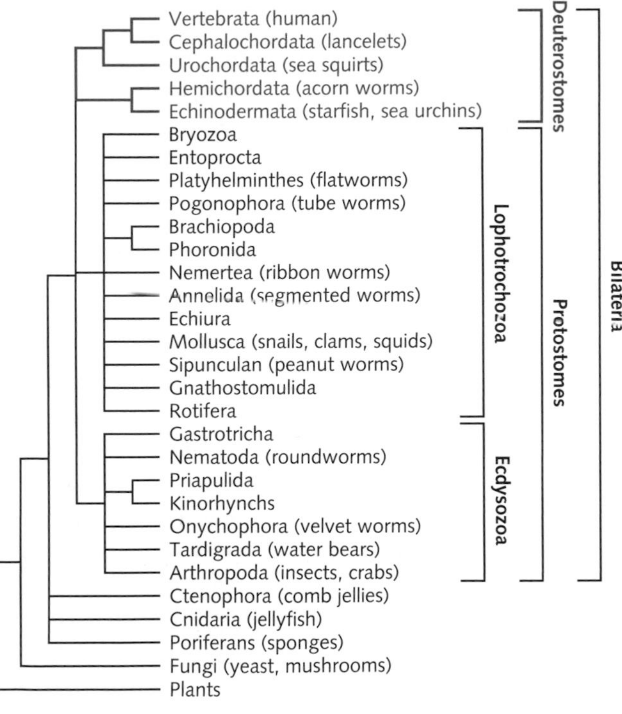

Figure 4.2 Phylogenetic tree of metazoa (multicellular animals). **Bilateria** include all animals that share a left–right symmetry of body plan. **Protostomes** and **deuterostomes** (red) are two major lineages that separated at an early stage of evolution, estimated at 670 million years ago. They show very different patterns of embryological development, including different early cleavage patterns, opposite orientations of the mature gut with respect to the earliest invagination of the blastula, and the origin of the skeleton from mesoderm (deuterostomes) or ectoderm (protostomes). Protostomes comprise two subgroups distinguished on the basis of the sequences of an RNA from the small ribosomal subunit and *HOX* genes. (*HOX* genes govern the development of body plans.) Morphologically, **ecdysozoa** have a moulting cuticle – a hard outer layer of organic material. **Lophotrochozoa** have soft bodies. Figure 4.3 shows in more detail the group that includes us – the deuterostomes (red).

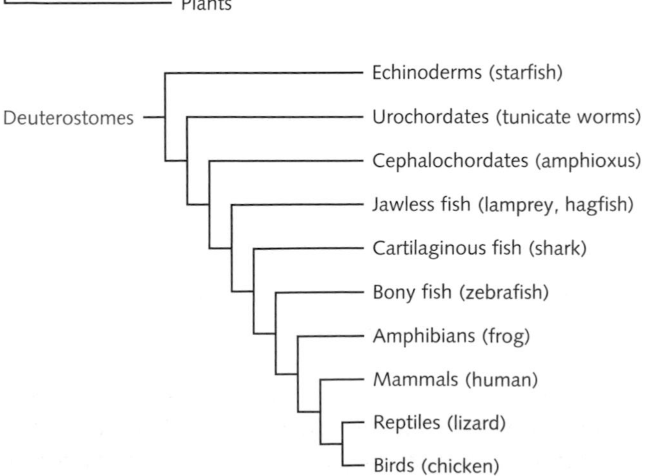

Figure 4.3 Phylogenetic tree of vertebrates and our closest relatives. Chordates, including vertebrates, and echinoderms are all deuterostomes. Examples of each are shown in blue.

- *The importance of horizontal gene transfer*. This is the acquisition of genetic material by one organism from another by natural rather than laboratory procedures through some means other than descent from a parent during replication or mating (see Box 4.3). Several mechanisms of horizontal gene transfer are known, including direct uptake, as in Griffith's pneumococcal transformation experiments, or via a viral carrier. Arrangements of species into phylogenetic trees, in contrast, assumes strict ancestor–descendant relationships between different organisms during evolution.

Horizontal gene transfer among different species has affected most genes in prokaryotes. It requires a change in our thinking from ordinary 'clonal' or parental models of heredity. Microorganisms do not easily fit into the structure of the 'tree' of life but require a more complex organizational chart.

Please pass the genes: horizontal gene transfer

On learning that *Streptomyces griseus* trypsin is more closely related to bovine trypsin than to other microbial proteinases, Brian Hartley commented in 1970 that '. . . the bacterium must have been infected by a cow'. This was a clear example of lateral or horizontal gene transfer – a bacterium picking up a gene from the soil in which it was growing, that an organism of another species had deposited there. The classic experiments on pneumococcal transformation by Griffiths, and those by O. Avery, C. MacLeod, and M. McCarthy that identified DNA as the genetic material, are another example.

Evidence for horizontal transfer includes (1) discrepancies among evolutionary trees constructed from different genes; and (2) direct sequence comparisons between genes from different species.

- In *Escherichia coli*, about 25% of the genes appear to have been acquired by transfer from other species.
- In microbial evolution, horizontal gene transfer is more prevalent among operational genes – those responsible for 'housekeeping' activities such as biosynthesis – than among informational genes – those responsible for organizational activities such as transcription and translation. For example:
  - *Bradyrhizobium japonicum*, a nitrogen-fixing bacterium, symbiotic with higher plants, has two glutamine synthetase genes: one is similar to those of its bacterial relatives; the other is 50% identical to those of higher plants;
  - rubisco (ribulose-1,5-bisphosphate carboxylase/oxygenase), the enzyme that first fixes carbon dioxide at entry to the Calvin cycle of photosynthesis, has been passed around between bacteria, mitochondria, and algal plastids, as well as undergoing gene duplication.
  - many phage genes appearing in the *E. coli* genome provide further examples and point to a mechanism of transfer.

Nor is the phenomenon of horizontal gene transfer limited to prokaryotes. Both eukaryotes and prokaryotes are chimaeras. Eukaryotes derive their informational genes primarily from an organism related to *Methanococcus*, and their operational genes primarily from proteobacteria, with some contributions from cyanobacteria and methanogens. Almost all informational genes from *Methanococcus* itself are similar to those in yeast. At least eight human genes appeared in the *Mycobacterium tuberculosis* genome. *S. griseus* trypsin is an example of eukaryote → prokaryote transfer.

The observations hint at the model of a 'global organism', or a genomic World Wide DNA Web from which organisms download genes at will! How can this be reconciled with the fact that the discreteness of species has been maintained? We offered the conventional explanation, that the living world contains ecological 'niches' to which individual species are adapted: the discreteness of niches explains the discreteness of species. But this explanation depends on the stability of normal heredity to maintain the fitness of the species. Why would the global organism not break down the lines of demarcation between species, just as global access to pop culture threatens to break down lines of demarcation among national and ethnic cultural heritages? Perhaps the answer is that it is the informational genes, which appear to be less subject to horizontal transfer, that determine the identity of the species.

Sizes and organization of genomes

We appeal to genomes to help us to understand ourselves as individuals, and our relationships with all of the other organisms that march in the pageant of life. To make progress, we must integrate several data streams, including:

- genome sequences;
- RNA and protein expression patterns;
- the spatial organization of individual macromolecules, their complexes, organelles, entire cells, tissues, and bodies; and
- regulatory networks, the internal structure and logic of adaptive control systems.

Even these may not be enough. History – sometimes observed but more usually inferred – provides essential additional clues. We see, today, a snapshot of one stage in a history of life that extends back in time for at least 3.5 billion years. We must try to read the past in contemporary genomes, which contain records of their own development.

US Supreme Court Justice Felix Frankfurter wrote that '. . . the American constitution is not just a document, it is a historical stream.' Like a genome!

This programme requires development of novel methods. New fields of study require new approaches. S.E. Luria once suggested that to determine common features of all life one should not try to survey everything, but, rather, identify the organism most different from us and see what we have in common with it. Let us combine this with a complementary idea: to take the most *closely related* organisms and identify the *differences*. That is:

- How do the human genome and the *E. coli* genome express our *common* heritage?
- How do genomes that are over 96% identical create the *differences* between humans and chimpanzees?

If we could answer these questions, we would have achieved a lot.

- We appeal to genomes to help us to understand evolutionary relationships. With a few exceptions, out access to genome sequences is limited to organisms alive today, that is, to a single snapshot in time. However, genomes contain records of their history which gives us a window onto the past.

Genome sizes

One reason for resistance to Darwin's theory of evolution was its denial to human beings of a special status relative to animals. Genomics threatens to do this all over again. Humans *do* have unique features. Many people, not excluding molecular biologists, expect these features to be reflected in the genome. And so they must be, although, frankly, not in any obvious way.

- The term C-value has been used to refer to the amount of DNA in a haploid cell, i.e. a gamete; the letter C refers to the *constancy* of the amount of DNA per cell in a species.

The overall size of the human genome is not special. Different organisms have different total amounts of DNA per cell (see Figure 4.4 and Table 4.2).

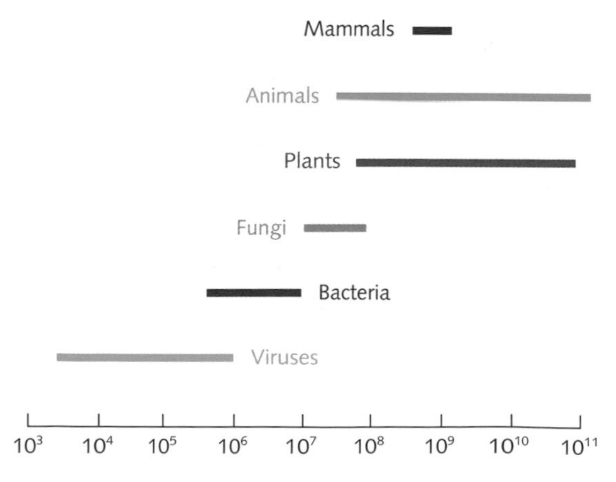

Figure 4.4 Distribution of genome sizes in different groups of living things. The horizontal scale gives the number of bases or base pairs.

Table 4.2 Genome sizes

Organism	Number of base pairs	Number of genes	Comment
ϕX-174	5 386	10	Virus infecting *E. coli*
Influenza A	13 590	10	Strain A/goose/guandong/1/96(H5N1)
Human mitochondrion	16 569	37	Subcellular organelle
Epstein–Barr virus (EBV)	172 282	80	Cause of mononucleosis
Nanoarchaeum equitans	490 885	552	Archaeon, smallest known genome of a cellular organism
Mycoplasma pneumoniae	816 394	680	Cause of cyclic pneumonia epidemics
Rickettsia prowazekii	1 111 523	834	Bacterium, cause of epidemic typhus
Mimivirus	1 181 404	1 262	Virus with the largest known genome
Borrelia burgdorferi	1 471 725	1 738	Bacterium, cause of Lyme disease
Aquifex aeolicus	1 551 335	1 749	Bacterium from hot spring
Thermoplasma acidophilum	1 564 905	1 509	Archaeal prokaryote, lacks cell wall
Helicobacter pylori	1 667 867	1 589	Chief cause of stomach ulcers
Methanococcus jannaschii	1 664 970	1 783	Archaeal prokaryote, thermophile
Haemophilus influenza	1 830 138	1 738	Bacterium, cause of middle-ear infections
Thermotoga maritime	1 860 725	1 879	Marine bacterium
Archaeoglobus fulgidus	2 178 400	2 437	Another archaeon
Deinococcus radiodurans	3 284 156	3 187	Radiation-resistant bacterium
Synechocystis	3 573 470	4 003	Cyanobacterium, 'blue-green alga'
Vibrio cholera	4 033 460	3 890	Cause of cholera
Mycobacterium tuberculosis	4 411 532	3 959	Cause of tuberculosis
Bacillus subtilis	4 214 814	4 779	Popular in molecular biology
Escherichia coli	4 639 221	4 485	Molecular biologists' all-time favourite
Saccharomyces cerevisiae	12 495 682	5 770	Yeast, first eukaryotic genome sequenced
Caenorhabditis elegans	100 258 171	19 099	'The worm'
Arabidopsis thaliana	135 000 000	25 498	Flowering plant (angiosperm), 'the weed'
Drosophila melanogaster	122 653 977	13 472	The fruit fly
Takifugu rubripes	3.65×10^8	23 000	Pufferfish (fugu fish)
Human	3.3×10^9	23 000	
Wheat	16×10^9	30 000	
Salamander	10^{11}	?	
Psilotum nudum	2.5×10^{11}	?	Whisk fern, a simple plant
Amoeba dubia	6.7×10^{11}	?	Protozoan

There is a general correlation between complexity of organism and amount of DNA per cell. Prokaryotes have less DNA per cell than eukaryotes, and yeast has less than mammals. However, although humans have more DNA per cell than certain other organisms popular in molecular biology, including *Caenorhabditis elegans* and the fruit fly, many organisms have even greater amounts than we do. The genome of *Amoeba dubia* is 200 times larger than the human genome. The genome of the marbled lungfish (*Protopterus aethiopicus*), a closer relative, is 43 times as large as ours.

Why the different amounts of DNA? As far as we know, most of the human genome does not encode protein or RNA. Regions of genomes without known function are often referred to as 'junk DNA'. Of course, the fact that we may not know the function of much of our genome does not mean that it has none. (Maybe it is junk, but it is certainly not all transcriptionally inert. A series of recent discoveries has revealed many new types of RNA molecules, mostly involved in control processes. It would be naïve to doubt that many more types will come to light.) Moreover, the amount of space between genes affects the rate of crossing over and recombination and, thereby, rates of evolution. Indeed, the large amount of repetitive sequence between our genes enhances recombination rates by promoting homologous recombination. Rate of evolutionary change is a characteristic of a species that is certainly subject to selective pressure. Features of the genome that affect rate of evolution cannot be dismissed entirely as junk.

If genome size *per se* does not single out humans, what about numbers of genes? Again there is a general correlation between complexity of organism and estimated numbers of genes. Viral genomes encode only a few proteins. Prokaryote genomes contain hundreds or thousands of genes. The simple eukaryote yeast has almost 6000 genes, fewer than twice as many as *E. coli*. Metazoa have tens of thousands of genes.

However, within groups of related organisms, including vertebrates, there is no simple correlation between apparent complexity of organism, or even genome size, and numbers of genes (see Table 4.3). Two vertebrates, the puffer fish and humans, appear to have roughly the same number of genes but differ by almost an order of magnitude in genome size. It was also unexpected to find that the worm *C. elegans* appears to have more genes than the fruit fly.

- Even taking alternative splicing and RNA editing into account, these figures give only a static idea of proteome complexity. Cells control gene expression patterns by complex and dynamic regulatory networks. Conclusion: it is difficult to correlate numbers of expressed genes with organismal complexity if one has no good way of measuring either.

The phenomena of alternative splicing and RNA editing show the situation to be more complicated than simple gene estimates make it appear. This is one reason why it has been difficult to get an accurate count of the number of genes in humans and other higher organisms. In eukaryotes, estimates of gene number refer to maximal sets of exons in units that are coordinately transcribed and translated. In fact, variation in splicing may create many proteins from each gene. As an extreme example, in the mammalian immune system, billions of distinct antibodies arise from regions in the genome containing fewer than ~100 exons. (The immune system is special: splicing occurs at the DNA, not the RNA, level.)

RNA editing is the alteration of bases in mRNA, after transcription. The changes are usually either C→U or A→I (I = inosine, has the coding properties of G). If only some mRNA from the same gene is edited, an extra degree of variability in the proteins arises. Investigation of RNA editing is a relatively new field, and many more implications of the process in health and disease remain to be revealed. However, it is known that defective RNA editing

Table 4.3 Distribution of genome sizes and gene densities

Species	Genome size (Mb)	Coding (%)	Approximate number of genes	Estimated gene density (kb/gene)
E. coli	4.64	88	4 485	1.03
Yeast	12.5	70	6 000	2.1
Puffer fish	365	15	23 000	10
A. thaliana	115	29	23 000	6
Human	3289	1.3	23 000	143

contributes to the pathology of sporadic amyotrophic lateral sclerosis (a neurodegenerative disease, of which the most famous sufferers have been Lou Gehrig and Stephen Hawking.)

The conclusion is that it is very different to estimate the size – to say nothing of the complexity – of a eukaryote's proteome from its genome.

The basis of the complexity of expression patterns, metabolic activity, and indeed all other phenotypic features is the organization of the genome itself. Different types of organism have experimented with different solutions of the problems of packaging long, narrow strands of DNA and of controlling access of transcriptional machinery to different regions.

Viral genomes

Viruses infect cells using specialized proteins on cell surfaces that effect attachment and invasion. Viral nucleic acid enters the host cell. In some cases, viral proteins required for replication also enter the host cell. Once inside the host cell, the invading viral molecules must (1) make multiple copies of the viral genome; (2) synthesize viral proteins, including enzymes active only within the host cell, and coat proteins (and others) to be assembled into the progeny virions; and then (3) 'pack up and leave'.

Viral genomes contain only relatively short stretches of nucleic acids. Some, such as the virus that causes hepatitis C, encode a single polyprotein, cleavage of which produces the few proteins the virus needs to take over the cell. Other viruses, such as human immunodeficiency virus type 1 (HIV-1), contain several genes. The HIV-1 genome is about 9.8 kb long, containing a total of nine genes (see Figure 4.5 and Box 4.4). One gene encodes the Gag–Pol fusion protein, which is cleaved to release Gag (the HIV-1 protease), reverse transcriptase, and integrase. Other mRNAs expressed by HIV-1 contain introns and are spliced to express Rev and Tat.

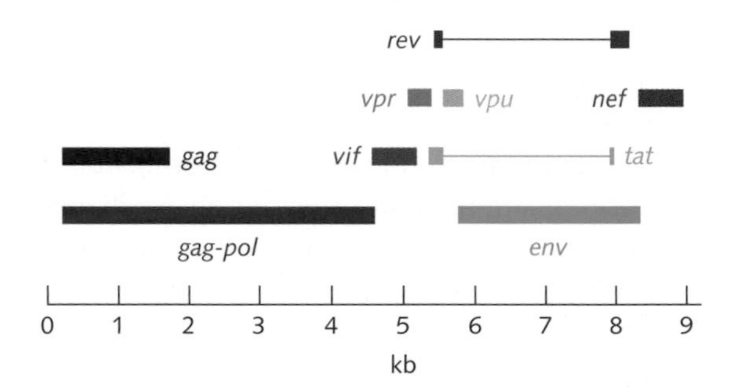

Figure 4.5 Diagram showing the sizes of the individual gene transcripts of HIV-1. The introns of the *rev* and *tat* genes are indicated by a thin line. The proteins encoded by these genes are:

Gene	Proteins	Function
gag	p24, p6, p7, p17	Structural proteins of capsid and matrix
pol	Reverse transcriptase, integrase, protease	Integration into host genome and cleavage of viral-encoded polyproteins
env	Precursors of gp120, gp41	Envelope proteins, active in attachment and fusion to host cells
tat	Tat	Facilitates transcription of viral RNA
rev	Rev	Enhances cytoplasmic export of transcripts
nef	Nef	Interferes with host immune function
vif	Vif	Interferes with host defence
vpr	Vpr	Needed for nuclear import of viral nucleic acid
vpu	Vpu	Promotes assembly and release of progeny virus; also stimulates degradation of host CD4 proteins, disabling the host immune system

Recombinant viruses

Mixed infections of a cell by different viruses permit genetic recombination. It is even possible to package *unaltered* nucleic acid from one strain into an envelope composed of protein from another strain. In that event, the absorption–penetration–surface antigenicity characters are those of the coat proteins, but the hereditary characteristics are those of the nucleic acid. (Infection with such a virus is a kind of natural Hershey–Chase experiment; see p. 81.) These effects can alter host specificity.

Types of viral genome

As assembled within the virion, a viral genome may consist of:

Nucleic acid	*Examples*
• Single-stranded DNA	Bacteriophages ϕX–174 and M13
• Double-stranded DNA	Adenoviruses, smallpox virus, Epstein–Barr virus, bacteriophage λ
• Single-stranded RNA	Bacteriophages MS2, Qβ, tobacco mosaic virus, HIV-1
• Double-stranded RNA	Bluetongue virus

Single-stranded DNA viral genomes are generally converted to double-stranded DNA by the host. Replication of RNA viruses is prone to mutation because the error-correction mechanisms active in host DNA replication do not apply. This helps viruses to evade host immune systems and facilitates their jumping between host species. (Emerging viral diseases, including but not limited to acquired immunodeficiency syndrome (AIDS) and avian flu, are usually based on viruses with RNA genomes.)

A viral genome consisting of single-stranded RNA can be:

(+)sense = same sequence as protein-translatable mRNA
(–)sense = complementary sequence to mRNA
ambisense = mixture of both.

Inside the cell, (+)sense viral RNAs present themselves as messenger RNA (mRNA) and are translated. (–)Sense viral RNAs and double-stranded viral RNAs require specialized polymerases for conversion to mRNA. Retroviral genomes contain (+)sense RNA, which is reverse transcribed into host DNA. These viral polymerases and reverse transcriptases are proteins that are contained in the infecting virion and enter the host cell along with the viral nucleic acid.

Some viral genomes are infectious on their own. For some RNA viruses, a DNA reverse transcript of the viral RNA is infectious (although at a lower rate than the natural virion). This permits preparation of large quantities of viral genomes for vaccines, avoiding the lability and high mutation rate of viral RNA replication.

- Both HIV-1 and influenza viruses have become major threats to human health after jumping from animal hosts. Their high mutation rates – with severe clinical consequences – reflect the fact that their genomes are RNA.

In the laboratory, a virus can be constructed as a vector to produce foreign proteins inside a cell. Two applications of this technique are as follows.

1. To produce a vaccine, insert a DNA sequence coding for the immunogen (perhaps the HIV-1 surface glycoprotein gp120) into the vaccinia virus genome. (The HIV-1 surface protein itself is of course not infectious: the Hershey–Chase experiment again!)* Infection by recombinant virus leads to expression of immunogen and elicitation of an immune response, giving the host protective immunity. The immunity created by such an *intracellular* exposure to the immunogen is much more powerful than that achievable simply by injecting the immunogen into the bloodstream.
2. A recombinant virus carrying a normal human gene can be useful for gene therapy. Can a retroviral vector reintroduce the normal variant into the patient's genome? A recent successful application has been the treatment of adrenoleukodystrophy (an inherited degenerative disorder leading to progressive brain damage, adrenal gland failure, and early death) by gene delivery using a virus derived from HIV. Treatments for many other diseases are at various stages of research. For instance, cystic fibrosis arises from a mutation in the cystic fibrosis transmembrane regulator (*CFTR*) gene. Gene therapy for cystic fibrosis is now in Phase I clinical trials.

* But HIV-1 does insert protein as well as RNA into the host cell. Fortunately for Hershey, Chase, and the field, the virus they worked with, bacteriophage T2, does not.

Influenza: a past and current threat

Influenza is a contagious disease caused by a virus that infects the respiratory tract. The virus is passed around a population in droplets created when an infected individual coughs or sneezes. Unlike HIV-1, the virus that causes AIDS, influenza virus can survive outside the host, greatly facilitating its transmission.

Every year, influenza seasonally affects many people worldwide. In a typical year, 38 000 people die in the USA from influenza or related complications, and 200 000 are hospitalized. The mortality rate is 0.8%, with most fatalities occurring in the very young or elderly. Worldwide, the usual annual fatality rate is 1–1.5 million.

However, in some years, influenza and associated complications attack more viciously. A famous pandemic occurred at the end of the First World War: within an 18 month period in 1918–1919, influenza killed an estimated 50–100 million people, far more than had died in the war. The mortality rate was 1% of those affected. Such an influenza pandemic today could have an even higher mortality in regions of the world containing many people immunocompromised by AIDS.

In most influenza seasons, fatalities occur as a result of bacterial infection of lungs weakened by the virus, to which the elderly are more vulnerable. The 1918–1919 epidemic was different, in the higher percentage of fatalities among *young* people. Several explanations have been offered including: (1) the high density of young soldiers in military camps and battlefields, leading to more effective transmission; (2) overcrowding and poor nutrition and health care among refugees; and (3) previous epidemics in the 1850s and 1889, leaving many elderly people with some immunity.

Three types of influenza virus are known, of which type A is the most dangerous. The virion contains a spherical lipoprotein coat enclosing eight nucleoproteins containing the RNA genome, encoding a total of ten proteins. Protruding from the envelope are several hundred 'spikes' containing the proteins haemagglutinin (80% of the spikes) and neuraminidase (about 20%) (see Figure 4.6). These proteins are essential to the reproduction of the virus. Haemagglutinin binds to host cell surface glycoprotein receptors to promote viral entry into cells.

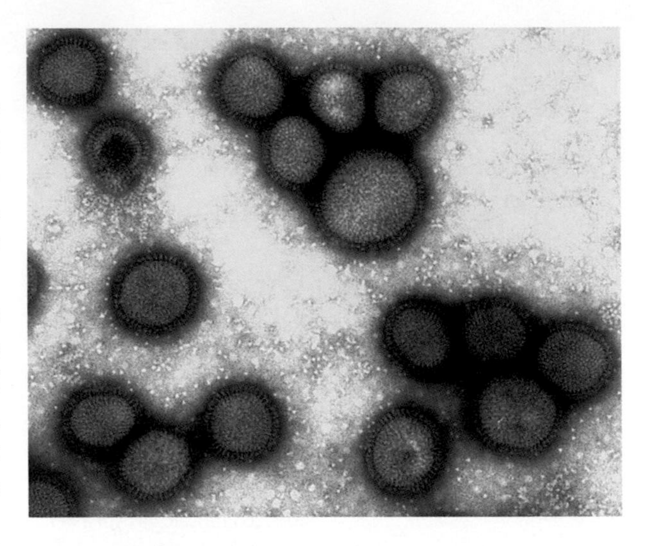

Figure 4.6 Influenza virus.

Picture courtesy Professor Y. Kawaoka, University of Wisconsin, USA, and University of Tokyo, Japan.

Neuraminidase helps progeny virions to get out. Both haemagglutinin and neuraminidase are targets of drugs.

The virus can evolve by point mutations, also called antigenic drift, or by genetic recombination. Immunologically distinct strains of viruses are called serotypes. Different serotypes vary both in the ease of their spreading and in the mortality of infection. For instance, binding of virus to mucosal respiratory surfaces may be affected by amino acid sequence polymorphisms of both viral proteins and host receptors. Contagion and mortality are also dependent on characteristics of the host population, including density and general health levels. A strain that is both highly contagious and has a high mortality rate would be very dangerous indeed. As part of an effort to understand why the 1918–1919 strain was so dangerous, scientists have recently reconstructed that virus, based on material recovered from contemporary postmortem specimens.

Different strains of influenza virus contain 1 of 13 recognized types of gene for haemagglutinin (H) and one of nine recognized types of gene for neuraminidase (N). These types identify different major strains of viruses. For instance, the strain that caused the 1918–1919 pandemic was H1N1.

Many strains of the virus infect only a restricted range of species and with different mortality rates.

These properties can change as the virus evolves. A strain that infects animals can potentially become infectious to humans. Species range depends on the different forms of sialic acid presented on viral glycoproteins. An important determinant is haemagglutinin residue 226, which is Gln in viruses infectious to birds and Leu in viruses infectious to humans.

Avian flu

In 2006, an H5N1 strain of avian flu characterized by very high mortality infected domestic poultry in several countries. It is considered a particularly dangerous threat to humans because it has a high mutation rate and recombines readily. One way for the virus to jump from birds to humans is for two strains to co-infect pigs and use them as a 'mixing vessel' for recombination.

Avian flu can normally infect only birds and, in some cases, pigs. Domestic poultry stocks raised under conditions of very high population density are particularly vulnerable. Often migratory birds are carriers but do not get sick. (The 2006 avian flu strain was spread from southeast Asia to Russia by migratory birds.)

The H5N1 strain prevalent in 2006 was first identified in Hong Kong in 1997 and traced to ducks from Guandong province. It jumped the species barrier to mammals, becoming infectious to pigs, in April 2004. It then became supervirulent, killing rodents, birds, and humans. This H5N1 strain is 100% fatal in domesticated chickens and in 54% of reported human cases. Human to human transmission is uncommon.

Compared with previous epidemics, the world today is particularly vulnerable because of:

- increased human population densities;
- widespread long-distance travel;
- intensive livestock production (including antibiotic feeding, which may create drug-resistant strains of infectious bacteria); and
- policies of various governments that do not adequately reimburse farmers who must sacrifice animals, creating a disincentive to report disease; the result is a delay or even default of an effective response.

Table 4.4 Population in China (millions)

Year	Humans	Pigs	Poultry
1968	790	5.2	12.3
2005	1300	508	13 000

Increased population densities of both humans and animals threaten a greater rate of spread of a dangerous strain of virus, even in comparison with the recent 1968–1969 epidemic (see Table 4.4).

Aggressive approaches to controlling avian flu have involved large-scale culling of stocks. In 1997, the H5N1 strain infected poultry in Hong Kong and caused six human fatalities. The entire poultry population of the island had to be destroyed: 1.5 million birds in three days. An H7N7 epidemic in the Netherlands in 2003 led to the killing of >30 million birds (approximately twice the human population of the country). In Asia in 2004, over 100 million birds were culled.

Drugs against influenza

Tamiflu (oseltamivir) and Relenza (zanamivir) are the two major drugs against influenza. Both are inhibitors of the viral neuraminidase.

Relenza (Figure 4.7) was designed at the Commonwealth Scientific and Industrial Research Organization (CSIRO) laboratory in Melbourne, Australia. Crystal structures of influenza neuraminidase showed that conserved sequences formed a cavity, suggesting a target site for drugs. By targeting the active site of the enzyme, it is harder for the virus to evolve resistance.

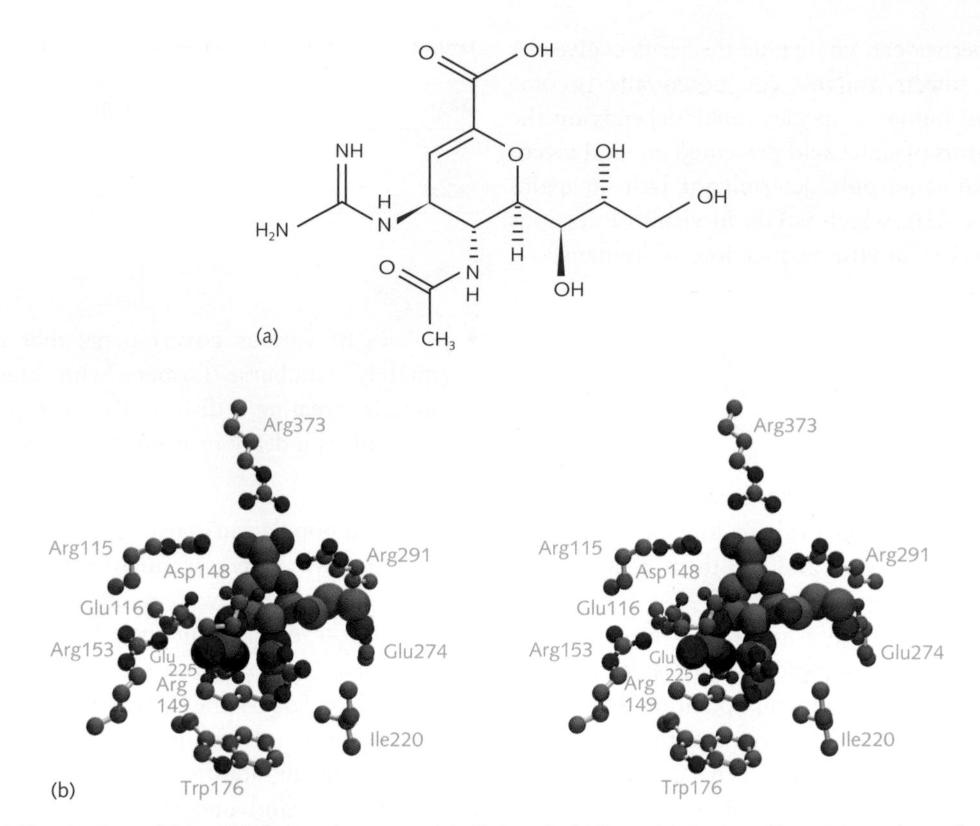

Figure 4.7 (a) The structure of the anti-influenza drug zanamivir (Relenza). (b) Zanamivir is a transition-state analogue that binds to the active site of influenza neuraminidase. Here the atoms of the drug are shown as large spheres and the residues from the neuraminidase are shown in ball-and-stick representation in stereo.

Ethical dilemma: publication of RNA sequence of the virulent 1918–1919 strain of influenza virus

Recently, scientists were able to recover and sequence the strain of influenza active in the 1918–1919 pandemic. The journal *Science* published the work and, consistent with editorial policy, required that the sequence be deposited in the nucleic acid databanks.

In *The New York Times* on 17 September 2005, R. Kurzweil and W. Joy wrote an article critical of the decision to make the sequence generally available in databanks on the grounds that terrorists might use the information to recreate the virus and use it as weapon.

Reactions to the publication of the reconstructed pandemic viral sequence illustrate the conflict between the recognition that free and open access to information is a benefit to the progress of science, and the dangers of its misuse. A precedent occurred before the Second World War when physicist Leo Szilard tried, unsuccessfully, to persuade colleagues not to publish results that might prove useful in the development of atomic weapons. He suggested that journals record dates of receipt and acceptance of manuscripts but then sequester the articles for the duration. This occurred well before the strict secrecy imposed after the Manhattan Project was organized.

Science did make an exception to its mandatory-deposition policy in publishing the Human Genome Draft Sequence by J.C. Venter and co-workers in 2001.

Genome organization in prokaryotes

A typical prokaryotic genome has the form of a single circular molecule of double-stranded DNA, between 0.6 and 10 million bp long. For instance, a cell of *E. coli* strain K12 contains a single molecule of double-stranded DNA 4 639 675 bp long, closed into a circle. The DNA is supercoiled and associated with histone-like proteins into a 'chromosome', appearing in a subcellular structure called the nucleoid. Some *E. coli* cells may contain plasmids: short, usually circular, double-stranded DNA molecules, ranging from 1 kb to several megabases in length.

Although single circular genomes containing most of the DNA are common in bacteria and archaea, many exceptions are known. Many prokaryotic cells contain plasmids. Some prokaryotes have linear DNA. *Borrelia burgdorferi*, the organism that causes Lyme disease, is an example. *B. burgdorferi* also contains numerous plasmids, some of which are circular and some linear. Other prokaryotes contain more than one chromosome. *Vibrio cholerae*, the organism that causes cholera, contains two circular DNA molecules of 2 961 146 and 1 072 314 bp.

Some but not all prokaryote genomes contain **insertion sequences**, mobile genetic elements similar to eukaryotic transposons.

The 4.6 Mb chromosome of *E. coli* encodes approximately 4500 genes, distributed on both strands. The absence of introns and the shorter intergenic regions account for the higher coding densities. A very large fraction of the DNA, 87.8%, codes for proteins, 0.8% codes for structural RNAs and only 0.7% has no known function (see Table 4.5 and Figure 4.8).

Many prokaryotic genomes have been sequenced. They illuminate, in a somewhat simpler context than the human genome, how these organisms solve problems common to all cellular life forms. In addition, there are good practical motives for studying features of prokaryotes. Differences between prokaryotic and eukaryotic metabolism – enzymes unique to prokaryotes – are appropriate targets for drugs against infection. Of great importance to clinical medicine is understanding how prokaryotes evolve to develop pathogenicity and antibiotic resistance.

Table 4.5 Coding percentage and average gene density

Species	Coding	Average gene density
E. coli	>90%	1 gene/kb
Pufferfish	15%	1 gene/10 kb
Human	5%	1 gene/30 kb

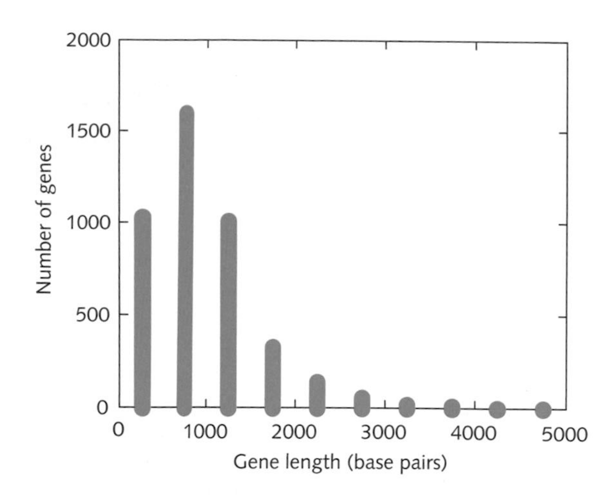

Figure 4.8 Distribution of gene lengths in *E. coli*. Two very long genes for hypothetical proteins, *yeeJ* and *ydbA*, of length 7152 and 8619 bp, respectively, are omitted. The average gene length is 960 bp. Most genes are less than 1500 bp long.

We visualize the contents of bacterial chromosomes as concentric circular diagrams, looking vaguely like 'tie-dyed' patterns (see Figure 4.9).

Replication and transcription

In *E. coli*, replication begins at a specific site called *oriC* and proceeds in both directions. This site is the calibration point from which the genome is indexed. Replication ends at the *terC* site, found almost, but not exactly, half way around the circle. (In contrast, archaea often have multiple sites of origin of replication.)

In prokaryotes, many mRNA transcripts contain several tandem genes, which require separate initiation of translation. (In this, they are unlike viral polyproteins, which are translated in one piece and then cleaved.) In bacteria, but less frequently in archaea,

(a)

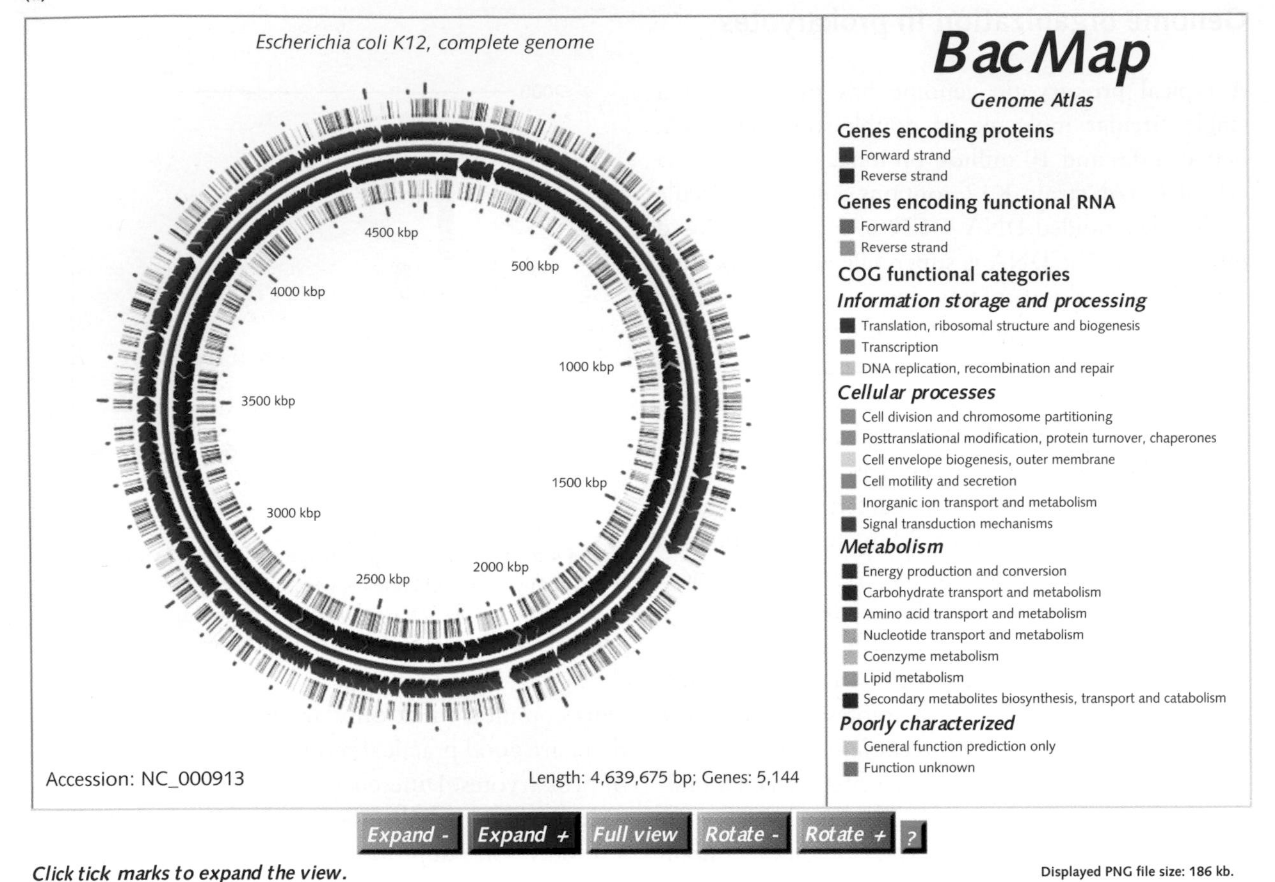

Figure 4.9 Map of the genome of *E. coli* K12. (a) Full view. Red arrows show protein-coding regions of the forward strand. Blue arrows show protein-coding regions of the reverse strand. Pink arrows show structural RNA-encoding regions of the forward strand. Cyan arrows show structural RNA-encoding regions of the reverse strand. Radial ticks identify individual gene products, colour coded according to function. COG categories refer to the Clusters of Orthologous Groups database (http://www.ncbi.nlm.nih.gov/COG/). (b) Expanded view of the region containing the *his* operon. The BacMap site provides access to genomes of bacteria and archaea (http://wishart.biology.ualberta.ca/BacMap/).

Pictures reproduced, by permission, from BacMap: An Interactive Atlas for Exploring Bacterial Genomes. See: Stothard, P., Van Domselaar, G., Shrivastava, S, Guo, A., O'Neill, B., Cruz, J., Ellison, M., & Wishart, D.S. (2005). BacMap: an interactive picture atlas of annotated bacterial genomes. Nucl. Acids Res. **33**, D317–D320.

co-transcribed genes have related functions, forming an 'operon'.

Timing illuminates the interrelationships among these processes. Under ordinary conditions, it takes *E. coli* 40 minutes to replicate its genome. The full generation time between cell divisions is about an hour. This explains why genes that require high rates of expression tend to be near the origin of replication: the availability for transcription of partially replicated DNA in effect increases the copy number of such genes. Conversely, the half-life of mRNA is only a few minutes. Therefore, translation must overlap transcription.

Gene transfer

There are three methods of transfer of DNA between prokaryotic cells.

(b)

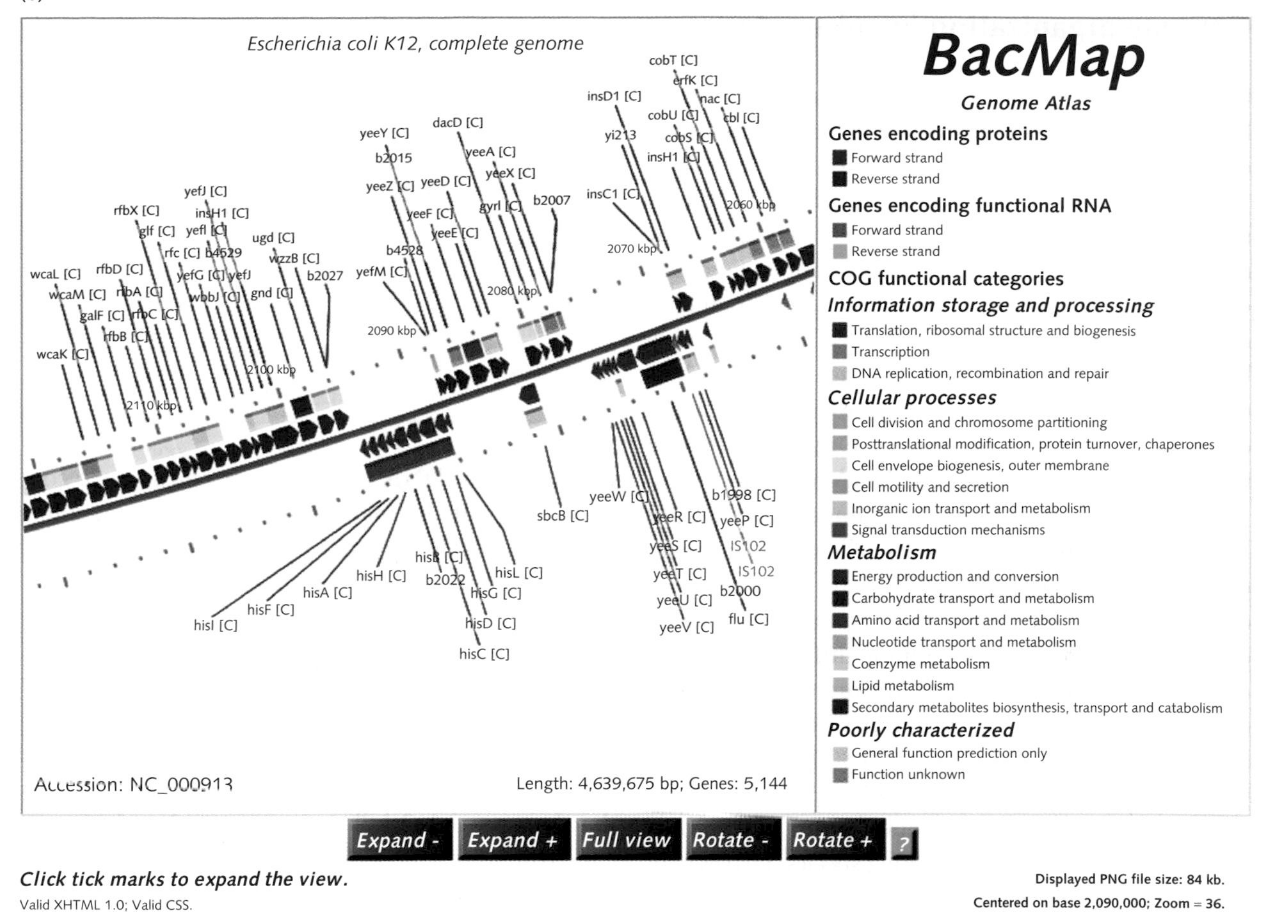

Figure 4.9 *(continued)*

- *Transformation.* The uptake of 'naked' DNA, as in the experiments of Griffith and of Avery, MacLeod, and McCarthy.
- *Conjugation.* Insertion of some or all of the DNA from one cell into another – the prokaryotic equivalent of 'mating', although there is no meiosis or zygote formation. Bacterial conjugation does permit formation of recombinants. The start point for DNA transfer varies with the position in the genome of a mobile site. This is *not* the same as the origin of replication, *oriC*.
- *Transduction.* Transfer of DNA from one cell to another via a bacteriophage. During replication in one cell, a phage can pick up fragments of bacterial DNA and transmit it to another cell subsequently infected by progeny virions.

Bacterial conjugation has proved very useful in genome mapping. Transfer of a complete genome takes 100 minutes. Interrupting the process at different times (by physical agitation) results in partial genome transfer. Identifying which genes have entered the recipient cell after different intervals revealed the order of the genes. Positions in the genetic map of *E. coli*, for example, were classically expressed in minutes. Now, of course, they are specified in terms of the DNA sequence itself.

- Prokaryotes have several mechanisms for sharing genetic material : transformation by naked DNA, conjugation, and transfer via viruses.

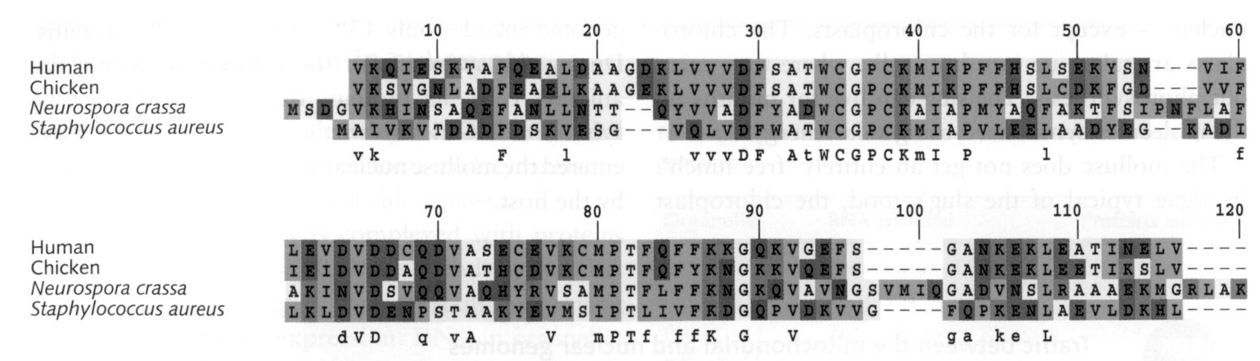

Figure 4.11 Thioredoxins are proteins that catalyse disulphide-exchange reactions, contributing to the speed and accuracy of the protein-folding process. The human thioredoxin gene extends over 13 kb and consists of five exons. This figure shows the alignment of amino acid sequences of thioredoxins from two vertebrates (human and chicken), a fungus (*Neurospora crassa*), and a bacterium (*Staphylococcus aureus*). Colour coding: green, amino acids with medium-sized and large hydrophobic side chains; yellow, small side chains; magenta, polar side chains; blue, positively charged side chains; red, negatively charged side chains. Upper-case letters in black on the line below the sequences indicate amino acids conserved in all four sequences. Lower-case letters in the line below the sequences indicate amino acids conserved in three of the four sequences.

mostly in the form of single-site mutations or insertions and deletions. Typically, there is reasonable correlation between overall species divergence and divergence of sequences of individual genes and the corresponding proteins. Comparisons of amino acid and gene sequences of thioredoxins provide a typical example (see Figure 4.11).

Compare these protein sequences with the corresponding gene sequences from human, chicken, and *Staphylococcus aureus* (see Figure 4.12). Note the large gap in the bacterial gene corresponding to the intron in the human and chicken genes. The asterisks under the sequences indicate positions containing the same base in all three genomes. The colons indicate positions containing two identical bases among the three; in most cases, two common bases appear in the human and chicken sequences, even in the non-coding regions. Note the frequent occurrence of patterns '**:' and '**-blank'. What is the likely reason for this?

- In most cases, divergence of the sequences of genes and proteins correlates well with the divergence of the species.

Duplications

Duplications of individual genes, of regions containing many genes, and of complete genomes have been an important mechanism of evolution. They are a prolific source of variation, the raw material of both selection and genetic drift.

BOX 4.6

What can happen to a gene?

During evolution:

1. A gene may pass to descendants, accumulating favourable (or unfavourable) mutations or drifting neutrally.
2. A gene may be lost.
3. A gene may be duplicated, followed by divergence or by loss of one of the pair.
4. A gene may undergo horizontal transfer to an organism of another species.
5. A gene may undergo complex patterns of fusion, fission, or rearrangement, perhaps involving regions encoding individual protein domains.

Duplications are seen in archaea, bacteria, and eukaryotes. Estimates of amounts of duplication vary, but there is agreement that it is substantial. The *A. thaliana* genome, for instance, contains over 60% duplications.

```
Human      actgcttttcaggaagccttggacgctgcaggtgataaacttgtagtagttgacttctca
Chicken    gctgattttgaggcagaactgaaagctgctggtgagaagcttgtagtagttgatttctct
S.aureus   gcagatttgattcaaaagtagaatctggtgtacaa---------ttagtagattttgg
            *:* **** *:: *:   *: * :***: *:::* ::  :::::::****:** **:*:

Human      gccacgtggtgtgggccttgcaaaatgatcaagcctttctttcat---------------
Chicken    gccacatggtgtggaccatgtaaaatgatcaagccatttttccatgtaagtagcctttgt
S.aureus   gcaacatggtgtggtccatgtaaaatgatcgctccggtattagaa---------------
           **:** ******** ** ** *********:::** :* ** :*:

Human      --------------gtgagtattaaacaatgtctgctttgtaagagatttgtgtttttg
Chicken    ttttcacagtaacagtaagtat-acacaaatacttctgtgcaacttgtcagtaatatg-g
S.aureus   ------------------------------------------------------------
                         :: ::::: : ::::    :: :: :: ::   :  ::  : :  :

Human      agttggtggtcacagtggtaggaaagaaagacagtt----aaaggatttggtttcggtg
Chicken    aggaaacctcctttgtctgtggggtggatggtatttcttgaaggagaatttgtagaagta
S.aureus   ------------------------------------------------------------
           ::        :   ::    ::   : : :  : ::    :: :    :: ::     ::

Human      gg-----gggatttctttggctccatctttggtctaaaagtagtagtataacaaataatt
Chicken    tgtgattggtaactattgaataaagtacttggatacacagcagggaacagacatgctgtt
S.aureus   ------------------------------------------------------------
            :     :: :  : ::        :  ::::    : :: ::       :::     ::

Human      taggtttgatacatgtagcccattgaaa-acaaattttagaagttaattttgtcttaaat
Chicken    gcatttgtctgtcctggtgctctgatgcacagtctgtaggtacctagcttccctcaaga
S.aureus   ------------------------------------------------------------
               ::::       : :  :  :::   :::   : :::      :  ::  :: ::

Human      agttctttttttccccacattgaaaca-----tgggcctta--tttgaaatcccagccta
Chicken    a---ctggtaacagtggagttgaaacagtgtgtgtacactggctctgattattaaaactg
S.aureus   ------------------------------------------------------------
           :   ::  :          ::::::::     ::  :  :   : :::      :  ::

Human      gaatttgatatgccaaactgtttt---atactaa--gaaaaatttgatttagagaaaatt
Chicken    cattcagataagctgtagcatctttctgtagtgatggggaggctgtaagggaaggaaagg
S.aureus   ------------------------------------------------------------
            : :  :::: ::   :   : ::     :: : : :   :  :  :     :: :::

Human      tatgtctcttagatctatgt-ctccaaaaga----tctaaattttggatctttaattag
Chicken    cctttcccttgtgcttaggtgcttcagcagactcttccaggggatggagctgaaaattaa
S.aureus   ------------------------------------------------------------
             : :: :::     :: :: :: ::  :::    :: :     : :        :::::

Human      tctctagttttattaagtttccatttaagaagcttaagcttgggtatgttgcattgccat
Chicken    tttctggtca-gtaaagtagctgctttaaggtaacgagtcaag---tctcacagcaccag
S.aureus   ------------------------------------------------------------
           : ::: ::    : ::::  :   :: :         ::   :   : :  ::    :::

Human      tacctagttctaaatctttt-------tggattttcatttaaattttccag-------
Chicken    tttctatgaatgcatcttttaaagaagtggctttcctggagcagtaactacataattttg
S.aureus   ------------------------------------------------------------
           :  :::    :  :::::::       ::: :::        :     : ::

Human      -------------tccctctctgaaaagtattccaac---gtgatattccttgaagtaga
Chicken    tttttcattctagagtctgtgtgacaagtttggtgat---gtggtgttcattgaaattga
S.aureus   -------------gaattagcagctgactatgaaggtaaagctgacattttaaaattaga
                           :* : :*: :*:* *    : :::*:: : :*: *::** * **

Human      tgt ggatgactgtcag
Chicken    tgt ggatgatgcccag
S.aureus   tgttgatgaaaatcca
           ***:*::::*    **:
```

Figure 4.12 Alignment of partial thioredoxin gene sequences from the genomes of human, chicken, and *Staphylococcus aureus*. The region shown contains exons 2 and 3 from the vertebrate genes.

Table 4.7 Percentages of duplicated genes

Species	Duplicate genes (%)
Bacteria	
Mycoplasma pneumoniae	44
Helicobacter pylori	17
Haemophilus influenzae	17
Archaea	
Archaeoglobus fulgidus	30
Eukarya	
Saccharomyces cerevisiae	30
Caenorhabditis elegans	49
Drosophila melanogaster	41
Arabidopsis thaliana	65
Homo sapiens	38

From: Zhang, J. (2003). Evolution by gene duplication: an update. Trends Ecol. Evol. **18**, 292–298.

Duplication of genes

Organisms in all three domains of life show duplication of individual genes.

After duplication, both copies of a gene may survive and diverge. Alternatively, one copy may turn into a pseudogene or be deleted, leaving only one functional copy.

As first proposed by S. Ohno in 1970, duplication followed by divergence is an important source of proteins with novel functions. It is generally easier to 'recruit' and adapt an already active molecule to a new function than to invent a new protein from scratch. The course of evolution of proteins descended from a common ancestor will differ, depending on whether they are retaining or changing their function (see Box 4.7).

> *'Walls supply stones more easily than quarries, and palaces and temples will be demolished to make stables of granite, and cottages of porphyry.' – Johnson,* Rasselas.

In analysing the divergence of related genes, how can we distinguish the effect of selection from genetic drift? Given two aligned gene sequences, we can calculate K_s, the number of synonymous substitutions, and K_a, the number of non-synonymous substitutions. Most but not all synonymous substitutions are changes in the third position of codons. (The calculation of K_a and K_s involves more than simple counting because of the need to estimate and correct for possible multiple changes.) The ratio of K_a/K_s distinguishes the role of selective pressure and drift in the divergence of genes after duplication:

$K_a/K_s \approx 1$ **Neutral evolution**: silent and substitution mutations have occurred to approximately equal extents.

$K_a/K_s \gg 1$ **Positive selection**: substitution mutations are more prevalent than silent mutations, implying that selective pressures are active and the substitutions are advantageous.

$K_a/K_s \ll 1$ **Purifying selection**: substitution mutations are underrepresented, implying that the sequence is optimized fairly rigidly, with relatively little tolerance for mutation.

BOX 4.7

Homologues, orthologues, and paralogues

Homologues are regions of genomes, or portions of proteins, that are derived from a common ancestor. Because only in rare cases can we actually observe the ancestor–descendant relationship, most assignments of homology are inferences from similarity in sequence, structure, and/or genomic context.

Paralogues are related genes that have diverged to provide *separate* functions in the *same* species. **Orthologues**, in contrast, are homologues that perform the *same* function in *different* species. (For instance, the α and β chains of human haemoglobin are paralogues, and human and horse myoglobin are orthologues.)

Other related sequences may be pseudogenes, which may have arisen by duplication or by retrotransposition from mRNA, followed by the accumulation of mutations to the point of loss of function or expression.

- A common mechanism of evolution is the duplication, of a gene corresponding to a protein subunit, of an entire gene, or even of a whole genome. Creating two copies of any of these entities means that one can continue to provide an essential function, while the other can diverge to explore other possibilities.

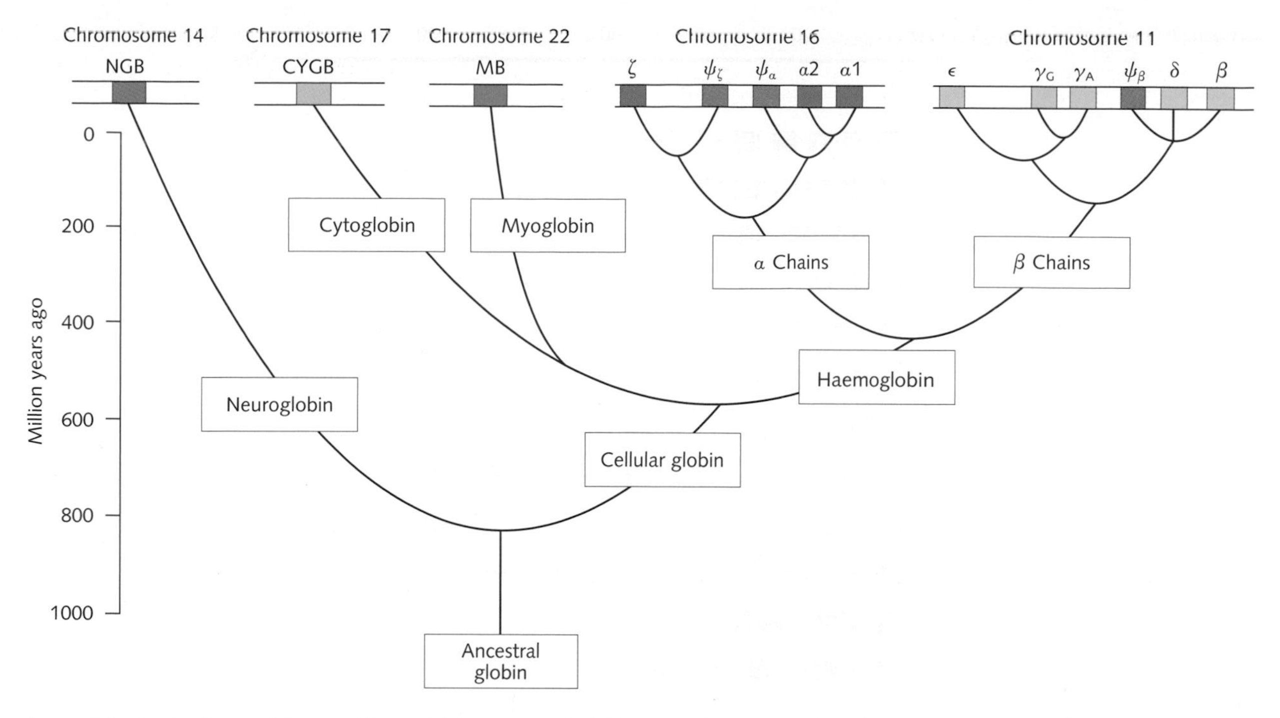

Figure 4.13 Duplication and dispersal through the genome of globin genes during animal evolution.

From: Burmester, T., Ebner, B., Weich, B., & Hankeln, T. (2002). Cytoglobin: a novel globin type ubiquitously expressed in vertebrate tissues. Mol. Biol. Evol. **19**, 416–421.

The many globins in the human genome provide a good example of gene duplication and divergence (see Figure 1.11). Genes for several versions of the haemoglobin α and β chains form clusters on chromosomes 16 and 11. Other, isolated, loci contain genes for neuroglobin, cytoglobin, and myoglobin. Closely linked genes, as in the α- and β-globin regions, suggest relatively recent divergence. Yet even within these clusters, the proteins encoded have diverged in function, showing small but significant variations in oxygen affinity and responses to allosteric effectors. Also, these proteins appear at different stages of our development, implying divergence in the control of their expression.

We can date the globin duplications by looking back into evolutionary history (see Figure 4.13). Neuroglobin split off from other globins before the last common ancestor of the vertebrates, perhaps 10^9 years ago. The divergence of myoglobin and cytoglobin from haemoglobin occurred before the emergence of the jawless fishes, during the Cambrian about 500 million years ago. The divergence of α- and β-globins occurred early in the vertebrate lineage, approximately 450 million years ago.

Within mammals, the α- and β-globin regions are quite variable in content and extent (see Figure 4.14). Even within the primate lineage, we can date a duplication of the γ-globin gene (see Figure 4.15).

Duplication can affect individual exons

Fibronectin, a large extracellular protein involved in cell adhesion and migration, is a **modular protein** (see Box 4.8) containing multiple tandem repeats of

Modular proteins

A **modular protein** contains a linear string of compact units, called **domains**. Domains appear to have independent stability and can be 'mixed and matched' with one another in different proteins. Domains sometimes, but by no means always, correspond to single exons.

Modular proteins are common in eukaryotes. Individual domains of eukaryotic modular proteins are often separately homologous to single-domain prokaryotic proteins (and, less-commonly, vice versa).

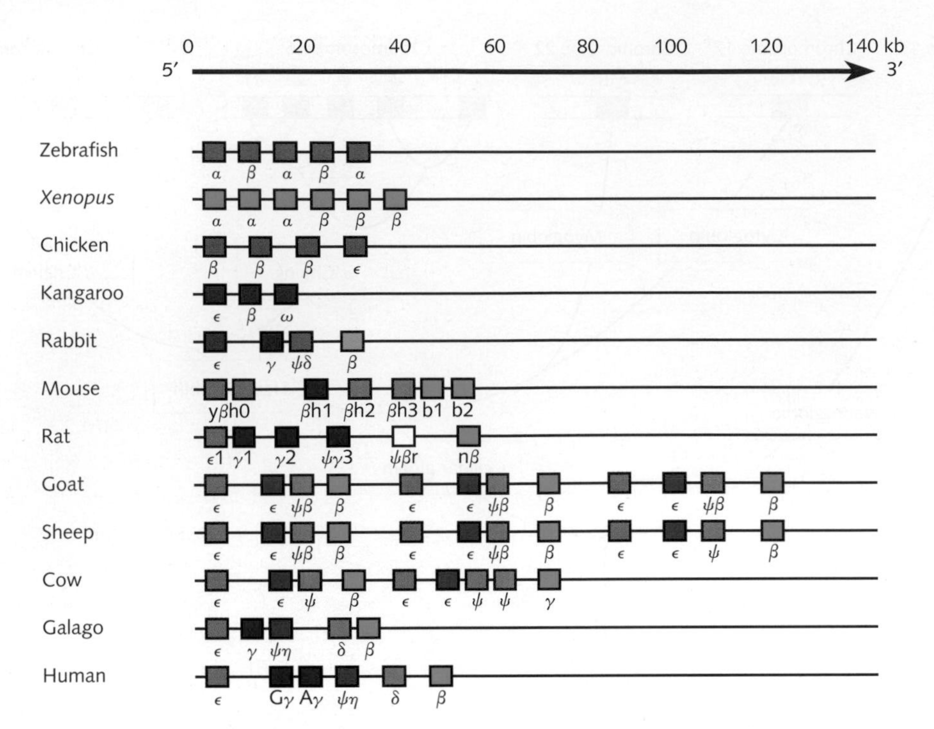

Figure 4.14 Layout of the β-globin locus in selected vertebrates. Colour coding: Light brown, fish; green, amphibian; purple, avian; magenta, marsupial; dark brown, ε-like; dark blue, γ-like; orange, δ-like; cyan, β-like; white, rat-specific pseudogene; red, η-like.

The zebrafish and *Xenopus* regions illustrate the organization of the region prior to the separation of α- and β-globins. They alone of the species illustrated here contain only α- and β-globin genes.

Goat and sheep are closely related species that show similar patterns. Rat and mouse are closely related species that show different patterns.

After: Aguileta, G., Bielawski, J.P., & Yang, Z. (2004). Gene conversion and functional divergence in the β-globin gene family. J. Mol. Evol. **59**, 177–189.

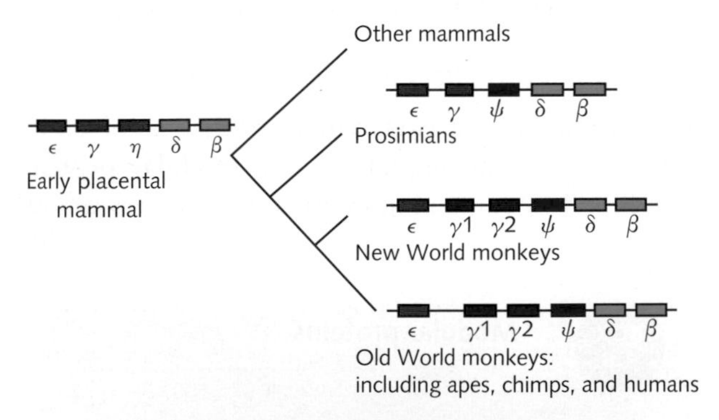

Figure 4.15 Evolution of primate β-globin region (not drawn to scale). Red boxes, embryonically expressed genes; green boxes, post-embryonically expressed genes; blue boxes, foetally expressed genes; black boxes, pseudogenes. Mammals ancestral to the groups in this chart had an embryonic ε-globin and a post-embryonic β-globin gene. Marsupials retain this pair (see Figure 4.14). By the time this diagram begins, the ε gene had duplicated to form three embryonic genes, ε, γ, and η, and the β gene had duplicated to form δ and β. The η gene fell into desuetude, mutating into a pseudogene. Subsequent duplication of the γ gene in anthropoids produced γ1 and γ2, and a change in the control of expression to convert the γ genes from embryonic to foetal expression.

three types of domain called F1, F2, and F3. It is a linear array of the form: $(F1)_6(F2)_2(F1)_3(F3)_{15}(F1)_3$ (see Figure 4.16). In the human genome, each domain of fibronectin is encoded by either one or two tandem exons.

Fibronectin domains also appear in other modular proteins. The duplication of the exon(s) encoding a domain, followed by transfer to another protein, is called 'exon shuffling'.

Family expansion: G-protein-coupled receptors

Repeated duplications can generate large numbers of homologues. G-protein-coupled receptors (GPCRs) are a large superfamily of eukaryotic cell-surface receptors active in signal recognition and processing, including the senses of sight, taste, and smell.

The human genome contains about 700 active GPCRs. They are integral membrane proteins with a common structure comprising seven transmembrane helices (see Figure 4.17). GPCRs interact with

in > 40 locations in mouse and >100 locations in humans. Formation of many of the clusters appears to antedate the divergence of humans and mice, but individual gene duplication and divergence within clusters has led to specialization.

- GPCRs are important in the pharmaceutical industry, for both therapeutic effectiveness and financial reward. About half of all prescription drugs target GPCRs.

Although mammals have of the order of 1000 odorant-receptor proteins, and each neuron in the nasal epithelium expresses only one odorant-receptor allele, over ten times as many scent molecules can be distinguished. How is this achieved? Each scent molecule interacts with several different receptors. Conversely, each receptor protein binds several related scent molecules. Each neuron signals the detection of one *group* of odours, and the brain compares the outputs and performs the required computation. Identification of a specific scent depends on its detection by a *combination* of receptors.

- We depend less on our sense of smell than mice and dogs do, but we have about half the number of expressed odorant receptor proteins. Would you have expected a larger discrepancy?

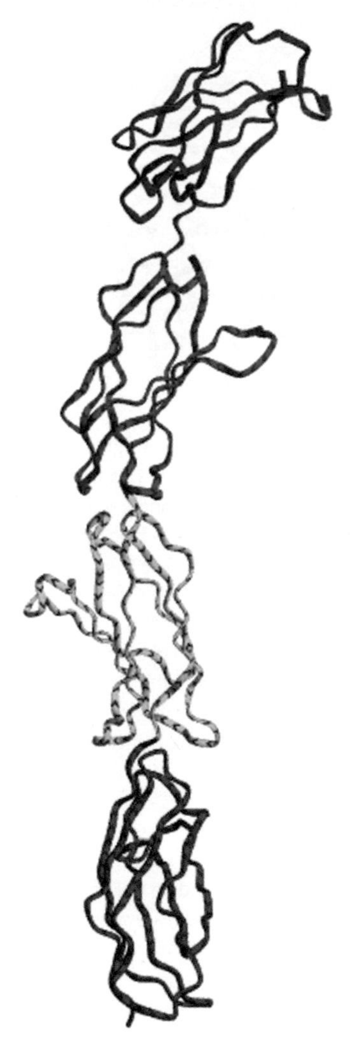

Figure 4.16 A fragment of fibronectin, a modular protein, showing four tandem domains.

G proteins within the cell. Some GPCRs mediate responses to extracellular chemical signals. Reception of the signal triggers *intracellular* signalling cascades, which may reach as far as the nucleus to affect gene expression.

The interaction patterns of large families of functionally related proteins can create great complexity. Odorant receptors are GPCRs expressed on sensory neurons in the nasal cavities of humans and animals. The human genome has about 1000 odorant-receptor genes, of which only 40% are active. The mouse genome has about 1300, of which 80% are active. These genes are distributed around mammalian genomes, arranged in clusters of up to 100 genes,

Large-scale duplications

The genomes of many species contain duplications of multigene regions, the length varying from species to species.

Large-scale segmental duplications are an important component of the difference between human and chimpanzee genomes, affecting about 2.7% of the genome. Some duplications are found in chimpanzees but not humans, some in humans but not chimpanzees, and some in both.

Some duplications that appear in the human but not in the chimpanzee genome involve segments associated with developmental disorders, including a region on human chromosome 15 involved in Prader–Willi and Angelman syndromes (see p. 85). These syndromes arise from microdeletions. In humans, the duplication contributes to the frequency of the

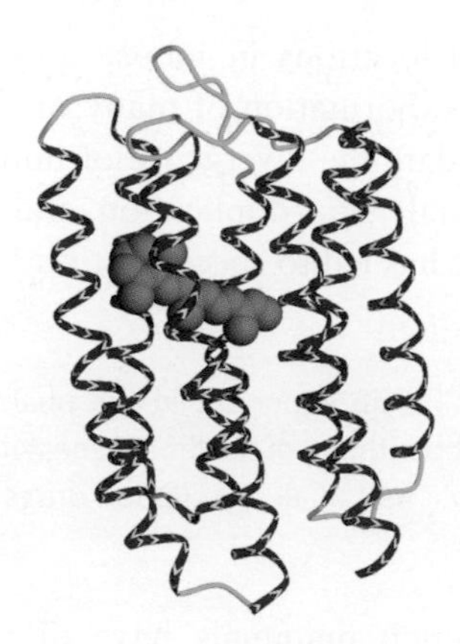

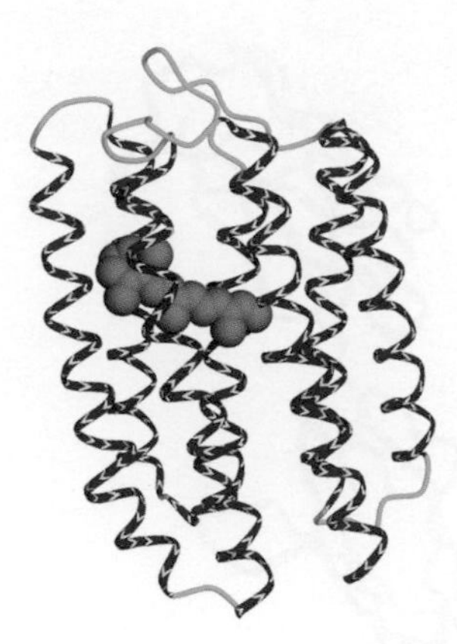

Figure 4.17 G-protein-coupled receptors (GPCRs) are a large family of transmembrane proteins involved in signal transduction into cells. They share a substructure containing seven transmembrane helices, arranged in a common topology. This figure shows the first experimentally determined mammalian GPCR structure, bovine opsin [1H68]. This molecule senses light and generates a nerve impulse.

The seven-helical structure is common to the family of GPCRs. The helices traverse the membrane, with loops protruding outside and inside the cell. This figure shows a view parallel to the membrane, with the extracellular side at the top. The transmembrane region is generally flanked by N- and C-terminal domains. The N-terminal domain is always outside the cell and the C-terminal domain always inside.

GPCRs constitute the largest known family of receptors. The family is as old as the eukaryotes and is large and diverse. Mammalian genomes contain ~1500–2000 GPCRs, accounting for about 3–5% of the genome. A similar fraction of the *C. elegans* genome codes for GPCRs.

Some GPCRs are involved in sensory reception, including vision, smell, and taste. Some, like opsin and bacteriorhodopsin, bind chromophores. (Bacteriorhodopsin is not a signalling molecule but a light-driven proton pump.) Others respond to extracellular ligands including hormones and neurotransmitters.

As expected from the structure, in many groups of GPCRs the sequences of the helical regions diverge less than the sequences of the loops. It is the loops that determine the specificity of the ligand, and of the G-protein partner.

The common mechanism of function of GPCRs is a conformational change, induced by receptor binding or light absorption. The activated state of the GPCR interacts with an intracellular G protein, triggering a signal cascade. As there are substantially more GPCRs than G proteins, many GPCRs must interact with a single G protein. For instance, all odorant receptors interact with the same G protein α-subunit.

GPCRs are the targets for many drugs used in the treatment of high blood pressure, asthma, allergies, and other conditions. The large number of related GPCRs is a challenge to the design of drugs that bind to a unique target. Many drugs have undesired side effects because of imperfect specificity.

disease, by presenting sites for homologous recombination during meiosis, which show up in some of the gametes as deletions. It is, therefore, likely that Prader–Willi and Angelman syndromes are *less* common in chimpanzees than in humans.

Whole-genome duplication

Genomes can duplicate if the chromosomes replicate but do not segregate properly into separate progeny cells upon mitosis.

The mere appearance of two copies of many genes does not prove whole-genome duplication. One must adduce (1) the genome-wide occurrence of pairs of homologous genes *appearing in the same order*; or (2) 'molecular clock' evidence showing equal divergence times in many pairs of homologues. (However, different genes diverge at different rates, and homologues may be under different selective pressures. Clock arguments are, therefore, relatively weak.)

The yeast genome underwent a duplication about 10^8 years ago. The effects are obscured by subsequent chromosomal rearrangements and by massive loss of duplicated material. The duplication has nevertheless left its traces in multiple homologues that retain their genomic order. The yeast genome contains 55 duplicated regions, on average 55 kb long, together covering ~50% of the genome and including 376 pairs of homologous genes.

In a seminal 1970 book, *Evolution by Gene Duplication*, S. Ohno proposed that the vertebrate genome is the product of one or more complete genome duplications. Genome sequences confirm his prescient insight. Nor are whole-genome duplications only limited to vertebrates. They have occurred frequently in plant lineages (see p. 219).

In the lineage leading to vertebrates, the genomes of the cephalochordate Amphioxus (*Branchiostoma floridae*), and the urochordate *Ciona intestinalis*, showed evidence for two rounds of whole genome duplication. Individual genes in these relatives corresponded to multiple genes in vertebrates, with enough synteny preserved to show that the process happened in parallel on a large scale. It appears that two whole gene duplications in the vertebrate line occurred after the split between the primitive chordate relatives, urochordates and cephalochordates, from vertebrates, about 400–600 million years ago. More recently, a third duplication occurred in the lineage leading to ray-finned fishes, such as zebrafish and medaka. (See p. 222.)

What happens to all those extra genes? Most metazoa have roughly the same number of protein-coding genes, in the range 20 000–25 000. Whole-genome duplication is followed by massive gene loss. Some genes do form paralogous groups, and can even duplicate further, individually, creating gene and protein families of various sizes. The globins and GPCRs are examples.

It is interesting to see which kinds of genes do take advantage of the duplication. The comparison of the genome of the primitive chordate *Branchiostoma floridae* with genomes of vertebrates shows that the set of duplicates that is retained after whole-genome duplication is enriched in genes for signal transduction, transcriptional regulation, neuronal activity, and development. Precisely the features with which early chordates were experimenting.

- Evolution subscribes to the advice, attributed to Yogi Berra: 'When you come to a fork in the road – take it.' We may even add: if you don't come to a fork in the road – take it anyway.

Plant genomes are very susceptible to duplication. The sequence of the *Arabidopsis* genome reveals at least two and possibly three successive duplication events.

Most plants are **polyploids**, i.e. they contain multiple sets of entire chromosomes. **Autopolyploids** contain multiple copies of genomes from the same parent. **Allopolyploids** contain multiple copies of genomes from different parents. Many crop species are polyploids, relative to the wild species from which they were domesticated, including wheat, alfalfa, oats, coffee, potatoes, sugar cane, cotton, peanuts, and bananas. Often polyploidy increases the size of the fruit or grain, a useful property for agriculture (see Box 4.9).

Polyploidy in wheat

The wheat first used in agriculture, in the Middle East at least 10 000–15 000 years ago, is a diploid called **einkorn** (*Triticum monococcum*), containing 14 pairs of chromosomes. Emmer wheat (*T. dicoccum*), also cultivated since palaeolithic times, and durum wheat (*T. turgidum*), are merged hybrids of relatives of einkorn with other wild grasses to form tetraploid species. Additional hybridizations, to different wild wheats, gave hexaploid forms, including spelt (*T. spelta*) and modern common wheat (*T. aestivum*). Triticale, a robust crop developed in modern agriculture and currently used primarily for animal feed, is an artificial genus arising from crossing durum wheat (*T. turgidum*) and rye (*Secale cereale*). Most triticale varieties are hexaploids.

Variety of wheat	Classification	Chromosome complement
Einkorn	*Triticum monococcum*	AA
Emmer wheat	*Triticum dicoccum*	AABB
Durum wheat	*Triticum turgidum*	AABB
Spelt	*Triticum spelta*	AABBDD
Common wheat	*Triticum aestivum*	AABBDD
Triticale	*Triticosecale*	AABBRR

A, genome of original diploid wheat or a relative; B, genome of a wild grass, *Aegilops speltoides* or a relative; D, genome of another wild grass, *T. tauschii* or a relative; R, genome of rye *S. cereale*.

All of these species are still cultivated – some to only minor extents – and have their individual uses in cooking. Spelt, or *farro* in Italian, is the basis of a well-known soup; pasta is made from durum wheat; and bread is made from *T. aestivum*.

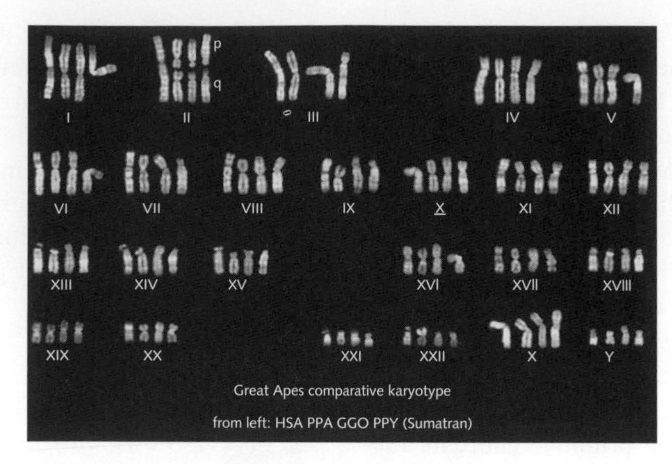

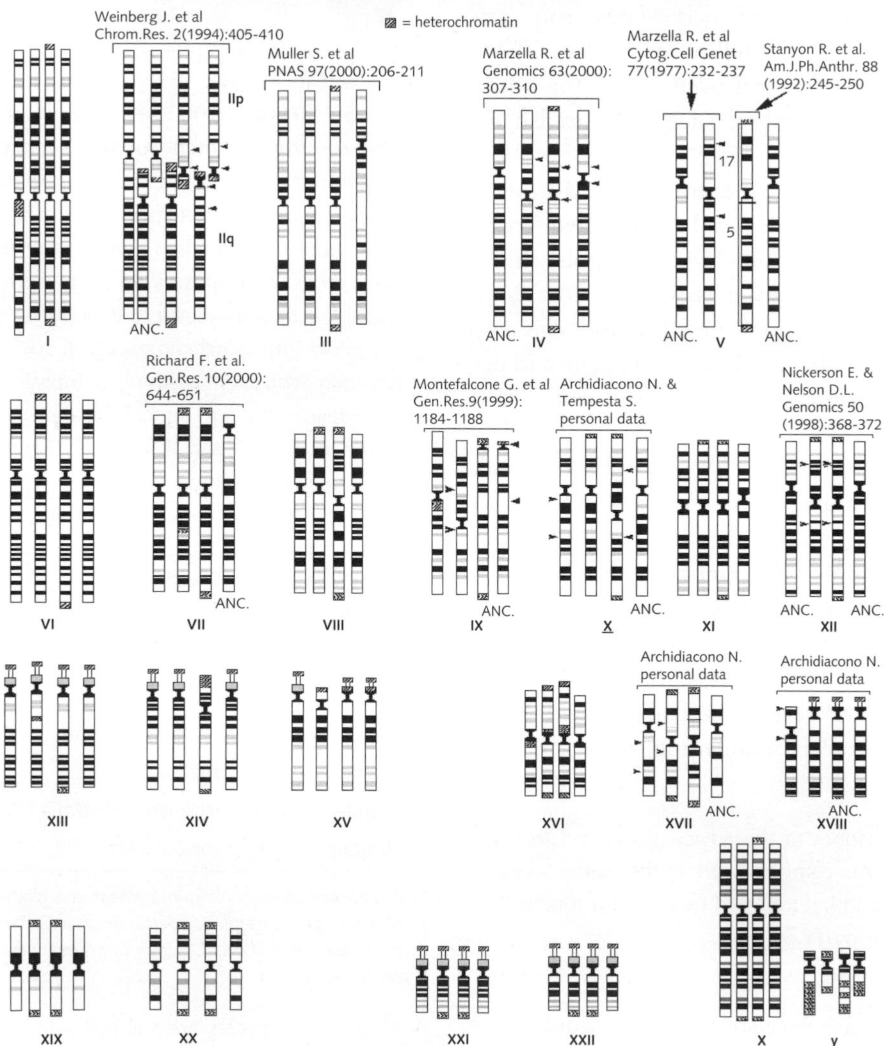

Figure 4.18 Top: photograph of banding patterns. Bottom: ideograms. HSA, *Homo sapiens*; PTR, *Pan troglodytes* (chimpanzee); GGO, *Gorilla gorilla*; PPY, *Pongo pygmaeus* (orang-utan).

Photographs courtesy of Prof. M. Rocchi, Università di Bari, Italy.

Polyploidy may have other advantages. In studies of Arctic flora, it is observed that the fraction of diploid and polyploid plant species increases towards higher latitudes. Many arctic plants tend to exist in small, separated populations and frequently go through 'bottlenecks' of marginal survival, for instance during glaciations. After recession of the ice, deglaciated areas may be repopulated by a few or even one dispersed seed. Carrying many copies of the genome in the cells of each individual may help to preserve genetic diversity, even in tiny populations.

Although polyploidization is much more common in plants than in animals, related species of frogs (genus *Xenopus*) are diploid, tetraploid, octaploid, and dodecaploid. One tetraploid mammal is known, the rat *Tympanoctomys barrerae* from the Monte Desert in west-central Argentina. These species provide a model for control of expression of duplicated genes. For example, in 'polyploid' frogs, silencing is non-syntenic. Each copy of the genome contains some expressed and some silenced genes. This is a different model from the silencing of an entire X chromosome in cells of mammalian females (see Chapter 1).

There are a number of examples of tissue-specific polyploidization. The endosperm of maize kernels undergoes repeated cycles of **endoreplication** (replication of nuclear DNA in the absence of mitosis) to produce cells that can have as many as 96 copies of the haploid genome. In mammals, it is observed that the number of polyploid cells in the liver increases with age, or in response to disease or surgery even in children. This may be a defence against oxidative stress. In the bone marrow of mammals, very large polyploid cells called megakaryocytes 'bud off' portions of their cytoplasm to form platelets. (Platelets are enucleate cells in the blood involved in clotting.) In a related condition, called **polyteny**, replicated chromosomes remain in alignment rather than separate as in polyploids. This is the origin of the giant salivary gland chromosomes of *Drosophila*, which played such an important role in the history of cytogenetics (see Figure 3.3).

Comparisons at the chromosome level: synteny

Comparison of chromosome banding patterns provides snapshots of similarities and differences in large-scale organization among eukaryotic genomes. Synteny literally means 'on the same band', that is, on the same chromosome. (The chromosome exchange that causes chronic myeloid leukaemia (see Chapter 3) is a *breaking* of synteny.) Closely related species generally show a correspondence between large syntenic blocks. The similarity of the banding patterns reveals the underlying similarity of the patterns in the DNA sequences themselves.

Figure 4.18 shows the relationships among the karyotypes of human, chimpanzee, gorilla, and orang-utan.

What makes us human?

It is too difficult to look only at the human genome and try to deduce . . . ourselves. Two approaches help in understanding the genome.

- *Comparative genomics*. We can compare the human and chimpanzee genomes and ask how *differences* between these genomes might give rise to *differences* between the species.
- *Study of human disease*. Many mutations cause disease and give clues to the functions of the affected regions. These regions may encode enzymes or regulatory proteins or RNAs, or they may be DNA sequences that are targets of regulatory mechanisms. Understanding the effects of mutations both illuminates human biology and, often, has immediate clinical applications.

Comparative genomics

The human and chimpanzee genomes are about 96% identical. To understand what makes us human – or at least what makes us not chimpanzees – we can focus on 13 Mb of different sequence, rather than the full 3.2 billion. There are even fewer differences in our amino acid sequences. Humans and chimpanzees express very similar sets of proteins, and most of the homologous proteins of chimpanzee and human are identical or very similar. About 30% of homologous human and chimpanzee proteins show no differences at all. On average, there are only two amino acid differences.

How, then, do humans and chimpanzees develop differently? The ultimate answer *must* lie within the static sequence of the genome. However, a satisfactory answer will require understanding of the dynamics, specifically of patterns of regulation of gene expression.

There is a paradox here. On the one hand, living systems are fairly robust to perturbations. Yeast, for example, survives individual knockout of 80% of its genes. On the other hand, the 4% differences between chimpanzee and human genomes make profound differences in phenotype. This suggests a *chaotic system*, one in which tiny perturbations can lead to large changes in the subsequent trajectory. *Superposed on the robustness are specific changes that exert immense leverage.*

There are two ways to find these crucial sequences. One is to look closely at the differences between human and chimpanzee genomes and try to figure out what the changed loci are doing. Another is to examine human mutations that affect phenotypic properties that chimpanzees do *not* share with humans, such as language and reasoning.

Combining the approaches: the *FOXP2* gene

Language is a unique feature of our species. It should show up as a genetic difference between our genomes and those of other species, including chimpanzee.

Many people suffer from diseases that interfere with production or comprehension of language, or both. Some of these are associated with trauma, or with complex genetics. However, one abnormality with simple Mendelian inheritance appears in a family in London. Members of the 'KE' family have a severe disorder affecting both the facial motor control involved in producing speech and also the mental processing of language.

The mutation responsible for this condition has been identified. It is a single-nucleotide polymorphism (SNP) in a gene called *FOXP2*, which encodes a transcription factor. The major protein encoded by *FOXP2* is 715 amino acids long. It is quite a stable protein from the evolutionary point of view, with only one substitution between mouse and the identical chimpanzee, rhesus macaque, and gorilla sequences. However, the human protein has two mutations relative to the other primate sequences.

This example illustrates the power of a combination of studies of human phenotypes and comparative sequence analysis. And yet the observation that the phenotype shown by members of the KE family arises from two SNPs in a single gene may be deceptively simple. We cannot conclude that the expression of only one gene is involved in creating the phenotype, as is the case for phenylketonuria, for example. The *FOXP2* gene product is a transcription factor. Its activity affects the expression of many genes. The effectiveness of their coordinated expression required co-evolution, i.e. sequence changes in other genes.

Genomes of chimpanzees and humans

The genome sequence of Clint, a male chimpanzee from the Yerkes National Primate Research Center at Emory University, in Atlanta, Georgia, USA, was reported in 2005. Clint represented the West African subspecies *Pan troglodytes verus*.

As expected from so closely related a species:

- There is close alignment of the genome: 96% is alignable with the human genome.
- The sequences of the alignable regions differ at 1.23% of the positions. Recognizing that there is intraspecies divergence among humans and among chimpanzees, it is likely that the true interspecies difference amounts to about 1% of the alignable sequence.
- The 4% non-alignable regions represent insertions and deletions. It is estimated that about 45 Mb of

human sequence do not correspond to chimpanzee sequence and a similar amount of chimpanzee sequence does not correspond to human sequence. More positions in the genomes differ as a result of insertions/deletions than differ as a result of base substitutions.

- The distribution of differences is variable across the genome. For all syntenic 1 Mb segments across the genomes, the range in difference is about 0.005–0.025%. Looking at the distribution with respect to the chromosomes, divergence tends to be higher near the telomeres. The divergence is lowest for the X chromosome and highest for the Y.
- The proteins encoded are also very similar in sequence. Of 13 454 orthologous proteins, 29% have identical sequences. On average, there are one to two amino acid residue differences between corresponding chimpanzee and human proteins.
- Although most proteins are very similar, a few show large K_a/K_s ratios, suggesting that they are under positive selection. These include two proteins, glycophorin C and granulysin (involved in combating infection) and other proteins involved in reproduction. (Selection can act most directly on reproduction itself, producing high rates of evolutionary change.)
- Changes in gene expression patterns show that genes active in the brain have changed more rapidly in humans.

Sadly and unexpectedly, Clint died a few weeks before the paper on his genome was submitted to Nature. *He was 24 years old. Even in the wild, chimpanzees can live for 40–45 years. Cheeta, a chimp that appeared in Tarzan movies in the 1930s, is almost 80 years old.*

Genomes of mice and rats

The mouse and rat are by far the most common mammalian laboratory animals. Knowledge of these species and correlation with human biology is encyclopaedic. Determination of mouse and rat complete genome sequences was clearly a high-priority goal. The genome of the laboratory mouse (*Mus musculus*) appeared in December 2002. The genome of the brown or Norway rat (*Rattus norvegicus*) appeared in April 2004.

Mice, rats, and humans are closely related mammals and illuminate one another. Laboratory studies on rodents are useful guides to the biochemistry and molecular biology of humans. Mice and rats provide the first test of tolerance to, and effectiveness of, novel drugs aimed ultimately at human therapy. Outside the laboratory, however, the close relationship has been tragic for humans: shared parasites permit rats to transmit disease. (An epidemic of bubonic plague in 1347–1352 killed a third of the population of Europe.) Shared diets make food supplies vulnerable to rodent infestation.

The last common ancestor of humans and rodents lived approximately 75 million years ago. Rats and mice separated much more recently: 12–24 million years ago. The genomes of all three species are approximately the same size. The rat genome is ~5% smaller than the human genome. The mouse genome is about ~15% smaller than the human genome. Sequence divergence and chromosome segment rearrangement appear to have been faster in the rodent lineage. The human genome shows more duplication – one reason why it is larger.

Human, mouse, and rat genomes encode similar numbers of genes. Most proteins have homologues in all three species, with very similar amino acid sequences (see Figure 4.19). (Transgenic animals – substituting rodent genes with human genes – can equip model organisms with exact human sequences if necessary.) The genes for most rodent–human homologues have a common exon–intron structure. Gene duplications create protein families, which may be of different sizes in different species. For instance, consistent with their greater dependence on a sense of smell, rodents have more odorant receptors than we do.

Some genomic variation is observable at the chromosomal level. The mouse has 19 chromosomes, plus X/Y. The rat has 20 chromosomes, plus X/Y. Synteny between the mouse and rat genomes is high.

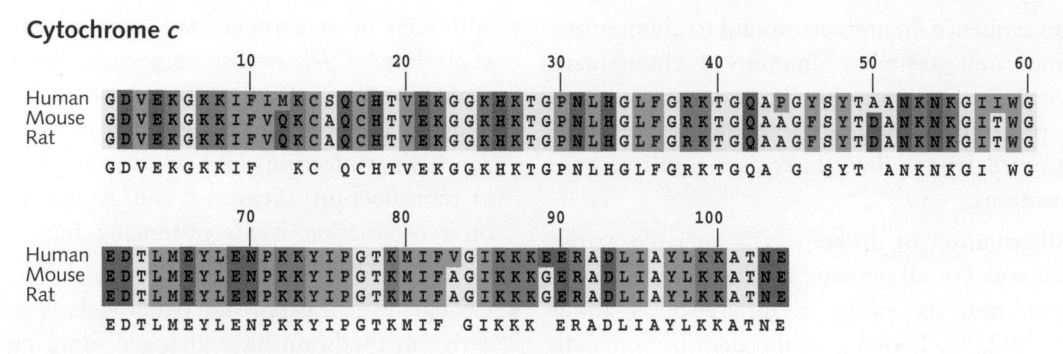

Figure 4.19 Alignment of the amino acid sequences of cytochrome *c* from human, mouse, and rat. All have 104 residues. The mouse and rat sequences are identical. Letters below the sequences show residues conserved in all three species. The human sequence shows nine substitutions, most of which are conservative. For colour coding, see Figure 4.11.

Synteny between the mouse and human genomes is variable. Most human chromosomes contain many small blocks that correspond separately to regions distributed among several mouse chromosomes. However, almost all of human chromosome 20 corresponds almost continuously with a region of mouse chromosome 2. Almost all of human chromosome X appears in mouse chromosome X, differing by rearrangement of nine contiguous blocks, including some reversals. Almost all of human chromosome 17 appears within a region of mouse chromosome 11; however, the sequence is broken into 16 segments that are rearranged, including some reversals.

The blocks are identified by matching sequences of genetic markers. Even though most of the correspondences are distributed among different chromosomes, most of the genomes can be partitioned into syntenic blocks, making up a total of 2.35 Gb (over 90% of the mouse genome). These regions include almost all known exons and regulatory regions.

Alignment of genetic maps is less stringent than alignment of sequences. The rearrangements within and among chromosomes make it impossible to align the three genomes as one long linear sequence. However, at the nucleotide level, 40% of human and mouse genomes are block-alignable, about 1 Gb in all.

The differences among the human, mouse, and rat genomes arise from selection or neutral drift. Regions changing under selection rather than drift are likely to be functional. This is a powerful way to search for regulatory regions, which are harder to identify in genomes than protein-encoding genes.

Genomics confirms the utility of the rat and mouse for clinical research. Of a set of ~1000 genes for which known mutations are associated with human disease, almost all have homologues in rodents. In certain interesting cases, the sequence of the human disease-associated mutant is identical to the mouse and rat wild type. There has probably been co-evolution, with a compensatory change in some other gene or genes. This can be a source of clues to the function and interactions involved in the disease.

Model organisms for study of human diseases

Above the molecular level, the differences between humans and flies and between humans and worms are more obvious than the similarities. At the biochemical and genomic level, the situation is reversed. The underlying common features of the structure, organization, and development of different species is of both academic interest and practical importance, as flies and worms provide models for human diseases.

Distribution of *C. elegans* genes

Chromosome	Size (Mb)	Number of protein genes	Density of protein genes (kb/gene)	Number of tRNA genes
I	7.9	2803	5.06	13
II	8.5	3259	3.65	6
III	7.6	2508	5.40	9
IV	9.2	3094	5.17	7
V	9.8	4082	4.15	5
X	10.1	2631	6.54	3

From: The *C. elegans* Sequencing Consortium (1998). Genome sequence of the nematode *C. elegans*: a platform for investigating biology. Science **282**, 2012–2018.

The genome of *Caenorhabditis elegans*

The nematode worm *Caenorhabditis elegans* entered molecular biology in 1963, at the invitation of Sydney Brenner. Brenner recognized its potential as a sufficiently complex organism to be interesting but simple enough to permit complete analysis of its development and neural circuitry, *at least* at the cellular level.

The *C. elegans* genome, completed in 1998, was the first full DNA sequence of a multicellular organism. *C. elegans* contains ~97 Mb of DNA distributed on six paired chromosomes (see Box 4.10). There is an X but no Y chromosome: different genders in *C. elegans* are a self-fertilizing hermaphrodite, genotype XX, and a male, genotype XO (i.e. a single unpaired X chromosome).

The *C. elegans* genome is about eight times larger than that of yeast, and its 19 099 predicted genes are approximately three times the number in yeast. Exons cover ~27% of the genome. The genes contain an average of five introns. The gene density is relatively low, for a eukaryote, with ~1 gene/5 kb of DNA. Approximately 25% are in clusters of related genes.

The genome of *Drosophila melanogaster*

The total chromosomal DNA of *Drosophila melanogaster* contains about 180 Mb. Approximately two-thirds is euchromatin, a relatively uncoiled and non-compact form, containing most of the active genes. The euchromatic portion, about 120 Mb, was the first segment of the sequence released. The other one-third of the *Drosophila* genome appears as heterochromatin, highly compact regions flanking the centromeres. Heterochromatin contains many tandem repeats of the sequence AATAACATAG and relatively few genes.

The genome is distributed over five chromosomes: three large autosomes, a tiny chromosome containing only ~1 Mb of euchromatin and an X/Y chromosome pair, of which the Y chromosome is heterochromatic and relatively gene poor. The fly's ~14 000 genes are approximately double the number in yeast, but fewer than in *C. elegans*, perhaps a surprise. The average density of genes in the euchromatin sequence is 1 gene/9 kb; about half that of *C. elegans* (see Box 4.11). The genes of the metacentric chromosomes 2 and 3 are reported separately for the two arms, arbitrarily designated left (L) and right (R). The other chromosomes are telocentric (see Figure 4.20).

Determination of the *D. melanogaster* genome sequence was a collaboration between industry (Celera Genomics) and the academic *Drosophila* Genome Projects based in Berkeley, California, USA, and in Europe. The project was a methodological testbed.

- First, it showed that a relatively large eukaryotic genome could be completed by the method of whole-genome shotgun sequencing (see Chapter 3).

BOX 4.11

Distribution of *D. melanogaster* genes

Chromosome arm	Size (Mb)	Number of protein genes	Density of protein genes (kb/gene)	Number of tRNA genes
X	22.2	2279	9.7	25
2L	22.4	2537	8.8	40
2R	20.8	2947	7.1	100
3L	23.8	2718	8.8	49
3R	27.9	3501	8.0	80
4	1.28	83	15.4	0

Data from Release 5.22: http://www.ncbi.nlm.nih.gov/mapview/map search.cgi?taxid=7227.

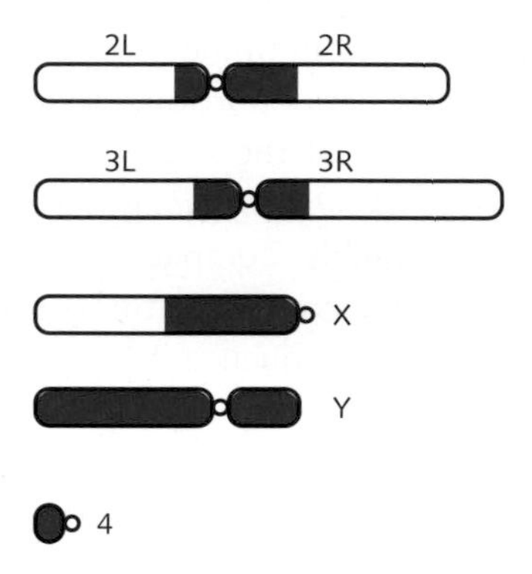

Figure 4.20 The chromosomes of *Drosophila melanogaster*. Heterochromatin is shown in red.

- Second, the annotation of the sequence took place in a burst of activity: the intensive sessions of an 11-day 'jamboree' meeting held at Celera in November 1999. The ~45 participants included experts representing the inherited knowledge of a century of fruit-fly biology and local computer experts involved in the sequencing. The flavour of the meeting is well described from the personal point of view by one of the participants, M. Ashburner.*

* Ashburner, M. (2006). *Won for All: How the* Drosophila *Genome Was Sequenced*. Cold Spring Harbor Laboratory Press, Cold Spring Harbor, New York, USA.

Homologous genes in humans, worms, and flies

Once the genomes were sequenced and annotated, comparisons showed that homologues of many genes appear in all three species. Forty-four per cent of protein-coding genes from *D. melanogaster* have human homologues, 25% of protein-coding genes from *C. elegans* have human homologues, and 23% of protein-coding genes from *D. melanogaster* have homologues in *C. elegans*.

Some proteins with common functions, such as cytochrome *c*, are expected to be quite similar in the different species (see Figure 4.21).

In other cases, different species have adapted homologous proteins to slightly different functions.

C. elegans and *D. melanogaster* are favourite subjects of developmental biologists. It is therefore of interest that a number of transcription regulators involved in developmental control are common to human, *D. melanogaster* and *C. elegans*. These include PAX (paired box domain) and HOX (homeobox domain) proteins.

Models of human disease

A model organism is a species in which an interesting feature of human biology – especially a disease – can be studied (Table 4.8). Ideally, a model organism is small and robust, has a relatively simple genome, is easy to maintain and manipulate (both physically and

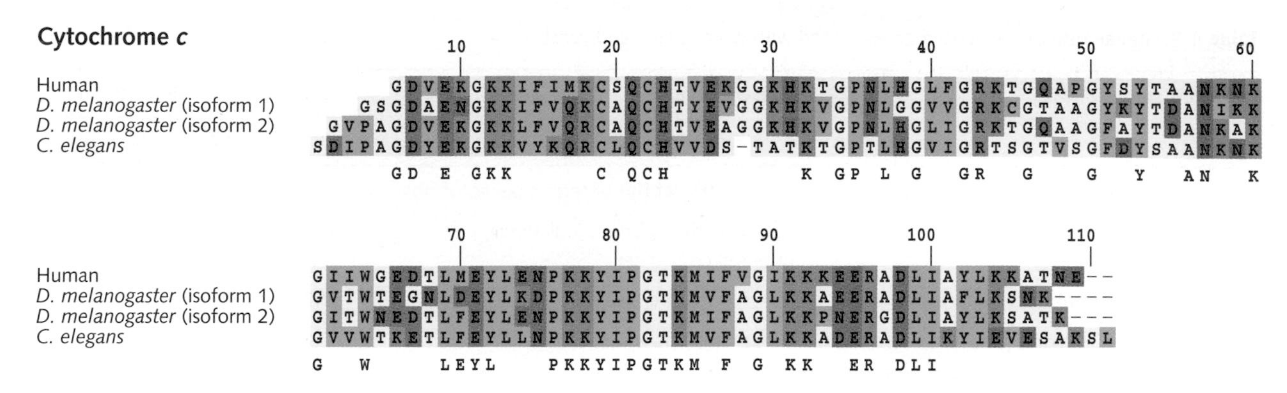

Figure 4.21 Alignments of the amino acid sequences of cytochrome *c* from human, *D. melanogaster* (two isoforms) and *C. elegans*. For colour coding, see Figure 4.12.

genetically) in the laboratory, has a short generation time, is safe to humans, and comes with an extensive knowledge of its biology. As each model organism has its own strengths and limitations, different organisms are useful for different investigations.

In principle, biologists will use the simplest organism that illustrates the human feature of interest. It is easier to do experiments with yeast than with fruit flies. However, sometimes there is no choice. The only animal other than humans that is susceptible to leprosy is the armadillo, which satisfies few, if any, of the criteria of an ideal laboratory organism.

There are two ways in which model organisms can contribute to understanding and treatment of human disease. The first is to observe homologues in the model organisms of genes implicated in human diseases. One can then study the effect on the model organism of mutation or knockout of the homologues. The second is to introduce a human gene into a model organism and discover its phenotypic effect. A model animal containing an active human gene makes it possible to screen libraries of compounds for potential drugs.

Table 4.8 shows some of the human disease-associated genes with homologues in *D. melanogaster*, *C. elegans*, and *Saccharomyces cerevisiae*. The database Homophila provides links between human disease-associated genes and *Drosophila* homologues.

Despite the fact that insects are not very closely related to mammals, fruit flies are useful in the study of human disease. The *D. melanogaster* genome contains homologues of human genes implicated in cancer and in cardiovascular, neurological, endocrinological, renal, metabolic, and haematological diseases. Some of these homologues have different functions in humans and flies. Other human disease-associated genes can be introduced into, and studied in, the fly. For instance, the gene for human spinocerebellar ataxia type 3, when expressed in the fly, produces similar neuronal cell degeneration. There are now fly models for Parkinson's disease and malaria.

C. elegans also provides human disease models. Mutations in the human gene for presenilin-1 (*PS1*) are associated with familial early onset Alzheimer's disease. Mutations in the homologous gene in *C. elegans*, *sel-12* (Figure 4.22) do show neurological defects, but in only a few neurons. Mutants do show more profound defects in egg laying, but this may be a secondary effect.

Although there are greater differences between the nervous systems of humans and *C. elegans* than between their machineries for respiratory energy transduction, the difference between the homologues shown here is not greater than the difference between the cytochrome *c* proteins. The relationship between sequence and function in proteins is full of surprises.

- We have discussed selected genomes – chimpanzee, mouse, rat, worm, and fly – the first because it the closest extant relative we have, and the others because of their importance as laboratory animals. Two aspects of comparative genomics are to study *differences* between genomes, to try to account for phenotypic divergence; and to study and apply *similarities*. One important application is to use laboratory animals as models for human diseases.

Table 4.8 Human disease-associated genes shared with worms, flies, and yeast

Affected area	Disease	Description	Gene	Similarity in		
				Worm	Fly	Yeast
Bones	Multiple exostoses	Ossification at tips of femur, pelvis, or ribs	*EXT1*	***	**	–
Blood	Leukaemia	Chronic myelogenous leukaemia, a blood cell cancer	*ABL1*	***	***	*
	Bruton agammaglobulinaemia	Lack of mature B cells	*BTK*	***	**	*
	Glucose-6-phosphate dehydrogenase deficiency	Drug- and stress-induced rupture of red blood cells	*G6PD*	****	****	****
Brain	Early-onset Alzheimer's disease	Common cause of mental retardation	*PS1*	**	**	–
	Fragile X syndrome		*FMR1*	**	–	–
	Juvenile Parkinson's disease		*PARK2*	***	**	*
Colon	Hereditary non-polyposis cancer	Polyps that become malignant	*MSH2*	***	***	***
	Adenomatous polyposis		*APC*	***	*	–
Ears	Hereditary deafness		*MYO15*	***	***	***
Eyes	Retinoblastoma	Cancer of the eye	*RB1*	*	*	–
Heart	Familial cardiac myopathy	Inherited cardiac disease	*MYH7*	***	***	***
	Long QT syndrome	Sometimes fatal cardiac arrhythmias	*3-SCN5A*	***	**	*
Kidney	Polycystic kidney disease 2		*PKD2*	**	**	–
Liver	Wilson's disease	Build-up of copper in cells, causing liver disease and other symptoms	*ATP7B*	***	***	***
Lung	Cystic fibrosis	Progressive disease of lungs and pancreas	*CFTR*	***	***	–
	Lung cancer	Caused by defects in *p53* gene, which can also cause cancer of the oesophagus, colon, brain, lung, breast, and skin	*p53*	*	–	–
Muscles	Duchenne's muscular dystrophy	Progressive atrophy of muscles	*DMD*	***	***	–
Pancreas	Pancreatic cancer		*MADH4*	***	*	–
	Pancreatic cancer		*RAS*	**	**	**
Prostate	Advanced cancer of the prostate	Caused by mutations in the *PTEN* gene, which can also cause cancer of the brain, endometrium, and breast	*PTEN*	**	**	*
Skin	Xeroderma pigmentosum D	Early-onset skin cancer	*XPD*	***	**	***
	Neurofibromatosis 1	Soft tumours at many sites, plus skeletal and neurological defects	*NF1*	***	*	**
Thyroid	Cancer of the thyroid	Multiple endocrine neoplasia type 2	*MEN2*	***	**	*

Based on data from Rubin, G.M., et al. (2000). Comparative genomics of the eukaryotes. Science **287**, 2204–2215. Presentation adapted from http://www.hhmi.org/genesweshare/e400.html.

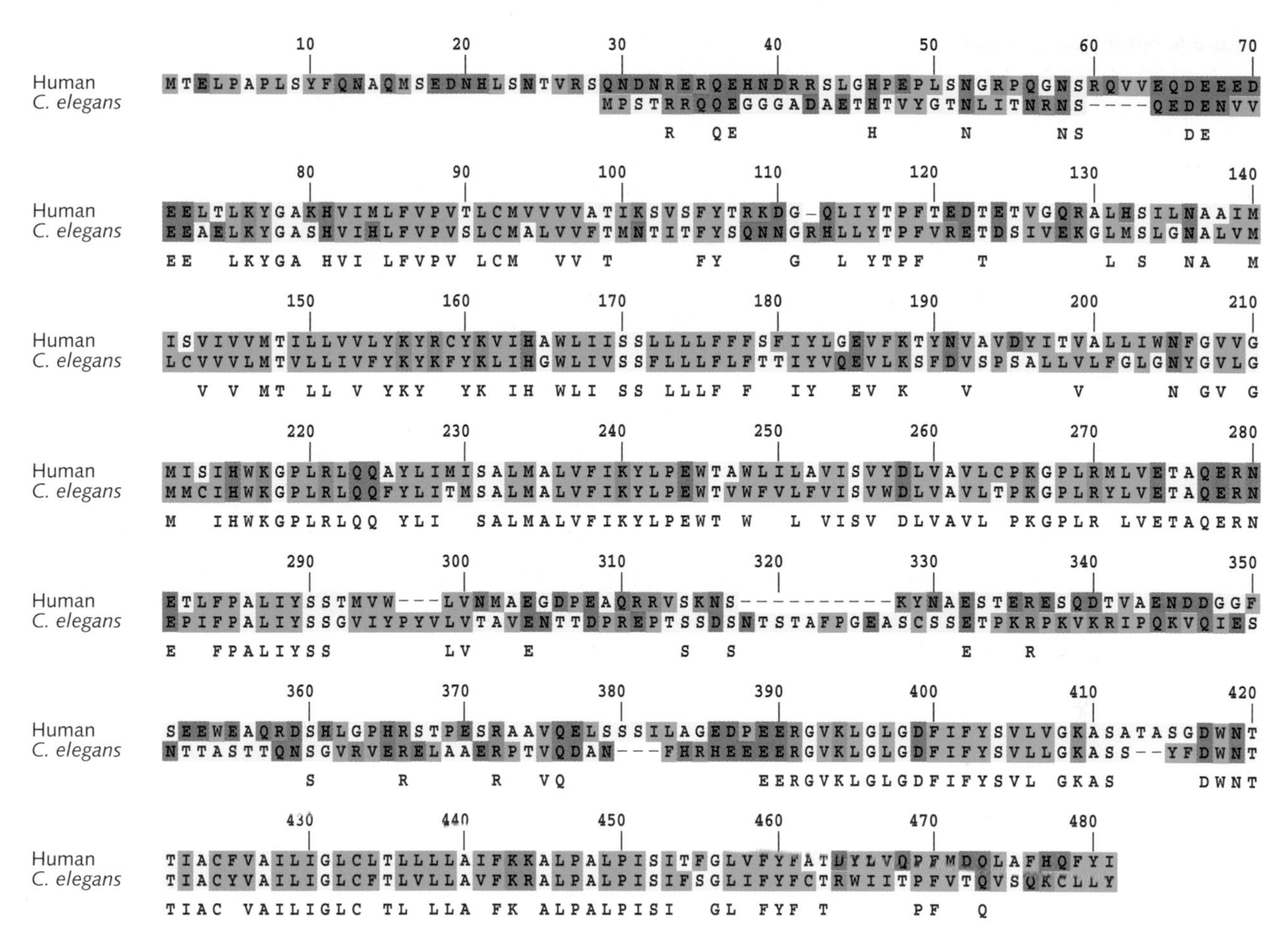

Figure 4.22 Alignment of the amino acid sequences of the human protein presenilin-1 and the *C. elegans* homologue SEL-12.

The ENCODE project

The ENCODE project (*Enc*yclopedia *o*f *D*NA *e*lements) is a systematic development and application of comparative genomics. It has the ultimate goal of developing methods for comprehensive identification of functional regions of the human genome, including coding and regulatory regions. A selected portion of the human genome – 1%, about 30 Mb – will be the initial focus. The basic approach will be comparative genomics and will involve both laboratory and computational analysis.

High-quality sequences will be finished to state-of-the-art standards, including resolving difficult regions. Medium-quality sequences will have >8-fold coverage, with manual refinement of assembly. Unfinished sequences are whole-genome shotguns; the coverage may vary and assembly may be incomplete.

Regions corresponding to the selected human genome segments from 29 vertebrates will be sequenced (see Table 4.9). These data will illuminate each other. The ENCODE project will apply, improve, and develop, as necessary, a variety of experimental and computational methods. Lessons learned from work with the selected subset will guide the scaling up of successful methods to analysis of entire genomes.

Coordinating with ENCODE, the HapMap Project (see Chapter 1) focuses on variations among humans in ten of the ENCODE regions. Sequences from 48 individuals from different geographic origins have yielded 30 000 SNPs.

Analysis of function involves two steps: deciding whether a segment has functional significance and, if

Table 4.9 Target species of the ENCODE project

		Quality of sequencing		
		High	**Medium**	**Unfinished**
Class:				
Actinopterygii		Zebrafish		
Amphibia		Frog		
Aves		Chicken		
Class: *Mammalia*				
Order:	**Suborder:**			
Monotremata				Platypus
Marsupialia			Opossum	
Proboscidia				African elephant
Insectivora				Tenrec
Xenarthra				Armadillo
Insectivora				Hedgehog
Insectivora				Shrew
Chiroptera				Bat
Artiodactyla		Cow		
Carnivora		Dog		
Carnivora				Cat
Rodentia		Mouse		
Rodentia		Rat		
Rodentia				Guinea pig
Lagomorpha				Rabbit
Primates	*Prosimii*			Galago
Primates	*Prosimii*			Mouse lemur
Primates	*Platyrrhini*			Duski titi
Primates	*Platyrrhini*			Owl monkey
Primates	*Platyrrhini*			Marmoset
Primates	*Catarrhini*			Colobus
Primates	*Catarrhini*	Macaque		
Primates	*Catarrhini*			Baboon
Primates	*Hominidae*		Orang-utan	
Primates	*Hominidae*	Chimpanzee		
Primates	*Hominidae*	Human		

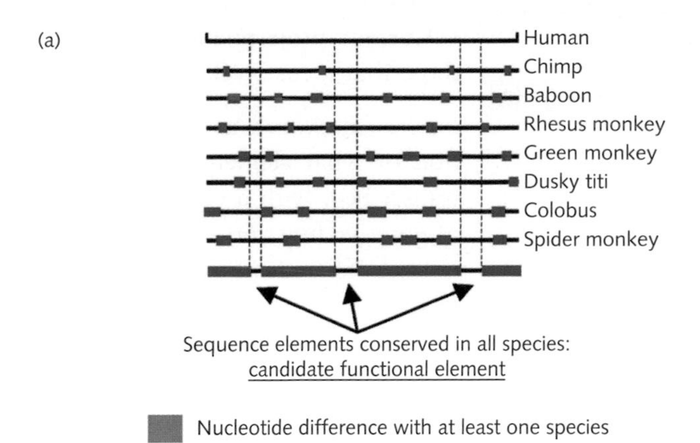

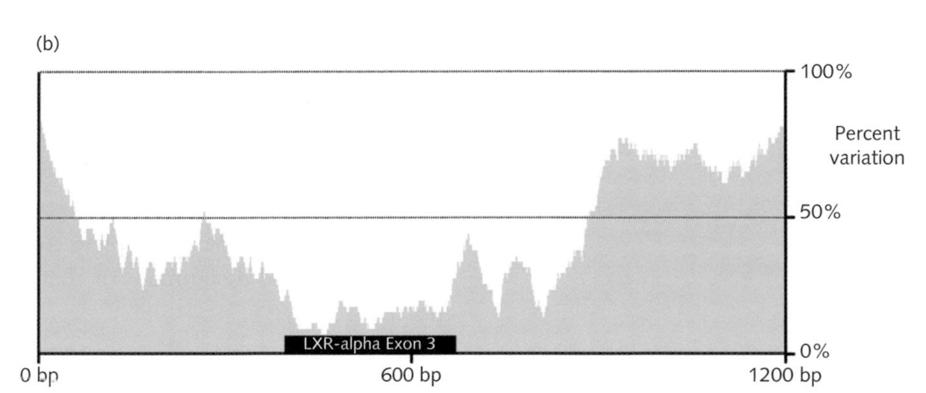

Figure 4.23 Patterns of variation in multiple sequence alignments can suggest regions of likely function. This diagram shows analysis of a 1200 bp region in primate genomes containing an exon of the liver X receptor α gene and flanking regions. This gene encodes a nuclear receptor responsive to elevated levels of intracellular cholesterol. (a) Human, reference sequence; purple, regions in which the sequence of the indicated region differs from at least one other species. The columns with no purple are conserved in all species and define regions likely to be functional. (b) Plot of % variation along the sequence. Regions of lowest variability correspond to the known exon. (A similar but not identical approach was used by E.A. Kabat and T.T. Wu in their classic work identifying complementarity-determining regions of antibodies from regions of hypervariability.)

From: Nobrega, M.A. & Pennacchio, L.A. (2004). Comparative genomic analysis as a tool for biological discovery. J. Physiol. **554**, 31–39.

so, identifying what it does (see Figure 4.23). Approximately 5% of the human genome is conserved with respect to mouse and rat sequences. This 5% should have interesting functions (without implying that the other 95% does not). Only about one-third of this 5% is predicted to encode protein. Analysis of function will require treatment of both protein-coding and non-protein-coding regions.

Accordingly, the criteria for selection of regions for the ENCODE project included choosing regions with ranges of gene density and of non-exonic conservation with respect to the mouse sequence. The result is a set of 44 discrete regions, spread around different human chromosomes and the syntenic regions in other species. These include well-studied regions such as the α- and β-globin loci and the region containing *CFTR*, the gene for the cystic fibrosis transmembrane conductance regulator, for which sequence information from different species is known. Sequences of the ENCODE target regions can be aligned and compared (see Figure 4.23).

In 2007, ENCODE moved into its second phase, and a companion project, modENCODE began (Table 4.10).

The modENCODE project

modENCODE extends the ENCODE project to model organisms. Its initial goal is to identify functional elements in the *C. elegans* and *D. melanogaster* genomes.

Table 4.10 Approximate sizes of ENCODE regions

Chromosome	Approximate sizes of ENCODE regions (Mb) (gene of interest)
1	0.5
2	0.5, 0.5, 0.5, 0.5
4	0.5
5	0.5, 0.5, 1.0 (interleukin)
6	0.5, 0.5, 0.5, 0.5
7	0.5, 1.0, 1.1, 1.2, 1.9 (CFTR)
8	0.5
9	0.5
10	0.5
11	0.5, 0.5, 0.6, 0.5 (Apo cluster), 1.0 (β-globin)
12	0.5
13	0.5, 0.5
14	0.5, 0.5
15	0.5
16	0.5, 0.5, 0.5 (α-globin)
18	0.5, 0.5
19	1.0
20	0.5
21	0.5, 1.7
22	1.7
X	0.5, 1.2

Current projects include but are not limited to those that focus on the following:

the transcriptome: as complete as possible a description of what elements of the genome are actually transcribed, with a classification, as far as possible, of putative function – at least whether the transcript is likely to code for protein or non-protein-coding RNA.

chromatin function and histone variants: genomic distribution of modifications to histones and other chromosome-associated protein regulatory elements.

the 3′ utr-ome: untranslated regions 3′ to coding sequences are important sites for post-transcriptional regulation of expression.

transcription factors: identification of the DNA-binding sites, and measurement of expression patterns at different life stages.

For *D. melanogaster*: a complete roster of small and microRNAs and assignment of their functions when possible.

Note the focus on regulatory elements. Papers appearing in *Science* in late 2010 report modENCODE results for worm and fly, respectively. These include reports of many new genes that encode proteins or RNAs.

RECOMMENDED READING

- Papers treating some of the current debate in biological taxonomy and the strengths and weaknesses of barcoding, and a description of the database:

Moritz, C. & Cicero, C. (2004). DNA barcoding: promise and pitfalls. PLoS Biol. **2**, e354.

Ratnasingham, S. & Hebert, P.D.N. (2007). BOLD: The Barcode of Life Data System. Molecular Ecology Notes **7**, 355–364.

Stoeckle, M.Y. & Hebert, P.D. (2008). Barcode of life. Sci. Am. **299**(4), 82–86, 88.

- Detailed review of work on genome comparisons and what they tell us about genome contents and evolution:

Miller, W., Makova, K.D., Nekrutenko, A., & Hardison, R.C. (2004). Comparative genomics. Annu. Rev. Genomics Hum. Genet. **5**, 15–56.

- How modern developments in biological data collection affect the use of model organisms in the study of human biology and disease:

Barr, M.M. (2003). Super models. Physiol. Genomics **13**, 15–24.

- Importance of duplications:

  Levasseur, A. & Pontarotti, P. (2011). The role of duplications in the evolution of genomes highlights the need for evolutionary-based approaches in comparative genomics. Biol. Direct. **18**, 11.

- Alternative splicing:

  Park, J.W. & Graveley, B.R. (2007). Complex alternative splicing. Adv. Exp. Med. Biol. **623**, 50–63.

- RNA editing:

  Maas, S., Kawahara, Y., Tamburro, K.M., & Nishikura, K. (2006). A-to-I RNA editing and human disease. RNA Biol. **3**, 1–9.

  Mass, S. (2010). Gene regulation through RNA editing. Discov Med. **10**, 379–386.

  Farajollahi, S. & Maas, S. (2010). Molecular diversity through RNA editing: a balancing act. Trends Genet. **26**, 221–230.

- ENCODE and modENCODE:

  The ENCODE Project Consortium (2011). A user's guide to the Encyclopedia of DNA Elements (ENCODE). PLoS Biol. **9**, e1001046.

  Elsner, M. & Mak, H.C. (2011). A modENCODE snapshot. Nat. Biotechnol. **29**, 238–240.

  Muers, M. (2011). Functional genomics: the modENCODE guide to the genome. Nat. Rev. Genet. **12**, 80.

EXERCISES, PROBLEMS, AND WEBLEMS

Exercises

Exercise 4.1 On a photocopy of Figure 4.4, (a) indicate the position of the human genome; (b) indicate the position of the pufferfish genome; (c) indicate the position of the rice (*Oryza sativa*) genome; (d) indicate the position of the yeast *S. cerevisiae* genome; (e) indicate the position of the *E. coli* genome; (f) indicate the range of sizes of mitochondrial genomes; and (g) indicate the range of sizes of chloroplast genomes.

Exercise 4.2 In H.G. Wells' 1898 novel *The War of the Worlds*, invaders from Mars are overcome by disease. Assuming that life on Mars developed independently of life on Earth, why is it unlikely that the Martians died of viral infections?

Exercise 4.3 Which of the following pairs are orthologues? Which are paralogues? Which are neither? (a) Human trypsin and horse trypsin; (b) human trypsin and horse chymotrypsin; (c) human trypsin and human elastase; (d) *Bacillus subtilis* subtilisin and horse chymotrypsin.

Exercise 4.4 On a photocopy of Figure 4.1, indicate estimates of the dates of the events at the points marked by asterisks.

Exercise 4.5 What RNA molecule is most closely linked to the *his* operon in *E. coli* (see Figure 4.9)?

Exercise 4.6 Human mitochondrial DNA is 16 569 bp long. A brain cell may contain 10 000 mitochondria. What fraction of the DNA in a brain cell is mitochondrial? Assume 1 mtDNA/mitochondrion.

Exercise 4.7 Some antibiotics, for example streptomycin, block protein synthesis in bacteria but not cytoplasmic protein synthesis in eukaryotes. Mitochondria and chloroplasts contain their own protein-synthesizing machinery. Would you expect streptomycin to block mitochondrial protein synthesis? Explain your answer.

Exercise 4.8 On a photocopy of Figure 4.1, indicate by an arrow linking them the approximate positions of the source and destination organisms for the endosymbiotic origins of mitochondria and chloroplasts.

Exercise 4.9 Why could a mollusc that extracts chloroplasts from algae that it eats not simply let the chloroplasts reside in some body cavity (like the symbiotic bacteria in the guts of ruminant animals) rather than endocytosing them?

Exercise 4.10 The main *E. coli* chromosome contains 4 639 221 bp. The cell is roughly a cylinder about 0.1 μm in diameter and 0.2 μm long. If the length of an extended segment of DNA in the B conformation is 3.4 Å per base pair (0.34 nm), what would be the diameter of the chromosome if it were geometrically a circle?

Exercise 4.11 On a photocopy of Figure 4.11, mark the positions that (a) contain the same amino acid in all three eukaryotes but differ in *S. aureus*; (b) contain the same amino acid in humans and chickens but are not the same in *Neurospora* and/or *S. aureus*; (c) contain the same amino acid in humans and *S. aureus*, but this amino acid does not appear at this position in both of the other two species.

Exercise 4.12 On a photocopy of Figure 4.11, indicate which regions of the amino acid sequences are encoded by which blocks of the nucleotide sequences in Figure 4.12. (This exercise is similar to Exercise 1.12.)

Exercise 4.13 On a photocopy of Figure 4.13, draw horizontal lines at the beginning and end of the Devonian (see Figure 2.15). What is earliest type of species to emerge after the split between neuroglobin and cytoglobin?

Exercise 4.14 Which animals in Figure 4.14 have the largest number of functional β-globin genes?

Exercise 4.15 Which animals in Figure 4.14 show a triplication of part of the β-region? Which genes make up the repeating unit?

Exercise 4.16 Which animals in Figure 4.14 show a pseudogene most closely related to a δ-globin?

Exercise 4.17 The human and chimpanzee genomes are 96% identical. (a) How many individual bases of the human genome differ from the corresponding positions of the chimp genome? (b) Assume that the human genome is 3% coding and contains about 20 000 genes, and that all of the sequence differences are independent, single-base changes distributed randomly throughout the genome (these assumptions are definitely *not true*). Estimate the fraction of genes mutated between humans and chimpanzees.

Exercise 4.18 On a photocopy of Figure 4.1, indicate where three whole-genome duplications are believed to have occurred.

Problems

Problem 4.1 On a copy of Figure 4.7(b), indicate the following interactions: (a) the positively charged guanidino group in zanamivir (left in Figure 4.7b) forms salt bridges with Glu116 and Glu225 in the neuraminidase active site; (b) hydroxyl groups of the glycerol moiety of zanamivir (at the right in Figure 4.7b) are hydrogen bonded to Glu274; (c) the carbonyl oxygen of the *N*-acetyl sidechain is hydrogen bonded to Arg149; (d) the methyl group of the *N*-acetyl sidechain makes hydrophobic interactions with Ile220 and Trp176.

Problem 4.2 Read the letters to *The New York Times* (17 September 2005) discussing the decision to publish the genome sequence of the 1918 pandemic strain of influenza virus. Summarize the arguments for and against publication.

Problem 4.3 From the partial sequences in Figure 4.12, (a) how many positions (necessarily in the coding regions) contain the same base in all three genomes? What percentage of positions contain the same base in all three genomes? (b) How many positions in the coding regions are common to human and chicken but different in *S. aureus*? To what percentage of coding positions does this correspond? (c) How many positions in the non-coding regions are common to human and chicken? To what percentage of non-coding positions does this correspond?

Problem 4.4 For the coding regions of the genes for human and chicken thioredoxins (p. 135), calculate the K_a/K_s ratio. What can you infer about the relative importance of selection and drift in accounting for the difference in the corresponding amino acid sequences?

Problem 4.5 To what time frame can you date the duplication of the γ gene in the β-globin locus of primates? (See Figures 2.15, 4.13, and 4.14.)

Problem 4.6 Figure 4.24 shows an evolutionary tree of hominids. With reference to Figure 4.18, can you find similarities in the chromosome structures that confirm these relationships between human, common chimpanzee, gorilla, and orang-utan?

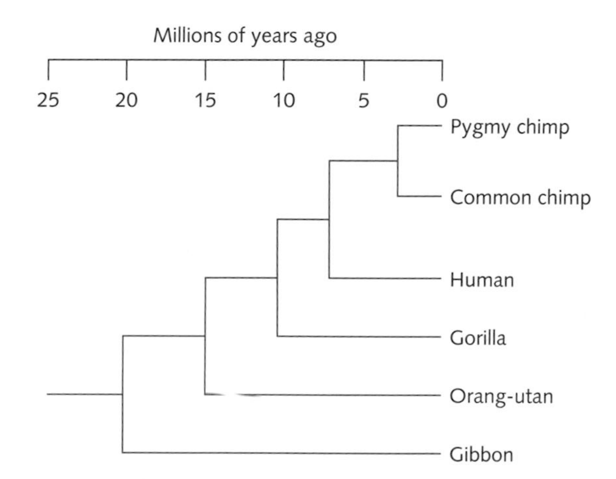

Figure 4.24 Phylogenetic tree of hominids.

Problem 4.7 Which substitutions between rodent and human cytochrome *c* would *not* be considered conservative mutations?

Problem 4.8 For the following pairs of homologous proteins, what are the percentages of identical amino acids in an optimal alignment and what are the percentages of identical residues or conservative substitutions in an optimal alignment: (a) the cytochrome *c* proteins of humans and *C. elegans*; and (b) presenilin-1 homologues of humans and *C. elegans*?

Weblems

Weblem 4.1 Identify a virus and a prokaryote such that the genome of the virus is larger than the genome of the prokaryote (see Figure 4.4).

Weblem 4.2 Find examples of viruses with (a) a double-stranded circular DNA genome; (b) a double-stranded linear DNA genome; (c) a single-stranded (+)sense RNA genome; and (d) a single-stranded (–)sense RNA genome.

Weblem 4.3 What was the most common serotype of influenza in the USA during the 2010–2011 season (www.cdc.gov/flu/weekly)?

Weblem 4.4 Give an example of a species that represents the simplest known form of (a) metazoan; (b) deuterostome; (c) placoderm; and (d) eutherian.

Weblem 4.5 Mitochondrial DNA is often edited before translation. When a mitochondrial gene is transferred to the nucleus, there are two possibilities: (1) the copying is DNA→DNA. In this case, the nuclear version will initially have the mitochondrial DNA sequence and will require mutation to effect the changes introduced in the mitochondrion by the editing; or (2) the nuclear gene is reverse transcribed from the edited mitochondrial mRNA. Compare the sequences of the nuclear gene for cytochrome oxidase II from mung bean (*Vigna radiata*) with mitochondrial sequences in related legumes, before and after editing. Do the data support either hypothesis?

Weblem 4.6 Align the following sequences: (1) the nuclear-encoded *rps14* gene from rice (*Oryza sativa*); (2) the mitochondrial-encoded *rps14* gene from broadbean (*Vicia faba*); (3) the nuclear-encoded *sdhB* gene from rice. Identify the leader sequence in the rice *rps14* gene, targeting mitochondrial import, that appears to have been borrowed from the rice *sdhB* gene.

Weblem 4.7 Make two histograms of *E. coli* genes, similar to that of Figure 4.8, showing in one of them the size distribution of genes appearing clockwise in the genome and in the other the size distribution of genes appearing counter-clockwise in the genome. Describe any systematic differences that appear.

Weblem 4.8 Many people believe that Rickettsiae are the closest extant relatives of the organism that, after endosymbiosis, gave rise to mitochondria. Rickettsiae are obligate aerobes. (a) Can you identify an enzyme that (1) catalyses an anaerobic function in mitochondria and (2) lacks a known homologue in Rickettsiae. (b) If you can find one, and if the rickettsial origin of mitochondria is a valid hypothesis, what are reasonable explanations of the presence of the anaerobic enzymatic function in mitochondria? (c) How would you test these explanations?

Weblem 4.9 In eukaryotes, the recombination rate per kilobase – the physical distance that corresponds to the genetic distance – varies among species by several orders of magnitude overall. (It also varies within each genome.) It depends primarily on overall genome size. You can find the data in computer-readable form through this book's Online Resource Centre (www.oxfordtextbooks.co.uk/orc/leskgenomics2e/). Draw graphs of recombination rate against genome size, distinguishing data from different groups of organisms. (a) What relationship do you observe, i.e. colloquially, what is the shape of the curve? (b) Can you plot the data in a way that gives a linear relationship? (c) Do the data from the different groups of organisms follow the same relationship? If not, how do they differ? (d) What general conclusions can you draw?

Weblem 4.10 Add to Figure 4.13 the corresponding partial gene sequence from *N. crassa*. Describe its relationship to human, chicken, and *S. aureus* sequences. In particular, answer questions analogous to those in Exercise 4.11 for the DNA sequences.

Weblem 4.11 What full-genome project is in progress, but not yet complete, for an organism in each of following categories: (a) fungus; (b) amphibian; (c) land plant; (d) insect (*not* a species of *Drosophila*); (e) primate.

Weblem 4.12 Collect and align sequences of the protein HSP70 from about six (of each) Gram-positive bacteria, proteobacteria, other Gram-negative bacteria, and archaea. Identify an insertion common to proteobacteria and other Gram-negative bacteria but absent from Gram-postive bacteria and archaea. On this basis, sketch the topology of a phylogenetic tree relating Gram-positive bacteria, proteobacteria, other Gram-negative bacteria, and archaea. Where do these results suggest placing the root of the prokaryotic tree?

Weblem 4.13 The genetic code used in translation of genes in animal mitochondria differs from the standard one. Was this feature simply inherited from the original symbiont that gave rise to the organelle, or did it arise subseqently by divergence? Determine (a) what genetic code is used by the living organism that appears to be the closest extant relative of the original symbiont, *Rickettsia prowazekii*. (b) Do all animal mitochondria use the *same* variant of the standard genetic code? Based on these data, suggest an answer to the question.

Weblem 4.14 What is significant about the following species that justified sequencing their entire genome sequences? (a) *Plasmodium falciparum*; (b) *Aspergillus fumigatus*; (c) *Tropheryma whipplei*; (d) *Sulfolobus tokodaii*; (e) sea urchin; (f) *Ciona intestinalis*.

Weblem 4.15 In 1976, swine flu killed Private David Lewis, a soldier at Fort Dix, in central New Jersey South of Princeton, USA. What was the response of the US Government, under President Gerald Ford? What was the course of the potential epidemic? In retrospect, does the response appear to have been the right one? Could you have advised President Ford to make a different response? If so, based on what facts known at the time of Private Lewis's illness?

Weblem 4.16 Compute K_a/K_s ratios for the genes for the two isoforms of *D. melanogaster* cytochrome *c*. Does the divergence appear to have arisen from selection or drift?

Weblem 4.17 Compute K_a/K_s ratios for the genes for human presenilin-1 and the *C. elegans* homologue *sel-12*. Does the divergence appear to have arisen from selection or drift?

Weblem 4.18 The classification of tarsiers among primates is ambiguous. Palaeontology places tarsiers with lemurs, lorises, and galagos, in the suborder prosimians. Some genetic relationships place tarsiers with monkeys, apes, and humans. It is also possible that tarsiers form a separate prosimian infraorder. Find the structure of the β-globin region of the tarsier genome and explain what these data suggest about where, in Figure 4.15, the tarsiers belong.

Weblem 4.19 The human homologues of *C. elegans* NPR-1 are the neuropeptide Y receptors NPY1R, NYP2R, . . . What mutations are known in these human receptors and what, according to Online Mendelian Inheritance in Man (OMIM™), are their effects?

Weblem 4.20 In humans, cholesteryl ester transfer protein is important in controlling blood levels of high-density lipoproteins. Do homologues of this protein exist in (a) mouse; (b) rat; and (c) hamster?

Weblem 4.21 Mouse and rat cytochrome *c* have identical amino acid sequences. Do they have identical gene sequences also? If not, show the differences between the cytochrome *c* genes in mouse and rat. Distinguish between exons and introns.

Weblem 4.22 How many cytochrome *c* pseudogenes are there in the human genome?

Weblem 4.23 Which of the following genes have the same number of exons in human and mouse orthologues? (a) Haemoglobin A (human gene *HBA*); (b) cytochrome *c* (human gene *HCS*); (c) spermidine synthase (human gene *SRM*).

Weblem 4.24 Are any regions containing globin genes, other than haemoglobin α and β, included in the choices of ENCODE regions studied by the HapMap project? If so, which ones?

CHAPTER 5

Evolution and Genomic Change

LEARNING GOALS

- To understand the coordination of changes in genotype and phenotype during evolution.
- To know the principles of biological classification and the grammar of biological nomenclature.
- To appreciate the distinction between similarity and homology, and that homology, usually unobservable, is an inference from similarity.
- To recognize that measurements of similarity between gene or protein sequences offer our best insight into relationships between individuals or between species.
- To understand the general idea of pattern recognition, and be able to use tools such as the dotplot to recognize similarities among sequences.
- To understand the basis of constructing phylogenetic trees, methods for calculating them, and the information different types of tree contain.

In the preceding chapter, we described some of the differences observed in comparing closely related species, and distantly related ones, at various levels from molecules to karyotypes. In this chapter we seek to understand how these changes came about. The general answer is of course evolution. The availability of the data from genomics and related fields – such as protein structure determinations – challenges us to probe into the detailed mechanism by which evolutionary changes occur. Tools for analysing similarities include sequence-alignment algorithms, and, from the similarities, methods for generating phylogenetic trees. These tools are part of the essential skill set for anyone working in the field of genomics.

Evolution is exploration

Evolution is exploration. Exploration leads to discovery. Discovery leads to change. Change can appear as creativity. (Let us avoid the word progress, much-abused in this context.) But it is all based on exploration.

By exploration, we mean that life can probe the vicinity of its current state, generating and testing variations. Evolution involves exploration and change at many levels. What makes biology so complex is that the changes at different levels are intimately linked.

- The most fundamental level of exploration is *mutation of genome sequences*. It is through mutations and, in sexually reproducing organisms, allelic reassortment, that life explores the neighbourhood of a current genotypic and phenotypic state.

  A mutation can affect a transcribed molecule, changing either the amino acid sequence of a protein, or a base in a non-protein-coding RNA. Alternatively, changes in splice sites or regulatory sequences can change protein expression levels. A mutation that causes loss of an essential function will be lethal – a blind alley of evolutionary exploration. Conversely, other mutations might be expected to have little or no effect. Such putatively 'silent mutations' include changes among synonymous triplets in protein-coding genes, or mutations in pseudogenes, or – presumably – mutations in the large portions of some genomes currently described as junk. However, even mutations producing synonymous codons in protein-coding genes could interact with transfer RNA (tRNA) levels to affect translation rates, and thereby affect protein structure (see pp. 9–10).

- Altered *proteins* explore possibilities of altered structure, including altered post-translational modification; and altered function, including changes in enzymatic activity or changes in regulatory signals.

  A single conservative amino acid substitution at a site on the surface of a protein distant from the active site would be expected to have only localized, self-contained effects on protein structure and function.

  However, some mutations send ripples through the system and have far-reaching effects. Small changes in single *HOX* genes have immense leverage in creating the overall body plans of animals. For example, a major change in body plan in metazoans occurred about 400 million years ago, with the emergence of insects with six legs from arthropod ancestors with large numbers of legs. Experiments of W. McGinnis and co-workers showed that changes in one protein, Ubx, a *HOX* homologue, are sufficient to achieve this large-scale anatomical transition.

- *At the chromosomal level*, evolution can explore distributions of genes. This can involve local or global gene duplication and transposition of either small segments of chromosomes or large-scale blocks. Degradation of synteny can lead to infertility. This is one of the mechanisms of speciation.

- *At the cellular level*, evolution has explored different kinds of organization, notably the prokaryote–eukaryote division. Some but not all cells have cell walls. Some, but not all, have chloroplasts. Complex organisms develop many types of specialized cell, tissue and organs.

- *Individuals* explore different possible life histories. The context and interactions of our lives shape our development. For humans more than other species, cultural heritage and experience have a great effect on physical as well as on mental development.

  We as individuals also have greater control over how we explore the potential inherent in our genomes. This freedom is, of course, incomplete, for societies are ecosystems and constrain our development and activities.

- *Within populations of individuals of the same species, evolution explores varying distributions of allele frequencies*. Natural populations show genotypic and phenotypic variation. Even in the absence of mutations, populations can 'react' to changing conditions by varying gene frequencies: **Industrial melanism** is a classic example (see Box 5.1).

- *At the level of body plan*, a visit to a zoo or botanical garden reveals life's stunning variety. Comparative anatomy reveals the underlying similarities among

BOX 5.1 Industrial melanism and its reversal

One variety of the British pepper moth, *Biston betularia* (variety *typica*), has a mottled black-and-white colouring. The moths are nocturnal; during the day, they roost on tree trunks. Before the rise of industrial pollution, light-coloured lichens encrusted the trees and the moths were protected from birds by camouflage.

Another variety, *carbonaria*, has a uniformly dark colour. It was first observed, as a mutant, in the mid-19th century, near Manchester in the north of England. Historical collections show the rarity of the dark variety at that time. The difference between the two varieties is controlled by a single gene that controls the amount of the black pigment, melanin, produced.

As the industrial revolution advanced, soot killed the lichens and blackened the bark of the trees. Against the darkened trees, the dark moths were better camouflaged. Within a century, the population had become 90% variety *carbonaria*. Figure 5.1 shows, quite convincingly, the differences in appearance of both trees and moths. The difference is a shift within the population of the allelic frequency distribution of a single gene.

Let it not be said that England did not take steps to curb air pollution. Coal fires were banned in London by Edward I in 1273, albeit only temporarily. Parliament followed up with the Clean Air Act of 1956. At present, *B. betularia* populations in areas recovering from soot deposits are shifting *back* to higher proportions of the mottled *typica* variety.

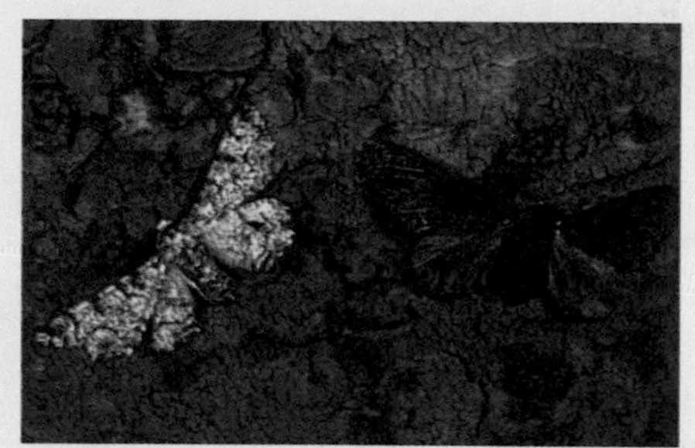

Figure 5.1 Industrial melanism. Left: light- and dark-coloured pepper moths (*Biston betularia*) on normal trees, with lichens growing on the bark. Right: light- and dark-coloured pepper moths on lichen-free trees encrusted with soot. Each picture contains two moths, one camouflaged and the other easily visible. Can you spot the camouflaged moths?

Reproduced with permission from: Kettlewell, H.B.D. (1956). Further selection experiments on industrial melanism in the *Lepidoptera*. Heredity **10**, 287–301.

different animals and plants, but also the different design solutions for structural, locomotory, and sensory systems between vertebrates and invertebrates.

- *At the level of ecosystems*, different populations explore their modes of interaction. Many pairs of species co-evolve. Some examples of co-evolution of species are known from *cooperating species*; for instance, the correlation between the anatomy of flowers and their insect pollinators – a subject studied by Darwin himself – or the correlation between colour changes in fruit ripening and the development of colour vision in animals. Other examples of co-evolution involve *species in competition or conflict*; for example, predator–prey relationships. These include the wars between humans and pathogenic bacteria and viruses.

The mechanism of evolutionary change is now understood, at least in general terms. Genetic reassortment and mutation generate inheritable phenotypic variation. Phenotype-dependent differential rates of reproduction – that is, **natural selection** – governs which alleles, at which frequencies, are passed on to succeeding generations. Alternatively, even in the absence of selection, **genetic drift** can lead to alterations in genome contents and distributions in populations.

The two elements of exploration – variation from the current state of the system and change to a new one – occur at many levels, from individual genomes to proteins to cells to ecosystems. A long-standing challenge of biology is to understand the relationships among these different levels of evolution.

Biological systematics

Classically, the unit of large-scale evolution is the species. Species represent nature's experiments in structures and lifestyles. Both Linnaeus and Darwin recognized the importance of species, and made them the focus of their work. The concept of species remains essential, despite its attendant difficulties (see p. 66). In view of the theoretical problems, biologists present the analysis of known species, and higher-order taxa, in terms based on tradition and convention. We shall explore how modern analytic methods based on genomic and structural data mesh with the classical approaches.

Study of the vast variety of living organisms requires that we organize what we observe and measure. We have to agree on what we call things. **Biological taxonomy** encompasses identifying new life forms (new in the sense of new to the scientific literature), deciding where they fit in, and assigning them a name – based on some 'real or fancied characteristic of the form described' (A.S. Romer) and equipped with a proper description and deposition of specimen material.

Biological nomenclature

Two problems in organizing biological nomenclature are what to name and how to assign the names. The taxonomic hierarchy – kingdom, phylum, class, order, family, genus, and species – introduced by Linnaeus is still in use, although with modifications. (Another legacy is the continued use of classical languages.) Members of more restrictive categories, for instance, several species in the same genus, have more shared features and higher degrees of similarity than the members of more inclusive categories, such as phylum or class. As Nature is not as 'neat' as the 18th century scientists would have liked, boundaries between taxa are fuzzy.

Descriptions of new species are more common than descriptions of new genera, families, etc. During the years 1970–1998, five times as many new species and subspecies were described than new genera. New families, orders, classes, and phyla appear much more infrequently (Table 5.1).

Biological names usually describe features of a species (e.g. giant kangaroo, *Macropus giganteus*, which means large-foot gigantic, not to be confused with the 'bigfoot' primate alleged to inhabit the northwest USA and adjacent regions of Canada). Names may indicate location (e.g. Virginia opossum, *Didelphis virginiana*), or recognize the discoverer (e.g. Darwin's rhea, *Rhea darwinii*, a large bird encountered by Darwin on his visit to South America). J. Gould named the species in Darwin's honour in 1837. (P.H.G. Mohring had named the genus in 1752, to reflect the large size of the birds: the Greek goddess Rhea, mother of Zeus, was a female titan, member of a mythological race of giants.) The scientific name of Père David's deer, *Elaphurus davidianus*, is another example of an animal named after its discoverer.

Table 5.1 New taxa described from 1970 to 1998

Taxa	Numbers described
New phyla	11
New classes	44
New orders	100
New families	731
New genera	8 579
New species	38 590
New subspecies	3 231

From: Winston, J.E. & Metzger, K. (1998). Trends in taxonomy revealed by published literature. BioScience **48**, 125–128.

Other sources of names include expedition sponsors, thesis supervisors, and public figures such as kings and queens, politicians, artists, musicians, and sporting figures. A marine mollusc, *Rotaovula hirohitoi*, was named for the former Emperor of Japan, who was himself a serious marine biologist. Some scientists have named creatures with loathsome features after rivals, as insults. Finally, J.E. Winston has written: 'These days, it would be considered pretty tacky . . . to name a species after yourself'.* For unusual and, in some cases, amusing examples, see http://home.earthlink.net/~misaak/taxonomy.html, http://cache.ucr.edu/~heraty/menke.html and http://cache.ucr.edu/~heraty/yanega.html#ECOLOGY.

Biological nomenclature is governed by international agreements, adopted by consent by professional scientists. The International Codes of Zoological and Botanical Nomenclature separately offer rules for the naming of animals and plants. Nomenclature of bacteria grew out of, and eventually split off from, the botanical code. Virologists have developed their own classification. Currently the International Union of Biological Sciences, a member of the International Council of Scientific Unions, concerns itself with biological nomenclature. It is a sponsor of the Species 2000 project, an effort to curate a complete and integrated database of the world's species, including plants, animals, fungi, and microbes. The Species 2000 project coordinates its activities with other similar international efforts, including the Interagency Taxonomic Information System (ITIS) and the Global Biodiversity Information Facility (GBIF). The related projects of the Tree of Life (http://tolweb.org/tree/) and ARKive (http://www.arkive.org/) include pictorial databases.

The World Conservation Union, usually known by its former name, the International Union for the Conservation of Nature and Natural Resources (IUCN), maintains a 'Red list' of endangered species.

Measurement of biological similarities and differences

Ultimately, comparisons in biology involve observations of differences between individual organisms. With access to living populations, it is possible to get a sense of the variability among individuals and to observe a variety of features, including physiology and lifestyle, in addition to 'static' anatomy.

In contrast, for extinct species, palaeontologists are often limited to fragmentary samples of hard parts – bones and teeth – sometimes from a single individual. Indeed, in the classical era of natural history exploration, a museum in Europe would often receive only a preserved body, even from species that still thrived elsewhere in the world. (Père David's deer is a typical example.) Despite these handicaps in data collection, biologists in the pre-molecular era built their taxonomic edifice on studies of comparative anatomy, embryology, and stratigraphy for dates. They developed spectacular expertise: it was said that Cuvier, the founder of vertebrate palaeontology, could reconstruct the entire skeleton of an animal from a single bone.

Understanding biological diversity requires observation and measurement of similarities and differences. What features should one compare? Classically, the choice depended on expertise and experience. W.E. Le Gros Clark wrote,

> While it may be broadly accepted that, as a general proposition, degrees of genetic relationship can be assessed by noting degrees of resemblance in anatomical details, it needs to be emphasized that morphological characters vary considerably in their significance for this assessment. Consequently it is of the utmost importance that particular attention should be given to those characters whose taxonomic relevance has been duly established by comparative anatomical and palaeontological studies.
>
> – Le Gros Clark, W.E. (1971). *The Antecedents of Man*, 3rd edn. Quadrangle Books, Chicago, pp. 11–12.

This approach works fine in the hands of a professional with expertise and as distinguished as Le Gros Clark, but it has also elicited attempts to make classification methods more quantitative and objective. These attempts include the development of computational methods for interpreting similarities of a wide spectrum of features, some but not all based on sequence data.

Molecular techniques

Many molecular properties have been used for phylogenetic studies, some surprisingly long ago.

* Winston, J.E. (1999). *Describing Species*. Columbia University Press, New York, p. 165.

Serological cross-reactivity was applied to detect relationships from the beginning of the last century until superseded by the direct use of sequences. E.T. Reichert and A.P. Brown published, over a century ago (in 1909), a phylogenetic analysis of fishes based on haemoglobin crystals. Their work was based on Stenö's law (1669), which states that although different crystals of the same substance have different dimensions – some are big, some small – they have the same interfacial angles. We now understand that this law reflects the similarity in microscopic arrangement and packing of the atomic or molecular units within the crystals. Reichert and Brown showed that the interfacial angles of crystals of haemoglobins isolated from different species showed patterns of similarity and divergence parallel to the species' taxonomic relationships.

Reichert and Brown's results are replete with significant implications, which can be appreciated only in retrospect. They demonstrate that proteins have definite, fixed shapes, an idea by no means recognized at the time. They imply that, as species progressively diverge, the structures of their haemoglobins progressively diverge also. In 1909, no one had a clue about nucleic acid or protein sequences. In principle, therefore, the recognition of evolution of protein structures preceded, by half a century, the idea of evolution of nucleotide and amino acid sequences. Reichert and Brown even saw a structural difference between oxy- and deoxyhaemoglobin.

- The Reichert and Brown work has my vote for the most premature scientific result ever.

Today, DNA sequences provide the best measures of similarities among species for phylogenetic analysis. Many genes are available for comparison. This is fortunate, because, given a set of species to be studied, it is necessary to find genes that vary at an appropriate rate. Genes that remain almost constant among the species of interest provide no discrimination. Genes that vary too much cannot be aligned.

Fortunately genes vary widely in their rates of change. The mammalian mitochondrial genome, a circular, double-stranded DNA molecule approximately 16 000 bp long, provides a useful fast-changing set of sequences for the study of evolution among closely related species. In contrast, slowly changing ribosomal RNA (rRNA) sequences were used by C. Woese to identify the three major divisions of life: archaea, bacteria, and eukarya (see Figure 4.1).

- In order to develop a clear picture of the relationships between species, it is necessary to pick a molecule that is changing at a reasonable rate. There must be enough change such that the signal does not sink below the noise level, but not too much change as to obscure common features.

Homologues and families

Products of evolution retain similarities. The similarities appear at many levels – related people, recently diverged species, tissues within an organism containing related cell types but varying protein expression patterns, amino acid sequences and structures of proteins, and DNA sequences. A major theme of biology traditionally has been to recognize and classify such similarities, with a view to understanding how they arose and, when appropriate, to what purpose (i.e. with what selective advantage).

To trace the course of evolution, we must quantitatively measure such similarities. There are many possible objects of such analysis – sequences of individual genes, full-genome sequences, sequences and structures of proteins, anatomical features, patterns of development, and any other phenotypic character one might choose. In many cases, the patterns of similarity between different features of a set of species give corresponding results, bolstering our confidence in their significance.

However, it is necessary to keep clearly in mind that similarity, which is observable, is a surrogate for relationship, which usually is not. Related biological objects are homologues, or families. In many cases, such as the globins, the similarities are sufficient to give us confidence that we are analysing a family of related molecules. Ideally, we have a spectrum of similarities, including close relatives and some distant ones, with the distant relatives linked by chains of close ones. For the comparisons of globins from different species, the congruence of the degrees of similarity of the molecules with measurements of similarities of other sets of molecules and with the classical taxonomic relationships between species is reassuring. For globins within a single species, the common conserved features argue that we are dealing with a single diverged family.

But divergence does not stop within the scope of our ability to detect homologues, and there are many cases where (a) a tantalizing tenuous degree of similarity suggests homology, but we remain unsure whether or not the inference of relationship is valid, or (b) there is sufficient dissimilarity between two molecules or structures that homology is unsuspected, but a series of missing links clearly connects the two.

Our most precise tools measure similarity between sequences or between molecular structures. These tools are mature methods. They have been calibrated to allow us to decide, in all but the hardest cases, whether or not we are dealing with homologues.

Pattern matching – the basic tool of bioinformatics

Given suitable data, computer programs can measure similarities and extract common patterns. Programs to extract patterns in sequences are powerful and readily available. Indeed, sequence comparisons are a problem common to many fields, including the text editors available in all computer systems. For protein structures also, it is possible to detect and measure similarities and common patterns. This makes it possible to study sequence–structure relationships quantitatively.

Other types of biological information – such as protein function, expression patterns, information about characteristics that distinguish species – do not present themselves quite so naturally in forms adapted for computational analysis. They require foundation work to create models of the information, including identification of the important categories of data, and controlled and carefully defined vocabularies for their description. The Gene Ontology Consortium's classifications of protein functions allowed development of tools for quantitative measurement of similarity and divergence of function. (See Chapter 11.)

Such rules and regulations governing how to express data are also essential for database integration. They provide the basis on which independent databases in related or overlapping fields can communicate and cooperate with one another. They allow information retrieval software to handle queries requiring coordinated access to several databases.

- To measure similarity between two sequences, find their optimal alignment – the best matching up of the individual characters – and produce a cumulative score of the similarities between the characters at each position.

Sequence alignment

Given two or more sequences, we wish to:

- measure their similarity;
- understand how the residues match up;
- observe patterns of conservation and variability; and
- infer evolutionary relationships.

If we can do this, we will be in a good position to go fishing in databanks for related sequences, and measuring relative degrees of similarity among genes or proteins. A major application of sequence alignment is to the annotation of genes, through identification of homologues, in order to assign structure and function to as many genes as possible.

Sequence alignment is the identification of residue–residue correspondences. Any assignment of correspondences that preserves the *order* of the residues within the sequences is an alignment. Alignments may contain gaps. For example,

Given two text strings: first string = a b c d e
second string = a c d e f

a reasonable alignment would be:
```
a b c d e -
a - c d e f
```

Some alignments are better than others. For the sequences gctgaacg and ctataatc:

An uninformative alignment:
```
        - - - - - - - g c t g a a c g
        c t a t a a t c - - - - - - -
```

An alignment without gaps:
```
g c t g a a c g
c t a t a a t c
```

An alignment with gaps:
```
g c t g a - a - - c g
- - c t - a t a a t c
```

And another:
```
g c t g - a a - c g
- c t a t a a t c -
```

Most readers would consider the last of these alignments the best of the four. To decide whether it is the best of *all* possibilities, we need a way of examining all possible alignments systematically. We need to compute a score reflecting the quality of each possible alignment and to identify an alignment with the optimal score. The optimal alignment may not be unique: several different alignments may give the same best score. Moreover, even minor variations in the scoring scheme may change the ranking of alignments, causing a different one to emerge as the best.

The dot plot

The **dot plot** is a simple picture that gives an overview of pairwise sequence similarity. Less obvious is its close relationship to alignments.

The dot plot is a table or matrix. The rows correspond to the residues of one sequence and the columns to the residues of the other sequence. In its simplest form, the positions in the dot plot are left blank if the residues are different and filled if they match. Stretches of similar residues show up as diagonals in the upper left–lower right (northwest–southeast) direction (see Figure 5.2).

Dot plots gives quick pictorial statements of the relationship between two sequences. Obvious features of similarity stand out. Figure 5.3 shows a dot plot of a sequence containing internal repetitions. Figure 5.4 shows a dot plot of a palindromic sequence (a sequence that is identical to its reversal).

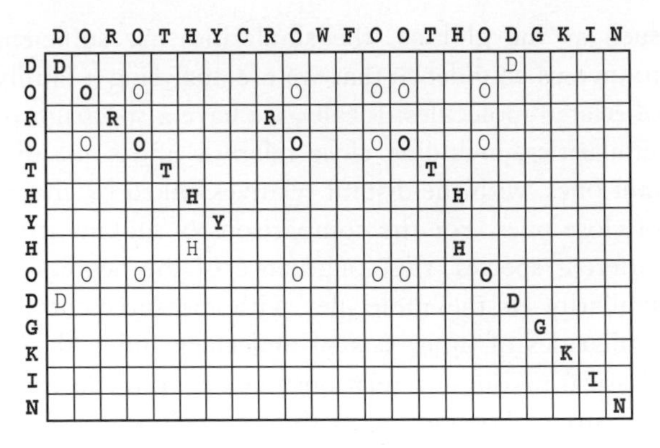

Figure 5.2 Dot plot showing identities between the short name (DOROTHYHODGKIN) and full name (DOROTHYCROWFOOTHODGKIN) of a famous protein crystallographer.

Letters corresponding to *isolated* matches are shown in non-bold type. The longest matching regions, shown in red, are the first and last names DOROTHY and HODGKIN. Shorter matching regions, such as the OTH of dorOTHy and crowfoOTHodgkin, or the RO of doROthy and cROwfoot, are noise. Note the effect of the 'insertion' of Crowfoot in interrupting and displacing the matching.

	A	B	R	A	C	A	D	A	B	R	A	C	A	D	A	B	R	A
A	A			A		A		A			A		A		A			A
B		B							B							B		
R			R							R							R	
A	A			A		A		A			A		A		A			A
C					C							C						
A	A			A		A		A			A		A		A			A
D							D							D				
A	A			A		A		A			A		A		A			A
B		B							B							B		
R			R							R							R	
A	A			A		A		A			A		A		A			A
C					C							C						
A	A			A		A		A			A		A		A			A
D							D							D				
A	A			A		A		A			A		A		A			A
B		B							B							B		
R			R							R							R	
A	A			A		A		A			A		A		A			A

Figure 5.3 Dot plot showing identities between a repetitive sequence (ABRACADABRACADABRA) and itself. The repeats appear on several subsidiary diagonals parallel to the main diagonal.

A dot plot relating real amino acid sequences – human and *Xenopus laevis* ephrin B3 – shows that the similarity is stronger in the N-terminal part of the protein. Figure 5.5 shows the dot plot and the corresponding sequence alignment. It is useful to look at these together and see how the regions of high and low similarity correspond in the two figures.

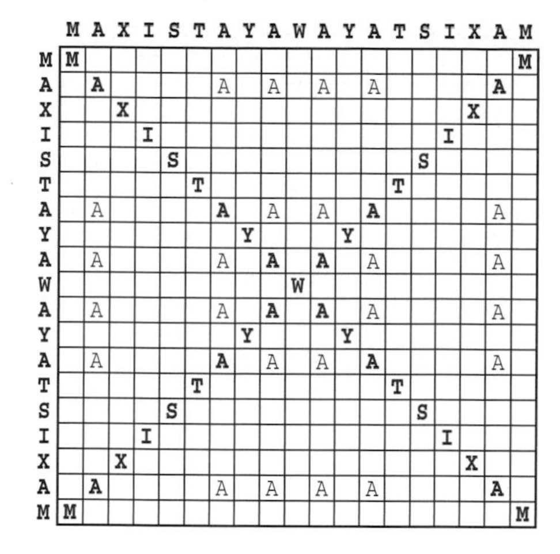

Figure 5.4 Dot plot showing identities between the palindromic sequence MAX I STAY AWAY AT SIX AM and itself. The palindrome reveals itself as a stretch of matches *perpendicular* to the main diagonal.

This is not just word play – regions in DNA recognized by transcriptional regulators or restriction enzymes have sequences related to palindromes. Longer regions of DNA or RNA containing inverted repeats of this form can form stem–loop structures.

- Ephrins are proteins that guide axons in the developing nervous system, and play a number of other important roles in development.

A disadvantage of the dot plot is that its 'reach' into the realm of distantly related sequences is poor. In analysing sequences, one should always look at a dot plot to be sure of not missing anything obvious, but be prepared to apply more subtle tools.

Dot plots and alignments

How can we derive an optimal alignment of two sequences? Conceptually, any alignment – that is, any assignment of residue–residue correspondences – is equivalent to a path through a dot plot. A diagonal move corresponds to an equivalence between two residues. Horizontal and vertical moves correspond to insertions and deletions. Because any allowable alignment assigns residues uniquely, and in order along the sequences, the *only* allowable moves are southeast (diagonal), east (horizontal), and south (vertical).

Figure 5.6 shows the optimal path through the Dorothy Hodgkin dot plot. The path passes through the largest number of matching residues. The horizontal segment of the path corresponds to the insertion of Crowfoot.

In this example, the optimal alignment and optimal path are obvious. In general, a computer program must examine all possibilities. How to do that effectively is a matter of some delicacy. Without explaining the methods in detail, the trick is to decide, for each *partial* path, what its best *extension* is. Algorithms for relating *locally* optimal moves to integrated optimal pathways – that is, for constructing full alignments – depend on a mathematical technique called dynamic programming.[1]

- A dotplot shows perspicuously the quality and distribution of the pattern of similarity between two sequences. Each possible alignment of the two sequences corresponds to a path through the dotplot, from upper left to lower right.

Varieties and extensions

Global alignment assigns correspondences to *all* residues in the sequences. If one sequence is shorter than the other, the difference in length must be made up by insertions/deletions.

Local alignment is a pattern-matching technique for identifying a match for a short probe sequence within a much longer text. Gaps outside the local match are not penalized. This is a common task in text searching and editing. Finding all instances of the word 'dream' in Hamlet is an example of local pattern matching. (Allowing a mismatch and a gap would pick up the word 'drum' as well as 'dream'.) If you consider the DNA sequence of an entire human chromosome as a long string of characters, searching for a particular gene sequence within the chromosome is a local matching problem.

A very important extension of pairwise sequence alignment is **multiple sequence alignment**, the mutual alignment of three or more sequences. Usually we can find large families of similar sequences by identifying homologues in many different species.

[1] For further details, see Lesk, A.M. (2008). *Introduction to Bioinformatics*, 3rd ed. Oxford University Press, Oxford.

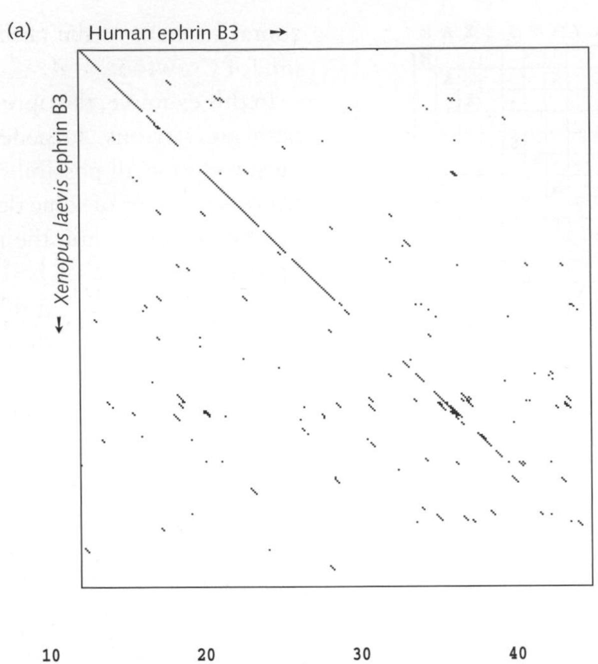

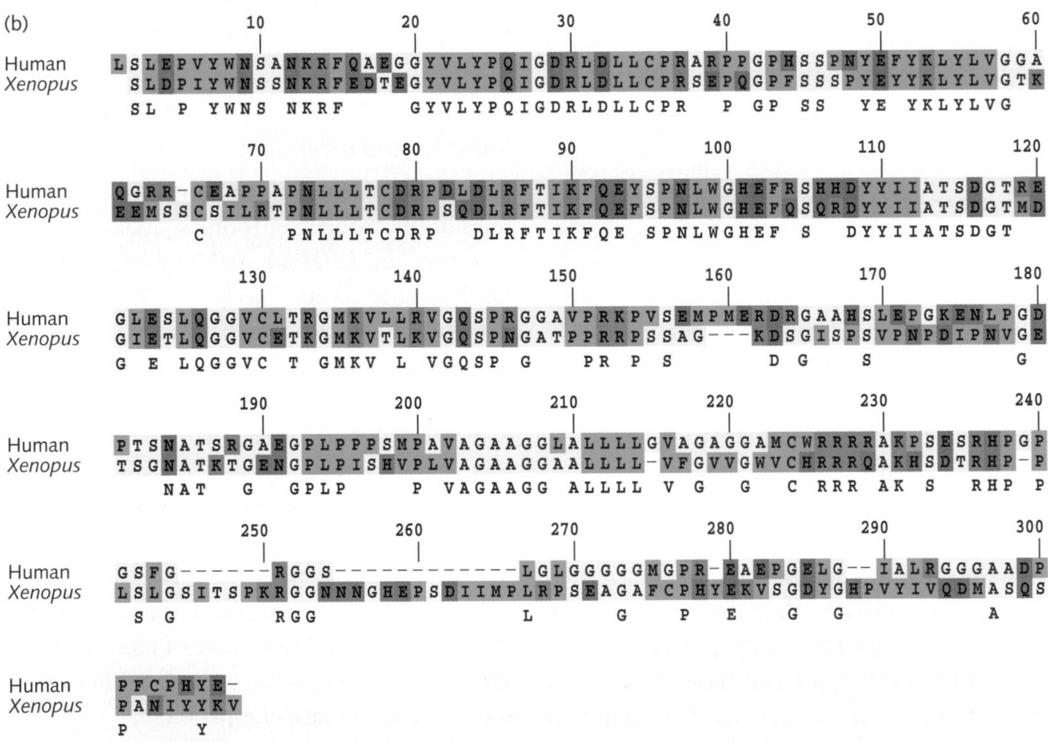

Figure 5.5 Relationships between the sequences of ephrin B3 proteins from human and *Xenopus laevis*. (a) Dot plot. The major signal is along the main diagonal, interrupted by occasional divergent regions, and showing the substantially weaker similarity near the C terminus. (b) Sequence alignment. Amino acids are colour coded by physicochemical type. Letters under the sequences indicate positions occupied by the same residue in both sequences.

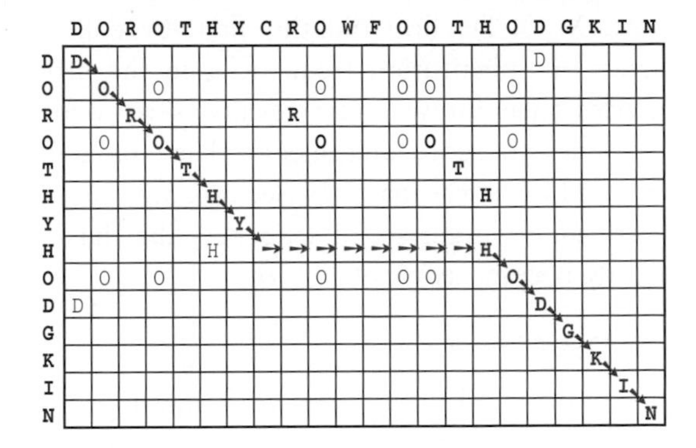

Figure 5.6 A path through the Dorothy Hodgkin dot plot. Diagonal arrows correspond to aligned residues, horizontal arrows to gap insertions. The corresponding alignment is the obvious one:

```
DOROTHY--------HODGKIN
DOROTHYCROWFOOTHODGKIN
```

Multiple sequence alignment reveals the underlying patterns contained in a set of related sequences much more clearly than pairwise sequence alignments.

Programs for all of these different alignment problems are available on the web:

Global alignment (pairwise and multiple)

CLUSTAL W	`http://www.ebi.ac.uk/clustalw`
T-Coffee	`http://www.igs.cnrs-mrs.fr/Tcoffee/tcoffeecgi/index.cgi`
EMBOSS	`http://www.ebi.ac.uk/emboss/align`

Local alignment

SSEARCH	`http://pir.georgetown.edu/pirwww/search/pairwise.shtml`
EMBOSS	`http://www.ebi.ac.uk/emboss/align/`

Defining the optimum alignment

To go beyond 'alignment by eyeball' via dot plots, we must define quantitative measures of sequence similarity and difference.

Given two character strings, two measures of the distance between them are as follows:

- the **Hamming** distance, defined between two strings of equal length, is the number of positions with mismatching characters.
- the **Levenshtein**, or edit, distance between two strings of not necessarily equal length is the minimal number of 'edit operations' required to change one string into the other, where an edit operation is a deletion, insertion, or alteration of a single character in either sequence.

For example:

```
agtc     Hamming distance = 2
cgta

ag-tcc   Levenshtein distance = 3
cgctca
```

A given sequence of edit operations induces a unique alignment, but not *vice versa*.

For applications to molecular biology, we wish to assign variable weights to different edit operations. For nucleic acids, we know that **transition** mutations (purine↔purine and pyrimidine↔pyrimidine; i.e. A↔G and T↔C) are more common than **transversions** (purine↔pyrimidine; i.e. (A or G)↔(T or C)). For proteins, amino acid substitutions tend to be conservative: the replacement of one amino acid by another with similar size or physicochemical properties is more likely to occur than its replacement by another amino acid with dissimilar properties. Similarly, the deletion of several contiguous bases or amino acids is more probable than the independent deletion of the same number of isolated bases.

A computer program can score each path through the dot plot by adding up the scores of the individual steps. For each substitution, it adds the score of the mutation, depending on the pair of residues involved. For horizontal and vertical moves, it adds a suitable gap penalty.

Scoring schemes

A scoring system must account for residue substitutions and insertions or deletions. (An insertion from one sequence's point of view is a deletion as seen by the other.) Deletions, or gaps in a sequence, will have scores that depend on their lengths.

For nucleic acid sequences, it is common to use a simple scheme for substitutions, +1 for a match, −1 for a mismatch, or a more complicated scheme based on the higher frequency of transition mutations than transversion mutations. One possibility is:

	A	G	T	C
A	20	10	5	5
G	10	20	5	5
T	5	5	20	10
C	5	5	10	20

For proteins, a variety of scoring schemes have been proposed. We might group the amino acids into classes of similar physicochemical type and score +1 for a match within a residue class and −1 for residues in different classes. We might try to devise a more precise substitution score from a combination of properties of the amino acids. Alternatively, we might try to let the proteins teach us an appropriate scoring scheme. M.O. Dayhoff did this first, by collecting statistics on substitution frequencies in the protein sequences then known. Her results were used for many years to score alignments. They have been superseded by newer matrices (see Box 5.2) based on the very much larger set of sequences that has subsequently become available.

The BLOSUM matrices

S. Henikoff and J.G. Henikoff developed the BLOSUM matrices for scoring substitutions in amino acid sequence comparisons. The BLOSUM matrices are based on the BLOCKS database of aligned protein sequences; hence the name: BLOcks SUbstitution Matrix. From regions of closely related proteins alignable without gaps, Henikoff and Henikoff

The BLOSUM62 matrix used for scoring amino acid sequence similarity

Rows and columns are in alphabetical order of the three-letter amino acid names. Only the lower triangle of the matrix is shown, as the substitution probabilities are taken as symmetric (not because we are sure that the rate of any substitution is the same as the rate of its reverse, but because it is difficult to determine the differences between the two rates).

Ala (A)	4																			
Arg (R)	−1	5																		
Asn (N)	−2	0	6																	
Asp (D)	−2	−2	1	6																
Cys (C)	0	−3	−3	−3	9															
Gln (Q)	−1	1	0	0	−3	5														
Glu (E)	−1	0	0	2	−4	2	5													
Gly (G)	0	−2	0	−1	−3	−2	−2	6												
His (H)	−2	0	1	−1	−3	0	0	−2	8											
Ile (I)	−1	−3	−3	−3	−1	−3	−3	−4	−3	4										
Leu (L)	−1	−2	−3	−4	−1	−2	−3	−4	−3	2	4									
Lys (K)	−1	2	0	−1	−3	1	1	−2	−1	−3	−2	5								
Met (M)	−1	−1	−2	−3	−1	0	−2	−3	−2	1	2	−1	5							
Phe (F)	−2	−3	−3	−3	−2	−3	−3	−3	−1	0	0	−3	0	6						
Pro (P)	−1	−2	−2	−1	−3	−1	−1	−2	−2	−3	−3	−1	−2	−4	7					
Ser (S)	1	−1	1	0	−1	0	0	0	−1	−2	−2	0	−1	−2	−1	4				
Thr (T)	0	−1	0	−1	−1	−1	−1	−2	−2	−1	−1	−1	−1	−2	−1	1	5			
Trp (W)	−3	−3	−4	−4	−2	−2	−3	−2	−2	−3	−2	−3	−1	1	−4	−3	−2	11		
Tyr (Y)	−2	−2	−2	−3	−2	−1	−2	−3	2	−1	−1	−2	−1	3	−3	−2	−2	2	7	
Val (V)	0	−3	−3	−3	−1	−2	−2	−3	−3	3	1	−2	1	−1	−2	−2	0	−3	−1	4
	A	R	N	D	C	Q	E	G	H	I	L	K	M	F	P	S	T	W	Y	V

calculated the ratio of the number of observed pairs of amino acids at any position to the number of pairs expected from the overall amino acid frequencies. In order to avoid overweighting closely related sequences, the Henikoffs replaced groups of proteins that had sequence identities higher than a threshold by either a single representative or a weighted average. The threshold of 62% similarity produces the commonly used BLOSUM62 substitution matrix. This is offered by all programs as an option and is the default in most.

The BLOSUM62 matrix is shown in Box 5.2. It expresses scores as *log-odds* values:

$$\text{Score of mutation } i \leftrightarrow j = \log_{10} \frac{\text{observed } i \leftrightarrow j \text{ mutation rate}}{\text{mutation rate expected from amino acid frequencies}}$$

The numbers are multiplied by 10, to avoid decimal points. The matrix entries reflect the probabilities of mutational events. A value of +2 (e.g. leucine↔isoleucine) implies that in related sequences the mutation would be expected to occur 1.6 times more frequently than random. The calculation is as follows: the matrix entry 2 corresponds to the actual value 0.2 because of the scaling. The value 0.2 is $\log_{10}$ of the relative expectation value of the mutation. As $\log_{10}(1.6) = 0.2$, the expectation value is 1.6.

The probability of two independent mutational events is the product of their probabilities. By using logarithms, we have scores that we can add up rather than multiply, a computational convenience.

Scoring insertions and deletions, or 'gap weighting'

To form a complete scoring scheme for alignments, we need, in addition to the substitution matrix, a way of scoring gaps. How important are insertions and deletions, relative to substitutions? We need to distinguish gap initiation:

```
aaagaaa
aaa-aaa
```

from gap extension:

```
aaaggggaaa
aaa----aaa
```

For aligning DNA sequences, the popular alignment software package CLUSTALW recommends use of the identity matrix for substitution (+1 for a match, 0 for a mismatch) and gap penalties of 10 for gap initiation and 0.1 for gap extension by one residue. For aligning protein sequences, the recommendations are to use the BLOSUM62 matrix for substitutions, with gap penalties of 11 for gap initiation and 1 for gap extension by one residue.

- To define optimal alignment, we must assign scores for each possible substitution and corresponding scores for gap initiation and extension.

Approximate methods for quick screening of databases

It is routine to screen genes from a new genome against databases, to find similarities to other sequences. Databases have grown so large that programs based on exact local alignments are too slow. Approximate methods can detect close relationships well and quickly but are inferior to the exact ones in picking up very distant relationships. In practice, they give satisfactory performance in the many cases in which the probe sequence is fairly similar to one or more sequences in a databank, and they are, therefore, certainly worth trying first.

The original paper on BLAST: Altschul, S.F., et al. (1990). Basic local alignment search tool. J. Mol. Biol. ***215****, 403–410, was the field's most highly cited paper published in the 1990s.*

A typical approximation approach such as BLAST (basic local alignment search tool) takes a small integer k and determines all instances of each 'word' of length k (i.e. each set of k consecutive characters, with no gaps) of the probe sequence that occur in any sequence in the database. A candidate sequence is a sequence in the databank containing a large number of matching k-tuples, with equivalent spacing in probe and candidate sequences. For a selected set of candidate sequences, *approximate* optimal alignment calculations are then carried out, with the time- and space-saving restriction that the paths through the matrix considered are restricted to bands around the diagonals containing many matching k-tuples. It is clearest to show the procedure in terms of a dot plot (see Figure 5.7).

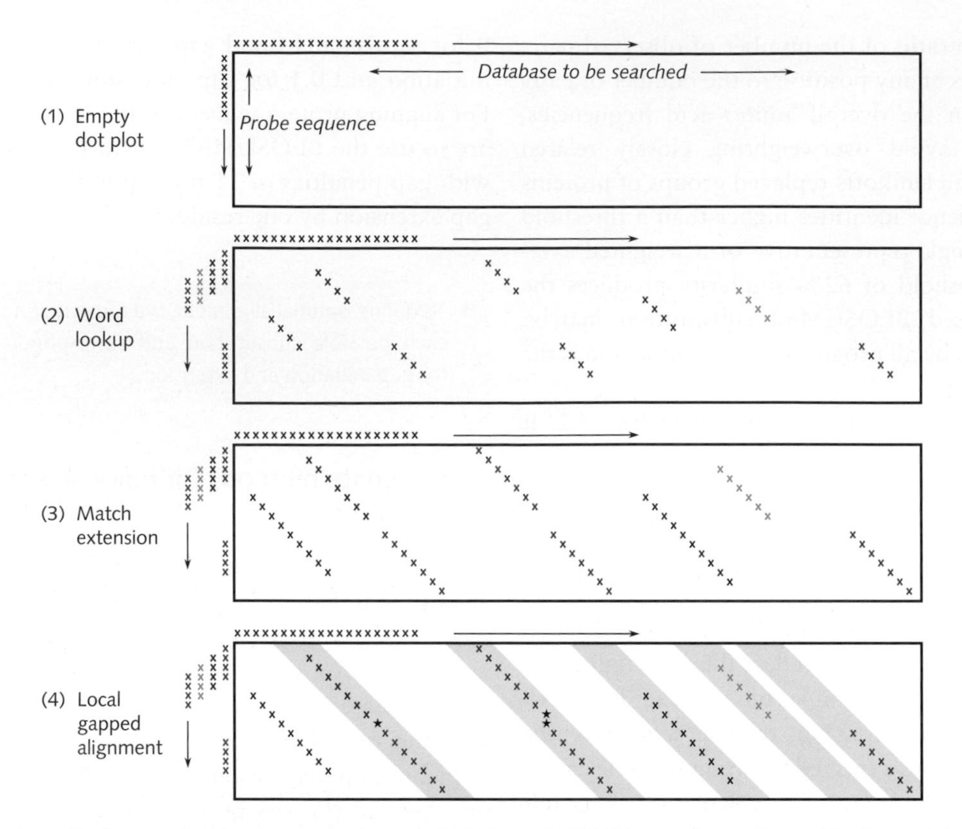

Figure 5.7 Schematic diagram showing the mechanism of a BLAST search. BLAST solves the problem of finding matches of a probe sequence in a full genome or a full database that are much longer than the probe sequence.

(1) The 'playing field' of the algorithm is the outline of a dot plot, just as if the problem were going to be solved by application of an exact-alignment method.

(2) BLAST first divides the probe sequence into fixed-length words of length k; here $k = 4$. It then identifies all exact occurrences of these words in the full database, with no mismatches or gaps. Note that the same four-letter word may occur several times in the probe sequence (shown here in red), and of course each four-letter word may match many times within the database. It is possible to do this step quickly after pre-processing the database to record the sites of appearance of all four-letter words.

(3) Starting with each match, BLAST tries to extend the match in both directions, still with no mismatches or gaps allowed.

(4) Given the extended matches, BLAST tries to put them together by doing alignments *allowing* mismatches and gaps, but only within limited regions containing the preliminary matches (grey areas). The result of this step is to add to the matches the positions shown as ★. This produces longer matching regions.

It is the restriction of the more complex matching procedure to relatively small regions, rather than applying it to the entire matrix, that gives the method its speed. The price to pay is that the method will miss a combined match lying outside the grey area. In the example illustrated, the matching regions coloured red and green, at the right of the matrix, will not be combined but reported as separate hits.

There are several variations on this theme, including the original BLAST program and its variants (see Box 5.3).

Multiple sequence alignments and pattern detection

Multiple sequence alignments are rich in information about patterns of conservation. They helps us to understand the common features of structure and function of a family of sequences, by showing us which residues are crucial (and therefore conserved). They also help us to identify distant homologues with greater confidence than a pairwise sequence alignment could.

The patterns inherent in a multiple sequence alignment are not merely inferences from the alignment table – this is leaving it too late – but can actively contribute to creating a high-quality alignment. The idea is for an algorithm to learn the underlying patterns *while* it is assembling the multiple sequence alignment.

BOX 5.3

Different 'flavours' of BLAST search different databases

Program	Searches for:	In:
BLASTN	Nucleotide sequence	Nucleotide sequence database
BLASTX	Six-frame translations of a nucleotide sequence	Protein sequence database
BLASTP	Protein sequence	Protein sequence database
TBLASTN	Protein sequence	Six-frame translations of a nucleotide sequence database
TBLASTX	Six-frame translations of a nucleotide sequence	Six-frame translations of a nucleotide sequence database

One very powerful program based on this approach is PSI-BLAST, an extension of BLAST for multiple sequence alignment (see Figure 5.8). PSI-BLAST constructs a **profile,** i.e. a conservation pattern, in an initial multiple alignment of the 'hits' from a preliminary BLAST search. Armed with the profile, the method returns to the database and does a more sensitive search, giving higher weight to well-conserved positions; it then realigns what it finds and refines the profile. Several such cycles of refinement of the profile give PSI-BLAST the power both to detect distant relationships and to create high-quality multiple sequence alignments.

Perhaps the most powerful pattern analysis algorithms are based on **hidden Markov models.** A hidden Markov model is a mathematical construct that generates sequences according to internal probabilistic rules. Successive rolls of a pair of dice generate a sequence of numbers between 2 and 12, with different probabilities. However, there is not even a probabilistic link between the successive numbers that come up. Typing with one finger at random generates a sequence of characters, again with no correlation between successive letters. However, if you insist on typing ten keys per second, so that you can only move your hand a limited extent between successive keystrokes, then Q is more likely to be followed by W than by M. The result would be a probabilistically generated sequence with the observed distribution of each character dependent on what preceded it.

To represent a set of sequences with a hidden Markov model, imagine a computer program that

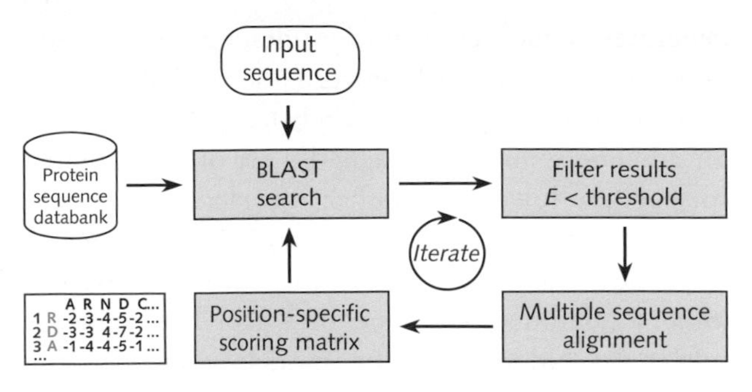

Figure 5.8 Schematic flowchart of a PSI-BLAST calculation to detect protein sequences in a database that are similar to a probe sequence. The user submits an input sequence and chooses a protein sequence databank to probe.

First, using the input sequence and a standard substitution matrix such as BLOSUM62, an ordinary BLAST calculation identifies similar sequences in the database and assigns a statistical measure of significance, E, to each 'hit'. For each sequence retrieved from the database, E is the number of sequences of equal or higher similarity to the probe sequence that would be expected to be found in the database, just by chance.

The program will select those sequences for which E is no greater than a specified threshold, often chosen as 0.005, and perform a multiple sequence alignment of them.

By counting the relative frequencies of different amino acids in each column of the multiple sequence alignment, the program will derive a position-specific scoring matrix. The red box at the lower left shows part of a position-specific scoring matrix. The columns are labelled by the 20 natural amino acids, shown in blue. The rows are labelled by the sequence to be scored by the matrix, residue numbers in red and amino acids in green. In this case, the N-terminal sequence of the sequence to be scored is RDA . . . The entries in the column are the log-odd scores of finding any amino acid at any position in the multiple alignment. For instance, the entry under A in row 3 is –1; therefore, the probability of finding an A at the third position is proportional to 10^{-1}.

To find the score of the sequence, add up the values in the R column of the first row, the D column of the second row, the A column of the third row, etc., to give: $10^{-3} + 10^{-7} + 10^{-1}$. In this example, the probabilities are expressed unscaled and as logarithms to the base 10. Note that the sequence being scored may contain gaps.

This matrix can be used as an alternative to the input sequence and substitution matrix in a BLAST search. Each subsequent BLAST search, based on the matrix derived in the previous step, will return a different set of 'hits'. With a sensible choice of input parameters, the procedure will usually converge to produce a more reliable set of similar sequences than would be returned by the simple BLAST search of the input sequence performed in the first step.

generates sequences of nucleotides, or amino acids, according to rules that govern the probability distributions of *successors* to each letter. For example, an adenine would be assigned a set of probabilities for being followed by another adenine, or a thymine, or a cytosine, or a guanine, or a gap. A different set of probabilities would govern the successors of a thymine, a guanine, a cytosine, or a gap. The enhanced power of hidden Markov models over position-specific scoring matrices stems from the correlation between successive positions.

For example, it is observed that the dinucleotide frequency CpG is lower in higher organisms than would be expected from the overall mole fractions of C and G in the genome. A hidden Markov model would reflect this in a lowered probability of G in a position following a C, relative to the probabilities of a G following A, T, or another G.

Given a set of sequences, the process of *training* a hidden Markov model involves adjusting all of the probability distributions so that the sequences generated by the model have a high probability of reproducing the set of sequences analysed.

- A multiple sequence alignment is much richer in information than a pairwise sequence alignment. A hidden Markov model is a method for capturing the information.

Pattern matching in three-dimensional structures

Given two or more structures – perhaps of several homologous proteins – we can frame questions generalized from sequence alignment. How can we measure structural similarity quantitatively? Can we derive a sequence alignment from structural comparisons?

In the native state of a protein, the mainchain follows a curve in space. The general spatial layout of this curve defines a **folding pattern**. The backbones of related proteins show recognizably similar but not identical folding patterns. A letter of the alphabet in different type fonts – for instance, b and *b* – illustrate the topological similarities, and differences, in detail seen among proteins related by evolutionary divergence that share a common folding pattern. A better analogy for widely divergent proteins might be the letters B and R, which share the letter P as a common core substructure but in addition have either a loop (B) or a stroke (R) that differ. Homologous protein structures typically contain fairly large, well-fitting substructures. Figure 5.9 shows a superposition of local regions of two proteins and an overall superposition of two entire structures.

Extraction of the maximum common substructure induces an alignment of the sequences. This is called a **structural alignment**. Because structure changes more conservatively than sequence during evolution, for distantly related proteins it may be possible to align the sequences on the basis of the structures even if methods based purely on sequences cannot recognize the relationship.

- A structure alignment is nevertheless an alignment = an assignment of residue–residue correspondences. Instead of assigning the correspondence by matching the characters in two or more sequences, a structural alignment assigns the correspondence to residues that occupy similar positions in space, relative to the molecular framework.

Evolution of protein sequences, structures, and functions

Extending Crick's classic 'central dogma' gives us the basic paradigm:

DNA → RNA → amino acid sequence of a protein
→ protein structure → protein function

During evolution, selection acts on protein function to alter gene frequencies in populations, closing the loop back to DNA.

Transcription of DNA to RNA, and translation of mRNA by ribosomes, takes us as far as the amino acid sequence. The amino acid sequence dictates the protein structure by a spontaneous folding process (see Chapter 1). Folding produces a native state. For most proteins, the native state structure contains an active site with the proper geometry, charge distribution, and hydrogen-bonding potential to interact specifically

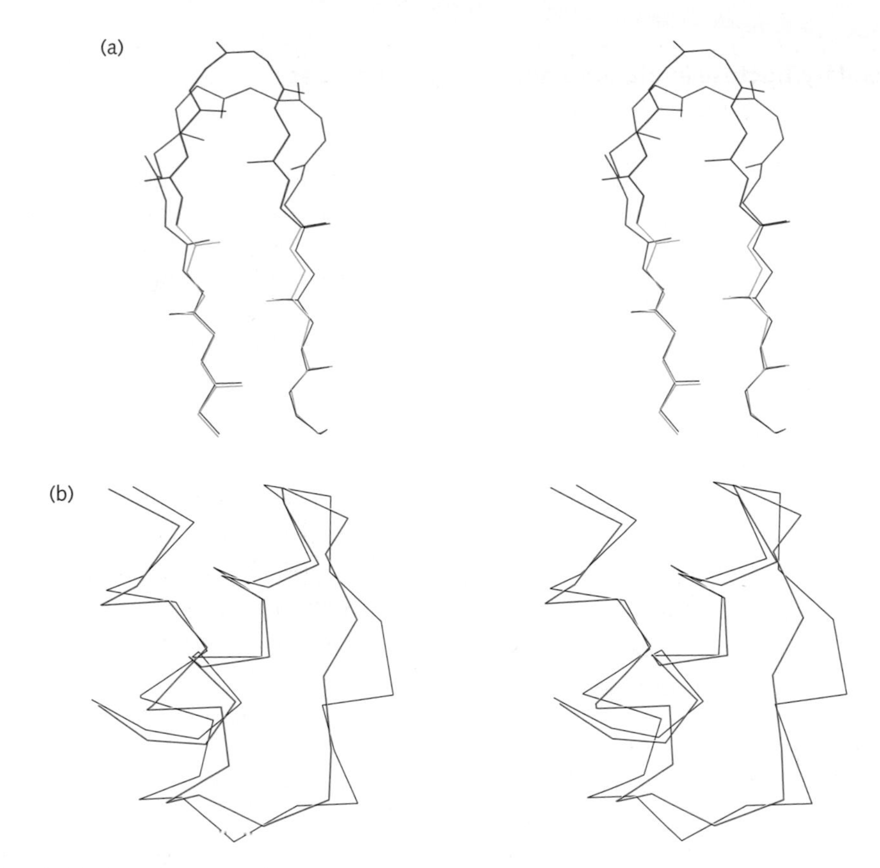

Figure 5.9 Local and global superpositions of protein structures.

(a) Two β-hairpins from the antigen-binding site of an antibody [1VFA, 2FBJ]. Only main-chain atoms are shown. The 'stems' of the loops, parts of strands of β-sheet, superpose well (black and cyan regions, at bottom of picture). The connections have different lengths and conformations, and do not superpose well (red and blue regions, at top of picture). This is an example of a *local* well-fitting substructure. It involves only a small contiguous region of the chains.

(b) Superposition of regions with a folding pattern called the HeH (helix–extended loop–helix). Black: domain from RNA-binding domain of transcriptional terminator protein ρ (from *E. coli*) [1A62]. Red: domain from KU heterodimer (human) [1JEQ]. This figure shows a 'chain trace,' a polygon in space connecting one point from each residue.

The helices at either end of the chains superpose well. The extended regions between the helices do not. The sequence alignment induced by the structural superposition is this:

```
RNA-binding domain   TPVSELITLGENMGLEN–LARMRKQDIIFAILKQH
KU heterodimer       FTVPMLKEACRAYGL–KSG-L-KKQELLEALTKHF
```

with other proteins, or with small-molecule ligands. In many cases, the active site contains catalytic residues that produce enzymatic activity.

The effects of single-site mutations

The native states of proteins are the cumulative effect of many inter-residue interactions. What then should we expect to be the result of a perturbation in the amino acid sequence?

Consider a SNP leading to a single amino acid substitution. Will the structure stay the same? Changing one amino acid without otherwise altering the structure would leave most interactions intact, except for those involving the mutated residue itself (and conservative mutations may preserve even these). Nevertheless, sometimes changing a single residue is enough to blow the original structure apart. An example is the mutation A174D in human aldolase (See Box 5.4). In other cases changes to residues providing specific interactions with ligands may alter activity. Some mutations do not alter the structure but destabilize it; frequently this is enough to cause disease. Box 5.4, treating human

BOX 5.4

Hereditary fructose intolerance and mutants of aldolase B

The enzyme aldolase catalyses the cleavage of two substrates:

fructose-1,6-bisphosphate → glyceraldehyde-3-phosphate + dihydroxyacetone phosphate

fructose-1-phosphate → glyceraldehyde + dihydroxyacetone phosphate

Fructose-1,6-bisphosphate is a mainstream metabolite in glycolysis and gluconeogenesis, classic pathways of glucose metabolism. Fructose-1-phosphate arises in metabolism of dietary fructose. Different isozymes of aldolase have different relative activities towards fructose-1,6-bisphosphate and fructose-1-phosphate.

Approximately 1 in 20 000 people suffer from *hereditary fructose intolerance*, a defect in the liver isozyme, aldolase B. The gene encoding this protein maps to locus 9q22.3 in the human, giving the trait an autosomal recessive inheritance pattern. For affected individuals, ingestion of fructose, a monosaccharide common in fruits and honey, leads to vomiting, discomfort, and hypoglycaemia. Problems often first appear in infancy as fructose and sucrose are added to the diet upon weaning. The condition can be fatal if unrecognized and untreated; however, for most patients it is sufficient to adopt a diet free of fructose and sucrose.

Numerous mutations have been associated with aldolase B dysfunction, including amino acid substitutions, nonsense mutations producing truncated protein, insertions and deletions including frameshifts, and changes in splice sites. The most common mutations are A149P (over 50% of cases worldwide) and A174D.

T. Cox and co-workers* characterized the proteins corresponding to several known mutants. Normal aldolase B is a tetramer of four 363-amino-acid subunits. Because all mutants were discovered in patients presenting with Hereditary Fructose Intolerance, all had reduced or absent enzymatic activity. Two classes of mutants were:

- *Catalytic mutants:* these can be expressed as intact tetramers, retaining some activity at 37°C. These include W147R and R303W.
- *Structural mutants:* these are destabilized, and show catalytic activity (if at all) only after expression at 22–23°C. These include N334K, A149P, L256P, and A174D.

The substitutions in the catalytic mutants occur in or near the active site. The substitutions in the structural mutants occur either in a residue buried in the monomeric structure (A174D, which does not fold at all, as a result of burying a charged sidechain), or in the subunit interface, which causes the protein to dissociate into monomers (N334K, L256P, and A149P).

* Rellos, P., Sygusch, J., & Cox, T.M. (2000). Expression, purification and characterization of natural mutants of human aldolase B.J. Biol. Chem. **275**, 1145–1151.

aldolase, illustrates several effects of single amino acid substitutions.

Many small changes in amino acid sequence leave the basic structure intact, producing only small conformational changes. In this sense, protein structures are *robust* to mutation – not to all mutations, but to enough mutations to allow variability. This is essential – and sufficient – for evolution.

Z. Wang and J. Moult have described some of the kinds of structural effects of single amino acid substitutions related to human diseases[2] (see Figures 5.10, 5.11, and 5.12).

[2] Wang, Z. & Moult, J. (2001). SNPs, protein structure, and disease. Hum. Mut. 17, 263–270.

- A sequence that so lacked robustness that any mutation would destroy it, could not exist. It could have no neighbouring precursor and processes of evolution could never find its sequence.

Evolution of protein structure and function

Sequences and structures of related proteins show coordinated evolutionary divergence. As sequences progressively diverge, structures progressively deform. Typically, a core of the structure, including the major elements of secondary structure and, usually, the active site, retains its folding pattern. Other, peripheral regions of the structure can refold entirely.

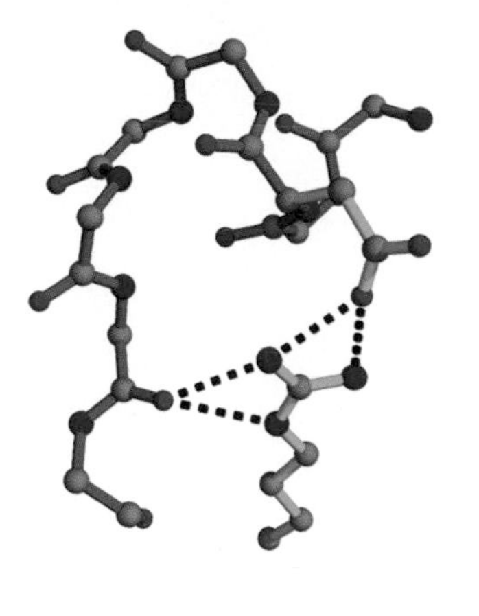
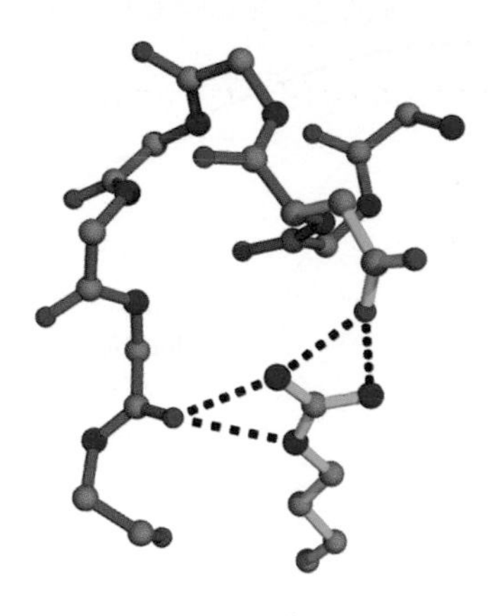

Figure 5.10 Factor XIIIa is the enzyme at the final step of the blood coagulation cascade. It cross-links fibrin molecules, stabilizing clots. The normal protein contains an arginine sidechain that forms a salt bridge and multiple hydrogen bonds, to a neighbouring asparate sidechain and to main-chain carbonyl atoms. The arginine and aspartate sidechains are shown in green. Mutation of the arginine to an isoleucine (not shown) removes these interactions, destabilizing the protein and resulting in poor clot formation [1F13].

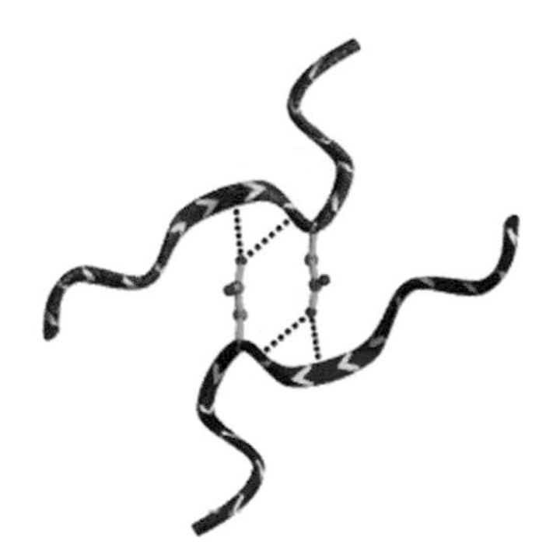
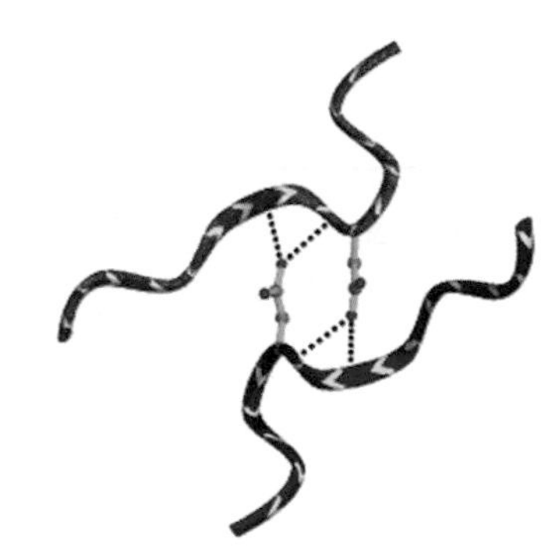

Figure 5.11 Aldolase A is an enzyme in the glycolytic pathway. It is an isozyme of aldolase B, the protein involved in Hereditary Fructose Intolerance. Aldolase is normally a tetramer, stabilized by hydrogen bonds involving an aspartate residue at the inter-subunit interface. Mutation of this aspartate to a glycine destabilizes the tetramer. Although this has no detectable effect on most cell types, red blood cells show weakened cell membranes, causing a congenital form of haemolytic anaemia [2ALD].

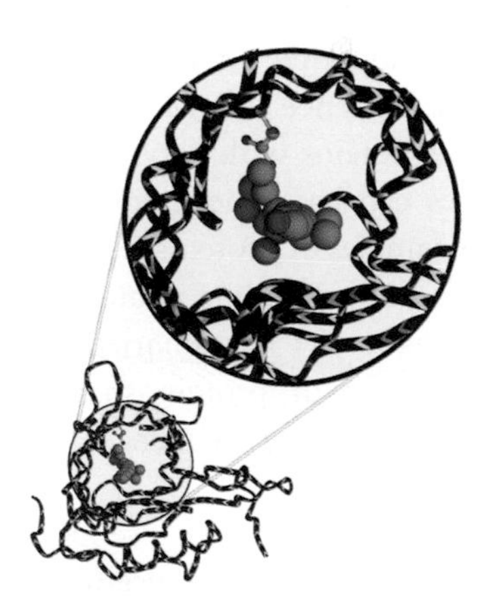

Figure 5.12 Retinol-binding protein transports vitamin A around the bloodstream, bound in a deep hydrophobic cavity within the protein. This figure shows a model of mutant 75Gly→Asp of retinol-binding protein. Note that the model was built *solely* by inserting the sidechain, to observe the structural consequences. No attempt was made to try to predict the structural deformation produced. What the model shows is that there are steric and electrostatic incompatibilities between the sidechain of 75Asp and the ligand. This explains the observation of decreased affinity for retinol, producing vitamin A deficiency and night blindness.

In this process, structure changes more conservatively than sequence. In many families of proteins, we can recognize structural similarity in relatives so distant that there is no easily visible signal of the similarity in the sequence.

The reason for retention of protein conformation in general, and the structure of the active site in particular, is selection for maintenance of function. A need to retain function imposes constraints on protein stability and structural change during evolution.

It is easier to see the effects of these constraints than to understand their mechanism. In many cases certain specific residues are directly involved in function – for example the iron-linked histidine of the globins – and these are immutable. In contrast, constraints that maintain the overall folding pattern are dispersed around the sequence, and it is only by studying patterns of residue conservation in large-scale alignments of homologous proteins that we can begin to understand the constraints imposed by structure on sequence.

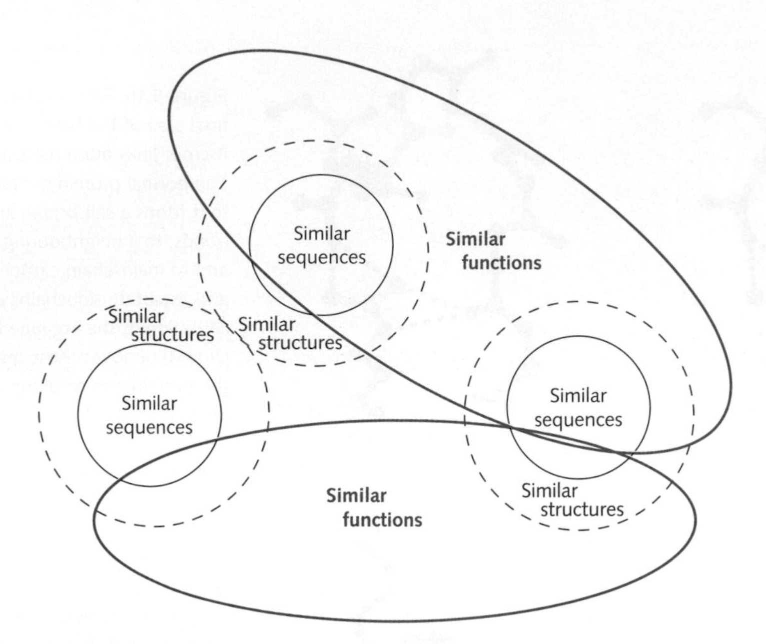

Figure 5.13 Relationships among sequence, structure, and function:

- *similar sequences can be relied on to produce similar protein structures*, with divergence in structure increasing progressively with the divergence in sequence;
- conversely, *similar structures are often found with very different sequences*: in many cases, the relationships in a family of proteins can be detected *only* in the structures, the sequences having diverged beyond the point of our being able to detect the underlying common features;
- *similar sequences and structures often produce proteins with similar functions*, but exceptions abound;
- conversely, *similar functions are often carried out by non-homologous proteins with dissimilar structures*, e.g. the different families of proteinases, sugar kinases, and lysyl-tRNA synthetases.

When a protein evolves to change its function, many of these constraints are released – or, more precisely, replaced by alternative constraints required by the new function. The relationship between sequence and function is much more complex than the relationship between sequence and structure. Small changes in sequence, during evolution, usually make only small changes in structure. Often they make only small changes in function also. But changes in function do not necessarily require large changes in sequence or structure – function can jump.

Indeed, a protein can change function without any sequence changes at all. For instance, in the duck, an active lactate dehydrogenase and an enolase serve as crystallins in the eye lens, although they do not encounter the substrates *in situ*. In other birds, crystallins are closely related to enzymes, but some divergence has already occurred, with loss of catalytic activity. (This proves that the enzymatic activity is not necessary in the eye lens.) Many other such examples are known of 'recruitment', or 'moonlighting', by proteins – adapting to a novel function with relatively little sequence change.

Conversely, proteins with very different sequences and structures can have the *same* function. For instance, many families of proteinases differ in sequence and structure, sharing only a common general catalytic activity. Figure 5.13 summarizes, in a schematic way, the landscape of protein space with respect to the relations among sequence, structure, and function.

All three features of proteins – sequence, structure, and function – are potentially useful in interpreting new genome sequences. We expect that many regions of the new genome encode proteins similar to relatives known in other species. We can find them by looking for similar patterns in the sequences. We can expect that the structures will be similar, and indeed can calibrate the expected difference in structure from the extent of divergence in sequence. As Figure 5.13 shows, however, we cannot be as confident in assuming that function will be conserved.

Phylogeny

Once we have measured the similarity of one or more properties of a set of individuals, or species, we can try to arrange them according to their apparent pattern of divergence. If in fact the similarities do arise during descent from a common ancestor, it should be possible to depict the relationships in a family tree (for individuals), or phylogenetic tree (for species and higher taxa). The goal is to present the pattern of similarities and divergences in a consistent diagram such that close relationships within the diagram correspond to high degrees of similarity.

The assumption is that such a diagram will have the form of a tree. (We saw an example of a tree in Figure 4.1.) The computations that derive the optimal phylogenetic tree from a matrix of similarities are not trivial, and the problem has been a challenge in research for some time.

The goal of phylogeny is a *logical arrangement* of a set of species, populations, individuals, and genes. The arrangement is derived from observed similarities. The basic principle is that *the origin of similarity is common ancestry*. Although there are many exceptions, arising from convergent evolution or horizontal gene transfer, this basic principle is crucial both for rationalizing contemporary observations and for opening a window onto the history of life.

From phylogeny, we infer relationships – among species, populations, individuals, or genes. Relationship is taken in the literal sense of kinship or genealogy, i.e. assignment of a scheme of ancestors and descendants (see Box 5.5).

- A phylogenetic tree is a diagram showing ancestor–descendant relationships, that captures a pattern of similarities, in that individuals or species closely linked in the tree have high similarity.

BOX 5.5

Concepts related to biological classification and phylogeny

- **Homology** means, specifically, descent from a common ancestor.
- **Similarity** is the measurement of resemblance or difference, independent of the source of the resemblance. Similarity is observable *now* and involves no historical hypotheses. In contrast, assertions of homology require inferences about historical events, which are almost always unobservable.
- **Similarity and dissimilarity**. Data suitable for phylogenetic analysis may be specified equivalently in terms of similarities between objects or by dissimilarities. In comparing two DNA sequences, we may count the percentage of identical residues in an optimal alignment. This is a measure of similarity – the higher the value, the more similar the sequences. Alternatively, we could count the number of mutations separating the sequences. This is a measure of dissimilarity.
- **Clustering** is bringing together similar items, distinguishing classes made up of objects that are more similar to one another than they are to other objects outside the classes. Most people would agree about degrees of similarity, but clustering is more subjective. When classifying objects, some people prefer larger classes, tolerating wider variation; others prefer smaller, tighter classes. They are called, respectively, **groupers** and **splitters**.
- **Hierarchical clustering** is the formation of clusters of clusters of . . .
- The distinction between clustering and **classification**. Clustering is the determination of the set of classes into which a group of samples should be divided. Classification is the assignment of a sample to its proper place in a known set of classes.
- **Phylogeny** is the description of biological ancestor–descendant relationships, usually expressed as a tree. A statement of phylogeny among objects *assumes* homology and *depends* on classification.

Relationships between species are rarely directly observable, even in such an 'obvious' case as Darwin's finches. However, from genomics, species relationships can usually be deduced reliably.

Evolutionary relationships give us an historical glimpse of the development of life (see Figure 4.1). Although molecules themselves cannot be dated, evolutionary events observed on the molecular level can be calibrated with the fossil record.

The results of phylogenetic analyses are usually presented in the form of an **evolutionary tree** (see Box 5.6).

Figure 5.14 shows the relationships among the ratites – large flightless birds, such as the ostrich. The ancestor of the ratites is believed to be a bird that could fly, probably related to the extant tinamous. Such a tree, showing descendants of a single original ancestral species, is said to be **rooted**. (The root of the tree usually appears at the top or the side; botanists will have to get used to this.)

Alternatively, we may be able to specify relationships but not order them according to a history.

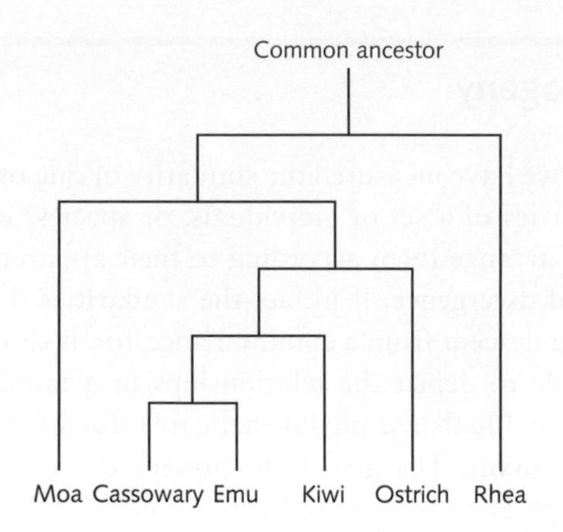

Figure 5.14 Phylogenetic tree of ratites (large flightless birds), based on mitochondrial DNA sequences. The common ancestor is at the **root** of this tree, appearing at the top of the graph. A surprising implication of these DNA sequences is that the moa and kiwi are not the closest relatives and, therefore, New Zealand must have been colonized twice by ratites or their ancestors. In terms of geography, this is less surprising if one looks at a map of the ancient continent Gondwanaland prior to its break-up, rather than at a contemporary map on which the distance between Africa and New Zealand is large.

BOX 5.6

Structure and contents of an evolutionary tree

In computer science, a tree is a particular kind of **graph**. A graph is a structure containing nodes (abstract points) connected by edges (represented as lines between the points). A path between two nodes in a graph is a series of consecutive edges that begins at one node and ends in the other. In a general graph, there may be many paths between any two nodes. (In Chapters 11 and 12, we discuss graphs in more detail.)

A tree is a special kind of graph. First of all, a tree must be connected, meaning that there is a path through the graph between any two points. Second, in a tree there is *only* one path between every two points. We have already seen several trees; for example, Figure 4.1.

A particular node may be selected as a **root** of a tree. However, this is not necessary – abstract trees may be rooted (for instance, Figure 5.14) or unrooted (for instance, Figure 5.15). In phylogenetic trees, the root is the earliest common ancestor of all of the other nodes. Rooted phylogenetic trees explicitly show ancestor–descendant relationships, as from any node there is a connected path up through successive ancestors terminating at the root. Unrooted trees show the topology of relationship but not the pattern of descent.

It may be possible to assign numbers to the edges of a graph to indicate some kind of 'length' of the edges, corresponding to a 'distance' between the nodes that the edges connect. These lengths are not necessarily geometric distances, but may be abstract values. Given edge lengths, the graph may be drawn to scale, with the sizes of the edges proportional to the assigned lengths.

In phylogenetic trees, edge lengths signify either some measure of the dissimilarity between two taxa, or the length of time since their separation. The assumption that differences between properties of living species reflects their divergence times will be true only if the rates of divergence are the same in all branches of the tree. Many exceptions are known. For instance, among mammals, many proteins from rodents show relatively fast evolutionary rates.

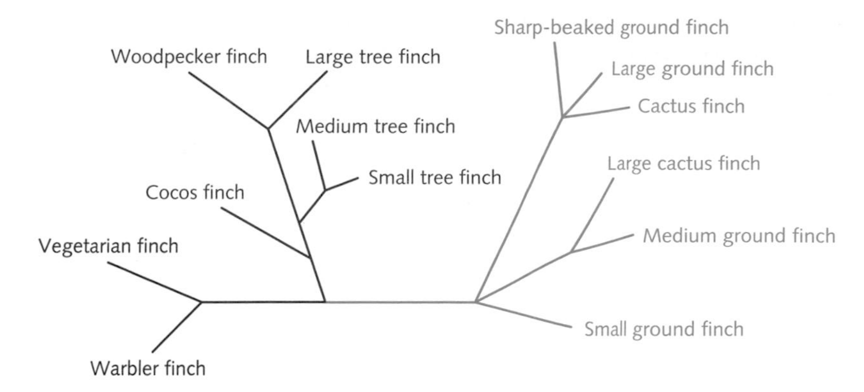

Figure 5.15 Unrooted tree of relationships among finches from the Galapagos and Cocos Islands. Darwin studied the Galapagos finches in 1835, noting the differences in the shapes of their beaks and the correlation of beak shape with diet. Finches that ate fruits had beaks like those of parrots, whereas finches that ate insects had narrow, prying beaks. These observations were seminal to the development of Darwin's ideas. As early as 1839 he wrote, in *The Voyage of the Beagle*, 'Seeing this gradation and diversity of structure in one small, intimately related group of birds, one might really fancy that from an original paucity of birds in this archipelago, one species had been taken and modified for different ends'.

The relationships among the finches of the Galapagos Islands, studied by Darwin, plus a related species from the nearby Cocos Island are shown in an **unrooted tree** (Figure 5.15). Addition of data from a species on the South American mainland ancestral to the island finches would allow us to root the tree.

- The idea of phylogeny is to observe different degrees of similarity among species or higher taxa, assume that the species are related by descent from a common ancestor and that higher degrees of similarity correspond to closer relationships, and to try to capture the relationships in a tree diagram showing ancestor–descendant relationships such that species more closely related according to the tree do have higher degrees of similarity.

The statement of a tree of relationships may reveal only the connectivity or topology of the tree, in which case the lengths of the branches are arbitrary. A more ambitious goal is to show the distances between taxa quantitatively, for instance to label the branches with the time since divergence from a common ancestor.

A phylogenetic tree tells us the organization of a set of taxa. It does not tell us how they should be grouped or partitioned. For example, the tree of Darwin's finches *looks* as if it could reasonably be partitioned into two or three clusters. The guiding principle in selecting a partition is that the intracluster similarities be relatively high and the intercluster similarities be relatively low. If the data are intrinsically well grouped, then the clustering is obvious. If there is no clear separation, proper clustering is ambiguous and difficult.

- The PHYLIP package (PHYLogeny Inference Package) of J. Felsenstein is an integrated set of links to tools for phylogenetics, and sources of software (http://evolution.genetics.washington.edu/phylip/software.html). Some multiple sequence alignment packages, such as CLUSTAL W, provide facilities to convert their alignments to phylogenetic trees.

Phylogenetic trees

Given a set of data that characterize different groups of organisms – for example, DNA or protein sequences, or protein structures, or shapes of teeth from different species of animals – how can we derive information about the relationships among the organisms in which they were observed? To what extent does the topology of the relationships depend on the choice of character? In particular, are there any *systematic* discrepancies between the implications of molecular and palaeontological analysis?

Broadly, there are two approaches to deriving phylogenetic trees. One approach makes no reference to any historical model of the relationships. Measure a set of distances between species and generate the tree by a **hierarchical clustering procedure**. This is called the **phenetic** approach. The alternative, the **cladistic** approach, is to consider possible pathways of evolution, infer the features of the ancestor at each node and choose an optimal tree according to some model of evolutionary change. Phenetics is based on similarity; cladistics is based on genealogy.

Clustering methods

Phenetic, or clustering, approaches to determination of phylogenetic relationships are explicitly non-historical. Indeed, hierarchical clustering is perfectly capable of producing a tree even in the absence of evolutionary relationships. A departmental store has goods clustered into sections according to the type of product – for instance, clothing or furniture – and subclustered into more closely related subdepartments, such as men's and women's shoes. Men's and women's shoes have a common ancestor, but there is no implication that shoes and furniture do.

A simple clustering procedure works as follows: given a set of species, determine for all pairs a measure of the similarity or difference between them. This could depend on a physical body trait, such as the difference between the average adult height of members of two species, or one could use the number of different bases in alignments of mitochondrial DNA.

To create a tree from the set of dissimilarities:

- First, choose the two most closely related species and insert a node to represent their common ancestor.
- Then replace the two selected species by a set containing both, and replace the distances from the pair to the others by the average of the distances of the two selected species to the others. Now we have a set of pairwise dissimilarities, not between individual species, but between sets of species. (Regard each remaining individual species as a set containing only one element.)
- Then repeat the process.

This method of tree building is called the UPGMA method (unweighted pair group method with arithmetic mean; see Box 5.7).

Calculation of phylogenetic trees by clustering

Consider four species characterized by homologous sequences ATCC, ATGC, TTCG, and TCGG. Taking the number of differences as the measure of dissimilarity between each pair of species, we will use a simple clustering procedure to derive a phylogenetic tree.

The distance matrix is:

	ATCC	ATGC	TTCG	TCGG
ATCC	0	**1**	2	4
ATGC		0	3	3
TTCG			0	2
TCGG				0

(As the matrix is symmetric, we need fill in only the upper half.)

The smallest distance is **1** (in boldface), between ATCC and ATGC. Therefore, our first cluster is {ATCC, ATGC}. The tree will contain the fragment:

The reduced distance matrix is:

	{ATCC, ATGC}	TTCG	TCGG
{ATCC, ATGC}	0	$\frac{1}{2}(2 + 3) = 2.5$	$\frac{1}{2}(4 + 3) = 3.5$
TTCG		0	**2**
TCGG			0

The number 3.5 in the upper right was calculated by averaging the distances between ATCC and TCGG = 4 and between ATGC and TCGG = 3.

The next cluster is {TTCG, TCGG}, with distance **2**. Finally, linking the clusters {ATCC, ATGC} and {TTCG, TCGG} gives the tree:

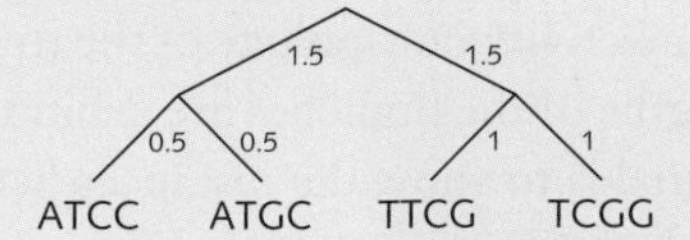

Branch lengths have been assigned according to the rule:

Branch length of edge between nodes X and $Y = \frac{1}{2}$ distance between X and Y

Whether the branch lengths are truly proportional to the divergence times of the taxa represented by the nodes must be determined from external evidence.

Cladistic methods

Cladistic methods deal explicitly with the patterns of ancestry implied by the possible trees relating a set of taxa. Their aim is to select the correct tree by utilizing an explicit model of the evolutionary process. The most popular cladistic methods in molecular phylogeny are the **maximum parsimony** and **maximum likelihood** approaches. They are specialized to sequence data, starting from a multiple sequence alignment. Neither maximum parsimony nor maximum likelihood could be applied to anatomic characters such as average adult height.

The maximum parsimony method of W. Fitch defines an optimal tree as the one that postulates the fewest mutations (see Box 5.8).

Calculation of phylogenetic trees by maximum parsimony

Given species characterized by homologous sequences ATCG, ATGG, TCCA, and TTCA, the tree:

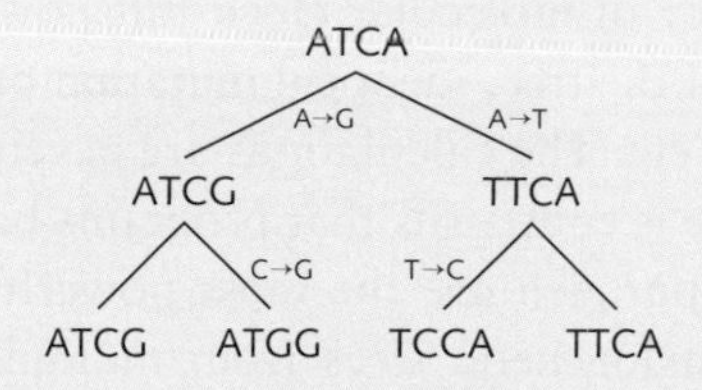

postulates four mutations. Note that the ancestral sequences, ATCG, TTCA, and ATCA, are not part of the observable data.

An alternative tree:

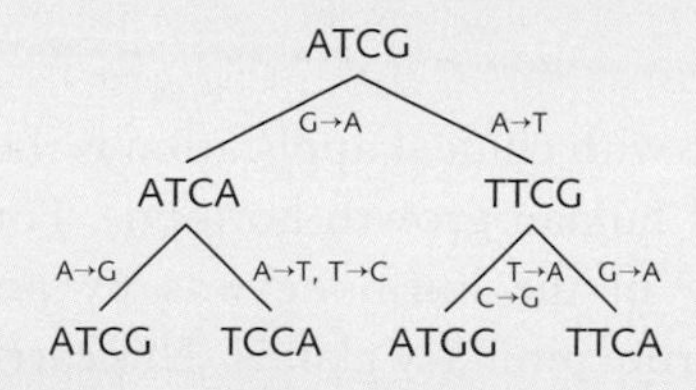

postulates seven mutations. Note that the second tree implies that the G → A mutation in the fourth position occurred twice independently. The former tree is optimal according to the maximum parsimony method, because no other tree involves fewer mutations. In many cases, several trees may postulate the same number of mutations, fewer than any other tree. For such cases, the maximum parsimony approach does not give a unique answer.

The maximum likelihood method assigns quantitative probabilities to mutational events, rather than merely counting them. Like maximum parsimony, maximum likelihood reconstructs ancestors at all nodes of each tree considered; however, it also assigns branch lengths based on the probabilities of the mutational events postulated. For each possible tree topology, the assumed substitution rates are varied to find the parameters that give the highest likelihood of producing the observed sequences. The optimal tree is the one with the highest likelihood of generating the observed data.

Both maximum parsimony and maximum likelihood methods are superior to clustering techniques. This has been demonstrated with cases where independent evidence – for instance, from palaeontology – provides a correct answer, and also with simulated data – computer generation of evolving sequences.

The problem of varying rates of evolution

Suppose that four species, A, B, C, and D, have the phylogenetic tree:

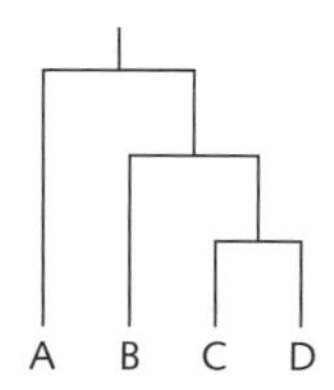

This tree is consistent with the dissimilarity matrix:

	A	B	C	D
A	0	3	3	3
B		0	2	2
C			0	1
D				0

Suppose, however, that taxon D is changing very fast, although the phylogeny is unaltered. The dissimilarity matrix might then be observed to be:

	A	B	C	D
A	0	3	3	20
B		0	2	20
C			0	20
D				0

from which we would derive the *incorrect* phylogenetic tree:

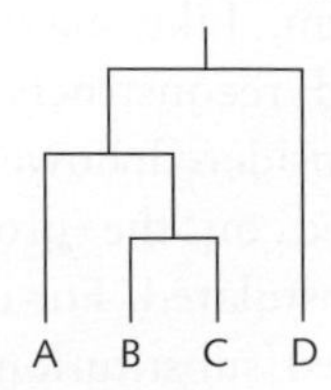

All of the methods discussed here are subject to errors of this kind if the rates of evolutionary change vary along different branches of the tree. To test for varying rates, compare the species under consideration with an **outgroup** – a species more distantly related to all of the species in question than any pair of them is to each other. For instance, if we are studying species of primates, a non-primate mammal such as the cow would be a suitable outgroup. If the rates of evolution among the primate species were constant, we would expect to observe approximately equal dissimilarity measures between all primate species and the cow. If this is not observed, the suggestion is that evolutionary rates have varied among the primates, and the character being used may well not provide the correct phylogenetic tree.

Bayesian methods

The problem we are trying to solve is:

Of all possible phylogenetic trees organizing the relationships among different species, based on a given multiple sequence alignment, which one has the highest probability of generating the observed multiple sequence alignment, under some model of evolutionary change? The model might be specified in terms of probabilities of mutation rates, etc., and for the moment it would seem that a weakness of the approach is the difficulty of knowing how to specify the model explicitly and accurately.

Nevertheless, from any such model of evolutionary change, we can compute the probability that any tree would produce the observed multiple sequence alignment. Suppose we begin the problem in a state of complete ignorance, meaning that we consider that initially – for all we know – all potential phylogenetic trees must be regarded as equally probable. Then Bayes' rule states that we want to choose the tree with the highest probability of producing the observed multiple sequence alignment.

What makes this approach so powerful is that we can optimize the probability of producing the observed data not only over possible trees, but over different models of evolutionary change. This releases us from making overly constricting assumptions such as constancy of molecular clock rates over different branches of the tree, identical mutation probabilities at all sites, etc. The calculations are nevertheless feasible. There is consensus that programs based on the Bayesian approach are the most powerful tools for deriving phylogenetic trees from multiple sequence alignments.

Short-circuiting evolution: genetic engineering

Evolutionary divergence arises in nature through generation of variation by random mutation, followed by either selection or genetic drift to alter allele frequencies in populations or to create novel species. Contemporary techniques allow deliberate transfer of genes, to create organisms with altered characters directly. In addition to gene therapy for disease (see p. 125) – and genetically modified crop plants – many applications are available or under development:

Use of microorganisms as protein factories. Microorganisms are routinely used in the laboratory to express proteins – human or otherwise – for research. An example with clinical application is the microbial synthesis of human growth hormone. Formerly, the only source of the hormone was by post-mortem extraction from pituitary glands. This carried the risk of transmitting prion disease. Other microbiologically produced human proteins with clinical applications include insulin, and many monoclonal antibodies. Still other applications include manufacture of fuels, or plastics, or dissolving oil spills.

In the USA the attempt to patent genetically modified bacteria that could break down hydrocarbons was a landmark case, decided in favour of granting the patent by a 5–4 decision of the United States Supreme

Court in 1980. Of course, novel varieties of flowers produced by classical methods of breeding and selection have been protectable for many years. The International Union for the Protection of New Varieties of Plants (UPOV) is an intergovernmental organization established by treaty in 1961. It is now, appropriately, turning its attention to biotechnology, and legal and intellectual property issues.

Genetically modified animals. Higher animals are also used as protein factories, in cases where the active protein requires postranslational modifications of which microorganisms are incapable. Production of drugs by this route is called 'pharming'. Genetically engineered goats secrete an anticoagulant, human antithrombin III, in their milk. This product has been approved for clinical use in the United States.

Other goals of genetically modified animals include:

(a) enhancing the nutritional value of food; e.g. pork enriched in ω-3 fatty acids.

(b) pigs lacking the surface antigens that produce rejection by the human immune system, as a source of organs for transplant.

(c) animals that grow faster and/or require less expensive feed; for example fast-growing salmon.

(d) protecting livestock against disease; e.g. cows lacking prion proteins and therefore immune to BSE.

(e) allergen-free pets.

(f) fish that glow in colours by virtue of genes for fluorescent proteins.

Genetically modified plants. Many crop plants are targets for genetic modification. Goals include:

(a) pesticide-resistant plants, that allow treatments to kill weeds or insect pests without damaging the plants.

(b) a related approach is a plant that makes its own insecticide. Bt-corn (maize) contains a natural insect-killing gene transferred from *Bacillus thuringiensis*.

(c) crops with enhanced nutritional value. An example is 'golden rice' enriched in vitamin A (see p. 245).

(d) fruit with longer shelf life, such as the 'flavr-savr' tomato.

(e) crops that produce only sterile seeds.

There are a number of controversial aspects to these activities. In the case of genetically modified plants, there is concern over the spreading of genes from the crops to undesired hosts. For instance, if a gene for herbicide resistance is introduced into a crop plant, it would make it easier selectively to kill weeds without affecting the crop plant. However, it has been observed that the gene can spread to the weeds. Another concern is economic. Use of sterile seeds requires farmers to purchase new seeds each year. It precludes traditional agricultural practice of holding back a portion of a crop for replanting.

In addition to the specific economic implications, there is a widespread feeling that biotechnology might alter the relationship between people and Nature that have been a common cultural heritage for thousands of years. It would be wrong to dismiss these feelings as irrational or as characterizing only a fringe.

RECOMMENDED READING

- Two general articles on phylogeny:

  Whelan, S., Liò, P. & Goldman, N. (2001). Molecular phylogenetics: state-of-the-art methods for looking into the past. Trends Genet. **17**, 262–272.

  Baldauf, S.L. (2003). Phylogeny for the faint of heart: a tutorial. Trends Genet. **19**, 345–351.

- Phylogenetic relationships in eukaryotes:

  Baldauf, S.L. (2003). The deep roots of eukaryotes. Science **300**, 1703–1706.

- Estimates of the history of diversity of living species:

  Jackson, J.B.C. & Johnson, K.G. (2001). Paleoecology: measuring past biodiversity. Science **293**, 2401–2404.

- Discussions of sequence analysis:

  Gusfeld, D. (1997). *Algorithms on Strings, Trees and Sequences* Cambridge University Press, Cambridge.

  Doolittle, R.F. (1986). *Of URFS and ORFS/A Primer on How to Analyze Derived Amino Acid Sequences.* University Science Books, Mill Valley, CA, USA.

  Li, H. & Homer, N. (2010). A survey of sequence alignment algorithms for next-generation sequencing. Brief. Bioinf. **11**, 473–483.

- It is possible to represent phylogenetic relationships in forms more general than tree structures:

  Bandelt, H.-J. & Dress, A.W.M. (1992). Split Decomposition: A new and useful approach to phylogenetic analysis of distance data. Molec. Phy. Evol. **1**, 242–252.

  Huson, D.H. & Scornavacca, C. (2011). A survey of combinatorial methods for phylogenetic networks. Genome Biol. Evol. **3**, 23–35.

- Discussion of applications of genetic engineering:

  Arnold, F.H. (2008). The race for new biofuels. Eng. Sci. **71**, 12–19.

  Brustad, E.M. & Arnold, F.H. (2011). Optimizing non-natural protein function with directed evolution. *Curr. Opin. Chem. Biol.* **15**, 201–210.

EXERCISES, PROBLEMS, AND WEBLEMS

Exercises

Exercise 5.1 On two photocopies of Figure 5.15, indicate a reasonable division of the species into (a) three clusters; (b) five clusters.

Exercise 5.2 What is the Hamming distance between the words DECLENSION and RECREATION?

Exercise 5.3 What is the Levenshtein distance between the words BIOINFORMATICS and CONFORMATION?

Exercise 5.4 The Levenshtein distance between the strings `agtcc` and `cgctca` is 3, consistent with the following alignment:

```
ag-tcc
cgctca
```

Provide a sequence of three edit operations that convert `agtcc` to `cgctca`.

Exercise 5.5 To what alignment does the path through the following dot plot correspond?

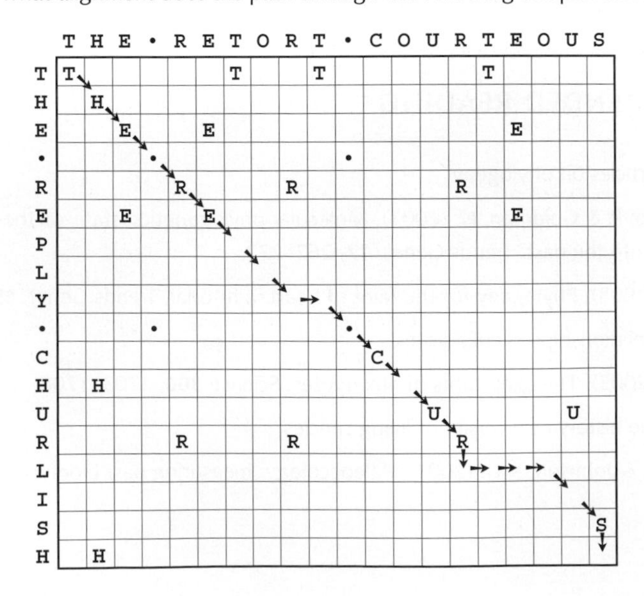

Exercise 5.6 In the dot plot appearing in Figure 5.5, there is an interruption of the matching at a height approximately at the level of the downward-pointing arrow at the left that precedes the words *Xenopus laevis*. On a photocopy of Figure 5.5(b), indicate where in the sequence this region appears.

Exercise 5.7 How would you use a dot plot to pick up palindromic DNA sequences of the type that appear partly on each strand, as in the specificity sites of restriction endonucleases?

Exercise 5.8 According to the BLOSUM62 matrix: (a) is a histidine (H) more likely to change to an asparagine (N) or to an aspartic acid (D)? (b) What is the ratio of the probability that a histidine will be *observed* to change to an asparagine to the probability *expected* on the basis of the amino acid composition of the protein that it will change to an asparagine?

Exercise 5.9 Consider the box (red outline) showing a part of a position-specific scoring matrix in Figure 5.8. Suppose you were scoring a protein with 225 residues according to this matrix. (a) How many columns would you expect there to be in the position-specific scoring matrix? (b) How many rows would you expect there to be?

Exercise 5.10 On photocopies of Figure 5.13, indicate points (a) where a pair of highly diverged homologous proteins with similar structure but without obvious sequence similarities might lie; (b) where a pair of non-homologous proteins with similar structure might lie; (c) where a pair of enzymes that share a function but not a structure (for instance, serine and cysteine proteinases) might lie.

Problems

Problem 5.1 Draw a dot plot of the following sequence from the wheat dwarf virus genome: `ttttcgtgagtgcgcggaggctttt` against itself. In what respects is it not a perfect palindrome?

Problem 5.2 How would you adapt the dot plot formalism to search for regions of DNA or RNA that form local double-helical regions? Assume that the two hydrogen-bonded regions are separated by only a short unpaired loop, as for example in tRNA (see Figure 1.4).

Problem 5.3 (a) How might the course of the BLAST calculation shown in Figure 5.7 differ if the word length were chosen as 3 instead of 4? (b) How might the course of the BLAST calculation shown in Figure 5.7 differ if the word length were chosen as 7 instead of 4?

Problem 5.4 The phylogenetic tree (p. 184) is derived from a complete dissimilarity matrix, i.e. a specification of a measure of the dissimilarity between *every* pair of tetranucleotides. The numbers associated with each edge reproduce the measures of dissimilarity between connected nodes: i.e. the sum of the edges in the path between ATCC and ATGC is $0.5 + 0.5 = 1$, which is the value in the matrix corresponding to row ATCC and column ATGC. For every pair of tetranucleotides, calculate the sum of the numbers associated with the edges in the path between them. For which pairs do the results agree with the original dissimilarity matrix? For which pairs do the results disagree?

Problem 5.5 Examples in the chapter derived a phylogenetic tree for the four sequences ATCC, ATGC, TTCG, and TCGG by the UPGMA method (unweighted pair group method with arithmetic mean) and a phylogenetic tree for the sequences ATCG, ATGG, TCCA, and TTCA by the maximum-parsimony method. Derive phylogenetic trees for the sequences ATCC, ATGC, TTCG, and TCGG by the maximum-parsimony method and for the sequences ATCG, ATGG, TCCA, and TTCA by the UPGMA method. Show all intermediate steps. Compare the results with the trees derived in the chapter.

Weblems

Weblem 5.1 Draw a picture of the human aldolase B monomer. Indicate the sites of the mutations N334K, L256P, A149P, N334K, and A174D. Indicate the region of the active site and the regions of intersubunit contacts. Comment on the possible severity of the effects of these mutations on structure and function of the protein.

Weblem 5.2 Retrieve the globin sequences shown in Figure 1.17(b). Perform a multiple sequence alignment, and draw the phylogenetic tree. Comment on ways in which the tree seems biologically reasonable; and – if any – ways in which it does not.

CHAPTER 6

Genomes of Prokaryotes

LEARNING GOALS

- To know the features that distinguish the major divisions of life and to appreciate how differences of lifestyle reflect differences in genomes and structures.
- To understand the molecular basis of adaptations; for example, to life at high temperatures, or different ocean depths.
- To appreciate, at the molecular level, the genomic and phenotypic differences among selected related species of prokaryotes.
- To face the problem of bacterial pathogenicity, and the development of antibiotic resistance.
- To recognize the vast variety of different microorganisms that inhabit, and mutually interact in, environmental samples. These habitats include oceans and soils, and internal environments such as the human (or animal) gut.

Evolution and phylogenetic relationships in prokaryotes

Prokaryotes have several claims on our interest.

- They cause infectious diseases. Some diseases, such as tuberculosis, are major public health problems. It is a challenge to control these diseases in the face of the development of antibiotic resistance.
- Molecular biologists study prokaryotes as examples of relatively simple cells, to understand fundamental principles of metabolism, genetics, and development.
- Historically, prokaryotes represent the earliest forms of life, from which all others are derived. They had the biosphere to themselves for over 2 billion years.
- Prokaryotes are important mediators of ecological processes and geological cycles. Indeed, geological and biological phenomena are linked in an intimate marriage, which has seen its turbulent episodes. Purely geological events such as asteroid impacts have caused mass extinctions. Purely biological events, such as the development of photosynthetic processes that released large quantities of O_2 into the atmosphere, and respiration that released CO_2, have altered general flows of matter and energy, affecting the development of the Earth's geochemistry and climate. Microbes respond to human-caused environmental damage. They can aggravate ecological problems, but also hold out hope of ameliorating them (Table 6.1).

The major habitats of prokaryotes are the open ocean, surface soils, and subsurface sediments beneath both ocean and soil. The total carbon content of prokaryotes is between 60 and 100% of the total carbon found in plants, but the total nitrogen and phosphorus in prokaryotes is probably ten times that of plants. The bodies of humans and other animals harbour many microbes, but – important as the consequences for health and disease may be – as an overall reservoir we are a minor player.

Table 6.1 Landmarks in history of life

Formation of Earth	~4.5×10^9 years ago
Origin of life	>3.8×10^9 years ago
Cyanobacterial photosynthesis	>2.7×10^9 years ago
Rise of atmospheric O_2	$2.3–1 \times 10^9$ years ago
First metazoan	~1×10^9 years ago
Cambrian	~0.5×10^9 years ago

- The oceans also contain viruses in very great abundance and variety. Most of these are uncharacterized. They have been called the 'dark matter of the biosphere'. It is likely that viruses are an important mediator of gene transfer between marine prokaryotes.

The exploration of potential habitats by prokaryotes approaches saturation. Prokaryote cells divide actively. Production is estimated at 1.7×10^{30} cells per year, the open ocean being the highest contributor. This fecundity gives prokaryotes the opportunity to evolve quickly. The resulting variety of prokaryotes includes the colonists of inhospitable habitats such as hot springs and very salty lakes. It also includes almost continuous local variations, adaptions to microniches (Table 6.2).

Major types of prokaryotes

C. Woese divided prokaryotes into archaea and bacteria, on the basis of 16S rRNA gene sequences. Figure 6.1 shows the secondary structures of a region within the 16S rRNA that differs in bacteria, archaea, and eukaryotes. In context, Figure 6.2 shows the tertiary structure of this region within the full *Escherichia coli* 16S rRNA structure in the ribosome.

Numerous other differences between archaea and bacteria have subsequently emerged, involving genomic, structural, and metabolic features:

- some genes in archaea but none in bacteria contain introns;
- there are systematic differences in tRNA sequences between archaea and bacteria;
- enzymes involved in DNA replication, such as DNA polymerases and some of the tRNA synthetases involved in protein synthesis, differ between archaea and bacteria;

Table 6.2 Distribution of prokaryotic cells

Habitat	Number of prokaryotic cells ($\times 10^{28}$)	Total carbon in prokaryotes ($\times 10^{15}$ g)
Ocean subsurface	355	303
Terrestrial subsurface	25–250	22–215
Soil	26	26
Oceans, lakes, and rivers	12	2.2
Within all human bodies	0.00004	

From: Whitman, W.B., Coleman, D.C., & Wiebe, W.J. (1998). Prokaryotes: the unseen majority. Proc. Natl. Acad. Sci. USA **95**, 6578–6583.

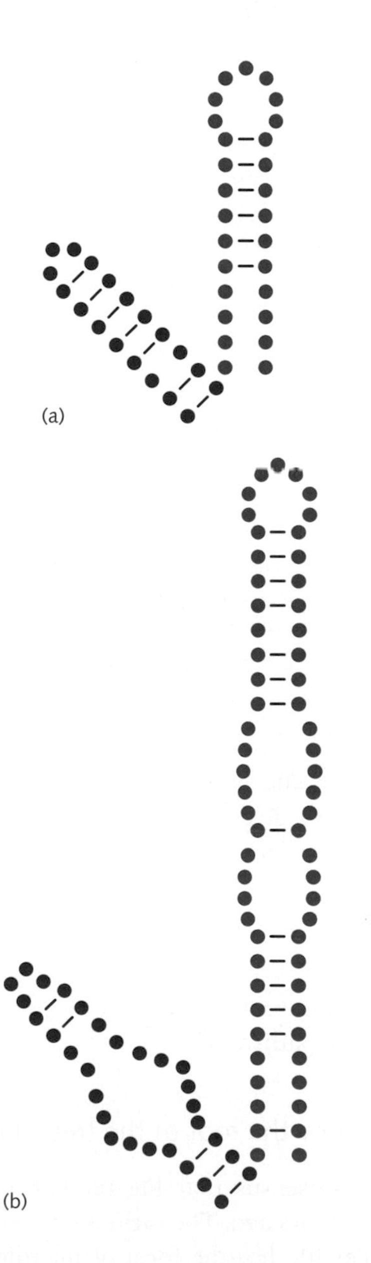

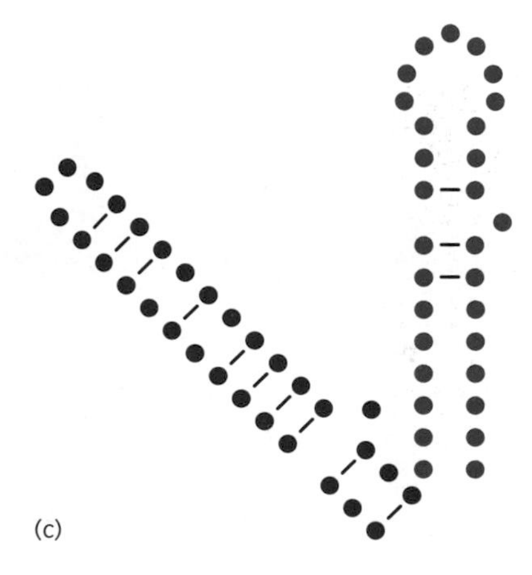

Figure 6.1 Secondary structure patterns of regions of 16S rRNA that differ among (a) bacteria (*Escherichia coli*); (b) archaea (*Methanococcus vannielii*); and (c) eukaryotes (*Saccharomyces cerevisiae*). Dots represent individual residues. Lines indicate complementary base pairing. The systematic differences in the lengths of the helical regions and the constraints imposed by the complementarity contributed to the patterns that Woese detected in the alignment of the sequences and in the derived phylogenetic trees. (rRNA = ribosomal RNA)

These diagrams provide only a two-dimensional view. Figure 6.2 shows the actual three-dimensional structure of this region within the entire 16S RNA structure of *E. coli*.

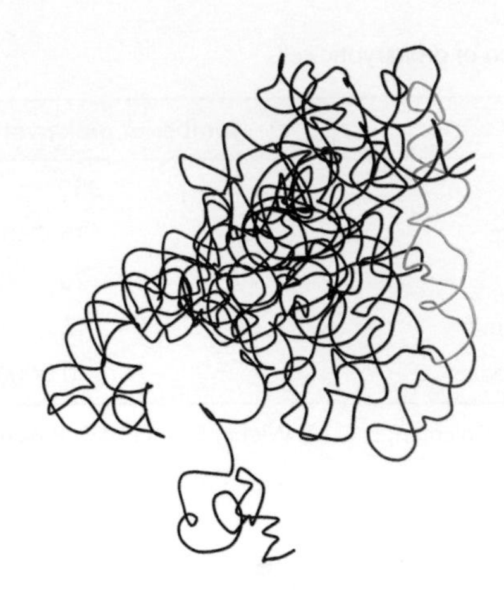
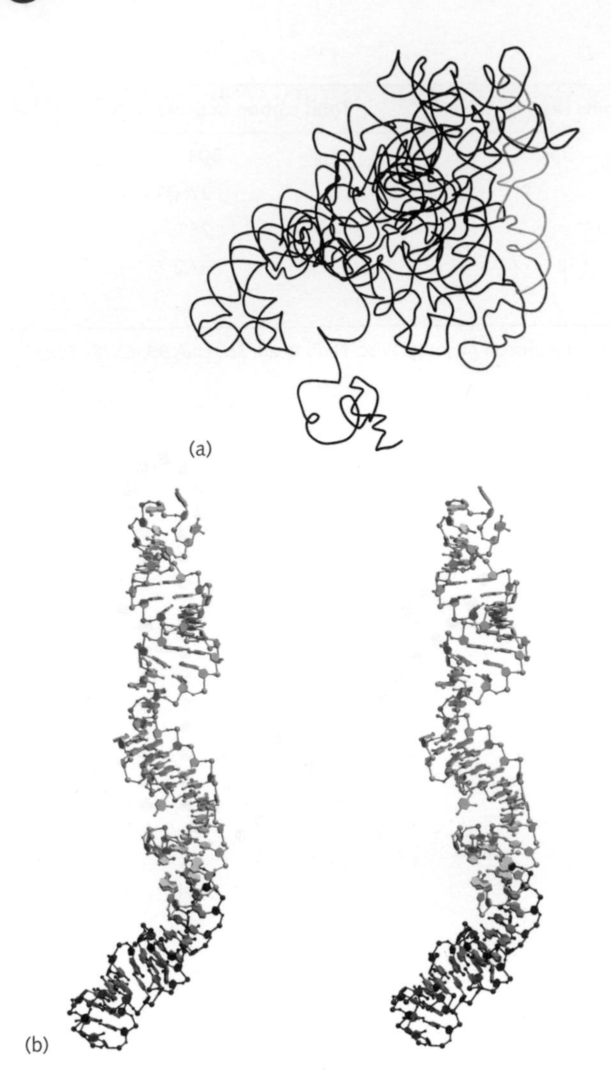

Figure 6.2 (a) Three-dimensional structure of 16S rRNA from the *Escherichia coli* ribosome [2AVY], showing the region of Figure 6.1(a) highlighted in red and blue. (b) Detailed structure of the region shown in Figure 6.1(a) Note that the acute angle between two helical regions drawn in the conventional representation of the secondary structure in the preceding figure is merely a drafting convention and does not correspond to the true three-dimensional structure.

- archaea but not bacteria contain DNA-associated proteins resembling histones;
- membranes of all cells contain phospholipids: compounds combining a glycerol molecule with long-chain organic molecules (see Figure 6.3); however:
  - bacteria and eukaryotes build cell membranes from phospholipids containing D-glycerol; archaea use L-glycerol
  - the organic chains in bacteria and eukaryotes are fatty acids, typically 16–18 carbon atoms long, while archaea instead use polyisoprenes (the branching of the isoprene chains permits the formation of links between different phospholipids in archaeal membranes; this allows the membrane to develop a higher-order structure)
  - bacteria and eukaryotes link the organic side-chains to glycerol with an ester linkage while archaea prefer an ether linkage;
- cell wall structures: bacterial but not archaeal cell walls contain peptidoglycan, a combination of sugar derivatives and peptides; and
- archaea and bacteria differ in their complement of metabolic pathways.

Do we know the root of the tree of life?

There is consensus that life on Earth began over 3.5 billion years ago. The earliest remaining evidence for cellular life has the form of microfossils, called stromatolites, from South Africa and Australia.

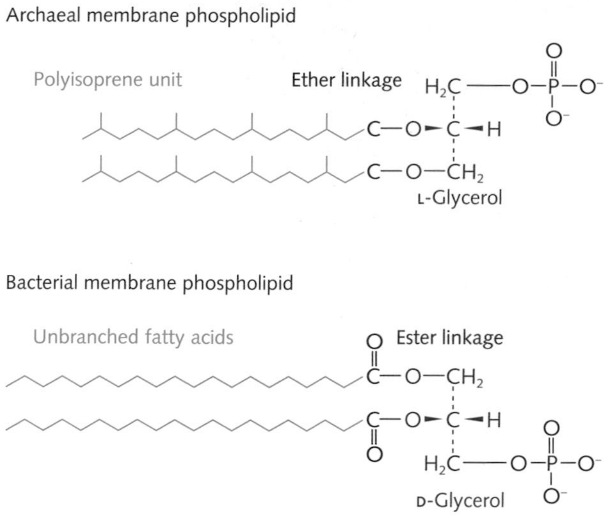

Archaeal membrane phospholipid

Polyisoprene unit

Ether linkage

L-Glycerol

Bacterial membrane phospholipid

Unbranched fatty acids

Ester linkage

D-Glycerol

Figure 6.3 The chemical structure of the cell membrane differs between archaea and bacteria. A phospholipid is a combination of glycerol (a three-carbon alcohol), a phosphate group, and long hydrocarbon moieties. In archaea, the hydrocarbons are terpenes, or polyisoprenes, attached to the glycerol with an ether linkage. (Double bonds are not shown for simplicity.) In bacteria, they are fatty acids, esterified with the glycerol. Glycerol has two mirror-image forms, and archaeal and bacterial membranes contain different enantiomers. In the glycerol moieties in the figure (shown in red), triangles indicate bonds to groups in front of the central carbon, whereas broken lines indicate bonds to groups farther away than the central carbon. These differences must have correlates in the genomes, which contain genes that encode alternative sets of synthetic enzymes to produce these structures.

These arose from cyanobacteria related to modern prokaryotes. What preceded them has not left physical remnants and can only be inferred from what traces they have left in contemporary molecular biology. It is widely believed that forms of life based on RNA – as both information archive and catalysts – existed before proteins took over the 'executive branch'.

The name archaea suggests that they represent the oldest forms of life. However, there has been extensive gene transfer between archaea and bacteria. It is not possible to assign LUCA – the last universal common ancestor of all known life forms – to either the archaeal or bacterial branch of the evolutionary tree. It is thought that the two lineages split very soon after the origin of cellular life. A branch of the archaea, the Korarchaeota, may be the closest extant relatives of LUCA.

- C. Woese divided all living things into three groups: archaea, bacteria, and eukaryotes. Although we do not have reliable knowledge of the earliest events in life history, it is likely that archaea are closest to LUCA, the last universal common ancestor of us all.

Archaea

The first archaea discovered lived at high temperatures near sea-floor hydrothermal vents or in lakes containing very high concentrations of salt, such as the Dead Sea. However, not all archaea are adapted to extreme environments. Indeed, there is some evidence that mesophilic archaea came first and that thermophiles were a later adaptation. Conversely, not all thermophiles are archaea; *Thermus aquaticus* (the source of *Taq* polymerase, an enzyme in common use for polymerase chain reaction amplification of DNA) is a bacterium.

Archaea are an abundant component of life in the open ocean, making up ~20% of all marine microbes. They also associate with a variety of metazoan hosts.

The major groupings of archaea (Figure 6.4) are as follows.

- *Crenarchaeota*. Many but not all of these are thermophiles. They include *Sulfolobus* and *Thermoproteus*.
- *Euryarchaeota*. These include methanogens, sulphate reducers, and many extreme halophiles, thermophiles, and acidophiles, including:
  - *Halobacter salinarum*, which can grow in salt concentrations above 4 M! Many people find its photosynthetic abilities even more interesting: *H. salinarum* contains a bacteriorhodopsin with which it captures sunlight energy as ATP without involving chlorophyll.
  - *Picrophilus torridus*, an extreme acidophile first isolated from the sulphurous volcanic springs of northern Japan. It can grow at pH 0.7!

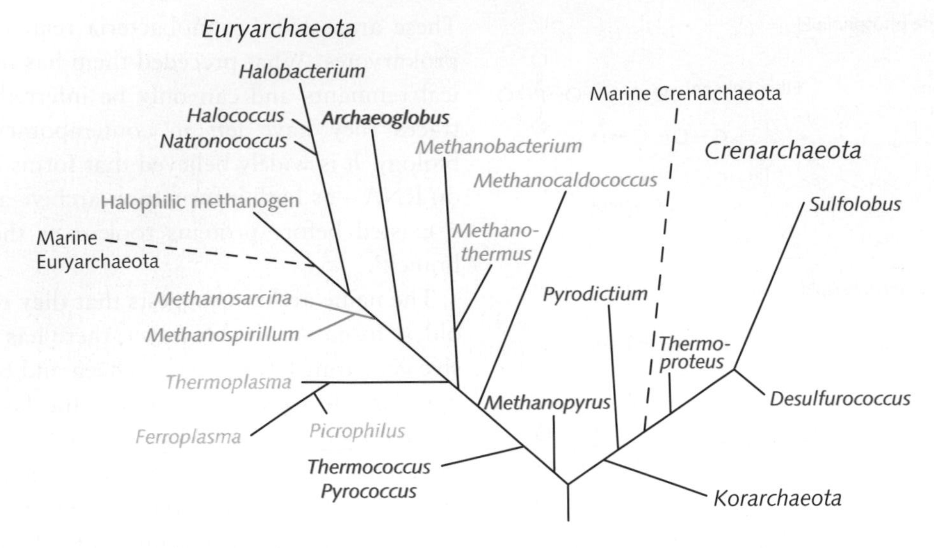

Figure 6.4 Phylogenetic tree of archaea, based on analysis of 16S rRNA sequences. Major archaeal groupings are coloured as follows:

Euryarchaeota:	***Crenarchaeota:***
Archaeoglobali	*Desulfurococcales*
Halobacteria	
Methanobacteria	***Korarchaeota***
Methanococci	
Methanomicrobia	***Nanoarchaeota*** (not shown)
Methanopyri	
Thermococci	
Thermoplasmata	

BOX 6.1

Methanogens as sources of greenhouse gas emission: the case of New Zealand

New Zealand is home to 4 million people, 10 million cattle, and 45 million sheep. The sheep and cattle host methanogenic archaea in their stomachs to help to digest fodder.

In the USA and European countries, animals make only a relatively small contribution to greenhouse gas emissions. In contrast, in New Zealand, ruminant-produced methane accounts for approximately half of the country's total greenhouse gas production. When New Zealand signed the Kyoto protocol, the government proposed to tax farmers to the tune of \$NZ11 per ton of carbon emitted. (At that time \$NZ1 ≈ UK£0.47 ≈ \$US0.76.) This would amount to an annual charge of about \$NZ0.09 per sheep and \$NZ0.72 per cow.

The proposal met determined resistance from the pastoral community, some of it couched in surprisingly ribald terms. The New Zealand government ultimately abandoned the idea. Research into the effects of different fodders on internal flora and even antibiotics specifically targeting archaea is now under way.

- Methanogens. These are strict anaerobes, dependent on the reaction:

$$CO_2 + 4H_2 \rightarrow CH_4 + 2H_2O$$

Methanogenic archaea live in the guts of ruminant animals and help to digest cellulose. Cellulases hydrolyse plant fodder to simple sugars, from which CO_2 and H_2 are produced by fermentation. A cow can produce hundreds of litres of methane per day! (See Box 6.1.)

- *Korarchaeota*. These were discovered by environmental sampling of a hot spring in Yellowstone National Park, Wyoming, in the western USA. They are perhaps closest to the root of all archaea.
- *Nanoarchaeota*. These have been identified as a single small (~400 nm diameter) hyperthermophile from a submarine hot vent.

The last two phyla are minor, at least in terms of our current knowledge of them.

The genome of *Methanococcus jannaschii*

The microorganism *Methanococcus jannaschii* was collected from a hydrothermal vent 2600 m deep off the coast of Baja California, Mexico, in 1983. It is a thermophilic organism, surviving at temperatures from 48 to 94°C, with an optimum at 85°C. *M. jannaschii* is capable of self-reproduction from inorganic components. Its overall metabolic equation is to synthesize methane from H_2 and CO_2. It is a strict anaerobe.

- Hydrothermal vents are underwater volcanoes emitting hot lava and gases through cracks in the ocean floor. They create niches for living communities disconnected from the surface; these use minerals from the vent as nutrients. These communities of microorganisms, and some animals, are the only forms of life not dependent on sunlight, directly or indirectly, for their energy.

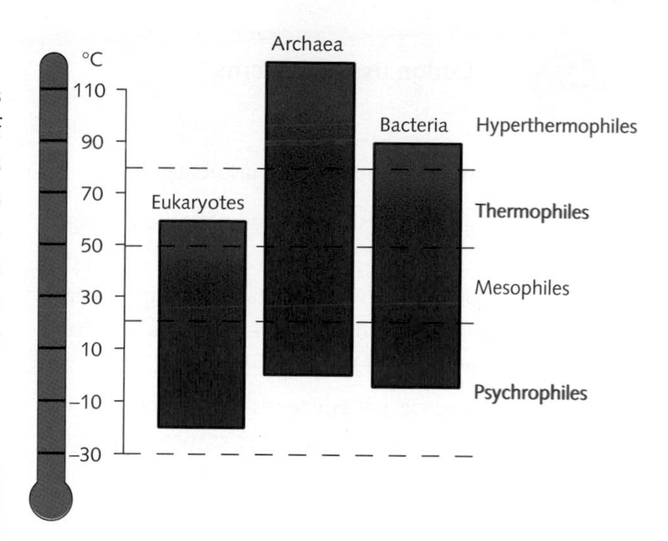

Figure 6.5 'Some like it hot, some like it cold . . .' Distribution of known growth temperatures of eukaryotes, archaea, and bacteria. Ranges defining hyperthermophiles, thermophiles, mesophiles, and psychrophiles are approximate.

The genome of *M. jannaschii* was sequenced in 1996 by The Institute for Genomic Research (TIGR). It was the first archaeal genome sequenced. It contains a large chromosome with a circular double-stranded DNA molecule 1 664 976 bp long and two extrachromosomal elements of 58 407 and 16 550 bp. There are 1743 predicted coding regions, of which 1682 are on the chromosome and 44 and 12 are on the large and small extrachromosomal elements, respectively. Some RNA genes contain introns. As in other prokaryotic genomes, there is little non-coding DNA.

M. jannaschii would appear to satisfy Luria's goal of finding our most distant extant relative. Comparison of its genome sequence with others shows that it is distantly related to other forms of life. Only 38% of the open reading frames could be assigned a function on the basis of homology to proteins known from other organisms. However, to everyone's great surprise, archaea are in some ways more closely related to eukaryotes than to bacteria! They are a complex mixture. In archaea, proteins involved in transcription, translation, and regulation are more similar to those of eukaryotes. Archaeal proteins involved in metabolism are more similar to those of bacteria.

Life at extreme temperatures

Organisms from the three major divisions of life – archaea, bacteria, and eukaryotes – show wide, overlapping ranges of optimal growth temperatures (see Figure 6.5). To survive at elevated temperatures, thermophiles and hyperthermophiles must synthesize molecules that are stable to heat denaturation. Accordingly, adaptations to high-temperature survival might be observed in DNA, in RNA, and in proteins. Enzymes from hyperthermophiles have applications in laboratory molecular biology and in industry.

What choices do organisms have to adjust the thermal stability of their constituents?

Thermophiles can evolve proteins with enhanced stability. Moreover, any set of favourable amino acid sequence is compatible with many gene sequences. In principle, organisms can take advantage of the redundancy of the genetic code to adjust the thermal stability of their nucleic acids. For double-stranded DNA and RNA, the thermal stability increases linearly with the G+C content. (However, this is only one aspect of the application of the redundancy in the code. See Box 6.2.)

Singer & Hickey* compared genome sequences from 40 prokaryotes. The organisms included eight

* Singer, G.A. & Hickey, D.A. (2003). Thermophilic prokaryotes have characteristic patterns of codon usage, amino acid composition and nucleotide content. Gene **317**, 39–47; Hickey, D.A. & Singer, G.A. (2004). Genomic and proteomic adaptations to growth at high temperature. Genome Biol. **5**, 117.

Codon usage patterns

General observations about codon distributions are:

- Different genomes show different codon usage patterns.
- The variation in codon usage pattern among different genes within the genome of one species is less than the variation between species.
- Codon preference pattern tends to be preserved in closely related species, but diverges as species diverge. The similarity of the pattern in archaeal and bacterial thermophiles is evidence that codon usage patterns can be determined by selection.
- Within genomes, highly expressed proteins show stronger bias in codon usage. Genes for highly expressed proteins are enriched in sets of 'preferred' codons, and these preferred codons are often correlated with greater tRNA abundances. Matching the pattern of codon usage and tRNA abundance can make protein synthetic throughput higher.

archaeal mesophiles and thermophiles and 32 bacterial mesophiles and thermophiles, with optimal growth temperatures ranging from 18 to 97°C. This permitted a study focusing on adaptations to high temperature, not biased by archaeal–bacterial differences. Their results show that:

- *DNA*. Perhaps surprisingly, overall genomic G+C content is *not* correlated with growth temperature. However, a correlation with growth temperature *is* observed in the distribution of *dinucleotides*, with thermophiles, mesophiles, and psychrophiles showing different characteristic patterns.

  Thermophiles and hyperthermophiles stabilize their DNA structures, not by increasing the G+C content, but by tight binding of special ligands.
- *RNA*. The G+C content of *non-protein coding RNA* is correlated with growth temperature, especially in double-stranded regions. High G+C content in double-stranded regions enhances their thermostability. *Single-stranded RNAs*, including messenger RNAs (mRNAs), are relatively rich in purines, notably adenine, for reasons that are not clear.

  Although the overall G+C content of DNA is not correlated with growth temperature, codon usage patterns are. Coding sequences in thermophiles are enriched relative to mesophiles in synonymous codons ending in C or G.
- *Proteins*. Comparisons of homologues show that proteins from thermophiles and hyperthermophiles:
  - tend to be shorter than their homologues from mesophiles, most of the residues lost coming from surface loops;
  - have more charged residues at their surfaces, both positive and negative (Asp, Lys, His, Asp, Glu); the formation of stabilizing salt bridges is a common feature of their structures (see Figure 6.6);

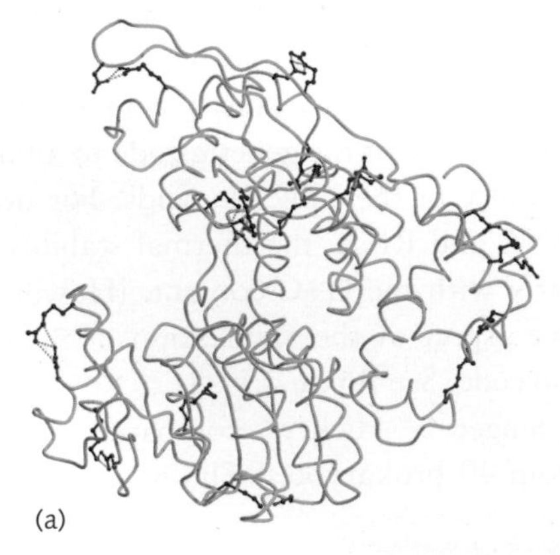
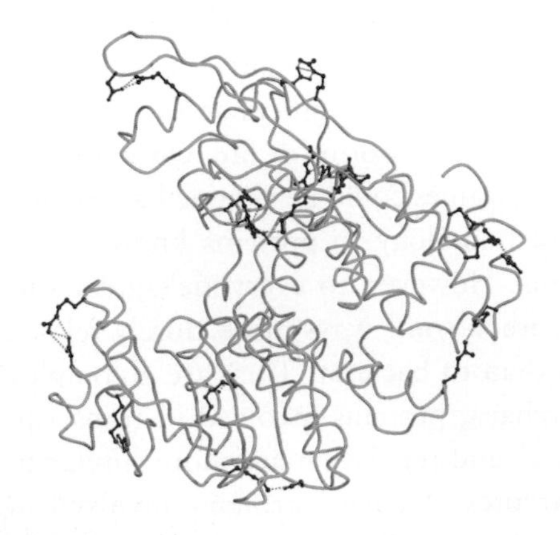

(a)

Figure 6.6 Proteins from thermophiles and hyperthermophiles are enriched in salt bridges relative to their mesophilic homologues. Positively charged sidechains are shown in blue. Negatively charged sidechains are shown in red. (a) Subunit of glutamate dehydrogenase from mesophilic archaeon *Clostridium symbiosum* [1HRD]. (b) Subunit of glutamate dehydrogenase from hyperthermophilic archaeon *Pyrococcus furiosus* [1GTM]. (c) Sequence alignment of archaeal hyperthermophilic and mesophilic glutamate dehydrogenase subunits.

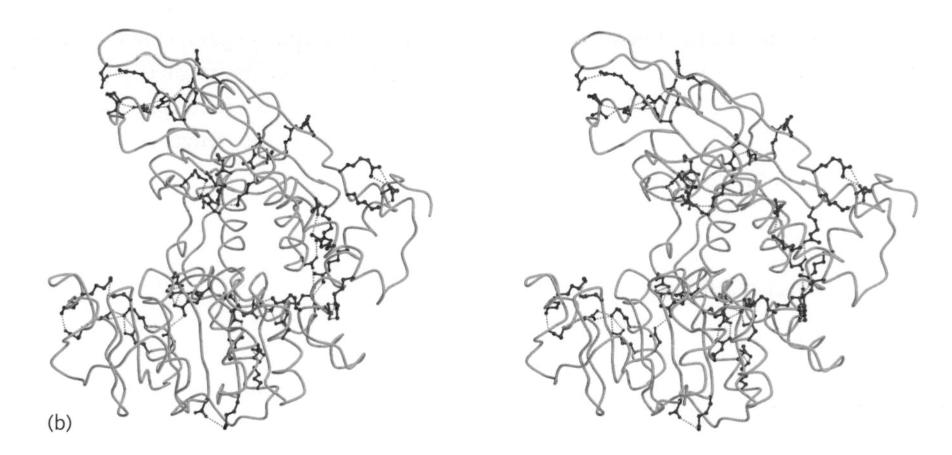

Glutamate dehydrogenase

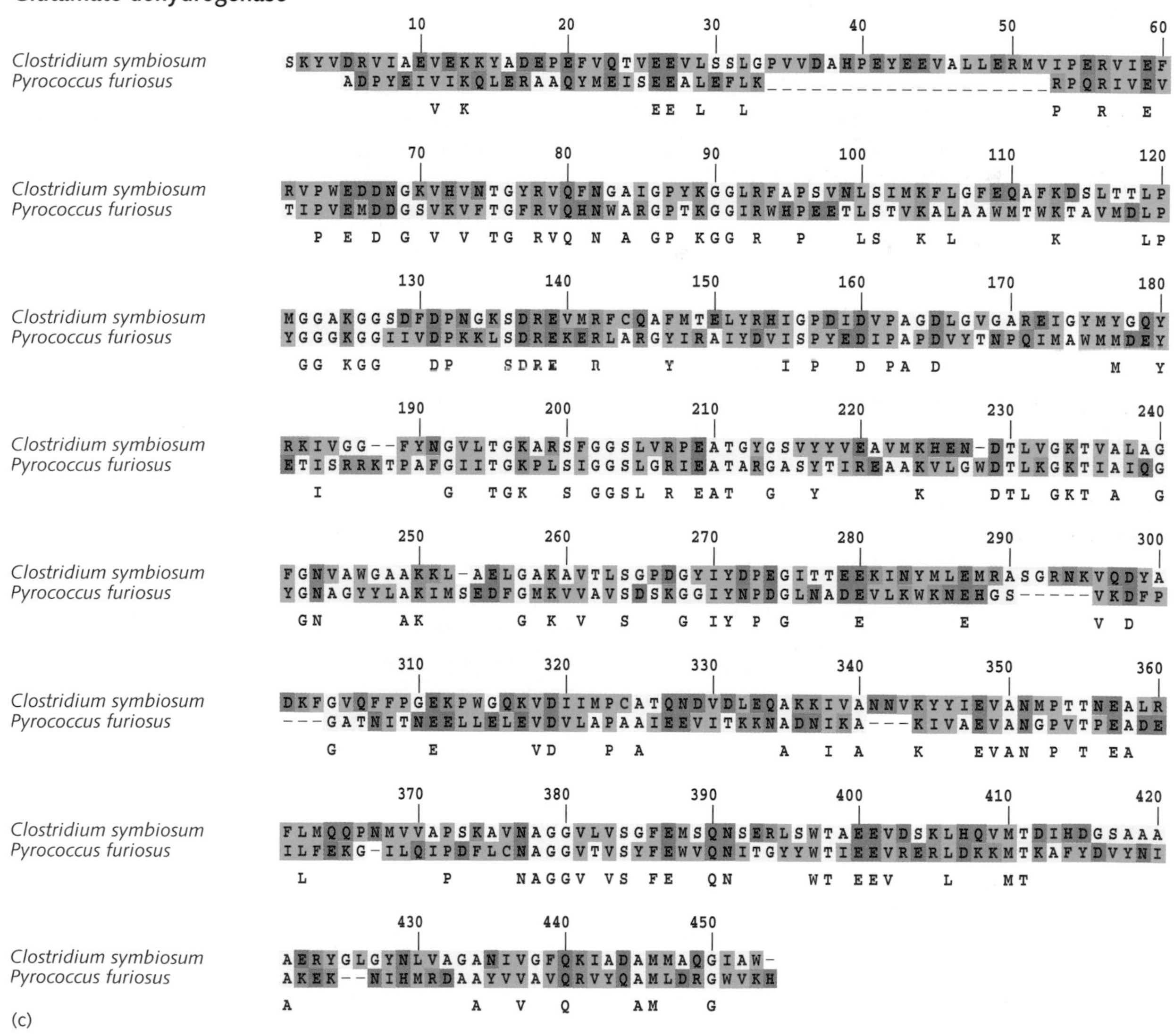

Figure 6.6 *(continued)*

- contain relatively fewer uncharged polar residues (Ser, Thr, Gln, Asn, Cys); some of these have thermolabile sidechains (His, Gln, Thr); and
- contain higher proportions of hydrophobic β-branched residues (see Box 6.3 and Figure 6.7).

Also, hyperthermophiles have special 'chaperones' – proteins that assist in the protein folding. It is likely that this is an adaptation to the challenge of high-temperature growth.

Effect of β-branched sidechains on protein stability

Proteins from thermophiles and hyperthermophiles are enriched in amino acids with β-branched sidechains. For example, compare leucine and isoleucine (see Figure 6.7).

How does this help to achieve high-temperature stability?

Folding of a protein to a unique native state is a compromise. Attractive inter-residue interactions favour formation of a compact native state. However, the greater conformational freedom of the polypeptide chain in the denatured state favours unfolding. To stabilize the native state, the attractive interactions must 'pay for' the loss of conformational freedom.

Thermodynamically, this is expressed by the criterion for stability:

$$G^{\text{Native}} - G^{\text{Denatured}} = \Delta G = \Delta H - T\Delta S < 0$$

where ΔH is the enthalpy change, ΔS the entropy change, and T the absolute temperature.

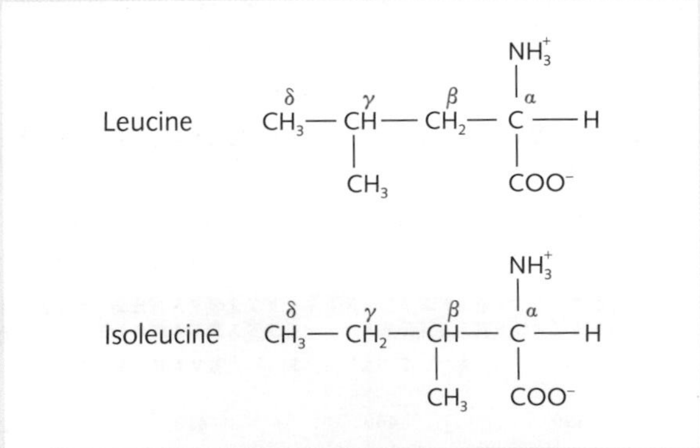

Leucine: $\overset{\delta}{CH_3}-\overset{\gamma}{CH}(CH_3)-\overset{\beta}{CH_2}-\overset{\alpha}{C}(NH_3^+)(COO^-)-H$

Isoleucine: $\overset{\delta}{CH_3}-\overset{\gamma}{CH_2}-\overset{\beta}{CH}(CH_3)-\overset{\alpha}{C}(NH_3^+)(COO^-)-H$

Figure 6.7 The amino acids leucine and isoleucine. The carbon atoms in the sidechain are labelled α, β, γ, and δ outwards from the carbon that will appear in the mainchain of a protein when the COO^- and NH_3^+ groups of the amino acid form peptide bonds. Isoleucine is said to be *β-branched* because, looking outwards along the sidechain from the C^β, there are two carbon substituents. The C^α of leucine, in contrast, has only one carbon substituent, looking outwards along the sidechain. (Leucine is branched at the γ-carbon.)

- The enthalpy, H, represents the attractive inter-residue interactions. Attractive interactions lower the enthalpy and make H more negative. The enthalpy of the native state is lower than that of the denatured state:

$$\Delta H = H^{\text{Native}} - H^{\text{Denatured}} < 0 \text{ favours folding.}$$

- The entropy, S, represents the conformational freedom. In the denatured state, the protein molecules adopt many different possible conformations, whereas in the native state, the conformations of many degrees of freedom are fixed; thus, the entropy of the denatured state is higher than the entropy of the native state:

$$\Delta S = S^{\text{Native}} - S^{\text{Denatured}} < 0 \text{ favours unfolding.}$$

- Because systems (at constant temperature and pressure) come to an equilibrium state of minimum Gibbs free energy (G), a protein will form the native state if and only if a favourable ΔH overcomes an unfavourable ΔS:

$$\Delta G = \Delta H - T\Delta S < 0$$

Because the entropy term is weighted by T, it assumes relatively higher importance at higher temperatures.

Assuming that a sidechain buried in the compact interior of a folded protein has a unique conformation, its loss of conformational freedom upon folding depends on the freedom it has in the denatured state. The lower the freedom in the denatured state, the lower the *loss* of entropy upon adopting a unique conformation in the native state. Because of the higher degree of crowding of atoms around the C^α in β-branched sidechains, they have less conformational freedom than β-unbranched sidechains, even in the denatured state. Therefore, β-branched sidechains contribute less to the unfavourable entropy change upon folding than β-unbranched sidechains. This is true at all temperatures but is more significant at the higher temperatures at which proteins from thermophiles and hyperthermophiles have to form their native states, because of the factor T in the entropy term.

Comparative genomics of hyperthermophilic archaea: *Thermococcus kodakarensis* and *Pyrococci*

A hyperthermophilic archaeon, *Thermococcus kodakarensis* strain KOD1, was isolated from a hot sulphur spring (102°C, pH 5.8) on the shore of Kodakara Island, in the Ryukyu archipelago between Kagoshima in southwest Japan, and Okinawa (29° 12′ N, 129° 19′ E). *T. kodakarensis* KOD1 is a strict anaerobe, normally growing by reducing elemental sulphur to H_2S.

General features of the genome

Fukui and co-workers reported the complete genome sequence of *T. kodakarensis* KOD1 (see Figure 6.8). The single, circular chromosome contains 2 088 737 bp, with a G+C content of 52 mole % (see Figure 6.8). A total of 2306 coding sequences were identified, with average length 833 bp, covering 92% of the genome. There are 46 genes for tRNA, two of which (for Trp and Met) contain introns.

Database searching suggested specific functions for half of the proteins (1165 out of 2306), and general functional classes for another 205. Of the proteins with known homologues, 240 are specific to the order Thermococcales. Of the remaining proteins, 261 appear to be unique to *T. kodakarensis*, as no homology to any other known protein was detectable.

Fifteen of the proteins are **inteins**, which catalyse the excision and splicing of intervening sequences *after* translation; i.e. the protein itself contains the self-splicing activity.

The genome contains numerous mobile elements, including four virus-related integrases and seven homologues of transposases – proteins that catalyse the movement of DNA segments around a genome. However, transposase activity was not observed, suggesting that these proteins have lost their function.

Molecular physiology of T. kodakarensis

By mapping the proteins of *T. kodakarensis* for which functions can be assigned, it is possible to reconstruct the metabolic and transport pathways of *T. kodakarensis*. (See Figure 6.9 for an overview.)

Comparative genomics of T. kodakarensis

The genome sequences of three close relatives of *T. kodakarensis* – *Pyrococcus abyssi*, *Pyrococcus horikoshii*, and *Pyrococcus furiosus* – allowed comparisons and descriptions of the evolutionary relationships between these archaea at the genome and protein levels.

Some of the differences are characteristic of the different genera – *Thermococcus* and *Pyrococcus*. These include the G+C content and the number of coding sequences (Table 6.3).

The loss of synteny between *Pyrococcus* and *Thermococcus* genera illuminates the origin of the difference in protein content (see Figure 6.10). The rearrangement among the *Pyrococcus* species themselves is substantial and involves a major inversion between *P. abyssi* and *P. horikoshii*. In contrast, between the genera, the genome has been completely shuffled and redealt. Indeed, there is no large contiguous region in the *T. kodakarensis* genome with no correspondence in pyrococci. This shows that the larger genome of *T. kodakarensis* is not the result of a recent horizontal transfer of a large block from a distant lineage.

Table 6.3 Characteristics of *T. kodakarensis*, *Pyrococcus abyssi*, *Pyrococcus horikoshii*, and *Pyrococcus furiosus*

Characteristic	*T. kodakarensis*	*P. abyssi*	*P. horikoshii*	*P. furiosus*
Genome size (bp)	2 088 737	1 765 118	1 738 505	1 908 256
G+C (mole %)	52.0	44.7	41.9	40.8
Coding sequences	2 306	1 784	2 065	2 065

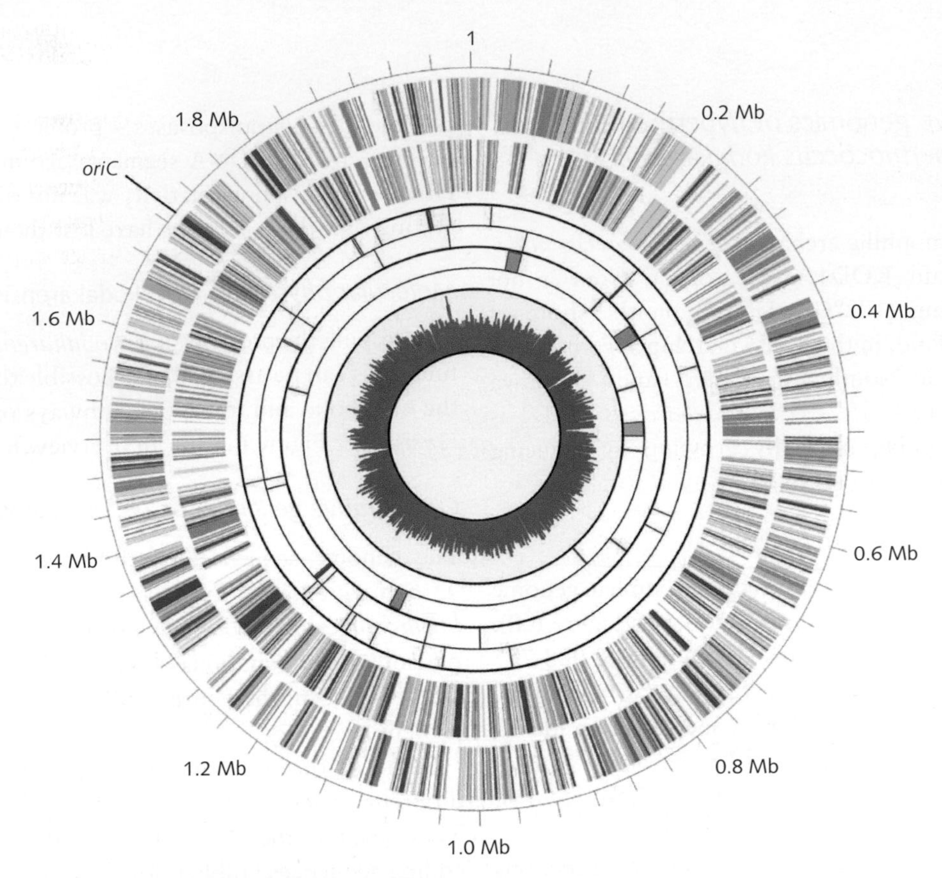

Figure 6.8 Diagram of the genome of *T. kodakarensis* strain KOD1. The contents of the consecutive circles, from the outermost, are:

(1) Scale in 0.2 Mb increments, plus the predicted origin of replication (*ori*C).
(2) Predicted protein-coding regions in clockwise direction.
(3) Predicted protein-coding regions in anticlockwise direction.
(4) Predicted tRNA coding regions in clockwise direction (red).
(5) Predicted tRNA coding regions in anticlockwise direction (blue).
(6) Predicted mobile elements in clockwise direction (red). Lines indicate transposase genes; boxes indicate virus-related regions.
(7) Predicted mobile elements in counter-clockwise direction (blue). Lines indicate transposase genes; boxes indicate virus-related regions.
(8) G+C content (mol. %) in 10 kb window.

Circles (2) and (3) are colour coded according to function:

Functional category	Colour
Translation, ribosomal structure, biogenesis	Magenta
Transcription	Pink
DNA replication, recombination, repair	Pale pink
Cell division, chromosome partitioning	Forest green
Post-translational modification, protein turnover, chaperones	Yellow
Cell envelope biogenesis, outer membrane	Light yellow
Cell motility, secretion	Light green
Inorganic ion transport/metabolism	Pale green
Signal transduction	Medium turquoise
Energy production/conversion	Purple
Carbohydrate transport/metabolism	Light blue
Amino acid transport/metabolism	Cyan
Nucleotide transport/metabolism	Violet
Co-enzyme metabolism	Pale turquoise
Lipid metabolism	Medium purple
Secondary metabolites biosynthesis/transport/catabolism	Light sky blue

From: Fukui, T., Atomi, H., Kanai, T., Matsumi, R., Fujiwara, S., & Imanaka, T. (2005). Complete genome sequence of the hyperthermophilic archaeon *Thermococcus kodakarensis* KOD1 and comparison with *Pyrococcus* genomes. Genome Res. **15**, 352–363.

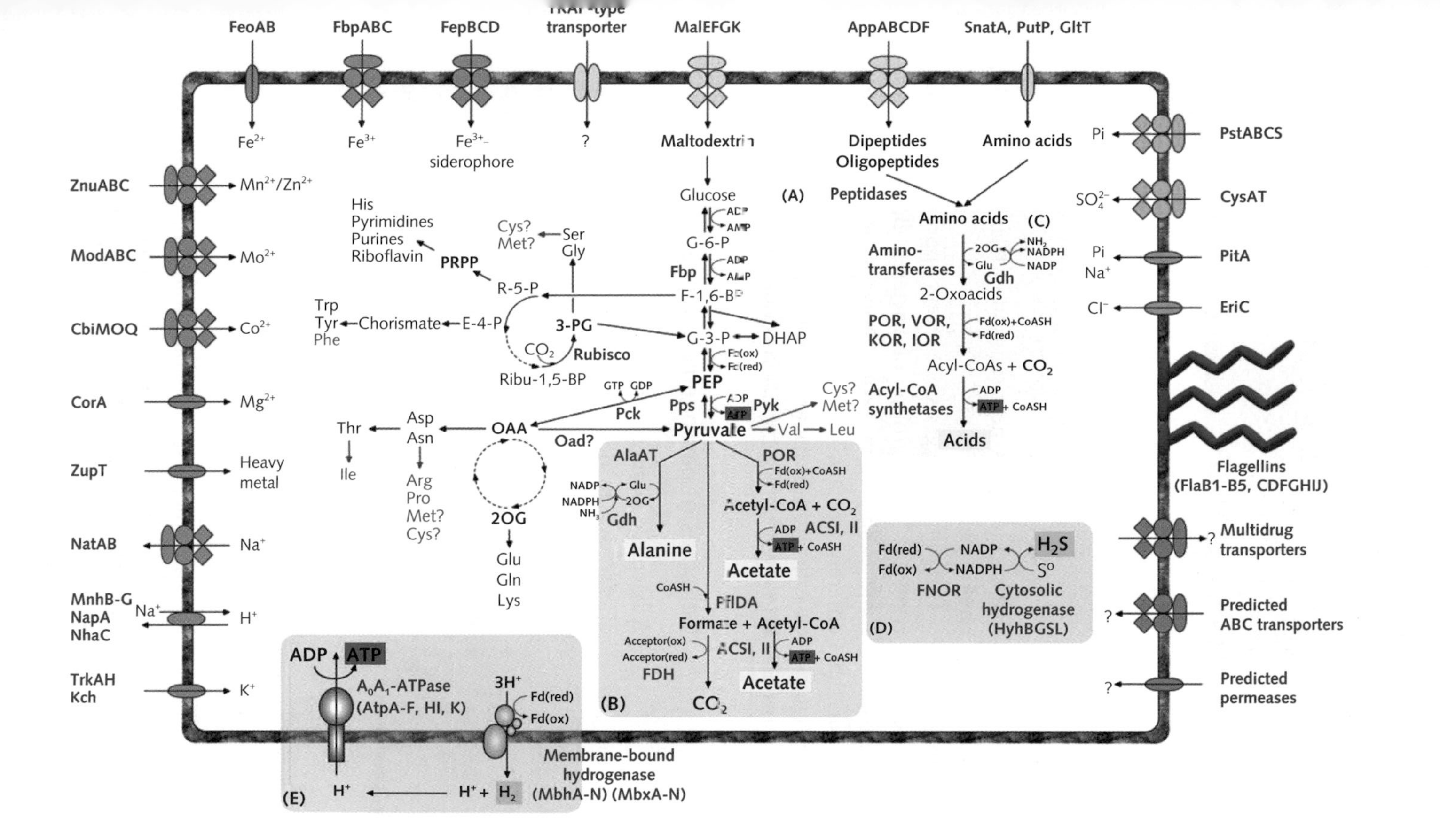

Figure 6.9 Reconstructed scheme of metabolism and solute transport in *Thermococcus kodakaraensis*. Components or pathways for which no predictable enzymes could be assigned appear in red.

Each gene product with a predicted function in ion or solute transport is illustrated on the membrane. The transporters and permeases are grouped by substrate specificity, as cations (violet), anions (green), carbohydrates/carboxylates/amino acids (yellow), and unknown (grey).

Metabolic pathways appear in the interior of the cell: (A) glycolysis (modified Embden–Meyerhof pathway); (B) pyruvate degradation; (C) amino acid degradation; (D) sulphur reduction; and (E) hydrogen evolution and formation of proton-motive force, coupled with ATP generation.

Abbreviations: DHAP, dihydroxyacetone phosphate; E-4-P, erythrose 4-phosphate; F-1,6-BP, fructose 1,6-bisphosphate; G-3-P, glyceraldehyde 3-phosphate; G-6-P, glucose 6-phosphate; OAA, oxaloacetate; 2OG, 2-oxoglutarate; PEP, phosphoenolpyruvate; 3-PG, 3-phosphoglycerate; PRPP, 5-phosphoribosyl 1-pyrophosphate; R-5-P, ribose 5-phosphate; Ribu-1,5-BP, ribulose 1,5-bisphosphate; ACS, acetyl-CoA synthetase (ADP-forming); AlaAT, alanine aminotransferase; Fbp, fructose 1,6-bisphosphatase; FDH, formate dehydrogenase; FNOR, ferredoxin:NADP oxidoreductase; Gdh, glutamate dehydrogenase; IOR, indolepyruvate:ferredoxin oxidoreductase; KOR, 2-oxoacid: ferredoxin oxidoreductase; PflDA, pyruvate formate lyase and its activating enzyme; Oad, oxaloacetate decarboxylase; Pck, phosphoenolpyruvate carboxykinase; POR, pyruvate:ferredoxin oxidoreductase; Pps, phosphoenolpyruvate synthase; Pyk, pyruvate kinase; VOR, 2-oxoisovalerate: ferredoxin oxidoreductase; AppABCDF, ABC-type dipeptide/oligopeptide transporter; CbiMOQ, ABC-type Co^{2+} transporter; CorA, Mg^{2+}/Co^{2+} transporter; CysAT, ABC-type sulphate transporter; EriC, voltage-gated Cl^- channel protein; FbpABC, ABC-type Fe^{3+} transporter; FeoAB, Fe^{2+} transporter; FepBCD, ABC-type Fe^{3+}-siderophore transporter; GltT, H^+/glutamate symporter; Kch, Ca^{2+}-gated K^+ channel protein; MalEFGK, ABC-type maltodextrin transporter; MnhB-G, multisubunit Na^+/H^+ antiporter; ModABC, ABC-type Mo^{2+} transporter; NapA and NhaC, Na^+/H^+ antiporter; NatAB, ABC-type Na^+ efflux pump; PitA, Na^+/phosphate symporter; PstABCS, ABC-type phosphate transporter; PutP, Na^+/proline symporter; SnatA, small neutral amino acid transporter; TrkAH, Trk-type K^+ transporter; ZnuABC, ABC-type Mn^{2+}/Zn^{2+} transporter; ZupT, heavy metal cation transporter.

From: Fukui et al. (2005) (see Figure 6.8).

Figure 6.10 Arrangement of homologous segments in the genomes of *Thermococcus kodakarensis*, *Pyrococcus abyssi*, *P. horikoshii*, and *P. furiosus*.

From: Fukui et al. (2005) (see Figure 6.8).

Bacteria

Bacteria form the other division of prokaryotes (Figure 6.11). Bacteria have been known for much longer than archaea (A. van Leeuwenhoek discovered bacteria in 1676). In consequence, bacterial taxonomy bears a considerable load of historical baggage. It has required genomes to sort out the phylogenetic relationships. However, the genomes also show large amounts of horizontal gene transfer that preclude any neat solution in terms of a simple phylogenetic tree.

Figure 6.11 suggests one recent approach to classifying the main groups of bacteria. Box 6.4 gives some examples of better-known organisms in the different groups, with some brief comments. For a classification of bacteria focusing on pathogens, see http://www.microbialrosettastone.com/.

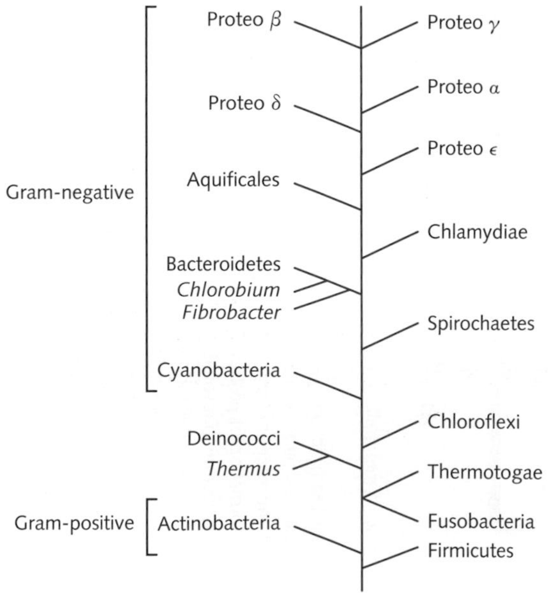

Gram-negative
Proteo β
Proteo γ
Proteo α
Proteo δ
Proteo ϵ
Aquificales
Chlamydiae
Bacteroidetes
Chlorobium
Fibrobacter
Spirochaetes
Cyanobacteria
Chloroflexi
Deinococci
Thermus
Thermotogae
Gram-positive
Actinobacteria
Fusobacteria
Firmicutes

Figure 6.11 Phylogenetic tree of some major bacterial types. This diagram reflects the topology of the tree but not the extent of divergence between and within groups. Some of the groups are phyla, others are genera.

Reference: Bacterial (Prokaryotic) Phylogeny Webpage (2006) (http://www.bacterialphylogeny.com/index.htm).

Genomes of pathogenic bacteria

A bacterial pathogen is a bacterium that can cause disease. It can do so by virtue of having **virulence factors**, which may include toxins, surface proteins that mediate attachment to cells, defensive shields (proteins and carbohydrates), and secreted enzymes. In many cases, closely related strains or species differ in pathogenicity. This suggests that comparison of their genomes could help to identify virulence factors. Knowledge of virulence factors would permit: (a) testing of foods for presence of pathogenic strains, for instance of *E. coli*, (b) choosing suitable drug targets, and (c) designing vaccines.

Examples include:

E. coli. Because a non-pathogenic strain (K-12MG1655) of *E. coli* occupies such a central role in molecular biology, it is easy to forget that it exists in nature, and that related strains are pathogenic. We discussed the *E. coli* genome in Chapter 4.

Strain *E. coli* 0157:H7 causes haemorrhagic colitis, which can be fatal. A comparison of the genomes of this strain with that of K12 strain MG1655 shows

Characteristics of major groups of bacteria

Group	Examples	Comments
Firmicutes	Bacilli, staphylococci, lactobacilli, Clostridia	*Listeria* and staphylococci can be infectious; *Clostridia* can cause food poisoning; lactobacilli are useful in yoghurt production
Actinobacteria	*Micrococcus, Streptomyces*	Decompose dead plant material; source of antibiotics
Fusobacteria	*Fusobacterium nucleatum*	Live in human gut, involved in periodontal infection
Thermotogae	*Thermotoga subterranea*	Thermophilic or hyperthermophilic; some are anaerobic
Thermus	*Thermus aquaticus*	Thermophilic; source of *Taq* polymerase
Deinococci	*Deinococcus radiodurans*	*D. radiodurans* is unusually radiation resistant
Chloroflexi	*Chloroflexus aurantiacus*	Photosynthetic, but do not produce O_2; may provide clue to early development of photosynthesis
Cyanobacteria	*Prochlorococcus marinus*	Chlorophyll-based photosynthesis; most split H_2O and produce O_2; give rise to chloroplasts via symbiosis
Spirochaetes	*Leptospira, Borrelia burgdorferi, Treponema pallidum*	Some are pathogenic (leptospirosis, syphilis, Lyme disease)
Fibrobacters	*Fibrobacter intestinalis*	Live in gut; help cattle to digest cellulose
Chlorobium	*Chlorobium tepidum*	Green sulphur bacteria; photosynthetic: reduce sulphide to sulphur
Bacteroidetes	*Bacteroides fragilis*	Some are marine plankton; others are anaerobic, live in the gut and can cause infection. *Porphyromonas gingivalis* causes gum disease
Chlamydiae	*Chlamydia trachomatis*	Grow intracellularly; major cause of blindness; also cause sexually transmitted infections of the urogenital system
Aquificales	*Aquifex aeolicus*	Extremophiles, autotrophs
ε-Proteobacteria	*Helicobacter pylori*	*H. pylori* lives in gut, cause of ulcers
δ-Proteobacteria	*Desulfovibrio desulfuricans*	Mostly aerobic; some anaerobic examples reduce sulphur or sulphate
α-Proteobacteria	*Rhodospirillum rubrum, Rhizobium, Rickettsia*	*Rhizobium* are symbiotic with legumes and fix nitrogen; *Rickettsia* cause typhus; give rise to mitochondria via symbiosis
β-Proteobacteria	*Burkholdia, Bordetella, Thiobacillus, Neisseria*	Some live on inorganic nutrients; others are infectious, different species causing pertussis, gonorrhoea and meningitis
γ-Proteobacteria	*Escherichia coli, Haemophilus influenzae, Pseudomonas aeruginosa, Yersinia pestis, Salmonella typhimurium*	Important in medicine and molecular biology: cause enteritis, typhoid, bubonic plague, and others

that the genome of K12MG1655, 4639221 bp, is shorter than that of 0157:H7, 5528445 bp. Regions amounting to 4.1 Mb are common to both strains. The unshared genes tend to cluster in strain-specific regions. Strain-specific regions in K12MG1655 contain 1.34 Mb, 1387 genes, and in 0157:H7 contain 0.53 Mb, 528 genes.

It is likely that the strains diverged about 4.5 million years ago. A clue to the origin of the differences between the strains is the atypical base composition of the strain-specific regions. This suggests that they entered the respective genomes by horizontal gene transfer.

- Horizontal gene transfer is a common theme in development of virulence and antibiotic resistance.

Helicobacter pylori. Half the world's population is infected with *H. pylori*. One out of 10 people develop clinical disease: gastritis, duodenal and gastric ulcers, and some cancers. Proof that *H. pylori* infection is the cause of ulcers was obtained – over the disbelief of the scientific and medical establishments at the time – by Barry Marshall, who swallowed a culture of *H. pylori*, and quickly developed symptoms of gastritis.

H. pylori strains are very diverse (they have been applied to tracking of patterns of human migration). Three strains have been sequenced completely. Strain 26695 contains about 1.7 Mbp, and about 1550 genes. Other sequenced strains differ by about 6%. Virulence appears to be associated with a common Cag 40 kb pathogenicity island containing >40 genes. The appearance of genes within this island is correlated with virulence. This pathogenicity island is common to many bacteria. It is likely that it has been circulated by horizontal gene transfer.

Staphylococcus aureus. *S. aureus* infections are a growing clinical problem because of the aggressive development of antibiotic resistance. (The development of resistance to vancomycin – the 'antibiotic of last resort' – is discussed in Chapter 9.)

The genomics of *S. aureus* has been pursued vigorously, in order to identify the mechanisms of development and spread of resistance. The *S. aureus* genome is about 2.8–2.9 Mb long. Assignment of approximately 2600 open reading frames accounts for almost 85% of the genome. There is a single plasmid containing about 25 000 bp. Genes for enhanced antibiotic resistance are encoded by a transposon inserted into the plasmid. Comparison of the sequences – in particular, observation of lack of synteny – has made it clear that the development of methicillin resistance was not a single event, producing a clone that was subsequently selected. Instead, the resistance elements were acquired many times by many strains, via horizontal gene transfer.

A comparison of the sequences of many *S. aureus* strains, encompassing different clinical phenotypes, showed that 78% of genes were common to all strains, including isolates from cow and sheep. The remaining 22%, that are at least partially strain-specific, tend to be localized within 18 large regions of difference (RDs), ranging from 3–50 kb long. Figure 6.12 shows the presence of these regions in the different strains, and the correlation of the pattern with methicillin resistance.

Genomics and the development of vaccines

Genomics and recombinant technology have made possible a new generation of approaches to vaccine design.

A vaccine against hepatitis B virus is expressed in yeast cells. It is a surface antigen, a viral envelope protein, the gene for which was cloned into yeast. A vaccine against *Bordetella pertussis* (the causative agent of whooping cough) is based on the toxin, a multi-subunit protein. By genetic engineering, the molecule was completely detoxified by introduction of mutants, which removed the enzymatic activity but left the immunological properties intact. That is, antibodies raised against the detoxified form protect against the native protein.

A more general approach to vaccine design involves comparing the genome sequences of pathogenic and nonpathogenic strains to identify virulence factors that might serve as the basis of vaccines.

Neisseria meningitidis serogroup B is the major cause of meningitis and septicaemia in children and young adults. From the 2 272 351 bp genome sequence, computational methods predicted 2158 genes. Algorithms predicted that 600 of them would be on the cell surface or secreted. These were candidates for vaccines. Of these, 350 were expressed in *E. coli*, and tested in mice for an immune response that produced

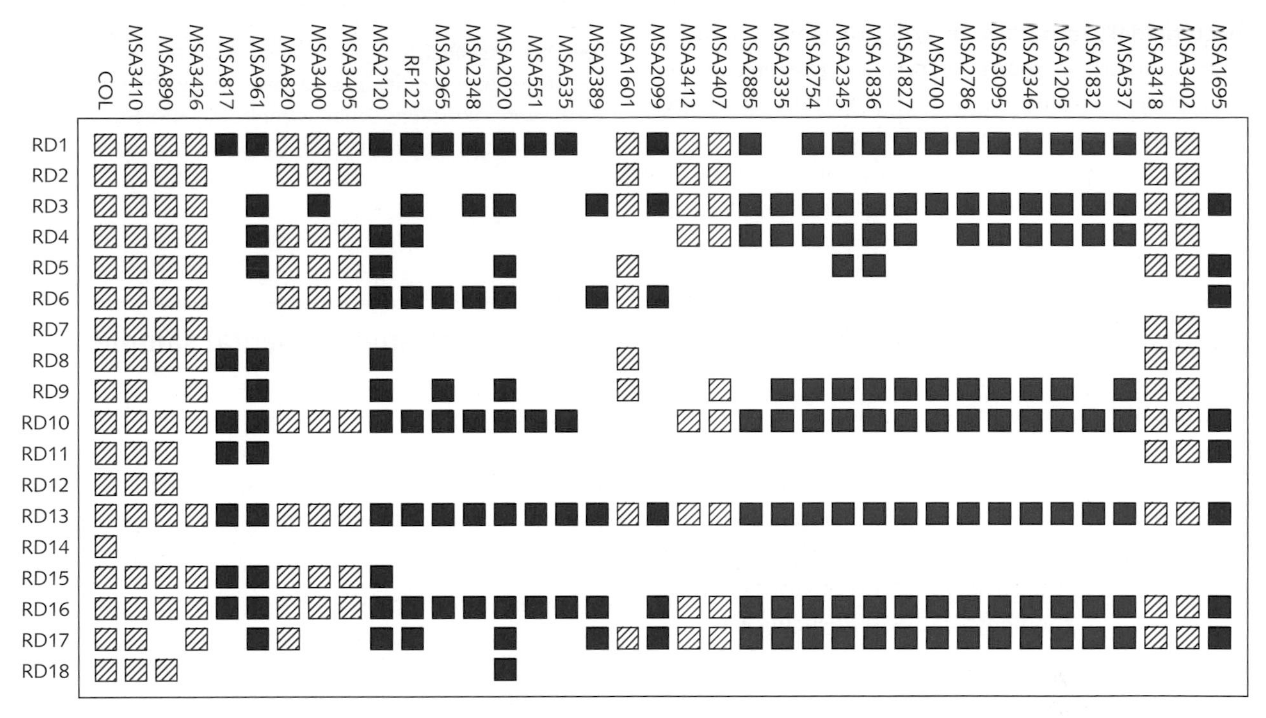

Figure 6.12 Results of comparison of sequences of 36 isolates of *S. aureus*. RD = regions of difference, localized segments of the genome of high variability among strains. Filled squares indicate RD present; empty squares, RD absent. Hatched squares correspond to methicillin-resistant strains. Red indicates isolates of electrophoretic type 234, the predominant type causing toxic shock syndrome.

From: Fitzgerald, J.R., Sturdevant, D.E., Mackie, S.M., Gill, S.R., & Musser, J.M. (2001). Evolutionary genomics of *Staphylococcus aureus*: Insights into the origin of methicillin-resistant strains and the toxic shock syndrome epidemic. Proc. Natl. Acad. Sci. USA **98**, 8821–8826. Reproduced by permission.

The future of antibiotic development

The development of resistant strains of pathogens presents a severe challenge to medicine. There is consensus that novel antibiotics will be needed. Some are already in the 'pipeline', currently in the clinical testing phase. However, the research that produced today's 'new' antibiotics was initiated in the early 1990s and pharmaceutical companies are reducing their emphasis on antibiotic research. It is likely that fewer new antibiotics will emerge in the current industrial climate. The paradox of a growing need for new discoveries coupled with the reduction in resources aimed at generating them creates a problem that may become a crisis.

The novel developments associated with genomics, the subject of this book, can in principle contribute to the development of antibiotics. It is possible to identify *targets* – specific proteins, essential for a pathogen, that differ from mammalian proteins sufficiently to suggest that drugs against these proteins would be effective against the pathogen but non-toxic to mammals. Unlike classical antibiotic research practice, experimental methods are now available that can define the mechanism of action of a drug while it is still under development.

A related prospect is to turn to biology as well as chemistry to discover new therapeutic agents, including revival of an old suggestion of using bacteriophages clinically. A number of small biotech companies have started up, funded by venture capital, to try to explore a number of non-traditional avenues. Given the decade 'lead time' required for a new drug to make its way from the laboratory to approval in clinical use, the process must be set in motion immediately. A problem with this approach is a line of recent US court decisions that impose stricter criteria on the patentability of procedures that might be considered 'natural processes'.

The problems are multidimensional, involving science, economics, long-term forecasting, and regulatory and patent law. Each field presents an individual set of difficulties. These difficulties are compounded by the necessity to solve them all simultaneously in the face of both genuine conflicts between different goals, and boundaries between professions that impede communication and cooperation. There is, however, consensus that the problems must be solved.

bactericidal antibodies. In order to achieve a vaccine that covered as many strains as possible, surveys of different strains for sequence variability of these candidate vaccines revealed which ones were relatively constant. The results of the work are promising candidates for vaccines now under development.

- Our descendants may well look back at the second half of the 20th century as a narrow window during which bacterial infections could be controlled, and before and after which they could not.

Metagenomics: the collection of genomes in a coherent environmental sample

Classically, microbiologists studied prokaryotes by growing them in culture, isolating pure strains for detailed study. Powerful as the methods were, and useful as they were for clinical applications and research, they were also blinders that prevented full appreciation of the variety and interactions of species in natural environments. DNA sequencing has made it possible to:

- clarify evolutionary relationships;
- use high-throughput sequencing methods to study a cross-section of the life in a natural sample;
- study the majority of strains that are difficult to grow in culture; and
- appreciate the relationships and interactions among different species that share an ecosystem.

- A millilitre of ocean water may contain 100–200 species. A gram of soil may contain 4000.

From natural samples containing complex mixtures, it is possible to amplify and determine sequences directly, without culturing individual strains. The molecule of choice has been 16S rRNA. This is partly because of its traditional role as a molecule that varies at the appropriate rate to distinguish ancient phylogenetic branching patterns. In addition, rRNA is not very prone to horizontal gene transfer. It thereby preserves the distinctions between taxa – perhaps, however, this disguises the mixing that has taken place with other genes. Another disadvantage of characterizing an organism by its rRNA is that rRNA does not reveal any details of the metabolism or other adaptations of the species.

An example of metagenomics is the sequencing of 16S rRNA genes from ocean water from the Sargasso Sea. A group led by J.C. Venter sequenced 10^9 non-redundant regions. Many novel sequences were found, although it is difficult to assemble complete genomes and avoid chimaeras.

Marine cyanobacteria – an in-depth study

The basic concepts of genome evolution are secure: organisms explore genome variations. Adaptations drive some changes – via divergence, allelic redistribution within a population, gene loss, or gene acquisition by horizontal gene transfer. Neutral genetic drift accounts for other changes, especially in small populations.

What is more difficult to understand is *how* populations make choices and adopt strategies. If organisms encounter environments varying in space and/or time, into how many populations, or even new species, will they split? Which proteins will diverge – in sequence, in function, or in expression pattern? What novel genes are needed and where will they come from?

In most field situations, the 'topology' of evolutionary space depends on a complicated interaction of physical and ecological variables, such as geographic barriers imposed by landscape, climate, and interspecies cooperation or competition. Complex environments give rise to complex biological communities.

The distribution of cyanobacteria in the open oceans in temperate regions offers a relatively simple ecological context. The distributions – both of environmental features and of species or strains – depend on a *single* variable, depth. There are many correlates of depth: light intensity and quality, temperature, pressure, ultraviolet light penetration, nutrient availability (notably, sources of nitrogen and iron), and the occurrence of predators and viruses. Yet, for all its complexity, the system is one-dimensional.

BOX 6.5 *Prochlorococcus* and *Synechococcus* genomes

Feature	*Prochlorococcus*		*Synechococcus*
	Strain MED4	Strain MIT9313	Strain WH8102
Preferred light level	High	Low	
Length (bp)	1 657 990	2 410 873	2 434 428
G+C (mol%)	30.8	50.7	59.4
Protein coding (%)	88	82	85.6
Protein coding genes	1 716	2 273	2 526
RNA genes	40	51	44

The major populations of cyanobacteria in temperate and tropical oceans belong to two related genera, *Prochlorococcus* and *Synechococcus*. They are responsible for a significant fraction of worldwide photosynthesis. *Prochlorococcus* is believed to have diverged from *Synechococcus* fairly recently.

Ocean environments are stratified. Studies of the distribution of *Prochlorococcus* ecotypes in a vertical column in the Sargasso Sea reveal a division into *two* types of strain. Closer to the surface than about 130 m depth, the majority of *Prochlorococcus* strains are adapted to high light levels. Strains prevalent below 130 m depth are adapted to low light levels (see Figure 6.13).

- *Prochlorococcus* strains are adapted to different ambient light levels, which decrease with increasing depth below the ocean surface. High light levels: ≥200 μmol photons m^{-2} s^{-1}. Low light levels: ≤30–50 μmol photons m^{-2} s^{-1}.

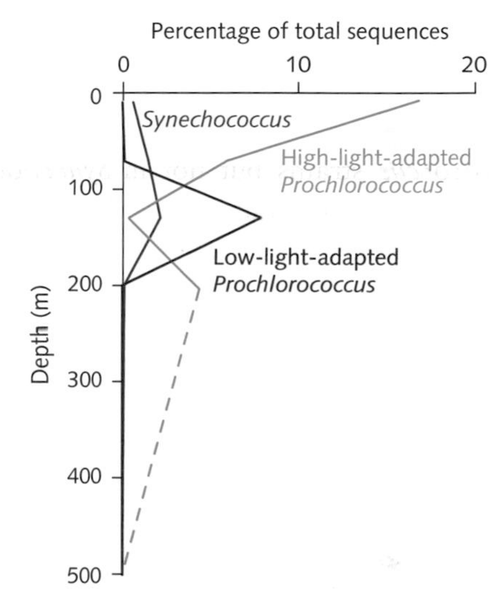

Figure 6.13 Distribution with depth of three types of cyanobacteria: *Synechococcus*, *Prochlorococcus* strains adapted to high-light-intensity habitats, and *Prochlorococcus* strains adapted to low-light-intensity habitats. The appearance of large amounts of low-light-adapted *Prochlorococcus* at 200 m, below the level where it seemed to have virtually disappeared, is puzzling. The broken line from 200–500 m depth is interpolated; measurements have been made at 200 and 500 m but not in between.

After: DeLong, Preston & Mincer, et al. (2006). Community genomics among stratified microbial assemblages in the ocean's interior. Science **311**, 496–503.

Rocap and co-workers compared the genomes of high-light- and low-light-adapted *Prochlorococcus* strains with *Synechococcus** (see Box 6.5). *Prochlorococcus* strain MIT9313 is adapted to growth in low light intensity. Both its distribution with depth (Figure 6.13) and the general features of its genome are similar to *Synechococcus*. In contrast, the high-light-adapted strain MED4 is 'lean and mean': its genome is unusually small and encodes fewer proteins. *Prochlorococcus* MED4 is the smallest known organism that generates oxygen.

* Rocap et al. (2003). Genome divergence in two *Prochlorococcus* ecotypes reflects oceanic niche differentiation. Nature **424**, 1042–1047.

BOX 6.6

Cyanobacterial photosynthesis

The photosynthetic apparatus of cyanobacteria contains large chromophore-containing macromolecular complexes:

- Two coupled photosystems that carry out energy transduction – the capture of light energy. These are called PSI and PSII.
- Antenna pigments that make light harvesting more efficient by absorbing light and transferring the excitation energy to the reactive chlorophylls.

Approximately half of the proteins of *Synechococcus* are common to all three species: 1314 out of 2526. Only 38 genus-specific proteins appear in both *Prochlorococcus* strains but not in *Synechococcus*. Many of these 38 proteins are involved in the synthesis of the light-harvesting complex of *Prochlorococcus*, which has a structure unusual among bacteria (see Box 6.6).

Within the *Prochlorococcus* genus, many genes are strain specific: MIT9313 has 923 proteins that do not appear in MED4 (about half of these do appear in *Synechococcus*). This, together with the observation that the MED4 genome and proteome are substantially smaller, implies that many genes have been lost in the differentiation of the strains.

How are different Prochlorococcus *strains adapted to differences in ambient light intensities and spectral distributions?*

Effective interactions with light require both efficient energy transduction and protection from photochemical damage caused by excitation energy spillover.

- Antenna complexes of photosystem II (PSII) of most cyanobacteria, including *Synechococcus*, contain protein complexes called phycobilisomes. *Prochlorococcus* is unusual among cyanobacteria in using, as PSII antennae, proteins binding unusual modified (divinyl) chlorophylls, called Pcb proteins.

  Where did the Pcb proteins come from? They appear to have been recruited from a family called **iron-stress-inducible** (ISI) proteins (see Weblem 6.4). ISI proteins are expressed in cyanobacteria under conditions of low iron concentrations. They provide an alternative to the iron-rich protein ferredoxin in the electron transport chain.
- High-light-adapted and low-light-adapted strains differ in the relative amounts of chlorophyll a_2 and b_2. High-light-adapted strains have mostly chlorophyll a_2; low-light-adapted strains have more chlorophyll b_2, which absorbs optimally in the blue region of the spectrum. This is an appropriate match to the colour of the ambient light below the ocean surface. The ratio of concentrations of chlorophylls a_2 and b_2 can vary with habitat conditions. In the low-light-adapted strain MIT9313, the ratio can change by at least a factor of two, producing more chlorophyll a_2 at higher light intensities. The chlorophyll a_2/b_2 ratio in the high-light-adapted strain MED4 is less sensitive to ambient light intensity. The nature of the control mechanism and the set of proteins that change expression patterns are not yet fully understood.

Another adaptation for 'scavenging' light in dark environments is to increase the number of genes for light-harvesting proteins. Low-light-adapted strains have more copies of the genes that code for the chlorophyll-binding antenna protein Pcb. In addition, MED4 contains a phycoerythrin, an antenna protein that binds a chromophore absorbing green light.

Protection against photochemical damage

Ultraviolet light can damage DNA. Major products include thymine dimers, the linkage of adjacent thymine residues in DNA. In response to the threat of mutation, cells contain repair enzymes, including **photolyase,** which recovers thymines from dimers.

Because ultraviolet light does not penetrate far into sea water, *Prochlorococcus* strain MED4, living nearer the surface, is in greater danger from photochemical damage than MIT9313. Indeed, MED4, but not MIT9313, contains a gene for photolyase. Another difference between the strains is also probably related to photo-oxidative stress: MED4 contains perhaps twice as many **high-light-inducible proteins** as MIT9313. From their distribution in the genome, some of these appear to have arisen by recent duplication events.

- 'Mad dogs and Englishmen go out in the midday sun,' sang Noël Coward. Plants and ocean-surface-dwelling cyanobacteria do too – they have no choice.

Utilization of nitrogen sources

In the ocean, the prevalent form of nitrogen near the surface is ammonium, produced in part by fixation of atmospheric nitrogen. In deep waters, nitrate is more common.* Nitrate is produced by degradation of dead organic matter: dead organisms sink.

Different organisms assimilate nitrogen from molecular nitrogen, nitrate (NO_3^-), nitrite (NO_2^-), or ammonium ion (NH_4^+).

The two *Prochlorococcus* strains differ in their ability to assimilate nitrogen from different sources (see Box 6.7). There has been a successive loss in nitrogen-assimilating ability with the divergence from *Synechococcus* to low-light-adapted *Prochlorococcus* (deeper-living) to high-light-adapted *Prochlorococcus* (living nearer the surface). *Synechococcus* has active nitrate and nitrite reductase genes and can use either ion, as well as ammonium, as a nitrogen source. *Prochlorococcus* MED4 has lost nitrate reductase but retains nitrite reductase; it can use nitrite but not nitrate as a nitrogen source. *Prochlorococcus* MIT9313 has neither reductase and must get its nitrogen from ammonium or from other reduced nitrogen compounds such as amino acids.

Protection against predators and viruses

Other cellular life forms graze on *Prochlorococcus* and *Synechococcus* strains. Viruses infect them. Defensive adaptations involve genes encoding proteins involved in the synthesis of lipopolysaccharides and polysaccharides, which form the basis of cell surface recognition. Both strains of *Prochlorococcus* have acquired, by horizontal gene transfer, clusters of genes for surface polysaccharides not shared by the other strain or by *Synechococcus*. As evidence of foreign origin, this 40.8 kb cluster in MIT9313 has a G+C content of 42 mole %, substantially lower than that of the MIT9313 genome as a whole, 50.7 mole %.

BOX 6.7

Assimilation of nitrogen

Reaction	Enzyme	*Synechococcus*	*Prochlorococcus*	
		WH8102	MED4	MIT9313
$N_2 \rightarrow NH_4^+$	Nitrogenase	Absent	Absent	Absent
$NO_3^- \rightarrow NO_2^-$	Nitrate reductase	Present	Absent	Absent
$NO_2^- \rightarrow NH_4^+$	Nitrite reductase	Present	Present	Absent
$NH_3^+ \rightarrow$ glutamine	Glutamine synthase	Present	Present	Present

* See: http://www.es.flinders.edu.au/~mattom/IntroOc/notes/figures/fig5a5.html.

RECOMMENDED READING

- General discussions of prokaryotic classification:

  Oren, A. & Papke, R.T. (2010). *Molecular phylogeny of microorganisms*. Caister Academic Press, Norfolk, UK.

- Last universal common ancestor (LUCA) and related topics:

  Mat, W.K., Xue, H., & Wong, J.T. (2008). The genomics of LUCA. Front. Biosci. **13**, 5605–5613.

  Puigbò, P., Wolf, Y.I., & Koonin, E.V. (2009). Search for a 'Tree of Life' in the thicket of the phylogenetic forest. J. Biol. **8**, 59.

- The history of the growth of oxygen in the atmosphere: an intersection between biochemistry and geology:

  Raymond, J. & Segrè, D. (2006). The effect of oxygen on biochemical networks and the evolution of complex life. Science **311**, 1764–1767.

  Holland, H.D. (2006). The oxygenation of the atmosphere and oceans. Phil. Trans. R. Soc. Lond. B: Biol. Sci. **361**, 903–915.

- Metagenomics:

  DeLong, E.F. & Karl, D.M. (2005). Genomic perspectives in microbial oceanography. Nature **437**, 336–342.

  Scanlan, D.J. et al. (2009). Ecological genomics of marine picocyanobacteria. *Microbiol. Mol. Rev.* **73**, 249–299.

  Wooley, J.C., Godzik, A., & Friedberg, I. (2010). A primer on metagenomics. PLoS Comput. Biol. **6**, e1000667.

  Bohlin, J. (2011). Genomic signatures in microbes – properties and applications. Sci. World J. **11**, 715–725.

- New approaches to vaccine design:

  Rossolini, G.M. & Thaller, M.C. (2010). Coping with antibiotic resistance: contributions from genomics. Genome Medicine **2**, 15.

  Scarselli, M., Giuliani, M.M., Adu-Bobie, J., & Rappuoli, R. (2005). The impact of genomics on vaccine design. Trends Biotech. **23**, 84–91.

EXERCISES, PROBLEMS, AND WEBLEMS

Exercises

Exercise 6.1 Which of the differences between archaea and bacteria, described on pp. 192–194, could you derive from genome sequence alone?

Exercise 6.2 Using the standard genetic code (Box 1.1), are there any amino acids for which a hyperthermophilic organism could *not* preferentially choose a codon with G or C in the third position?

Exercise 6.3 From the table in Box 6.4, what are two groups of bacteria that are (a) photosynthetic, (b) live in the human gut, (c) pathogenic?

Exercise 6.4 On a photocopy of Figure 6.6, circle salt bridges that appear at (approximately) common positions in both structures.

Exercise 6.5 On a photocopy of Figure 6.7, identify positions at which there is a charged residue in the *P. furiosus* sequence but an uncharged residue in the *C. symbiosum* sequence. (Charged residues = K, R (shown in blue), D, E, H (shown in red) uncharged residues = all the others.)

Exercise 6.6 On a photocopy of Figure 6.9, (a) circle the glycolysis/gluconeogenesis pathway; (b) circle the Calvin cycle.

Exercise 6.7 On a photocopy of Figure 6.10, circle a region in which synteny is maintained in *P. abyssi, P. horikoshii*, and *P. furiosis.*

Exercise 6.8 Why is the recombinant vaccine against hepatitis B expressed in yeast and not *E. coli*?

Exercise 6.9 If 1 ml of seawater contains 200 species, the metagenome would be how big?

Problems

Problem 6.1 Draw a Venn diagram showing the numbers of genes specific to one, common to pairs, and common to all three of the following: *Prochlorococcus* strain MED4, *Prochlorococcus* strain MIT9313, and *Synechococcus* strain WH8102.

Problem 6.2 In Figure 6.12, hatched squares indicate MRSA strains. (a) Find two RDs (regions of diversity) that are present in all methicillin-resistant strains studied. (b) Find two RDs that are present in some methicillin-resistant strains, but not every strain containing either of them is methicillin resistant. (c) Is there any RD that is present in every methicillin-resistant strain, and absent from every methicillin-sensitive strain?

Weblems

Weblem 6.1 Print out the complete secondary structures of 16S rRNAs from *E. coli*, *Methanococcus vannielii*, and *Saccharomyces cerevisiae* (http://www.rna.icmb.utexas.edu/). On each structure, indicate where the region illustrated in Figure 6.1 appears.

Weblem 6.2 Are any archaea implicated in human disease?

Weblem 6.3 (a) Identify a bacterium with a growth temperature below 10°C. (b) Identify an archaeon with a growth temperature below 10°C. (See Figure 6.5.)

Weblem 6.4 Where did *Prochlorococcus* get the chlorophyll-binding proteins of its photosystem II antenna? (a) Using either of the *Prochlorococcus* Pcb proteins (UniProt ID PCBA_PROMM or PCBA_PROMP), search using PSI-BLAST for homologous cyanobacterial proteins with different functions. You should find – among others – ISIA_SYNP6, ISIA_SYNP7, and ISIA_SYNP2. Present the results of this search, editing the output down to the most relevant information. (b) What is the function of the ISI family of proteins? (c) Align the full sequences of PCBA_PROMM, PCBA_PROMP, ISIA_SYNP6, ISIA_SYNP7, and ISIA_SYNP2, using CLUSTAL W or T-Coffee. Comment on the extent of the divergence. (d) Determine the fraction of identical residues between every pair of sequences from the multiple sequence alignment and draw a phylogenetic tree.

Weblem 6.5 The high-light-intensity-adapted *Prochlorococcus* strain MED4 contains a phycoerythrin, but the low-light-intensity-adapted strain MIT9313 does not. Do most *Synechococcus* strains contain phycoerythrins? If so, align the sequences of a prochlorococcal and synechococcal phycoerythrin and comment on the extent of the divergence compared with the divergence of prochlorococcal Pcb proteins from synechococcal ISI proteins (see Weblem 6.4).

CHAPTER 7

Genomes of Eukaryotes

LEARNING GOALS

- Having a clear sense of how eukaryotic cells differ from prokaryotic cells.
- Understanding the relationships among the major types of eukaryotes.
- Recognizing that fungi in general and yeasts in particular are among the simplest eukaryotes, and have served as model organisms in the molecular biology laboratory (in addition to applications in agriculture and cooking).
- Knowing the unique features of higher plants, and the focus on *Arabidopsis thaliana* – 'the fruit fly of botany'.
- For the animals, appreciating the evolutionary path from early organisms, more distantly related to humans, to mammals.
- Respecting the power – albeit limited – of the ability to recover and sequence DNA from extinct organisms.

We, like all other species that exist or have existed, are the product of a long evolutionary history. Genomes give us snapshots of landmarks along the route. They allow us to understand the topology of the path – where and when it branched. They reveal when certain features arose. In some cases they show us the experiments – some productive, some abortive – that preceded the mature result subsequently adopted.

The origin and evolution of eukaryotes

There is consensus that eukaryotes are descended from prokaryotes. Evidence includes the observations that both prokaryotes and eukaryotes use the same genetic code, and share many metabolic pathways. When did eukaryotes originate? There is consensus that prokaryotes had the world to themselves for many years since the origin of life (dated at no later than 3.5 billion years ago). The first eukaryotic fossil is approximately 2 billion years old. The discovery in datable oil deposits of biochemicals produced only by eukaryotes suggests an earlier origin, almost 3 billion years ago.

The first eukaryotes began as unicellular organisms. Multicellularity began with formation of colonies, followed by cell specialization, perhaps first by simple symbiosis. Developmental programmes allowed specialization within clonal clusters.

There followed exploration of a great variety of body plans, based on a variety of tissue types. Major landmarks, that we know about, in the development of higher organisms, include the division between animals and plants. Some animals adopted body structures with bilateral symmetry. Some of these became vertebrates. Eventually, some became human.

Until relatively recently, our view of the history of life was limited to organisms that have left either descendants or fossils. The Burgess Shale deposits show us that we have missed a lot of interesting alternatives. It is true that it has become possible to recover and sequence ancient DNA. This has opened a window on extinct species. But only in a limited way.

Our only real possibility of reconstructing our history is through the genomes of extant organisms that have been around for a long time, and to read the story that their sequences tell.

- In order to reconstruct evolutionary history from genomes, we must analyse genomes from species, the ancestors of which first arose in the distant past. What do they share with their close relatives, that might offer genomic characterization of a group of species; for instance, what are the defining genomic features of vertebrates? What do ancient species share with their precessors and successors? What innovations did they achieve? Which of those were dead ends and which did other species descended from them adopt and develop?

Evolution and phylogenetic relationships in eukaryotes

Genome sequences provide detailed information about evolutionary relationships among species. So many genomes are now known that we must pick and choose only a few of the interesting examples. The Genomes On Line Database (GOLD) lists 2007 completed or ongoing eukaryotic genome sequencing projects.

Figure 7.1 shows the major groups of eukaryotes. Note the 'star' topology – at this level of low resolution, eukaryotes are a bush rather than a tree. In this chapter, however, we shall follow a more directed path, roughly in the direction towards higher complexity. This correlates fairly well with date of origin.

From the comparative genomics of eukaryotes, we can ask questions about features of humans that we consider essential. For instance, we can look into the origin of some features of a body plan, or the immune system, or the endocrine system, or features of the nervous system. Phrasing the questions loosely: Who invented them? When? What if anything did they come from? What alternatives were experimented with and how well did they work?

This is an ambitious programme. It is appropriate to begin with one of the simplest eukaryotes, yeast.

The yeast genome

Yeast, like *E. coli*, is an organism better known to many of us from molecular biology labs than from nature. It has served as a model eukaryote, because of its relative simplicity, ease of growth, having both haploid and diploid states, and safety. Of course we

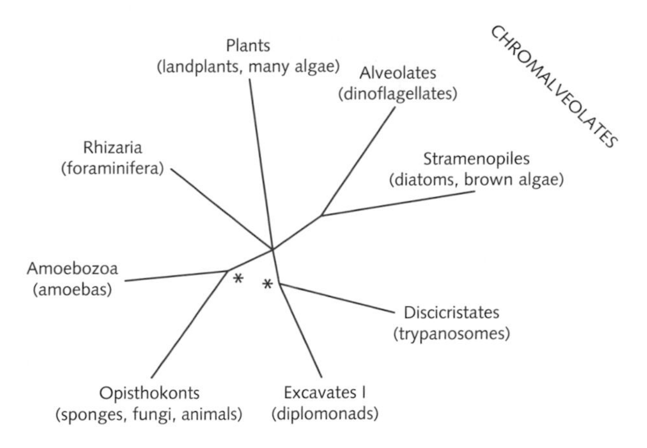

Figure 7.1 Major classes of eukaryote, with example species in parentheses. The asterisks mark possible positions of the root of the tree.

From: Baldauf, S.L. (2003). The deep roots of eukaryotes. Science **300**, 1703–1706, and personal communication.

are also grateful to yeast for bread, wine, and beer. On the other hand some related fungi are infectious.

Yeast is one of the simplest known eukaryotic organisms. Its cells, like our own, contain a nucleus and other specialized intracellular compartments. The sequencing of its genome, by an international consortium comprising ~100 laboratories, was completed in 1992.

The genome of baker's yeast, *Saccharomyces cerevisiae*, contains 12 052 000 bp, distributed among 16 chromosomes. The chromosomes range in size over an order of magnitude, from the 1352 kbp chromosome IV to the 230 kbp chromosome I.

The *S. cerevisiae* genome is 3.5 times the length of that of *E. coli*, and less than a tenth of the human genome. Many strains also contain one or more plasmids. The genome is relatively compact, with genes accounting for about 72% of the sequence. There are fewer repeat sequences compared with genomes of more complex eukarya.

A duplication of the entire yeast genome appears to have occurred ~150 million years ago. This was followed by translocations of pieces of the duplicated DNA and loss of one of the copies of most (~92%) of the genes.

Approximately 6000 protein-coding genes are predicted. Relatively few contain introns. However, the genes of a related yeast, *Schizosaccharomyces pombe*, are much richer in introns.

S. cerevisiae contains many genes for non-protein-coding RNAs. A single tandem array on chromosome XII encodes 120 copies of ribosomal RNA. There are 40 genes for small nuclear RNAs, and 275 genes for transfer RNA, about one-third of which contain introns.

Of the protein-coding genes, 4777 correspond to molecules to which a function can be assigned. About 1000 more contain some similarity to known proteins in other species. Another ~800 are similar to ORFs in other genomes that correspond to unknown proteins. Many of these homologues appear in prokaryotes. Only ~1/3 of yeast proteins have identifiable homologues in the human genome.

The classification of yeast protein functions shown in Table 7.1 is taken from the *Saccharomyces* genome

Table 7.1 Assignment of *Saccharomyces cerevisiae* genes products to different functional categories

Functional category	Number of proteins
metabolism	1514
energy	367
cell cycle and DNA processing	1012
transcription	1077
protein synthesis	480
protein fate (folding, modification, destination)	1154
protein with binding function or cofactor requirement (structural or catalytic)	1049
regulation of metabolism and protein function	253
cellular transport, transport facilities, and transport routes	1038
cellular communication/signal transduction mechanism	234
cell rescue, defence, and virulence	554
interaction with the environment	463
transposable elements, viral and plasmid proteins	120
cell fate	273
development (systemic)	69
biogenesis of cellular components	862
cell type differentiation	452
unclassified proteins	1393
functionally classified proteins	4777
functionally unclassified proteins	1394

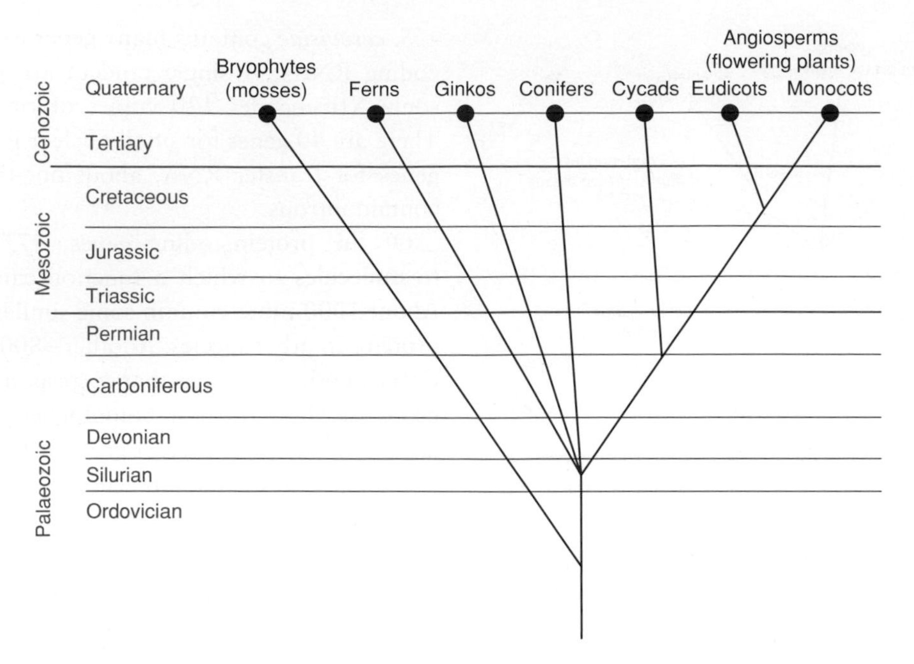

Figure 7.2 Phylogeny of land plants. The picture is limited to groups with extant examples with which readers may be familiar. Monocots and eudicots are named for the difference between single and double cotyledons in the embryo, but many other features separate them.

database, http://mips.gsf.de/genre/proj/yeast/Search/Catalogs/catalog.jsp.

The evolution of plants

Plants and animals parted company a long time ago. Although all life forms share much of their molecular biology, plants derive energy from sunlight via photosynthesis. The consequences for their structure, lifestyle, and developmental programmes have been profound. They require many proteins dedicated to their unique biophysical and metabolic activities. Plant genome sequences illuminate the similarities to and differences from other eukaryotes:

- Plants share some functions with animals. At the genomic and proteomic level, are they achieved in similar ways?
- Some functions are unique to plants. At the genomic and proteomic level, where did they come from? Were they invented, adapted, or borrowed?

From a common ancestor living about 800 million years ago, metazoa split into three major groups: fungi, animals, and plants. Higher plants evolved from single-celled organisms formerly classified as green algae. Green algae have now been split into streptophytes and chlorophytes; streptophytes are related to higher plants.

Plants came ashore to occupy land environments about 450 million years ago, in the mid-Ordovician. Most plants today are angiosperms, or plants with flowers (see Figure 7.2). Angiosperms arose about 140–190 million years ago.

The first complete nuclear genome of a higher plant to be sequenced was that of *Arabidopsis thaliana*, common name thale cress. It is related to turnip, cabbage, and broccoli. Its ease of handling and rapid generation time have made it a favoured subject for research in plant molecular biology. *A. thaliana* has been called 'the fruit fly of botany'.

The Arabidopsis thaliana *genome*

A. thaliana has a relatively small genome – 146 Mb – distributed over five chromosomes (see Box 7.1). (The maize genome is almost 20 times as large.)

- The compact genome was one reason why the research community adopted *Arabidopsis*.

BOX 7.1

The *Arabidopsis thaliana* genome

Genome size	146 Mb (estimated)
Sequenced nuclear DNA	115 936 794 bp
Predicted protein coding genes	26 732
Pseudogenes	3818
Alternately spliced genes	2330
Transposons	>10% of genome

Gene distribution	Length (bp)	Number of genes
Chromosome 1	30 432 563	6 905
Chromosome 2	19 705 359	4 178
Chromosome 3	23 403 063	5 313
Chromosome 4	18 585 042	4 088
Chromosome 5	23 810 767	6 248
Full nucleus	115 936 794	26 732
Mitochondrion	366 924	135
Chloroplast	154 478	122

The *Arabidopsis* nuclear genome is relatively compact. Protein-coding genes contain an average of 5.4 exons, of average length 276 bp, separated by relatively short introns about 165 bp long. The intergenic spacing is also short, about 4.6 kb. A feature of plant genes is that the G+C content of exons (44 mole %) is higher than that of introns (32 mole %).

The structure of the *A. thaliana* genome reveals both local and genome-wide duplications. There were probably *three* polyploidizations, estimates of the dates of which vary widely. The ranges 225–300 million years ago for the first, 150–170 million years ago for the second, and 25–40 million years ago for the most recent have been suggested. In addition, local duplications have affected ~17% of genes. Close relatives, such as cabbage and cauliflower, have undergone additional polyploidizations during the 12 million years since they diverged from *Arabidopsis*.

A. thaliana has a mitochondrial and a chloroplast sequence as well. Genome analysis must address questions of divisions of labour. Relative to animal cells, organelles in plant cells bear a greater metabolic burden, if only because of the activities of chloroplasts. Chloroplast genomes are relatively gene dense, with preserved gene order. In plant mitochondria, genes are more widely spaced and recombination is more common. Mitochondrial and chloroplast genes contain fewer introns (Table 7.2).

Table 7.2 Genes containing introns

	Genome		
	Nuclear	Chloroplast	Mitochondrial
Genes containing introns (%)	80	18.4	12

The *Arabidopsis* proteome contains many genes specific to plants, including those involved in photosynthesis and in the metabolism of components of cell walls. *Arabidopsis* is rich in genes that encode water-transporting channels, peptide-hormone transporters, metabolic and biosynthetic enzymes, and proteins involved in defence, detoxification, and environmental sensing.

Plants have many special metabolic pathways, for photosynthesis and for the metabolism of cell wall components, alkaloids, and growth regulators such as auxins and gibberellins. Complex metabolism requires the genome to encode a large and varied set of enzymes. In keeping with the essential role of light in plant life, *Arabidopsis* has many light sensors that regulate development and circadian responses.

Plants are also threatened by pathogens and have evolved defence mechanisms dissimilar from our immune system. One weapon that plants deploy against pathogens involves the production of reactive-oxygen species. Plants synthesize other defence molecules against animals, but, also, other molecules that attract pollinators. These attractants have provided useful sources of flavours, fragrances, and drugs, encompassing traditional 'herbal medicine' and modern pharmacology.

Comparing the proteins encoded in the nuclear genome of *Arabidopsis* with human proteins, the fraction of homologues observed varies with functional category. For protein synthesis, 60% of nuclear-encoded *Arabidopsis* genes have human homologues. For transcription regulation, the figure is only 30%. It is not that transcription is poorly represented in plant genomes; it is just that plants do it differently. In fact, plants have several times as many transcription factors as the fruit fly. Although many components of the signal-transduction pathways familiar from animals are absent in plants, plants have developed specific transcription factor families unknown in animals.

Many *Arabidopsis* genes are homologous to human genes implicated in disease. For instance, plants and animals have similar DNA repair systems, and *Arabidopsis* has a homologue of *BRCA2* (see Chapter 2). For some human disease-associated genes, the plant homologue is more similar to the human protein than those from fruit fly or *Caenorhabditis elegans*. Study of the function of the plant homologues will be illuminating, even though it is unlikely that *Arabidopsis* will be suitable for clinical trials of drugs intended for human use!

Plants are of course not ancestral to humans. We now turn to species that represent some important branching points in our own ancestry. That is, a succession of species with which we shared more and more recent common ancestors: urochordates, fishes, birds, monotremes, and other mammals.

The genome of the sea squirt (*Ciona intestinalis*)

Vertebrates arose within the chordate phylum, branching off from other lines that led to sea squirts and lancelets (see Figure 7.3). The sea squirt (*C. intestinalis*) represents one of the most primitive of our chordate relatives (Figure 7.4). Its genome provides insight into chordate and vertebrate origins.

- Chordates have a notochord, a rudimentary cartilaginous skeleton running dorsally from head to tail. A nerve cord lies parallel, adjacent, and dorsal to the notochord. Vertebrates retain the notochord during early embryonic development (vertebrates are also chordates) but replace it with the spinal cord.

The *C. intestinalis* genome has approximately 160 Mbp, about 1/20 the size of the human genome. From the initial sequence determined, approximately 16 000 proteins were adduced. This is comparable to invertebrates, and lower than typical vertebrate repertoires. The genes are tightly packed (7.5 kb/gene, compared to 100 kb/gene for humans).

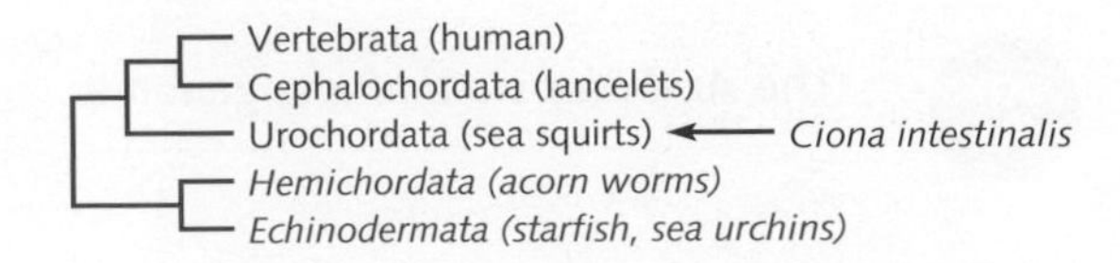

Figure 7.3 The evolutionary position of the sea squirt, *Ciona intestinalis*, compared to other chordates and our nearest non-chordate deuterostome relatives. (From Figure 4.2.)

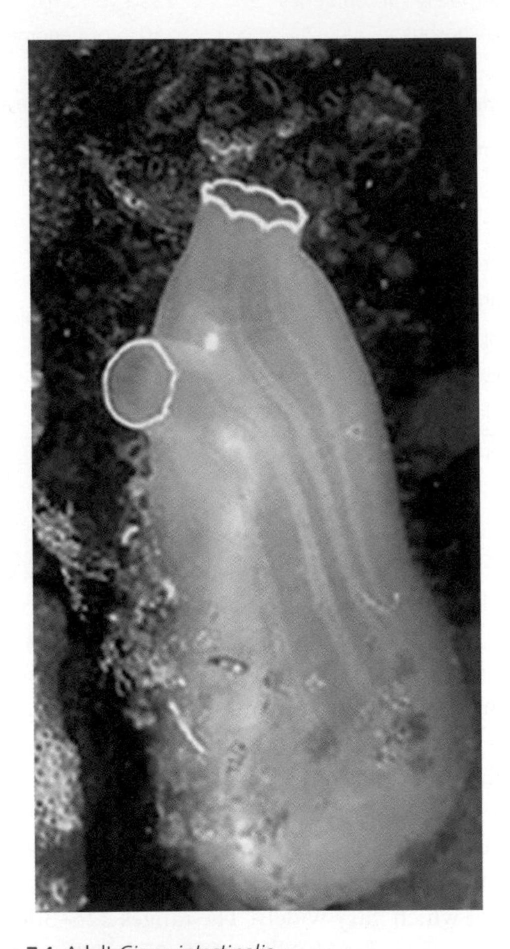

Figure 7.4 Adult *Ciona intestinalis.*

Cristian Cañestro, C., Bassham, S., & Postlethwait, J.H. (2003). Seeing chordate evolution through the Ciona genome sequence. Genome Biol. **4**, 208 (Photo by Andrew Martinez).

Because of the position of *Ciona* in the evolutionary tree, it is of interest to analyse its genes according to where homologues exist.

- Almost 60% of the genes have homologues in *C. elegans* and/or *D. melanogaster.* These represent genes shared by species ancestral to both invertebrates and chordates.

- A few genes look more similar to genes from worm and/or fly than to vertebrate genes. It is likely that these are vestiges of the common ancestor lost in the lineage leading to vertebrates. An example is the gene for haemocyanin, the oxygen carrier in many invertebrates. *Ciona* also contains genes for four globins, the vertebrate oxygen carrier.

  Some haemocyanins also have enzymatic activity, as phenoloxidases, enzymes which convert monophenols to diphenols, and/or diphenols to *o*-quinones. These reactions are involved in invertebrate immune responses: a defence reaction activates phenoloxidase activity, producing reactive quinones, which contribute to the inactivation of foreign organisms.

  Indeed, one system carefully looked for in *Ciona*, but conspicuous by its absence, is an adaptive immune system. *Ciona* does not appear to have genes for immunoglobulins, T-cell receptors, or MHC proteins. This appears then to be a later invention, by vertebrates. The characteristic molecules are absent from even the primitive jawless vertebrates – lamprey and hagfish.

- A fifth of the genes have no apparent homologue in vertebrates or invertebrates. It is likely that homologues will be discovered – perhaps when the protein structures are determined – or they may be very highly diverged within the urochordate lineage.

  Some *Ciona* proteins carry out functions specific to urochordates. The urochordate body is surrounded by a 'tunic', made of fibrous cellulose-like polysaccharide. (Tunicates is another name for this group.) *Ciona* contains enzymes for synthesis of cellulose, and endogluconases, which degrade cellulose. Of course cellulose as a structural material is unusual in organisms other than plants and bacteria. There is evidence that the last common ancestor of urochordates acquired the cellulose synthase gene by lateral transfer from bacteria. Cellulose *degradation* is more widespread. The endogluconases of *Ciona* are most similar to homologues in animals that digest cellulose, such as termites and some cockroaches.

The genome of the pufferfish (*Tetraodon nigroviridis*)

Tetraodon nigroviridis is a freshwater pufferfish. Ancestors of humans and fish parted company 450 million years ago. Comparing the genome of *T. nigroviridis* with other vertebrate genomes should therefore reveal defining properties of vertebrates.

The *T. nigroviridis* genome is about 340 Mb in length, an order of magnitude smaller than the human. Contributing to the compactness are a relative paucity of repetitive transposable elements, and shorter introns and intergenic regions. Forty per cent is protein coding! Approximately 28 000 protein-coding genes have been identified.

The *T. nigroviridis* genome strikingly illuminates the large-scale structure of vertebrate genomes. First, there is evidence for whole-genome duplication. Chromosome rearrangements have complicated what might originally have been a simple pattern. There remains, however, a considerable degree of synteny between pairs of groups of paralogous genes on different chromosomes. These common syntenic blocks *within* the *T. nigroviridis* genome arose by whole genome duplication followed by chromosome rearrangement.

Second, it is possible to map syntenic groups between *T. nigroviridis* and human. Figure 7.5 shows reciprocal maps. Consider Figure 7.5a. Imagine each human chromosome coloured a constant separate colour; for example, colour human chromosome 2 pink. Then for each gene on human chromosome 2, find homologues on *T. nigroviridis* chromosomes, and colour them pink also. The large blocks of pink in *T. nigroviridis* chromosome 2 indicate long blocks that are syntenic with human chromosome 2. Of course, some of human chromosome 2 appears elsewhere in the *T. nigroviridis* karyotype; for instance, at the top of *T. nigroviridis* chromosome 3. Figure 7.5b is the reciprocal map: colour the *T. nigroviridis* chromosomes a solid colour, and map them onto the human set.

It cannot be seen in Figure 7.5 directly, but typically one human region aligns with *two* regions in *T. nigroviridis*. The explanation is whole-genome duplication in *T. nigroviridis* but not in human.

Figure 7.6 shows this in more detail. Here Hsa = *Homo sapiens*; human chromosomes are numbered

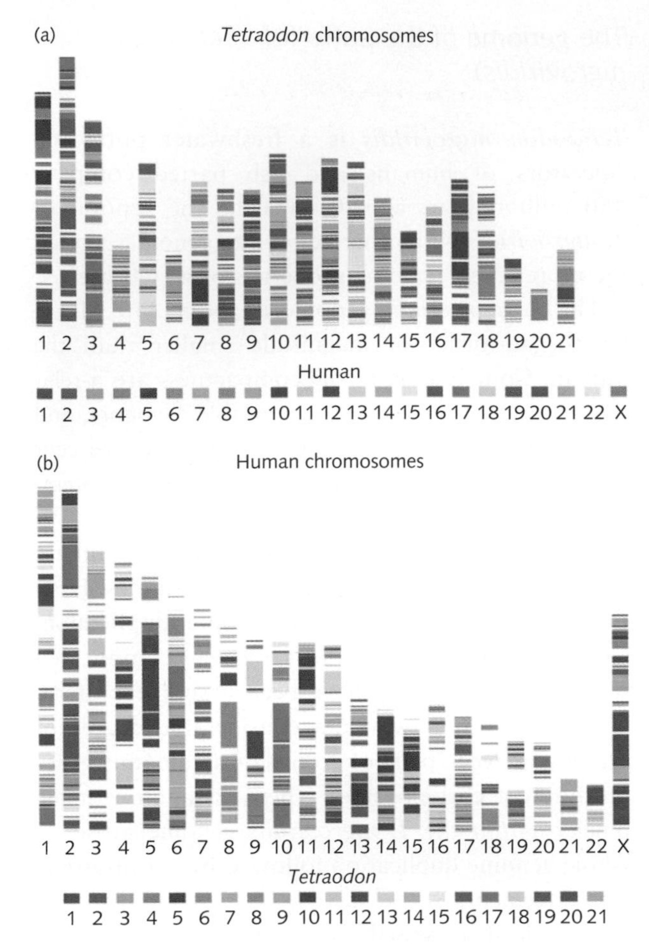

Figure 7.5 Mapping of syntenic blocks between human and *Tetraodon nigroviridis* chromosomes. The definition of synteny is not a very strict one. A synteny is recorded if a grouping of two or more genes in one species has an orthologue on the same chromosome in the other species, independent of order and orientation.

(a) Each coloured band in each of the *T. nigroviridis* chromosomes corresponds to a conserved syntenic block in the human chromosome of the same colour. For instance, *T. nigroviridis* chromosome 2 has many pink areas, indicating extensive relationship with human chromosome 2. *T. nigroviridis* chromosome 17 has many purple areas, indicating relationship with human chromosome 10.

(b) Reciprocal map, showing mapping of *T. nigroviridis* blocks onto human chromosomes. Here the close relationship between human chromosome 10 and *T. nigroviridis* chromosome 17 appears in green.

From: Jaillon, O., Aury, J.M., Brunet, F., Petit, J.L., Stange-Thomann, N., Mauceli, E., et al. (2004). Genome duplication in the teleost fish *Tetraodon nigroviridis* reveals the early vertebrate proto-karyotype. Nature **431**, 946–957.

Hsa1–Hsa22 plus HsaX. Tni = *T. nigroviridis*, and Anc stands for ancestral vertebrate chromosomes. This pattern is very interesting. First, as mentioned, to most regions of the human chromosomes there correspond *two* regions from *T. nigroviridis*.

Moreover, there are strange interleaving patterns within the mapping. This is expanded in two examples, from human chromosomes 16 and X. Each small box represents a gene. The expanded region of human chromosome 16 is a combination of *T. nigroviridis* chromosomes 13 and 15. It is the pattern you would expect from a pack of cards if you started with the red cards in one hand, the black cards in the other hand, and shuffled the deck.

The history that explains this pattern – and for which these observations provide compelling evidence – is that there was a whole-genome duplication in the *T. nigroviridis* lineage, but not in the common ancestor, nor in the human lineage after divergence. The chromosomes duplicated in an ancestor of *T. nigroviridis*, after divergence from the lineage leading to humans.

This produced what ultimately became *T. nigroviridis* chromosomes 5 and 13. If there were no chromosomal rearrangements or gene loss, the matchings of *T. nigroviridis* chromosomes 5 and 13 with human chromosome 16 would be the same. In Figure 7.6, in the upper expanded box, there would be green boxes above the Hsa16 line corresponding to *all* the red boxes below it, and red boxes below the Hsa16 line corresponding to *all* the green boxes above it.

But there have been rearrangements and loss of many of the duplicated genes. (As both human and *T. nigroviridis* have, to a first approximation, the same number of genes, clearly approximately half the genes present after the duplication must have been lost.) *Which* of the pair of duplicates was lost appears to be random. And there were also chromosomal rearrangements. For instance, *T. nigroviridis* chromosomes 5 and 13 also contribute to human chromosome 15.

The assumption of a whole-genome duplication can not only rationalize the pattern we see now, it can be run in reverse to infer the ancestral vertebrate karyotype. The alternating patterns seen on human chromosomes 16 and X in Figure 7.5 correspond to gene loss but not chromosomal rearrangement (within the expanded region). It is possible to ask for

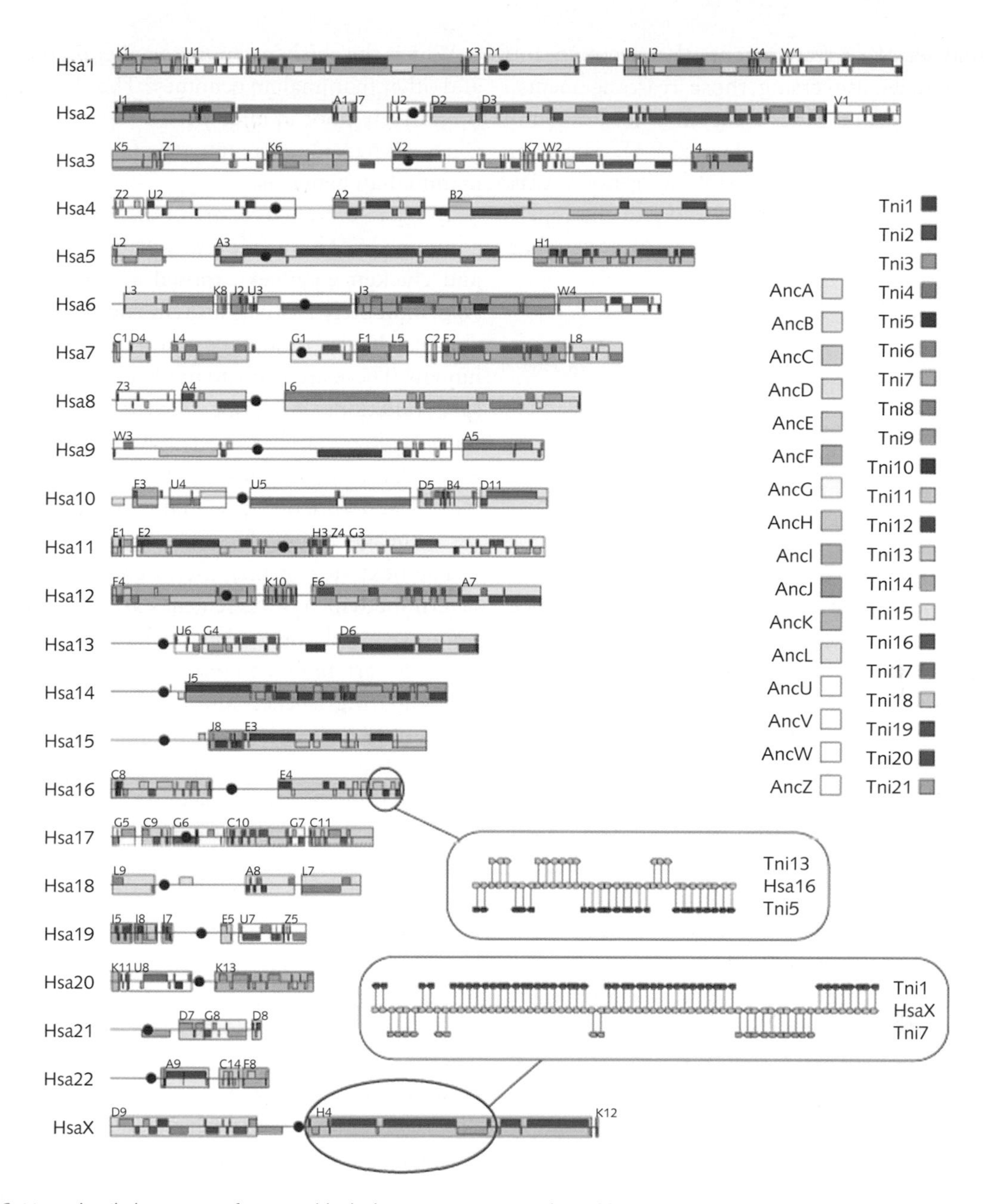

Figure 7.6 More-detailed mapping of synteny blocks between *T. nigroviridis* and human chromosomes. The two 'blown-up' regions show matches between individual genes. Notice that in general *two T. nigroviridis* regions map to one human one, evidence for whole genome duplication. In the detailed regions there is an alternation of matches, arising from random loss of one copy of each pair of genes produced by the whole genome duplication. Hsa = *Homo sapiens;* human chromosomes are numbered Hsa1–Hsa22 plus HsaX. Tni = *T. nigroviridis.* Anc = ancestral vertebrate. (Blocks AncU, AncV, AncW, and AncZ contain small amounts of sequence that could not be assigned to the twelve ancestral chromosomes.)

From: Jaillon, O., Aury, J.M., Brunet, F., Petit, J.L., Stange-Thomann, N., Mauceli, E., et al. (2004). Genome duplication in the teleost fish *Tetraodon nigroviridis* reveals the early vertebrate proto-karyotype. Nature **431**, 946–957.

the minimal set of rearrangements that accounts for the entire pattern. Reversing those rearrangements, on paper of course, provides a sketch of the ancestral vertebrate chromosomes (see Figure 7.6). The suggestion is that the ancestral vertebrate genome was distributed on 12 chromosomes, and had the repertoire of ~20 000–30 000 protein-coding genes that are typical of extant vertebrates.

- The pufferfish is a vertebrate. At least three-quarters of its genes have human homologues. By comparison of its genome with humans, it is possible to reconstruct the ancestral vertebrate karyotype.

The chicken genome

The chicken genome was the first complete sequence of a bird. It is a useful outgroup for study of the comparative genomics of mammals. The lineages leading to birds and mammals diverged about 310 million years ago.

The chicken is also important as an animal raised for food. The consumption of chickens in the UK amounts to 2.5 million birds per year, and about 11 billion eggs. Chickens were domesticated from the grey jungle fowl (*G. sonneratii*) in Asia, about 8000 years ago. Consequently, the genetics of chickens has been studied extensively. There are very many different breeds; for example, some are specialized for egg production, others for meat. It is anticipated that the genome will have practical applications in food production.

The chicken has also been a popular laboratory animal. It has contributed to research in developmental biology, virology, immunology, and cancer.

At ~1.2 Gbp, the chicken genome is substantially smaller than most mammalian genomes. However, it contains approximately the same number of genes. The chicken has 38 autosomes and one pair of sex chromosomes, called Z and W. Like other birds, but different from mammals, females are heterogametic (ZW) and males are homogametic (ZZ). About 90% of the 1.05 Gb of assembled sequence was anchored to its proper chromosome location. It is interesting that the synteny between human and chicken is more conserved than between human and mouse.

Why is the chicken genome smaller than the human and other mammalian genomes? The chicken genome is relatively poor in interspersed repeats and pseudogenes. Expansion of many gene families is greater in mammalian genomes.

To try to extract a 'core' set of vertebrate proteins, comparison of human, pufferfish (*Takifugu rubripes*), and chicken genomes exposed a common gene set. Approximately 7000 protein-coding genes from chicken have orthologues in both pufferfish and human. These are likely to implement common necessary functions, and one expects to find them in most higher vertebrates. These common genes are expressed in many different tissues. This is another typical signature of a gene that is not rapidly evolving for a lineage-specific function.

The next step in this line of enquiry would be to determine which of these genes are also expressed in primitive vertebrates and invertebrates. This will define a higher-vertebrate common core gene repertoire.

Comparing chicken and human only, about 60% of the ~23 000 chicken protein-coding genes have unique human homologues. Whereas such pairs between human and mouse or rat show an average of 88% sequence conservation, between human and chicken this drops to 75.3%. Proteins with different classes of function differ in conservation with human homologues: transport proteins are more highly conserved than average, and proteins of the immune response show only 60% sequence conservation.

The chicken has some avian-specific proteins. These include one family of keratins, which in chickens form feathers; mammals have expanded a different family, to form hair. Chickens have genes for avidin, a protein appearing in the egg whites of reptiles, amphibians, and birds. The very strong binding of biotin to avidin ($K_D \approx 10^{-15}$ M) has been applied in the laboratory for purification. What is its natural function? It is thought to protect eggs from bacteria, which require free biotin as a cofactor in numerous reactions.

Conversely, some human genes that chickens lack are:

- milk proteins such as casein
- enamel proteins, associated with loss of teeth in the bird lineage subsequent to Archaeopteryx, a primitive bird that did have teeth

- vomeronasal receptors. The vomeronasal system is a secondary chemosensory or odour detection system, that appears in many vertebrates, including humans, fish, reptiles, and others. Its absence from chicken and other birds signifies a loss in the avian lineage, rather than an invention in the mammalian one.

The platypus genome (*Ornithorhynchus anatinus*)

Extant mammals form a class divided into three orders:

monotremes: only the platypus and two species of echidna

marsupials: kangaroos, opossums, koalas, and many others, including all mammals native to Australia and New Guinea

placentals: all other mammals, including humans.

The platypus (*Ornithorhynchus anatinus*) is as distant a relative as we have among mammals. Startling to its discoverers, and to us now, is its mixture of mammalian and reptilian characteristics. Like mammals, the platypus has hair, and nurses its young (it has mammary glands, but not teats – the milk is released through localized pores in the skin, modified sweat glands). Like reptiles it lays eggs; and has venom, delivered by males through ankle spurs. Careful study of the anatomy revealed many other unusual characteristics. For instance, the name monotreme (= single aperture) refers to the single orifice serving both the urogenital and digestive systems.

An unusual sensory capacity is electroreception, the ability of the platypus to perceive electrical impulses. The platypus can locate and catch prey through use of a combination of mechano- and electroreceptors in its bill. A platypus will attack a battery immersed in water in the dark.

Study of the molecular biology of the platypus revealed other surprises, including 10 sex chromosomes (males are always XYXYXYXYXY). However, the sex determination system is closer to that of birds than of most mammals. In marsupials and placental mammals, the primary locus for sex determination is SRY, a gene on the Y chromosome that encodes a transcription factor. The platypus lacks this gene.

The closest to an extant reptile–mammal transitional form that we have, the platypus has offered us the chance to see how the basic distinctive features of mammals originated.

The platypus's was the first monotreme genome sequenced. It has 2.2 billion base pairs. 18 527 protein-coding genes were identified. Not unexpectedly, a majority of these have orthologues in opossum (a marsupial), human, dog, and mouse (placentals), and even chicken. Of particular interest are genes *not* found in other mammals. Like its anatomy, the platypus genome shows a mixture of mammalian and non-mammalian features. These include:

Odour receptors: the platypus odorant receptor genes are for the most part recognizable homologues of those in other mammals. The repertoire is more akin to that of other mammals than to reptiles. There are roughly half the number of odorant-receptor genes as in other mammals, but this may possibly be a reflection of the animal's aquatic lifestyle.

Milk: although true milk is unique to mammals, non-mammal animals that incubate their eggs secrete fluids that protect eggs from desiccation and/or infection. However, unlike those primitive precursors, platypus milk resembles that of other mammals. It is a complex mixture with both nutritive and anti-microbial functions.

Eggs: unlike the eggs of marsupials and placental mammals, which are nourished internally, the platypus lays eggs that contain yolk. Common to the yolks of eggs of fish, amphibians, reptiles, birds, and most invertebrates is the protein vitellogenin. Vitellogenin is the precursor of the lipoproteins and phosphoproteins that are major protein components of egg yolk. However, vitellogenins are not restricted to eggs. For instance, bees use it as food store also. Vitellogenin genes were lost in the lineages leading to marsupials and placental mammals, and retained in the monotremes. In other mammals, the placenta became the locus of embryonic development, and the mother supplied nutrients. Monotremes have a primitive form of placenta, called a yolk-sac placenta.

Venom: venom is one of those ideas that has proved useful to a variety of species, including monotremes

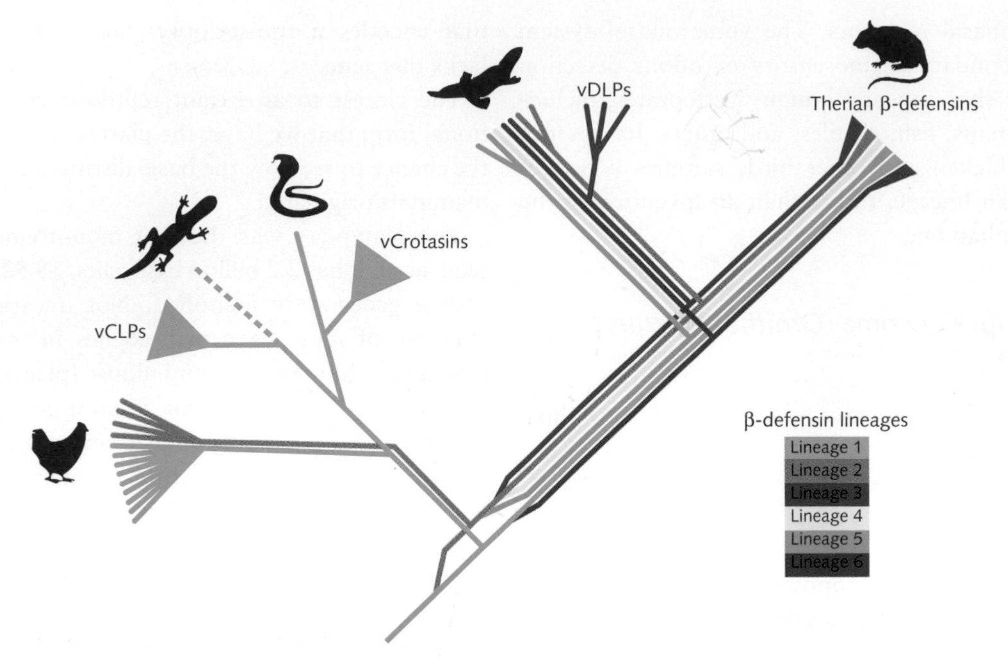

Figure 7.7 Evolutionary tree and points of gene duplication of defensins in birds, reptiles, platypus, and therians (= marsupial + placental mammal). Defensins are a group of families of small proteins found in a variety of vertebrate and invertebrate species. The therian molecules are not components of venom. They have antibacterial activity, generally functioning by forming pores within the microbial cell membrane, allowing cell contents to leak out.

A class of venom defensin-like proteins (vDLPs) has arisen independently in several lineages, including reptiles and platypus. In *Crotalus* snake venomes, the vDLPs are neurotoxins affecting voltage-gated sodium channels. The mechanism of action of the vDLP in platypus venom is still unknown.

From: Warren, W.C., Hillier, L.W., Graves, J.A.M., Birney, E., Ponting, C.P., et al. (2008). Genome analysis of the platypus reveals unique signatures of evolution. Nature **453**, 175–183.

and reptiles. Platypus venom is a complex cocktail containing proteins evolved by duplications of genes with other functions. However, although there are some biochemical features common to platypus and snake venoms, there is evidence that they developed independently. The enlistment of defensin-like peptides as components of venom is an example of convergent, or at least parallel, evolution (see Figure 7.7).

- In its macroscopic phenotypic characteristics, the platypus shows a combination of reptilian and mammalian features. Analysis of its genome also reveals these characteristics of a transitional form. However, there are some details that appear only from detailed sequence information; for instance, that the defensin homologue in platypus venom is not retained from reptiles, but convergently evolved.

The dog genome

Dogs and humans have lived and cared for each other for over 10 000 years. Dogs are work, sport, and companion animals. All readers will know of dog–human partnerships that rival human–human relationships in emotional intimacy.

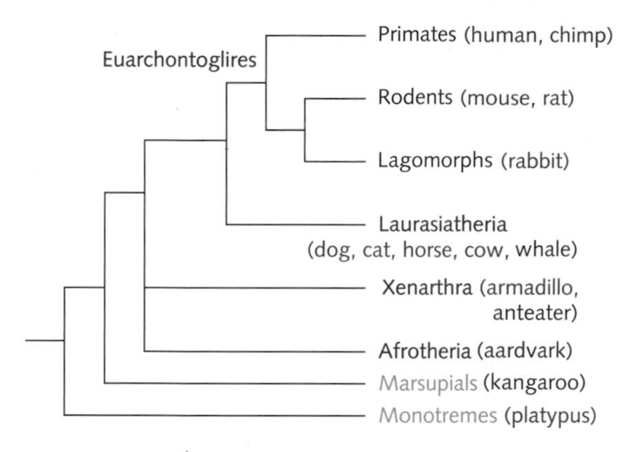

Figure 7.8 Phylogeny of mammals, showing monotremes and marsupials (green) and the four major groups of eutherian mammals: Euarchontoglires, Laurasiatheria, Xenarthra, and Afrotheria (blue). Human, chimpanzee, mouse, and rat are all Euarchontoglires. Dogs belong to the Laurasiatheria. Complete genome sequences are known for species shown in red.

Were those not sufficient reasons for interest in the dog genome, the biology of the dog presents numerous scientific challenges and opportunities.

- The dog is an outgroup of other mammals for which complete genome sequences have been determined (see Figure 7.8).
- Dogs are an ideal species in which to study domestication. To a far greater extent than other genera, dogs and their relatives offer both a variety of inbred populations – the different breeds – and corresponding wild populations. The genomes of dogs and wolves are much closer than those of humans and chimpanzees. The sequence divergences in chromosomal DNA between wolves and dogs is 0.04% in exons and 0.21% in introns. Unlike humans and chimpanzees, dogs and wolves can interbreed.

Romulus and Remus, founders of Rome, were, according to tradition, suckled by a wolf.

- Dogs share many human genetic diseases. Many are specific to individual breeds, and good genealogical and clinical records are available. The breeds are highly inbred: many have small founder populations and some have gone through bottlenecks. This simplifies the search for the gene or genes responsible for the disease.
- Because many drugs are tested in dogs, understanding of their molecular biology is useful. Dogs have also been used in research on gene therapy.
- Dogs show a vast morphological variation, notably in size. Information about genetic regulation of developmental pathways is implicit in the comparative genomics of different breeds. For instance, a single mutation controls breadth of skull and shortness of face. In humans, mutation in the homologous protein is responsible for Treacher Collins syndrome, a developmental disorder affecting the skull and face.

  Different breeds also vary generally in personality traits, providing an opportunity to identify genes for aggressiveness and passivity.

History of the dog

The order Carnivora, to which domestic dogs and cats belong, originated during the Palaeocene, ~60 million years ago (see Figure 7.9). Dog-like carnivores are known from fossils from 40 million years ago. The current closely related species – wolf, coyote, jackal, and red fox – split off about 3–4 million years ago. The wolf lineage gave rise to the domesticated dog, *Canis familiaris*.

Domestication of dogs is recorded in archaeological artefacts 14 000–15 000 years old, but probably took place much earlier. The first colonists of North America, who came across the Bering Strait about 20 000–15 000 years ago, brought domesticated dogs with them.

Evidence from the genome suggests that dogs went through two population bottlenecks. The first occurred ~9000 generations ago (~27 000 years) upon domestication. The second, ~30–90 generations ago, signals the origin of breed divergence. There are now about 300–1000 breeds of dogs. The American Kennel Club recognizes 150 as genetically separated populations, with closed gene pools.

Genome variation among breeds of dogs

The most complete canine genome is that of Tasha, a female boxer. Her genome was determined by the shotgun method, with 31.5 reads providing ~7.5-fold coverage (Table 7.3).

The dog genome is slightly smaller than that of humans, in part because dogs have fewer repeat

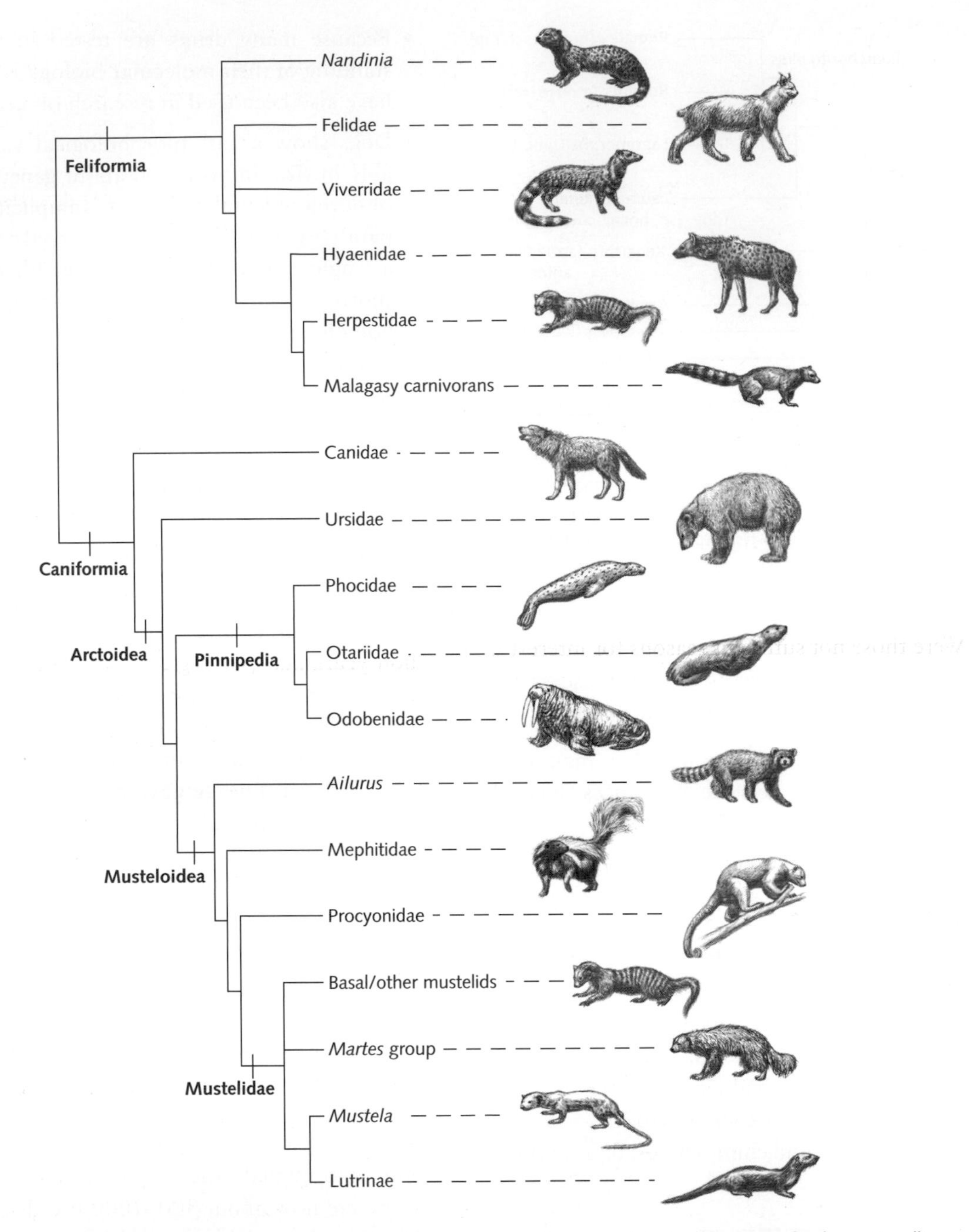

Figure 7.9 Domestic dogs and cats fall into the two main suborders of the order Carnivora. The two lineages split about 48 million years ago. Note that the pictures of the animals are not drawn to scale.

sequences. The short, interspersed element (SINE) is shorter in dogs than in humans (1 500 000 copies of SINEs make up 13% of the human genome).

In addition to determining the reference sequence, 2.5 million single-nucleotide polymorphisms (SNPs) were determined from 11 breeds of dogs. Given the inbred nature of the individual breeds, it is not surprising that fewer SNPs appear when comparing individuals within breeds than in comparisons amongst different breeds. Comparing individual

Table 7.3 The dog genome

Feature	Value	Comment
Number of chromosomes	39 pairs	More than humans
Genome length	2.4×10^9 bp	Slightly less than humans
Number of proteins identified	37 774	

boxers, there is ~1 SNP/1600 bases. Between breeds, there is ~1 SNP/900 bases.

Concomitant consequences of closer relationships within breeds are:

- *Greater interbreed than intrabreed sequence difference.* Over the entire species, dogs and humans show similar levels of nucleotide diversity between individuals: a frequency of different bases of $\sim 8 \times 10^{-4}$. However, the genetic homogeneity is much greater within breeds of dogs than within distinct human populations.
- *Longer haplotype blocks.* Within breeds, haplotype blocks may be as long as 100 kb. Haplotype blocks shared by different breeds are about 10 kb long. In comparison, the length of haplotypes in modern humans is about 20 kb.
- *Linkage disequilibrium within breeds extends over several megabases.* Across all breeds it is greatly reduced, extending only over tens of kilobases.

Comparison of dog, human, and mouse genomes

Dogs have 39 pairs of chromosomes compared with 23 pairs in human and 21 in mouse. Therefore, the human and mouse chromosomes must have been reassorted to make up the dog karyotype. Nevertheless, 94% of the dog genome appears in conserved synteny blocks with the human and mouse genomes.

Approximately 5% of the dog genome constitutes functional elements common to dog, human, and mouse. This is higher than the protein-coding fraction of the human genome. It includes regulatory elements and non-protein-coding RNAs, and further suggests that it is premature to dismiss as 'junk' the regions of the genome to which we cannot yet assign function.

Palaeosequencing – ancient DNA

Recovery of DNA from ancient samples

The recovery and sequencing of DNA from extinct species offers us a window onto evolutionary history. Source material includes fossils collected from their deposits, mummified samples, eggshells in the subfossil state (it was even possible, by testing DNA from the outer surface of moa shells, to show that *male* birds were responsible for incubating the eggs), preserved seeds or other plant material, specimens from museums, and clinical collections of pathogens. For instance, there are repositories of samples of influenza virus dating back almost a century.

Although often only minuscule amounts of material are available, PCR amplification can produce reasonable quantities for sequencing.

Like other biological material, DNA degrades after death, unless it is preserved. The best protection is to exclude liquid water. This can occur if the samples are frozen, or desiccated by heat, or if sequestered within compartments such as teeth, bone, or hair. Because DNA from ancient samples is usually present in only microscopic quantities, contamination is a serious danger. Contaminants can be microbial, or human from the scientists handling the specimens. Even without contamination, samples suffer from fragmentation, and from chemical change resulting in sequence changes. The most common is deamination of cytosine to uracil.

However, advances in isolation methods can reduce further damage during the extraction phase, and careful technique can exclude contamination by scientist DNA. Paleosequencers must 'rough it' during field work, but apply unusually painstaking care in handling samples. A speciality practised by few, but of interest to many.

DNA from extinct birds

The moas of New Zealand

In the absence of terrestrial mammals, New Zealand's largest animals were flightless birds. The moas ranged in size up to 3 m tall, 300 kg giants (Figure 7.10). They became extinct after the arrival of human settlers from Polynesia.

Moas are ratites, an order of birds that also includes the New Zealand kiwi, the ostriches of Africa, the emu and cassowary of Australia and New Guinea, the rhea in South America, and the extinct elephant bird of Madagascar (*Aepyornis maximus*). This monster was 3.3 m tall, and weighed 450 kg!

The taxonomy of moas has been difficult to resolve by classical methods, in the absence of living examples.

Figure 7.10 Photograph of an assembled skeleton of *Dinornis novazealandiae*, towering over celebrated 19th-century anatomist Sir Richard Owen. Based in London, Owen was the recipient of many interesting specimens discovered in the far reaches of the Empire, including the bones of moas, and a preserved specimen of the platypus. He was the driving force behind the establishment of the British Museum of Natural History in South Kensington.

The first specimen from a moa to reach Owen was a 15-cm fragment of bone. Owen correctly identified it as coming from the femur of a giant bird. In this photo Owen is holding the original fragment in his right hand. With his left, he is indicating its position within the full skeleton, discovered and assembled later.

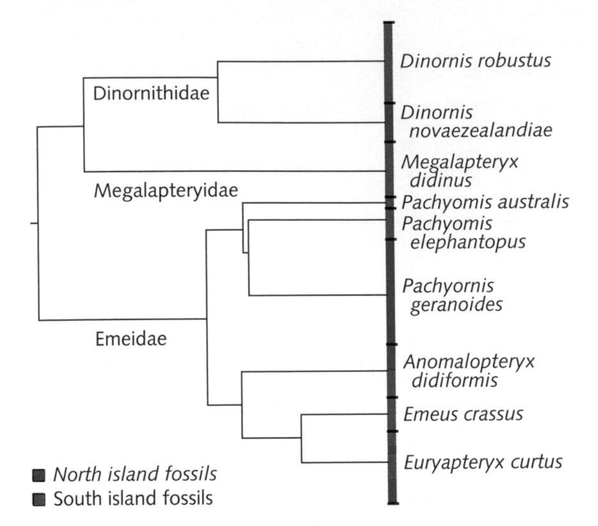

Figure 7.11 Phylogenetic tree of moa species, from mitochondrial DNA sequences. This classification divides moas into nine species, grouped into three families. Widths of coloured bands corresponding to species reflect intraspecies variation. North island specimens in red, South in blue. For *A. didiformis* and the two *Dinornis* species – but not *E. curtus* – the clustering separates specimens from the two islands.

After: Bunce. M, Worthy. T.H., Phillips. M.J., Holdaway, R.N. et al. (2009). The evolutionary history of the extinct ratite moa and New Zealand Neogene paleogeography. Proc. Nat. Acad. Sci. U.S.A. **106**, 20646–20651. Copyright (2009) National Academy of Sciences, U.S.A.

Mitochondrial DNA sequences from 29 specimens produced a phylogenetic tree comprising nine species, grouped into three families: the Dinornithidae, Megalapterygidae, and Emeidae (Figure 7.11). The data suggest a separation date of the Emediae of 5.27 million years, and more copious recent radiation within the last 2 million years.

It is interesting to correlate the estimated divergence times of the species with the geological history of New Zealand. Today, the North and South Islands are separated by Cook Strait, 23 km across at its narrowest. But, approximately 30–21 million years ago, sea levels reduced New Zealand to a few scattered islands. The somewhat larger South Island was isolated from the North, until about 2–1.5 million years ago.

Approximately 8.5–5 million years ago the New Zealand 'alps' formed. This mountain chain runs roughly parallel to the main axis of the South island. It divides the habitats into wet rainforest on the West, and dry, warmer regions on the East.

Family: Dinornithidae
Dinornis
Systematics: Two species D. robustus (South Island, blue) and D. novaezealandiae (North Island, red)
Dimensions: 56–249 kg and 90 to 200 cm in height – significant sexual dimorphism with females up to three times the mass of males.
Habitat: Browsing generalist – has been found in upland, lowland, and open forest habitats. The larger forms occupied low rainfall areas.
Family: Megalapterygidae
Megalapteryx
Systematics: Monotypic, M. didinus, (South Island).
Dimensions: 28–80 kg and 65 to 95 cm. Pleistocene specimens are significantly larger than Holocene forms.
Habitat: Subalpine scrub, grassland, and high country forests (usually >900 m).
Family: Emeidae
North Island
Anomalopteryx
Systematics: Monotypic, A. didiformis.
Dimensions: 26–64 kg and 50 to 90 cm.
Habitat: Non-coastal lowland forests with a continuous canopy.
Southern Alps
Cook Strait
South Island
Euryapteryx
Systematics: Monotypic, E. curtus (formally E. gravis and E. geranoides).
Dimensions: 12–109 kg and 51 to 103 cm.
Habitat: Drier climates – typically lowland open forest and coastal sites.
Emeus
Systematics: Monotypic, E. crassus (South Island).
Dimensions: 36–79 kg and 73 to 99 cm in height.
Habitat: Preference for lowland forest (usually <200 m) and swamps.
Pachyornis
Systematics: P. geranoides (North Island), P. elephantopus (blue), and P. australis (green) (South Island).
Dimensions: 17–163 kg and 54 to 121 cm in height.
Habitat: P. australis occupied subalpine grassland, P. geranoides and P. elephantopus preferred lowland forest edges and wetland vegetation.

Figure 7.12 Reconstructions, classification, and estimated geographical distributions of species of moa, extinct flightless birds from New Zealand.

From: Bunce, M., Worthy, T.H., Phillips, M.J., Holdaway, R.N., et al. (2009). The evolutionary history of the extinct ratite moa and New Zealand Neogene paleogeography. Proc. Nat. Acad. Sci. USA **106**, 20 646–20 651. Copyright (2009) National Academy of Sciences, USA.

The data on divergence of sequences is consistent with the historical geology. The suggested scenario is that the divergence of the major groups took place on the South Island, after the alps formed. When the land links arose, during glaciations, birds began to inhabit the North Island, taking advantage of the new surroundings to generate a new round of divergence (Figure 7.12).

The dodo and the solitaire

The dodo (*Raphus cucullatus*) was a large, flightless bird that inhabited the island of Mauritius, in the Indian Ocean, east of Madagascar (Figure 7.13). It was a large, robust bird, about a metre in height and weighing about 20 kg (about three times the size of a typical Thanksgiving turkey). It is common for island species to be either significantly larger, or significantly smaller than their mainland counterparts. Such substantial changes can obscure taxonomic positions.

Figure 7.13 Mauritius dodo.

From: 'A German Menagerie Being a Folio Collection of 1100 Illustrations of Mammals and Birds' by Edouard Poppig, 1841.

Sailors stopping at the island found the dodo easy prey – it could not fly, and lacked appropriate fear of human predators. As a result, the dodo became extinct. The last survivor was shot in 1681.

The solitaire (*Pezophaps solitaria*) was a related bird from a neighbouring island east of Mauritius, Rodrigues. It also became extinct, outliving the dodo by perhaps a century.

Even museum specimens of the dodo are rare. The Oxford University Museum of Natural History had one. It was seen by don Charles Dodgson, who used it as a character in Alice in Wonderland. Only partially saved from a fire during a tidying-up exercise, the Oxford specimen is the only known source of soft tissues from a dodo. What remains now comprises a head, and a leg and foot, each with some skin attached. There are many bones on the tropical island, but preservation conditions on the tropical island are not conducive to preservation of DNA. It was the Oxford remnants that provided the material for DNA sequencing.

From the Oxford sample of the dodo, and samples from Rodrigues of the solitaire, it was possible to amplify and sequence short overlapping fragments between 120 and 180 bp. For comparison, corresponding sequences were analysed from many extant species of putative relatives, including various species of pigeons and doves. From the extant species, sequences were determined of 1.4 kb of mitochondrial DNA, and regions of the genes for 12S ribosomal RNA (360 bp) and for cytochrome *b* (1050 bp).

A phylogenetic tree constructed from these data fixed the taxonomic position of the dodo. It is a pigeon, of the family Columbidae. The closest extant relative of the dodo and solitaire is the Nicobar pigeon (*Caloenas nicobarica*), which lives in Southeast Asia.

High-throughput sequencing of mammoth DNA

Mammoths are unusual among extinct organisms in the favourable conditions of their Arctic habitats for preservation of DNA. Even better than most specimens was a ~28 000-year-old jawbone, found on the shore of Baikura-turku, a bay extending from the south-eastern corner of Lake Taymyr. Extraction from 1 g of bone yielded ~0.73 μg DNA.

Fragments of the DNA attached to small sepharose beads were amplified in lipid vesicles by PCR. Six runs of a Roche/454 Life Sciences Genome Sequencer

20 System produced a total of 1 943 593 reads, with average length of ~95 bp.

The DNA in the sample contained about 50% mammoth DNA, a mixture of nuclear and mitochondrial; the rest was bacterial contaminant. Using reference sequences to sort out mammoth sequences from bacterial sequences, and mammoth nuclear sequences from mitochondrial ones, the aggregate harvest was ~95 Mb of mammoth sequence. This included 7.3-fold coverage of the 16 770 bp mitochondrial DNA. The remainder gave a partial view of the mammoth nuclear genome (~3%).

In addition to what it tells us about the mammoth, the significance of this work is its demonstration of the power of (1) the latest instrumentation and (2) resequencing in determining the organelle genome. *There is no need for selective amplification of the target sequence.* Instead, it is possible to assemble the mitochondrial DNA from the very large quantity of data available, given the scaffolding available from a related sequence, in this case that of the Indian elephant. Note that, with over sevenfold coverage, the reference sequence is not really needed for the assembly, but rather for identifying the reads corresponding to mitochondrial DNA.

Indeed, computational experiments have shown that a reasonably good assembly is possible using as a reference sequence the mitochondrial DNA of the dugong, a distant relative. Ancestors of mammoths and dugongs diverged around 65–70 million years ago. Mammoth and dugong mitochondrial DNA sequences are only 75.3% identical. In practice, a reference sequence from a closer relative than the dugong is to mammoth would in most cases be available. Therefore the dugong–mammoth assembly offers a 'worst-case' analysis.

For the sake of argument, suppose one's interest were limited to the mitochondrial DNA sequence. One might choose to amplify the mitochondrial DNA and sequence that only. Sequencing fragments from all of the DNA is, comparatively, very inefficient in its use of the data produced, as far as determining the mitochondrial sequence is concerned. However, comparisons with other ancient-DNA sequencing projects suggest that it *is* efficient in terms of the amount of precious sample used. For the study of extinct species, this is an overriding consideration.

The mammoth nuclear genome

The mammoth nuclear genome presented a harder problem. It is estimated to be approximately 4.17 Gbp in length, longer than the human genome.

DNA was extracted from hair samples from a Siberian animal, denoted M4, that died about 20 000 years ago. It is interesting that because the DNA is fragmented, the relatively short read lengths of the sequencer were not a great problem. The average read length produced was about 150 bp. This yielded 3.6 Gb of sequence. Combining this with additional sequences determined from other individuals produced a total of 4.17 Gb of sequence. Calibration of error rates suggest an average of 6 errors out of 10 000 bases arising from DNA damage, and 8/10 000 from sequencing.

The genome of the African elephant (*Loxodonta africana*) provided a reference sequence. The *L. africana* genome had been sequenced at 7× coverage, and assembled. Alignment of the reads from the mammoth samples, to *L. africana* and to other genomes representing potential contaminants, showed that over 90% of the reads were mammoth DNA, for a total of 3.3 Gb of mammoth sequence.

It was possible to compare amino acid sequences of mammoth proteins with the orthologues in elephants and other species. The results suggest that mammoth and African elephant differ, on average, in one residue per protein. It is difficult to assign selective or even functional significance to these, in general. Even in the cases of residues unique to mammoth, compared to a wide spectrum of other placental mammals, the sequence is only the starting point for investigation of the proteins thereby identified as interesting candidates for follow-up studies.

The phylogeny of elephants

Access to DNA from extinct species has allowed resolution of two problems in elephant phylogeny:

(a) How many species of African elephants are there?

There are two populations of elephants in Africa, living in the savannah and in the forest. Some authorities have described them as separate species: *Loxodonta africana* (savannah) and *L. cyclotis* (forest). Others have considered them a single species, or regard *L. cyclotis* as a subspecies of *L. africana*.

(b) Are mammoths more closely related to African elephants or Indian elephants (*Elephas maximus*)?

Both questions have engendered considerable debate in the relevant specialist literature.

In order to assess phylogenetic relationships among African and Indian elephants and mammoths, the extinct American mastodon provided an outgroup. DNA from a tooth, estimated to be between 50 000 and 130 000 years old, provided 1.76 Mb of mastodon sequence. The corresponding regions from elephants and mammoths were also sequenced. The data set for analysis contained, from each species, approximately 40 000 bp of sequence from 375 loci. These data allowed comparison of the divergences between the groups.

The results showed that the variation between *L. africana* and *L. cyclotis* is approximately the same as between mammoths and Asian elephant (*E. maximus*).

Note that mammoths and Asian elephants are not even in the same genus, but many people have believed that *L. africana* and *L. cyclotis* are the same species! Nevertheless, the conclusion is that mammoths are more closely related to Asian elephants than to African elephants, and that *L. africana* and *L. cyclotis* should be considered as separate species.

- Returning to the question of what defines species boundaries, it is clear that this distinction was drawn at least primarily on the basis of similarity of DNA sequence. What about the classical biological definition of whether the species hybridize in nature. (In captivity, even African and Indian elephants are fertile.) *L. Africana* and *L. cyclotis* do give rise to hybrids in the Uganda–Congo border region where their ranges overlap. However, as implied by the divergence of the DNA sequences, hybrids are in fact rare, and the two populations do maintain separate gene pools.

The splitting of a species into two has implications for conservation, because in principle splitting would require a separate decision about whether each were endangered. In fact all world elephant species are endangered.

RECOMMENDED READING

- General discussions of topics in eukaryotic evolution, many in the form of collections of papers:

Hirt, R.P. & Horner, D.S., eds. (2004). *Organelles, Genomes and Eukaryote Phylogeny: An Evolutionary Synthesis in the Age of Genomics*. CRC Press, Boca Raton, FL, USA.

- On June 29, 2006, The Royal Society held a discussion meeting, Major steps in cell evolution: palaeontological, molecular, and cellular evidence of their timing and global effects. The meeting was organized by T. Cavalier-Smith, M. Brasier, and T.M. Embley, and published in Philosophical Transactions of the Royal Society B, volume 361, issue 1470.

Katz, L.A. & Bhattacharya, D. (2006). *Genomics and Evolution of Microbial Eukaryotes*. Oxford University Press, Oxford.

Baldauf, S.L. (2008). An overview of the phylogeny and diversity of eukaryotes. Journal of Systematics and Evolution **46**, 263–273.

Telford, M.J. & Littlewood, D.T.J., eds. (2009). *Animal Evolution / Genomes, Fossils and Trees.* Oxford University Press, Oxford.

- An atlas of life forms, showing phylogenetic relationships and dates of divergence:

Hedges, S.B. & Kumar, S. (2009). *The Timetree of Life*. Oxford University Press, Oxford.

- Theory and applications of linkage disequilibrium:

Slatkin, M. (2008). Linkage disequilibrium – understanding the evolutionary past and mapping the medical future. Nature Reviews Genetics 9(6), 477–85.

EXERCISES, PROBLEMS, AND WEBLEMS

Exercises

Exercise 7.1 What fraction of the intergenic space in the nuclear genome of *A. thaliana* is occupied by transposons?

Exercise 7.2 On a photocopy of Figure 7.2, mark the approximate dates of the duplications in the *A. thaliana* genome on the branch leading to eudicots.

Exercise 7.3 Calculate the average gene density in the nuclear, mitochondrial, and chloroplast genomes of *A. thaliana*.

Exercise 7.4 Describe how you would test the assertion that the last common ancestor of urochordates acquired the cellulose synthase gene by lateral transfer from bacteria. What sequence information would you gather, and how would you analyse it?

Exercise 7.5 In Figure 7.5(b), which *T. nigroviridis* chromosomes have substantial regions of synteny with human chromosome 10?

Exercise 7.6 On a photocopy of Figure 7.5(a), indicate the regions in which *T. nigroviridis* chromosomes 5 and 13 contribute to human chromosome 15.

Exercise 7.7 Figure 7.12 shows that specimens of *Euryapteryx curtus* appear on both North and South Islands. On which island is it likely that the species arose? What reasoning leads you to this conclusion?

Exercise 7.8 In the dog, the variation in mitochondrial DNA sequences is lower than the variation in nuclear DNA sequences. What does this suggest about the breeding behaviour of domesticated dogs?

Exercise 7.9 For which of the following domesticated species could a population survive if released into the wild? Dog, cat, chicken, parakeet, maize, rice, and wheat.

Exercise 7.10 It is much easier to study mitochondrial DNA than nuclear – it is smaller, and more abundant in cells. What is the danger of assigning phylogeny by comparing populations through sequencing mitochondrial DNA of various individuals, in species that are matrilocal (that is, females remain within a herd, males leave)? An example was a study of elephants, based on mitochondrial DNA sequences. What criticism might be raised?

Problems

Problem 7.1 Figure 7.14 shows an alignment of globins from *Ciona intestinalis* and human haemoglobin α and β, myoglobin, cytoglobin, and neuroglobin. Some N- and C-terminal extensions have been trimmed. (a) Which pair of globins has the largest number of identical residues in this alignment? (b) Are the *Ciona* globins more similar to one another than the human globins are to one another? (c) Which human globin do the *Ciona* globins most resemble?

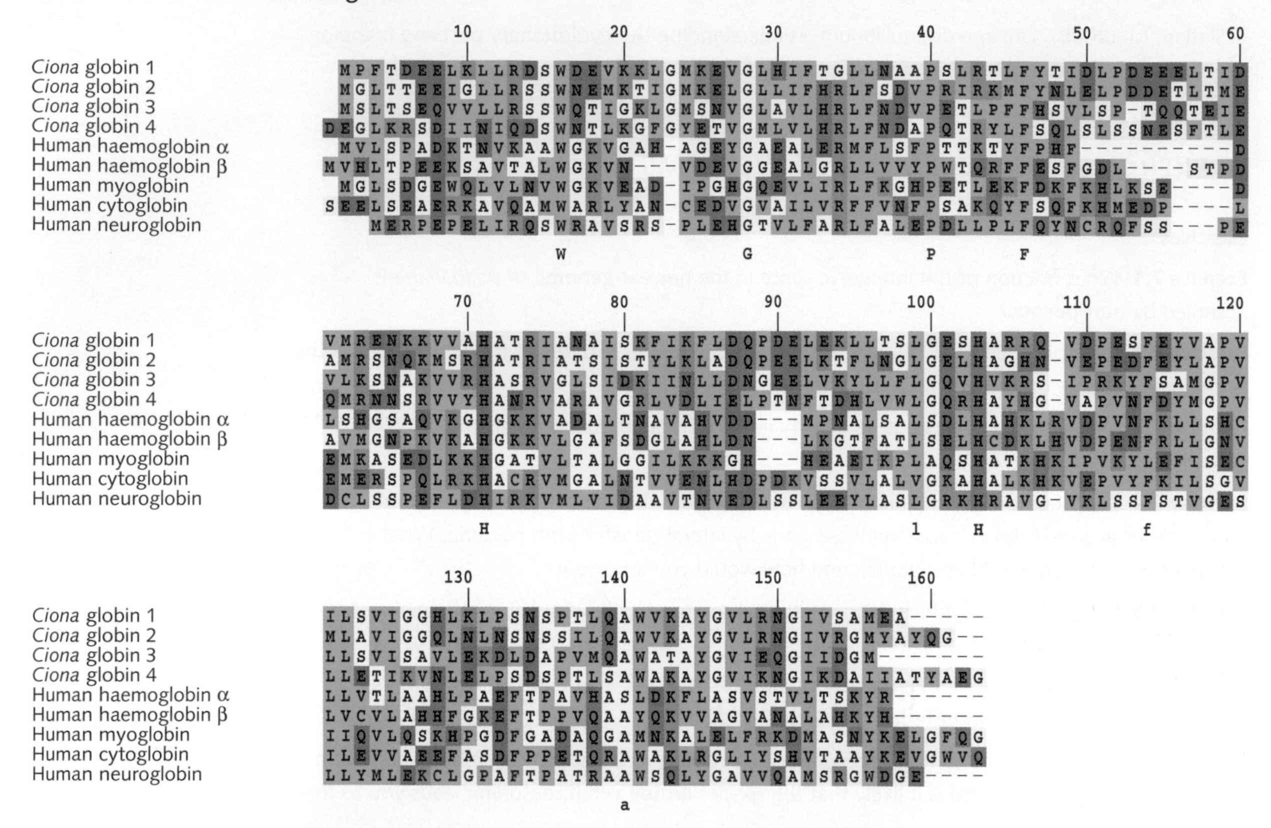

Figure 7.14 Alignment of globins from *Ciona intestinalis* and human haemoglobin α and β, myoglobin, cytoglobin, and neuroglobin. Some N- and C-terminal extensions have been trimmed.

Weblems

Weblem 7.1 (a) Has at least one organism from each of the major eukaryote classes shown in Figure 7.1 been the subject of a full-genome sequencing project? For each class, give an example of such a species if possible. (b) For which eukaryotic phyla has at least one species been the subject of a full-genome sequencing project? For each phylum, give an example of such a species if possible.

Weblem 7.2 Find a yeast protein in each of the first ten functional categories in Table 7.1.

Weblem 7.3 What is the latest common ancestor of the human and the aardvark? (Hint: compare the full taxonomy listings in any entry of a human and an aardvark sequence.)

Weblem 7.4 The UniProtKB entry for chicken ovocleidin 116 is Q9PUT1_CHICK. This protein is eggshell-specific in chickens. Does it have any mammalian homologues?

Weblem 7.5 Mammals excrete nitrogenous waste in the form of urea. Birds excrete uric acid. Urea in mammals is formed by the urea cycle, a metabolic pathway that forms urea ($H_2NC(=O)NH_2^+$) from ammonia and aspartate. The first enzyme in this pathway is carbamoyl phosphate synthetase 1, which catalyses the reaction:

$$2ATP + HCO_3^- + NH_4^+ \rightarrow 2ADP + H_2NC(=O)OPO_3^{2-} \text{ (carbamoyl phosphate)} + P_i$$

(a) Does the chicken contain a homologue of human carbamoyl phosphate synthetase I?
(b) Is there reason to believe it is not functional? (c) In what tissues in chicken is it expressed?
(d) What might be the function of this enzyme in the chicken?

CHAPTER 8

Genomics and Human Biology

LEARNING GOALS

- To understand the science underlying the use of genomics for personal identification.
- To recognize what characteristics of an unknown individual can be inferred from a sample of blood or saliva, and that the use of these inferences in criminal investigation remains controversial, with different jurisdictions adopting different regulations.
- To appreciate how mitochondrial DNA sequences were used to identify the remains of the Russian royal family.
- To see the domestication of crop plants as experiments in directed genome change, and to appreciate that we can now analyse the genetic differences between the wild progenitor and the varieties used in contemporary farming.
- To consider the relationship between humans and Neanderthals, based on a sequencing of the Neanderthal mitochondrial and nuclear genomes.
- To understand past patterns of human migration, as reflected in mitochondrial DNA haplotypes.

A prominent theme in our presentation of genomics has been the potential for applications that improve the health of humans, animals, and plants. In this short chapter we collect a few applications of genomics to some of the other human sciences.

Genomics in personal identification

Legal applications of DNA sequencing depend on several scientific facts:

1. The genomes of all individuals except identical siblings are unique. Like fingerprints, genomes provide a unique personal identification. A bloodstain at a crime scene, like a set of fingerprints, can be traced to a specific individual.
2. The genome of every person combines chromosomes from his or her parents. Unlike fingerprints, therefore, genomes can indicate familial relationships; notably, identification of paternity.
3. Each person's genome contains genes that influence, even if they do not inevitably determine, recognizable features, such as eye colour. In principle, DNA left by an unknown individual at a crime scene could be analysed to suggest a physical description of the source individual.
4. Thus, unlike fingerprints, genomes contain much more information about a person than simple identification. The treatment of this information by governmental authorities raises ethical and legal questions. We have already raised some of these questions in Chapter 1.

The use of molecular characteristics to identify people and relationships is a century old. The earliest methods applied the classical blood groups – A, B, AB, and O. A suspect with blood type O must be innocent of a crime committed by a person of blood type A. A person of blood type O could not be the parent of a child of type AB. However, many people share the same blood type. Therefore, blood typing can prove innocence but not guilt. In contrast, DNA sequences can provide *positive* proof of guilt or paternity.

A. Jeffreys at the University of Leicester discovered DNA 'fingerprinting' in 1984, when he and his colleagues compared the sizes of the restriction fragments of DNA samples including a human family group (father, mother, and child). Different individuals give different patterns – the gel provided a 'bar code' unique to each individual. (This bar code is different from the one used for species identification – see p. 117.) Moreover, the pattern from the child's DNA was a combination of those of the parents (see Figure 8.1).

Why do different individuals give different patterns of restriction fragment sizes? One possible cause of the difference is a mutation in a restriction site, causing that site not to be cleaved. In this case, two fragments from unmutated DNA will correspond to a single longer fragment from the mutated sample. (In terms of our analogy between restriction maps and distances between consecutive Starbucks cafes on Broadway in New York City, imagine the effect on the pattern if one of the cafes were to close (see p. 90.) Alternatively, somewhere between two restriction sites there may be a short repetitive stretch of DNA, the number of copies of which is unstable during replication, where the polymerase 'stutters'. Expansion of such a repeat will lengthen the restriction fragment in which it appears. The fragment will occupy a different position on a gel that separates fragments according to size.

Such a short repetitive segment of DNA is called a **variable number tandem repeat** (VNTR). VNTRs are

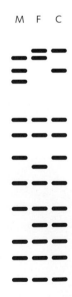

Figure 8.1 Pattern of a gel showing DNA fingerprints from a mother (M), father (F), and child (C). Every band in lane C matches one appearing in lane F or lane M or both. The bands on the gel correspond to restriction fragments from a complete digest by *Hin*fI. The gel separates fragments according to size.

generally flanked by recognition sites for the same restriction enzyme, which will neatly excise them, producing fragments of different lengths (see Box 8.1). It is these fragment lengths that vary between individuals, known as **restriction fragment length polymorphism** (RFLP). The fragments can be separated on a gel according to size and detected by Southern blotting.

- VNTRs are characteristics of genome sequences; RFLPs are artificial mixtures of short stretches of DNA created in the laboratory in order to identify VNTRs.

The patterns were easy to determine from a sample of DNA. Jeffreys and his co-workers quickly established that they were unique to individuals, providing a 'genetic fingerprint'.

The first legal application, in 1985, was to a case of disputed identification involving a family of UK citizens. A child in the family visited Ghana. When he returned to the UK, immigration authorities suspected him of being an impostor, not entitled to UK residency. None of the classical blood tests, including A, B, AB, O, and other blood groups, and even MHC haplotyping – which gives much higher discrimination – produced definitive results. Indeed, there was a possibility that the boy was related to the woman who claimed to be his mother, but perhaps he was her nephew rather than her son. Quite fine distinctions were therefore essential. Jeffreys' DNA fingerprints, comparing the patterns from the child's DNA with that of members of the UK family, proved his identity to the satisfaction of the Home Office. The family were reunited.

BOX 8.1 The two restriction enzymes in most common use in DNA profiling: *Hae*III and *Hin*fI

Sequence specificity			
	↓		↓
*Hae*III	5'...GGCC...3'	*Hin*fI	5'...GANTC...3'
	3'...CCGC...5'		3'...CTNAG...5'
	↑		↑

Jeffreys also applied his method to criminal identification. In the first case, DNA fingerprinting proved the innocence of a suspect who had actually confessed to two crimes. The true criminal was discovered after a survey of DNA samples from almost 4000 people living in the region. In this single case, DNA fingerprinting proved both the innocence of a man under arrest, in serious danger of conviction and punishment; and the guilt of the real criminal.

A substantial number of persons convicted and sent to jail before Jeffreys' discovery have subsequently been proved innocent by analysis of samples saved from the evidence presented at their trial.

Despite its successes, identification by gel separation of RFLPs has disadvantages in practice. It requires relatively large amounts of undegraded DNA (10–50 ng of material no shorter than 20 000–25 000 bp). Since the development of PCR, DNA-based identification methods have tested for the presence of selected regions known to vary in the population, using PCR to amplify those present. This greatly improves sensitivity. Subnanogram amounts suffice to identify 100 bp regions. It is possible to get a positive identification from a single hair (of a person, or in one case of the cat of a criminal's parents), or from the saliva on a licked envelope.

The method in common use now is to PCR amplify a short tandem repeat (STR) typically containing 2–5 bp, repeated between a few and a dozen times. Amplification produces fragments about 200–500 bp long. Loci in common use show 5–20 common alleles, and 8–15 loci are tested.

Jeffreys has recently introduced a newer identification method based on a single VNTR locus (see Problem 8.1).

- DNA fingerprinting has shown itself to be a very reliable and useful method of personal identification. We discussed ethical, legal, and social issues associated with databanks of DNA sequences in Chapter 1.

Mitochondrial DNA

Human mitochondrial DNA is 16 569 bp long. It contains a hypervariable 100 bp region, which varies by 1–2% between unrelated individuals. The mitochondrial DNA of unrelated people typically differs at eight positions. Mitochondrial DNA is very abundant and survives very well. It was used to identify the remains of the Russian royal family (see Box 8.2).

Gender identification

It is possible to decide whether a nuclear DNA sample came from a male or female. Obviously, detection of any sequence unique to the Y chromosome will prove male origin. Another technique in common use applies the appearance of different versions of the gene for angiogenin on the X and Y chromosomes. The X version contains a 6 bp deletion. PCR amplification of this region from a female will give one band from the two identical X copies of the gene; DNA from a male will give two bands, one from the X and one from the Y.

DNA identification has provided evidence in several very high-profile cases that readers will be familiar with.

- The trial of O.J. Simpson for the murder of his wife; he was acquitted despite presentation of evidence by the prosecution that he was the source of fresh bloodstains found at the scene of the crime.
- A stain on a White House intern's dress provided evidence against US President William J. Clinton.
- Comparison of DNA from descendants of early 19th-century US President Thomas Jefferson and Sally Hemmings, a slave on his Virginia plantation, proved Jefferson to be the father of Hemmings' children.

Less sensational, but more important in everyday law enforcement, is the fact that DNA evidence is sufficiently definitive and widely accepted to *avoid* many trials, by not indicting innocent people.

Applications of DNA identification techniques to animals include the proof of claims that Dolly the sheep was indeed a clone, testing of horses and dogs to confirm breeders' claims of pedigrees, testing of commercial whale meat to check for endangered species, and even a suggestion of creating a database

Identification of the remains of the family of Tsar Nicholas II from analysis of mitochondrial DNA

For most of us, all of our mitochondria are genetically identical, a condition called homoplasmy. However, in some individuals, different mitochondria contain different DNA sequences; this is called heteroplasmy. Such sequence variation in a disease gene in the mitochondrial genome can complicate the observed inheritance pattern of the disease.

The most famous case of heteroplasmy involved Tsar Nicholas II of Russia. After the revolution in 1917, the Tsar and his family were taken into exile in Yekaterinburg in Central Russia. During the night of 16–17 July 1918, the Tsar, Tsarina Alexandra, at least three of their five children, their physician, and three servants who had accompanied the family were killed and their bodies buried in a secret grave. When the remains were rediscovered, assembly of the bones and examination of the dental work suggested – and sequence analysis confirmed – that the remains included an expected family group. The identity of the remains of the Tsarina were proved by matching the mitochondrial DNA sequence with that of a maternal relative, Prince Philip, Chancellor of the University of Cambridge, Duke of Edinburgh – and grandnephew of the Tsarina. (Prince Philip's shared maternal line with Alexandra means that in principle his chances of suffering from haemophilia were 12.5%.)

However, comparisons of mitochondrial DNA sequences of the putative remains of Nicholas II with those of two maternal relatives revealed a difference at base 16 169: the Tsar had a C and the relatives a T. Extreme political and even religious sensitivities mandated that no doubts were tolerable. Further tests showed that the Tsar was heteroplasmic; T was a minor component of his mitochondrial DNA at position 16 169. To confirm the identity beyond any reasonable question, the body of Grand Duke Georgij, brother of the Tsar, was exhumed and was shown to have the same rare heteroplasmy.

to identify dogs whose owners do not clean up after them in municipal parks.

Physical characteristics

Suppose a sample containing DNA is collected at a crime scene, and there is reason to believe that a criminal deposited it. It is possible to use the sample for identification. But suppose the source individual is not represented in the forensic databanks, and is not one of the suspects – usual or unusual – rounded up. It is still technically feasible to make some inferences about the person the police are looking for.

It is possible to predict certain physical characteristics from analysis of DNA sequences. Gender, obviously, but also colour of hair, eyes, and skin, and ethnic background. Use of these inferences for suspect profiling is controversial.

In some cases, it is possible to infer the source individual's family name! Oxford don Brian Sykes discovered that all males named Sykes in the UK are descendants of a single founder individual, and all carry specific diagnostic features of their Y-chromosome sequences. Police could deduce, from a sample left at a crime scene, whether the source individual was named Sykes or not (unless, of course, he changed his name).

Other possible analysis of a blood sample left at a crime scene might provide an estimate of the time of day of deposition. Certain chemicals, for instance melatonin, vary in concentration in blood and saliva following regular circadian rhythms. Such tests do not involve DNA. DNA methylation correlates with age. It might not be too fanciful to imagine a police investigator asking:

Now then, Grandfather Sykes, where were you at 11 pm last night?

- In addition to matching a DNA sample with an individual, it is possible to analyse crime-scene samples to infer several characteristics, including eye and hair colour, complexion, and ethnicity. Use of these inferences in criminal investigation remains controversial, and there is substantial variation in what different jurisdictions permit.

The domestication of crops

The transition from hunting/gathering to agriculture represents a major change in human activity, diet, and social and economic organization. Domestications changed the biology of plants and animals, and even of humans – for instance, the ability to digest lactose past infancy is associated with domestication of cattle.

Many different plants were domesticated, in different regions around the world (Figure 8.2). Although these domestications were independent events, there are many common features:

- Characteristics that improve the product:
  - enlargement of fruit and/or seed.
  - improved flavour and/or nutrition.
- Characteristics that facilitate harvesting:
  - synchronization of ripening time.
  - larger central stalks relative to side shoots – technically, increased apical dominance. For instance, contemporary maize has a central stalk, with the ears growing at the tips of short branches. The ancestral species, teosinte, was highly branched (see Figure 8.3). This allows more plants per unit area tilled; it is analogous to building skyscrapers in cities.
  - seeds do not fall off the plant (called shattering). However, to facilitate harvesting the link of the seed to the plant should be relatively weak. The loss of seed dispersal can render the plant no longer viable in the wild.
- Tillering – shoots fill empty spaces between plants. This makes it unnecessary to plant seeds at specific intervals.
- Increased self-pollination.

These favourable properties are the result of genetic changes during domestication. Documented types of changes include: amino acid substitutions, deletion/truncations altering the functions of individual proteins,

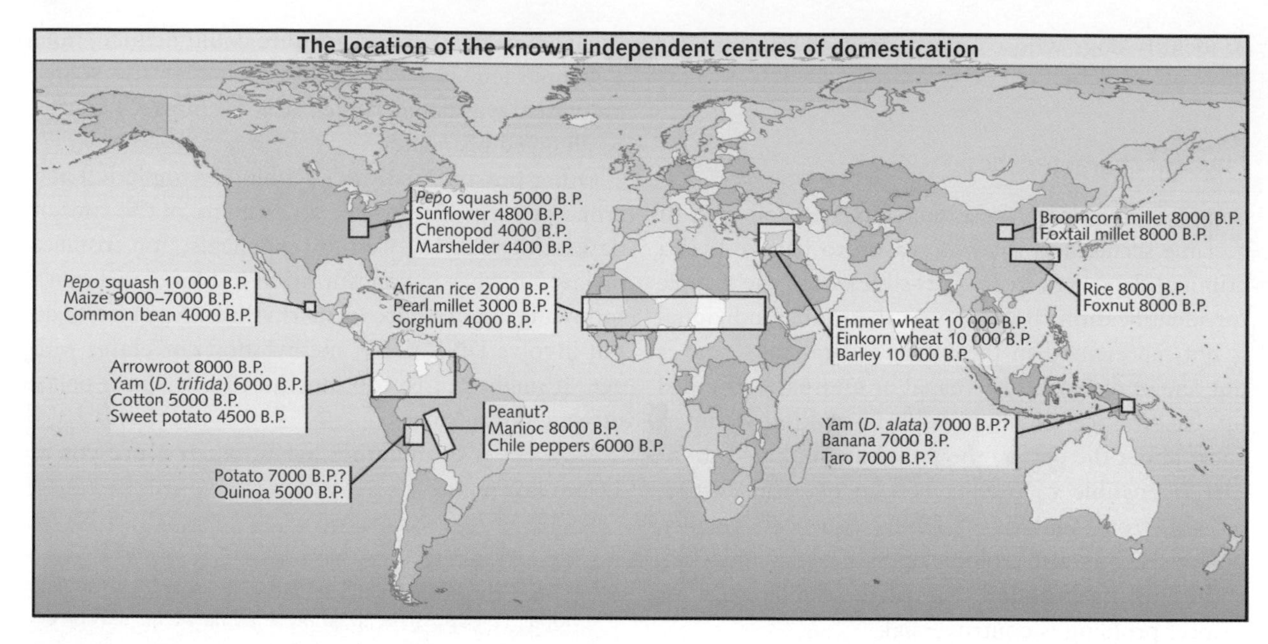

Figure 8.2 Populations in many regions of the world have domesticated plants. Dates shown are based on archaeological evidence, not DNA sequence analysis. (B.P. = before present.)

From: Doebley, J.F., Gaut, B.S., & Smith, B.D. (2006). Cell **127**, 1309–1321.

Figure 8.3 Comparison of modern maize with its teosinte progenitor (*Z. mays* subsp. *parviglumis*). (a) Teosinte grows many long, tasselled branches. (b) In modern maize, many short branches bear the ears at their tips. (c) The kernels of teosinte are encapsulated in a hard compartment. This picture shows both mature (left, dark) and immature (right) kernels. (d) Comparison of kernels of teosinte and modern maize.

Sources: (a) From US Department of Agriculture, Natural Resources Conservation Service, Plant Materials Program, Plant Release Photo Gallery. (b) Photo by David T. Webb, distributed by the Botanical Society of America. (c, d) Photographs by Hugh Iltis.

transposon insertion, regulatory changes, splice-site mutation, and gene duplications.

In general, domesticated plants show lower genetic diversity than their wild, progenitor species. Causes of loss of genetic diversity include (a) population bottlenecks and/or (b) selective sweeps. Comparisons with genes that are selectively neutral, with respect to domestication phenotypes, reveal the relative importance of these two effects. Selectively neutral genes lose genetic diversity only through bottlenecks.

A very interesting question about the genetics of domestication is: to what extent were the genes selected for in domestication present in progenitor populations, and to what extent are they *de novo* mutations? Examples of both types are known. The large genetic variability in the progenitor population makes it harder to find a rare allele even if it were present in the original population.

- Studies of the genomes of crop plants have applications to agriculture, including the search for varieties that produce yields improved in quality and quantity, require less fertilizer and pesticides, and are resistant to disease. If genomes of wild progenitor species are also available, it is possible to study the genomics of domestication.

Maize (*Zea mays*)

The cultivation of maize (*Zea mays*) supported the pre-Columbian civilizations of Central and South America. Maize was brought to Europe in the 15th and 16th centuries and quickly spread around the world. Maize is now the world's third-largest crop plant, after rice and wheat. The annual harvest amounts to 7×10^{11} kg, raised on 3.3×10^{11} km^2 of land. Maize is grown primarily as food, for humans and animals, but some is converted into ethanol for fuel.

In the late 20th century, the intensive development of high-yielding crop varieties began. This 'green revolution', in addition to improving yields, bred maize for a higher content of lysine. Corn was formerly lysine-poor. As a result, people with diets based primarily on maize were traditionally lysine-deficient.

- In the USA and Canada, **maize** is called **corn**.

Maize is a domesticated form of a grass called teosinte. Several varieties of wild teosinte survive in Mexico. Analyses of isozyme and microsatellite diversity – 'paternity tests' for species – have identified the progenitor of maize as a teosinte subspecies, *Z. mays* subsp. *parviglumis*. A combination of archaeological and molecular evidence suggests that teosinte was domesticated in southern Mexico between 6000 and 9000 years ago. The earliest artefacts showing domesticated maize are cobs found in a cave in Oaxaca, Mexico, dated 6250 years ago.

Rafael Guzmán, an undergraduate student at the Universidad de Guadalajara, discovered the teosinte progenitor strain of maize during field work in southwestern Mexico in late 1978. Guzmán was stimulated by a challenge contained in a New Year's card from botanist Hugh Iltis. The importance of his discovery for maize science cannot be overestimated. Access to the progenitor strain makes possible detailed comparisons of sequences. Maize and teosinte are still interfertile, permitting the reintroduction of specific alleles. The teosinte that Guzmán discovered contains unique virus-resistance genes that have been bred into the maize used in agriculture.

Notable differences between teosinte and modern maize include:

- Teosinte has many long branches tipped by tassels. The tassels correspond to male flowers. Modern maize has a single main stalk, with the tassel at the top. The many short lateral branches bear the ears at their tips. The ears, containing the seeds, of course develop from female flowers. (Separate male and female flowers are a feature of varieties of teosintes and maize, not shared by other grasses.)
- The teosinte ear contains 5–12 kernels. Their structure adapts them for dispersal by passage through the digestive tracts of birds and mammals. Individual seeds grow inside an encasing **glume** hardened by silica and lignin (see Figure 8.3c). For human consumption, the kernels would have to be ground, or 'popped' by heating. At maturity, teosinte seeds separate spontaneously, another aid to dispersal. In contrast, an ear of modern maize has several hundred kernels, with no hard casing, and the kernels remain fixed to the ear (see Box 8.3).

These very large phenotypic differences between teosinte and maize appear in plants that have very similar genomes. This presents a paradox. Its resolution must emerge from study of the genomes, the proteins, and the expression patterns. How did the genome change upon domestication? How many genes were involved? Which genes? How have the changes in the genome produced the phenotypic effects?

BOX 8.3 Glume and doom: modern maize could not survive in the wild

Loss of teosinte's hard and adhering seed casing in modern maize is fatal for its natural method of seed dispersal. If eaten by birds or other animals, the seeds would not pass through and be sown. They would be digested, as they are when we eat maize. As a result, modern maize could not survive in the wild. It depends on humans to plant seed each year.

There is a symmetry, literally a symbiosis: human populations – including Mesoamerican civilizations and the earliest pilgrim settlements in New England – have been dependent on maize for food, and maize is dependent on humans for sowing its seed. One could say – depending on one's point of view – that maize has domesticated humans. This is a characteristic of many but not all domestications.

- For maize we are fortunate to have both the modern varieties and the progenitor, teosinte.

The first clues to the extent of the genetic change came from classical genetics. G. Beadle – better known for his 'one-gene–one-enzyme' hypothesis – crossed a strain of maize with teosinte and examined the second generation (F_2). He concluded that domestication involved about five major loci. The experiment was not intended to pinpoint the number of altered genes – Beadle wanted primarily to distinguish whether a large or a small number of loci were involved.

- Some people believed that the phenotypic differences were so great as to preclude conversion of teosinte to maize by human selection. Beadle's point was that the genetic differences might be simpler than suspected.

The 2.3 Gb genome of maize was sequenced in 2009. With the availability of sequence information, it is possible to determine which genes in maize have been subject to selection. *Sites selected during domestication should show enhanced loss of genetic diversity*. There should be a signal from these sites, observable in comparisons of sequence diversity between maize and teosinte. This effect is amplified by the high natural variation in the progenitor population: in *Z. mays* subsp. *parviglumis*, the nucleotide diversity at silent sites is as high as 2–3%. (Of course, these numbers may not accurately reflect the characteristics of a small founder population that gave rise to maize.) The overall loss in diversity upon domestication is estimated at ~25–30%, attributable to the 'domestication bottleneck'. The signal of selection must stand out, at a higher conservation level, above this background.

- What are the signatures of selection? A decrease in nucleotide diversity, increased linkage disequilibrium, and altered population frequencies of polymorphic nucleotides in a gene and linked regions.

The results of such sequence comparisons suggest that about 3%, or about 1200 genes, were targets of selection during maize domestication. At least 50 contribute to agronomically significant traits. Many of the genes involved in morphological changes are clustered around loci that may correspond to Beadle's observations. Others affect biochemical characters that Beadle did not investigate.

Some of the genes involved have been identified as transcription regulators.

- One gene that differs between teosinte and maize is *tb1* (tb = teosinte branched). The maize version represses lateral shoot development and converts tassels to ears (compare Figures 8.3a and 8.3b). Maize has ~30% of the diversity in teosinte populations in the protein-coding region of this gene – approximately the background level of diversity changes upon domestication – but only ~2% of diversity of teosinte in the region 5′ to the gene. TB1 is a repressor, which binds to specific sites in promotors of cell-cycle genes. In this case, selection has been applied not to the coding region of the gene but to a regulatory region. Indeed, the transcription level of *tb1* in maize is higher than that of teosinte. This appears to be the primary mechanism of action of the genomic difference: there has been selection for expression level, rather than for a changed amino acid sequence of the protein expressed. For *tb1*, the alleles selected for domestication are present in wild teosinte.

- Another gene that differs between teosinte and maize is *tga1* (tga = teosinte glume architecture). This gene affects the structure of the glume, the surroundings of the seed. The teosinte allele produces the hard seed case. The maize allele reduces the glume to a soft membrane underneath the kernels. For *tga1*, the expression level is similar in teosinte and maize. However, there is a specific single nucleotide change in one exon, substituting a lysine in a teosinte protein with an asparagine in maize. The maize allele has not been found in wild teosintes.

Can we know whether these and other genetic differences appeared at the time of domestication? It has been possible to sequence DNA from cobs found in sites of a variety of ages, the oldest dated at 4400 years ago. The 4400-year-old cob shows modern-maize alleles for three genes tested, including *tb1*, showing that all three were present in the maize population at least that long ago. However, teosinte-specific alleles of some genes appear in 2000-year-old cobs from New Mexico, in the southwest USA. Selection was incomplete then, at least in regions distant from the site of original cultivation.

Rice (*Oryza sativa*)

Rice is a very important crop plant in much of the world. It forms the major component of agriculture in Asia and India. Rice accounts for about 20% of human calorie intake world-wide.

Rice was domesticated independently at least twice. *O. sativa japonica* originated in south China, at least 10 000 years ago. *O. sativa indica* originated in eastern India or Indonesia. The progenitor of both was *Oryza rufipogon*. Familiar from grocery shops are many different varieties of rice, used in different national cuisines. The characteristics of the major ones are:

Indica: long, slender grains; grown mainly in tropical areas

Japonica: rounder, shorter grains; temperate and tropical varieties

Minor varieties:

Aus: drought tolerant; Bangladesh and West Bengal

Aromatic: Iran, Pakistan, India, Nepal; including Basmati.

What was the course of rice domestication? Was there a sequential process:

$$O.\ \textit{rufiponga} \rightarrow O.\ \textit{sativa japonica} \rightarrow O.\ \textit{sativa indica}$$
$$(\text{or } \textit{rufiponga} \rightarrow \textit{indica} \rightarrow \textit{japonica})?$$

These simple models do not explain the observations that *japonica* and *india* share some domestication alleles, but not others. The currently accepted model involves originally separate *O. rufiponga* populations, separately domesticated to *O. sativa japonica* and *O. sativa indica*, followed by genetic exchange and subsequent specialization.

- Do not confuse the wild progenitor *Oryza rufipogon* of domesticated rice *O. sativa* with the vegetable known as 'wild rice', now grown primarily in the northern United States, *Zizania palustris* and certain other *Zizania* species. These are not close relatives of *Oryza*.

Golden rice

Vitamin A deficiency is an important public health problem, prevalent in regions where rice is an important dietary component. Golden rice is characterized macroscopically by yellow grains, and biochemically by high concentrations of β-carotene, precursor of vitamin A.

Golden rice was created by genetic engineering of rice by inserting two genes in the β-carotene synthesis pathway. These genes are under the control of an endosperm-specific promoter. The genes introduced are: phytoene synthase, from maize; and carotene desaturase from the soil bacterium *Erwinia uredovora*. It is estimated that an individual could achieve the minimum daily requirement of vitamin A by eating 75 g of golden rice per day. Whether cultivation of golden rice, like other genetically modified plants, will be allowed, is under discussion.

Chocolate (*Theobroma cacao*)

The source of chocolate is the seed of a tree, *Theobroma cacao*, that originated in the Amazon basin of western South America (Figure 8.4). It is a relatively

(a)

(b)

Figure 8.4 (a) *Theobroma cacao* tree showing fruit. (b) A ripe pod split open, showing the beans. To create chocolate from beans, the seeds are fermented and dried. Processing extracts cocoa butter (triacylglycerols) and cocoa powder (proteins and polysaccharides, plus small molecules such as flavonoids, terpenes, and theobromine).

Photo (a) by Paul Bolstad, University of Minnesota Bugwood.org, (b) by Keith Weller, USDA Agricultural Research Service, Bugwood.org.

finicky plant, unable to tolerate low temperatures or aridity. This restricts it, primarily, to within 10° latitude of the equator. Chocolate was an important drink in the major central American Olmec, Maya, and Aztec cultures (see Box 8.4). It became popular in Europe after the Spanish conquests. Today, beyond its native habitat in South and Central America, *T. cacao* is important in the agriculture of Brazil and tropical Africa. West Africa currently produces 70% of the world's chocolate, led by the Côte d'Ivoire and Ghana.

Some history of chocolate

In the Maya and Aztec civilizations in Central America chocolate was a medicinal and ceremonial drink, and an expensive one.[1] Fermentation could produce an alcoholic beverage. The Aztecs prepared chocolate, not as a sweet as we know it, but mixed with chilli pepper to form a spicy, bitter drink. A Maya vase in the collection of the Princeton University art museum shows a woman pouring chocolate from a height into a receptacle, to produce froth. Even more analogous to wine, in the Maya culture it served sacramental, nutritious, and medicinal purposes. The beans also served as currency: a turkey cost 100 cacao beans.

European contact with cacao beans began in 1502 when Columbus captured the cargo of two Mayan trading canoes on his fourth voyage. The beans made little impact. Wider appreciation of chocolate awaited Cortez's return from his conquest of Mexico. In 1528 Cortez presented to Charles V a sample of cacao beans, and the tools and recipe for preparing them. In 1544, a group of Maya nobles

[1] '. . . chocolate drinks occupied the same niche as expensive French champagne does in our own [American] culture.' S.D. Coe and M.D. Coe, *The History of Chocolate*, 2nd. ed., Thames and Hudson, London, 2007, p. 61.

accompanied Franciscan friars to the court of Philip II and demonstrated its preparation.

In contrast to the spicy, bitter Mayan and Aztec beverages, the Spanish added sugar. This initiated the transition of chocolate from a medicinal to a recreational substance. Chocolate beverages containing alcohol are now prepared by adding spirits to chocolate.

Chocolate was introduced to Europe before coffee or tea were. It was the first exposure of Europeans to stimulant alkaloids such as caffeine. The taste for chocolate grew and spread, first as a drink and later in the solid, familiar 'bar' form. It is widely believed that the marriage of Philip II's daughter Anne of Austria to Louis XIII of France in 1615 brought chocolate across the Pyrenees. The first chocolate house in London opened in 1657. The popularity of these beverages derived originally from a belief in their medicinal value, as well as their flavour. Samuel Pepys's diary mentions drinking chocolate on several occasions, alluding to its curative properties:

> Waked in the morning with my head in a sad taking through the last night's drink, which I am very sorry for; so rose and went out with Mr. Creed to drink our morning draft, which he did give me in chocolate to settle my stomach. (24 April 1661)

The odour and taste of natural chocolate derives from unique flavonoids. Pharmacologically active compounds in chocolate include:

- theobromine (10% by weight of dark chocolate): heart stimulant, cough suppressor, vasodilator, lowering blood pressure leading to feelings of relaxation, diuretic. Pet lovers be warned: theobromine is toxic to dogs and cats.
- phenethylamine: a stimulant, similar in effect to amphetamines.

The 18th-century physician and architect Sir Hans Sloane was the inventor of milk chocolate. Sloane is better known as the founding donor of the original collection of the British Museum. London's chic Sloane Square was named after him because his heirs owned the land developed. Less well known is that the classic red British phone box, now an endangered species, was based on his tomb, which Sloane himself designed.

T. cacao is susceptible to fungal infections, notably 'witches' broom'. This fungus has hit Brazil's plantations very hard, but has so far not affected trees grown in Africa. A search for resistant varieties has motivated many expeditions to the headwaters of the Amazon. These expeditions also provided an opportunity to map out the geographic distribution of genetic variation in *T. cacao* trees. Areas of the highest variability are in the upper reaches of the Amazon, in what are now parts of Ecuador. This identifies the probable region of origin of the species.

Fourteen genetic clusters have been identified, extending the known varieties, which have been differentiated according to the appearance of the beans – as well as their flavour.

Cultivars of *T. cacao* important in agriculture include: *criollo* (*T. cacao* ssp. *cacao*), the best tasting, and used for the finest products; *forestero* (*T. cacao* ssp. *sphaerocarpum*), inferior in taste but resistant to disease (90% of the world's cacao crop is *forestero*); and *trinitario*, a *criollo*–*forestero* hybrid, superior in taste to *forestero* and providing its hardiness.

- Unlike maize and rice, *Theobroma cacao* did not have to undergo a substantial genetic modification to domesticate it. Originally, the seeds from the native varieties were harvestable directly.

The *T. cacao* genome

Two *T. cacao* projects have sequenced the *criollo* and *forestero* varieties.

T. cacao is a diploid organism containing 10 chromosomes. The assembly of the 420 Mbp *criollo* genome at 16.7 X coverage includes 76% of the genome. Much of the remainder comprises repetitive regions. A linkage map allowed anchoring 67% of the 326 Mb sequenced within the 10 chromosomes.

28 798 protein-coding genes were identified, higher than *A. thaliana*. The average gene size was 3346 bp, with a mean of 5.03 introns/gene. Figure 8.5 shows the shared and unique genes from *T. cacao*.

Motivations for sequencing the *T. cacao* genome include the desire to understand:

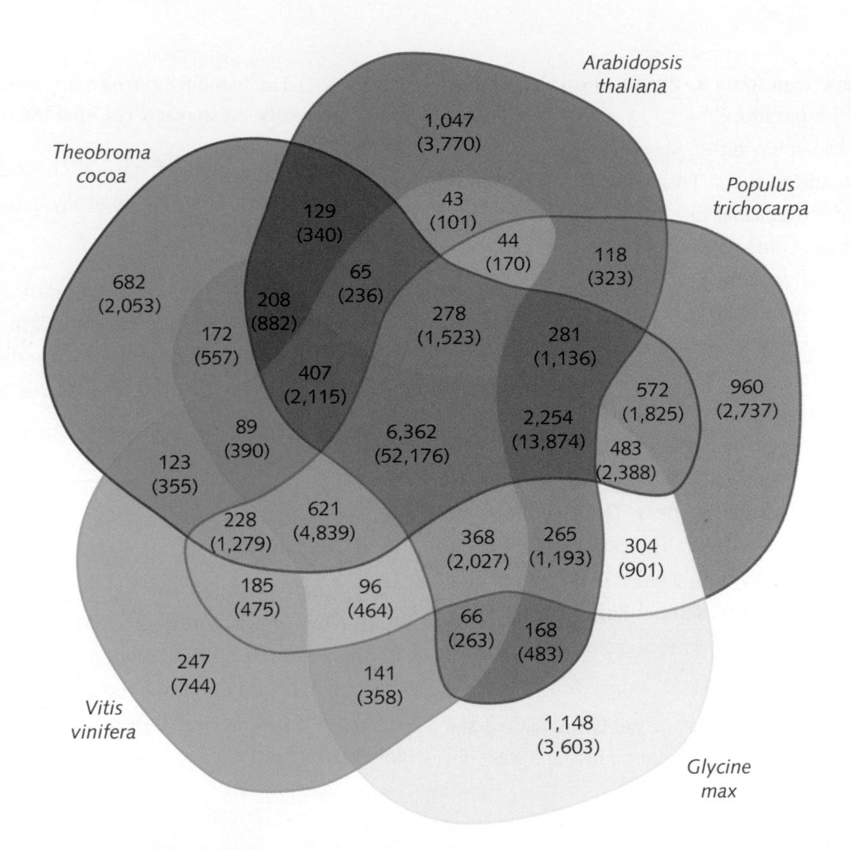

Figure 8.5 Numbers of shared and unique gene families and, in parentheses, numbers of genes from chocolate (*Theobroma cacao*), thale cress (*Arabidopsis thaliana*), black cottonwood (*Populus trichocarpa*), soybean (*Glycine max*), and wine grape (*Vitis vinifera*). From: Argout, X., et al. (2011). The genome of *Theobroma cacao*. Nat. Genetics **43**, 101–108.

- the determinants of resistance to witches' broom and other diseases
- the sources of the flavours in the chocolate produced
- the place of *T. cacao* in the general phylogeny of angiosperm plants.

Known disease-resistance genes in plants encode nucleotide-binding site/leucine-rich repeat (NBS-LRR) proteins and receptor protein kinase (RPK) proteins. The *T. cacao criollo* genome is relatively poor in one class of NBS-LRR genes, including those encoding the toll interleukin receptor (TIR) motif. In *T. cacao* only 4% of genes orthologous to NBS-LRR contain the TIR motif. The corresponding number for *A. thaliana* is 65%! In contrast, the *T. cacao criollo* genome contains orthologues of all for NPR1 subfamilies known in *A. thaliana*.

The molecules that give cocoa its distinctive flavour include flavonoids, alkaloids, and terpenoids. *T. cacao criollo* is rich in enzymes producing these compounds. *A. thaliana* has 36 genes for flavonoid biosynthesis pathway enzymes; *T. cacao* has 96. *T. cacao criollo* also has expanded families of gene encoding enzymes involved in terpenoid synthesis.

It would be tempting to hypothesize that the paucity of toll interleukin receptor genes might account for the lower disease resistance of *criollo* relative to *forestero*, and the enhancement in numbers of genes for flavonoid and terpenoid biosynthesis enzymes might account for the enhanced flavour of *criollo* relative to *forestero*. Comparison of the genomes of the two subspecies, which is so far only in the preliminary stages, suggests that this may be too simplistic an analysis.

Comparing the genomes of several flowering plants allows inference of the primordial angiosperm genome. The evolution of eudicot genomes is characterized by polyploidization events. Therefore synteny is an important clue to the phylogeny, in addition to the divergence of individual gene sequences.

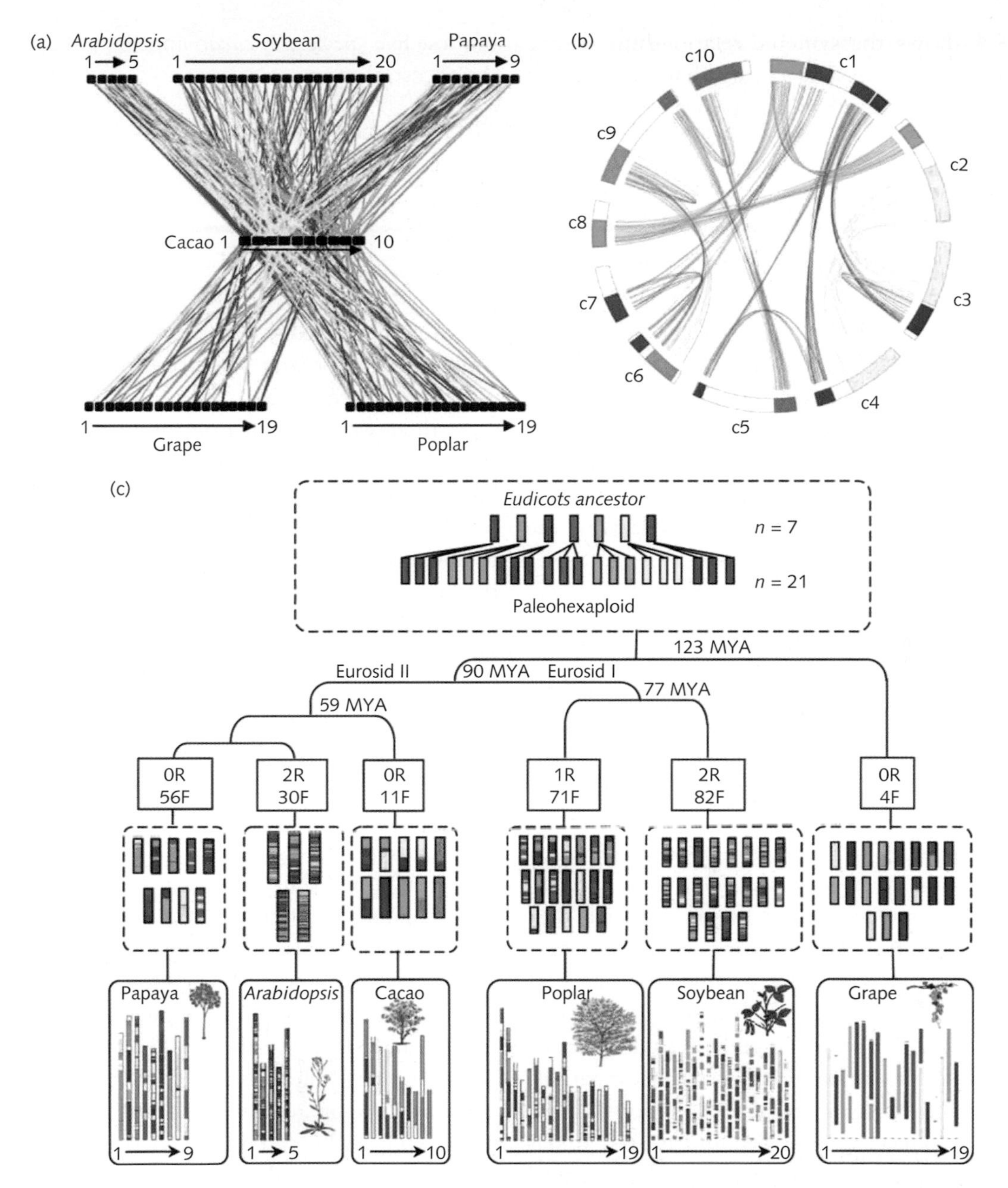

Figure 8.6 Syntenic relationships of eudicot genomes, comparing *T. cacao* with thale cress (*Arabidopsis thaliana*), soybean (*Glycine max*), black cottonwood (*Populus trichocarpa*), papaya (*Carica papaya*), and wine grape (*Vitis vinifera*), and inferring the chromosome structure of the common ancestor. (a) Mapping of orthologues from *T. cacao* to other five genomes. For each species chromosomes are numbered consecutively. Different colours represent seven ancestral eudicot linkage groups. (b) Syntenic relationships *within* the *T. cacao* genome. There is evidence for ancestral triplication of the genome. For example, light green links regions of chromosomes 1, 2, and 8. (c) A model for the evolutionary history of the karyotype. The six species studies derived from an original eudicot ancestor with 7 chromosomes. To produce the current karyotypes there were whole-genome duplications (R) and chromosome fusions (F). From comparisons of the chromosomes of the six species, it is possible to trace the path from the current chromosome structure of the six species (shown at the bottom), back to the common ancestor. The gene distribution among *T. cacao* chromosomes is the closest, of these six species, to the ancestral form.

From: Argout, X., et al. (2011). The genome of *Theobroma cacao.* Nat. Genetics **43**, 101–108.

Figure 8.6 shows the syntenic relationships of five eudicot species. The results suggest that the ancestral eudicot contained seven chromosomes. Interestingly, of these five species, *T. cacao* appears to be closest to the ancestral form.

Genomics in anthropology

Our genomes contain the history of the origins and development of our species. We have already seen how genomics can elucidate phylogenetic relationships, and how palaeosequencing of extinct species can widen the coverage of these studies. The palaeosequence that has most captured popular interest has been that of one of our closest relatives: the Neanderthal genome.

The Neanderthal genome

In 1856, about a dozen bones were found in a limestone quarry in the Neander Valley near Düsseldorf (Valley = Thal in German). Although, in retrospect, Neanderthal bones had appeared earlier, the 1856 discovery was the first to be recognized as remains from a new species.

Neanderthals were hominids closely related to modern humans. The earliest fossils in which they appear are about 130 000 years old. They were approximately the same size as humans, but substantially more robust. Their cranial capacity was as large as modern humans or perhaps slightly larger, although this in itself does not guarantee comparable cognitive abilities. Neanderthals did fashion and use tools, and produce at least decorative arts. It has been suggested that they wore makeup. They engaged in complex funerary practices.

Both modern humans and Neanderthals inhabited Europe and Central Asia, until the Neanderthals vanished, about 30 000 years ago. Articles in scientific journals inquire about their similarities to and differences from modern humans, and why they became extinct. The popular press focuses on the possibility of sexual congress between Neanderthals and humans.

The sequencing of Neanderthal mitochondrial DNA was completed in 2008. For the nuclear genome, DNA extracted from bones from three individuals from a site in Croatia produced a total of 4 Gb of Neanderthal sequence.

A three-way comparison among Neanderthal, human, and chimpanzee sequences allows analysis of human–Neanderthal divergence. There are 78 sites in protein coding genes containing amino acids specific to humans, that is, the amino acid at that position is the same in chimpanzee and Neanderthal, and different in human. In contrast, one result of interest is that Neanderthals have the human version of the FOXP2 gene, involved in language skills in humans.

The data suggest that the populations ancestral to Neanderthals and humans split about 270 000–400 000 years ago. Admittedly, this is a very rough estimate. But it antedates the migration of humans from Africa. It suggests the following scenario: the populations split in Africa 370 000 years ago. One population migrated to Europe to give rise to the Neanderthals. The other remained in Africa, and gave rise to humans. Then humans migrated to Europe about 40 000 years ago, where they coexisted with Neanderthals for about 6000 years.

Did humans interbreed with Neanderthals? The question is quite controversial. It is argued that by looking at human variation patterns from the HapMap data, Neanderthals are more closely related to humans from regions other than Africa, than to African populations. If there were no genetic admixture, one would expect no difference: Neanderthals should be equally closely related to all human populations. The scenario implied is that on their way out of Africa, humans interbred with Neanderthals and carried the genes thereby picked up around the rest of the world. The estimate is that 1–4% of the human genome is Neanderthal in origin. One might reasonably suppose that Neanderthals should be most closely related to European human populations – for that is where they coexisted for longest – but the data do not support this.

- It has been possible to sequence DNA from bones of Neanderthals, a species of hominid that has been extinct for about 30 000 years. Humans and Neanderthals both inhabited Europe for about 6000 years. One question addressed using the Neanderthal sequence is whether there was human–Neanderthal interbreeding. It has been estimated that 1–4% of the human genome is Neanderthal derived. However, this conclusion is controversial.

Ancient populations and migrations

When people move around, they take their DNA with them. This makes is possible to trace patterns of migration. Many studies focus on mitochondrial DNA sequences, which reveal lines of maternal inheritance (see Box 8.5). Y chromosomes provide complementary paternal information. Although studies of Y sequences are more sparse, in many cases they corroborate the implications of the mitochondrial data.

There is now a consensus that our species, *Homo sapiens*, arose in Africa approximately 100 000–150 000 years ago. Migrations beginning approximately 60 000 years ago took our ancestors around the world, and continue to do so. Unlike modern population flows, documented in historical records, we depend on archaeological relics, modern genomics, and linguistics to infer the timing, the routes, the numbers of individuals, and even perhaps the motivation of ancient migrations.

Other crucial transitions in human social organization, such as turning from hunting to agriculture, are reflected to some extent in domestications of other species such as maize and dog.

Human mitochondrial DNA haplogroups

Human mitochondrial DNA is a double-stranded, closed, circular molecule 16 569 bp long. It is inherited almost exclusively through maternal lines. A fertilized egg contains the mother's mitochondria. Although sperm contain mitochondria – essential to provide energy for their motility – the few paternal mitochondria that enter the egg are selectively eliminated. As a haploid entity, mitochondrial DNA is, therefore, not subject to recombination, and changes only by mutation.

Mitochondrial DNA is estimated to adopt one mutation every 25 000 years. This gives a reasonable rate of divergence to trace human migration patterns. (Nuclear DNA mutates approximately ten times more slowly than mitochondrial DNA because (1) histones protect it; (2) active repair mechanisms edit out some mutations; and (3) the activity of mitochondria in oxidative phosphorylation exposes the DNA to mutagenic oxygen radicals.)

Human mitochondrial DNA contains genes for 22 tRNAs, two ribosomal RNAs, and 13 proteins. The major non-coding region is the control region, or D-loop, involved in regulation and initiation of replication. This region is about 1 kb long. It shows a higher rate of substitution than the rest of the mitochondrial genome, by a factor of about four.

Different mitochondrial DNA sequences are associated with different populations. Mutations are referred to the first human mitochondrial DNA sequence determined, called the Cambridge Reference Sequence. Groups of related sequences are called haplogroups. (The distribution of the number of sequence differences between different individuals has a peak at ~70 for Africans and ~30 for non-Africans.) The original classification of sequence variants depended on changes in restriction sites (see Figure 8.7). This was followed by explicit sequencing of the control region, focusing on its two highly polymorphic segments. For finest resolution, contemporary studies are now more frequently determining full mitochondrial DNA sequences, except in cases of ancient DNA where the best recoverable material may be fragmentary.

Several databases focus on human mitochondrial genomes, including MITOMAP (http://www.mitomap.org) and mtDB (http://www.genpat.uu.se/mtDB).

→

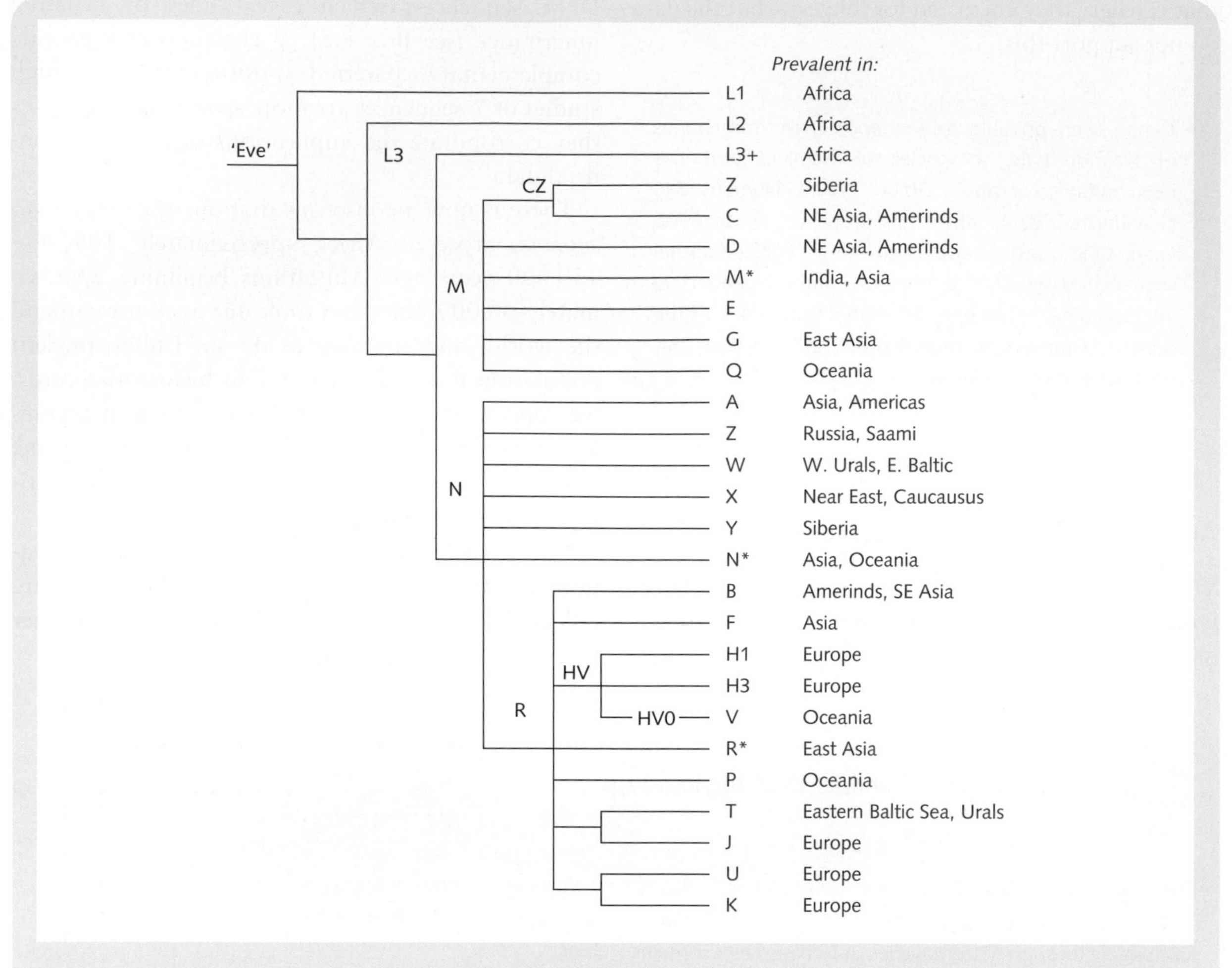

Figure 8.7 Phylogenetic tree of major mitochondrial haplogroups. The nomenclature began with a study of Native Americans, or Amerinds, and the letters A, B, C, and D were assigned to them. Other letters were introduced and were subdivided as needed as more detailed sequencing data appeared. HV0 was formerly called pre-V.

P. Forster has created a 'movie' showing successive stages of human dispersal (see Figure 8.8).

The evidence for human origins in Africa is that contemporary genetic diversity is highest there. The mitochondrial DNA haplogroup L1, believed to be the oldest haplotype that survives, is found in the KhoiSan of the Kalahari Desert in southern Africa and in the Biaka pygmies of the central African rainforest (see Figure 8.8a). An expansion of L2 and L3 haplogroups took place within Africa about 80 000–60 000 years ago (see Figure 8.8b). Over two-thirds of contemporary Africans belong to these groups.

The first emigration from Africa occurred about 85 000–55 000 years ago. The participants in this first dispersal carried the L3 mitochondrial DNA haplogroup, which arose ~84 000 years ago. Mutations in L3 gave rise to haplogroups M and N, ~60 000 years ago (see Figure 8.8c). Haplogroup M mitochondrial DNA appears in ancient populations from the Andaman Islands, southern continental India, and the Malaysian Peninsula. This suggests a dispersal via what is now southern Iran along the coast to India, the Malaysian Peninsula, and Australia. Archaeological evidence shows that humans reached Australia by 46 000 years ago. Dating of the earliest human remains in Australia is consistent with the extinction of many large mammals and birds shortly thereafter (see Figure 8.8d).

From 60 000 to 30 000 years ago, human populations expanded in southern Europe and Asia, accumulating mutations in mitochondrial DNA to form new haplogroups (see Figure 8.8e).

Expansion into Europe was delayed and interrupted. Between 20 000 and 30 000 years ago – an interglacial era – the first human Europeans encountered and replaced the Neanderthal population (see Figure 8.8f). The same climate conditions permitted a northwards expansion in Asia.

The closing of the Bering Strait allowed humans from northwest Asia to move across to North

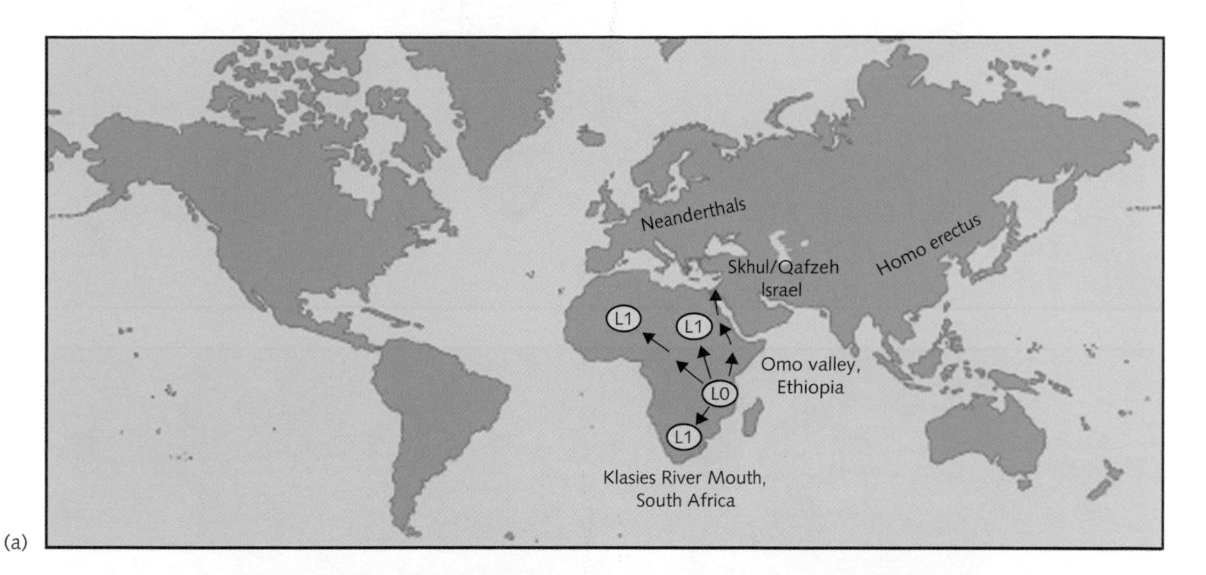

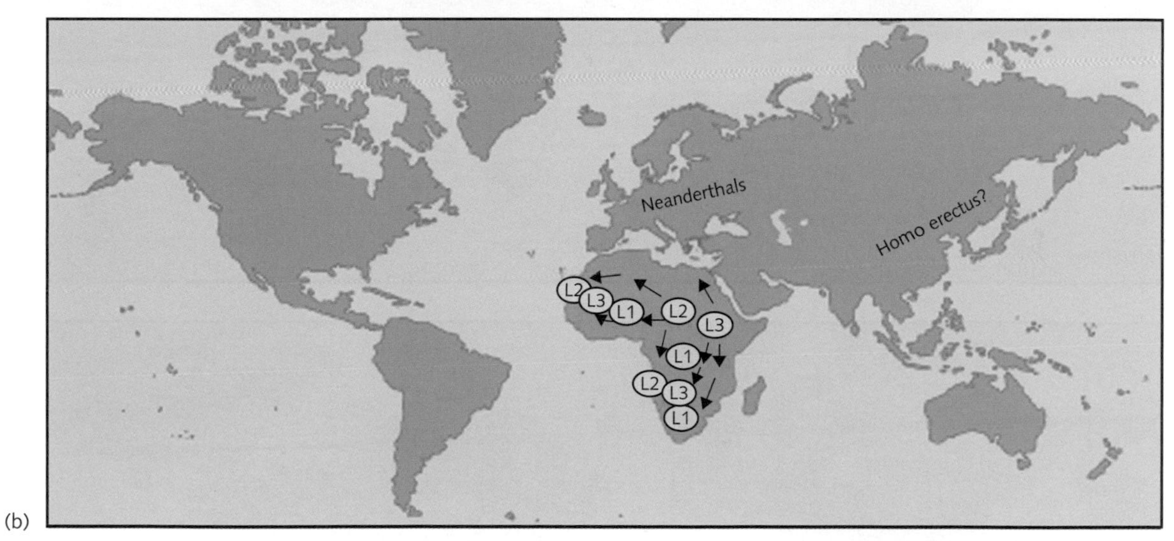

Figure 8.8 'Movie' of human migration patterns, based on mitochondrial DNA sequences. Letters indicate haplotypes. (a) The beginnings, ~150 000 years ago. (b) Expansion and divergence within Africa, 80 000–60 000 years ago. (c) Out of Africa, 60 000–50 000 years ago. (d) Spread through the south coast of the Indian Ocean, reaching Australia, 50 000–30 000 years ago. (e) Expansion in the eastern Mediterranean, India, southeast Asia and Australia, and central Asia, 30 000 years ago. (f) After 30 000 years ago, milder climate conditions permitted expansion northwards in Europe and Asia. (g) Crossing the Bering Strait, first human inhabitants of American continents, 20 000–15 000 years ago. (h) Ice Age, 20 000 years ago, with humans forced to retreat south. (i) Spread to South America, 18 000 years ago. (j) Warmer climate, with resettlement of northern latitudes, 15 000–13 000 years ago. (k) Sudden warming and subsequent stable climate, 11 400 years ago, allowing the spread of agriculture. (l) Expansion to islands, 2000 years ago. (m) The current picture.

From: www.rootsforreal.com. See also: Forster, P. (2004). Ice ages and the mitochondrial DNA chronology of human dispersals: a review. Phil. Trans. R. Soc. B: Biol. Sci. **359**, 255–264 (Figure 5).

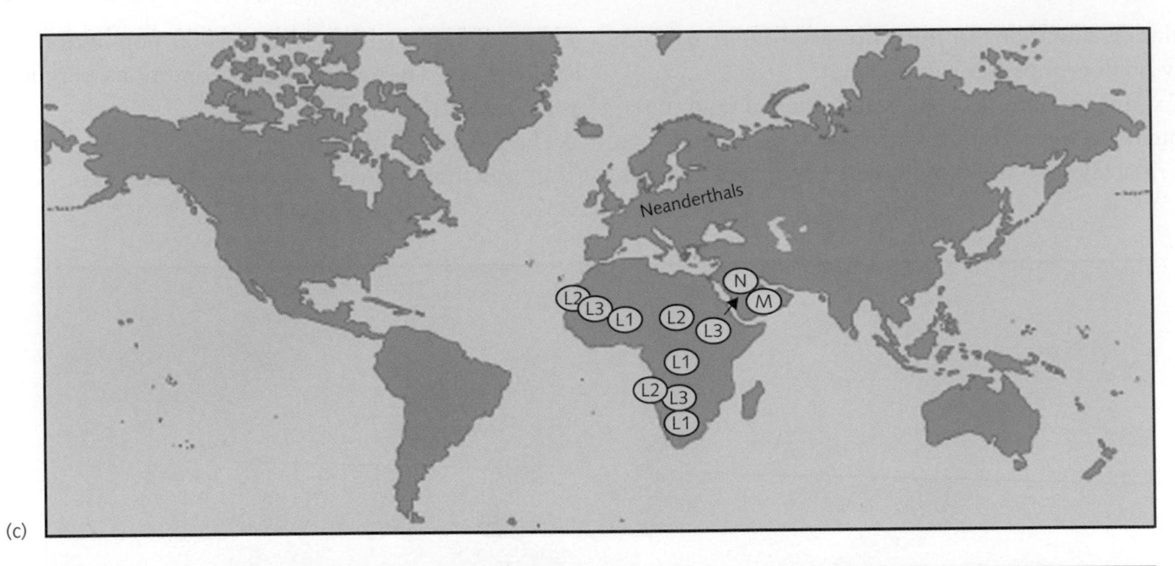

(c)

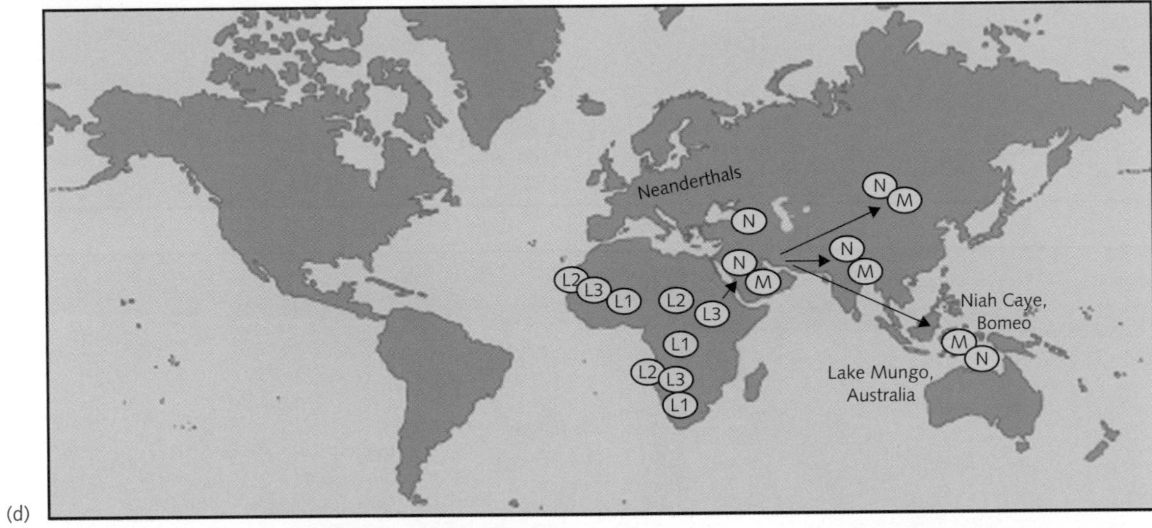

(d)

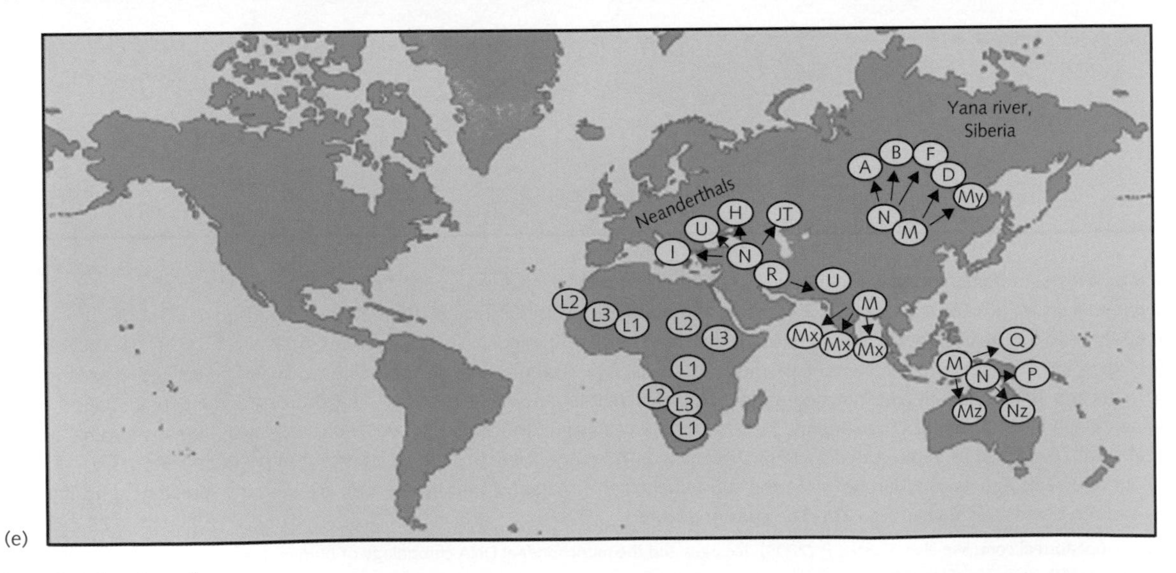

(e)

Figure 8.8 *(continued)*

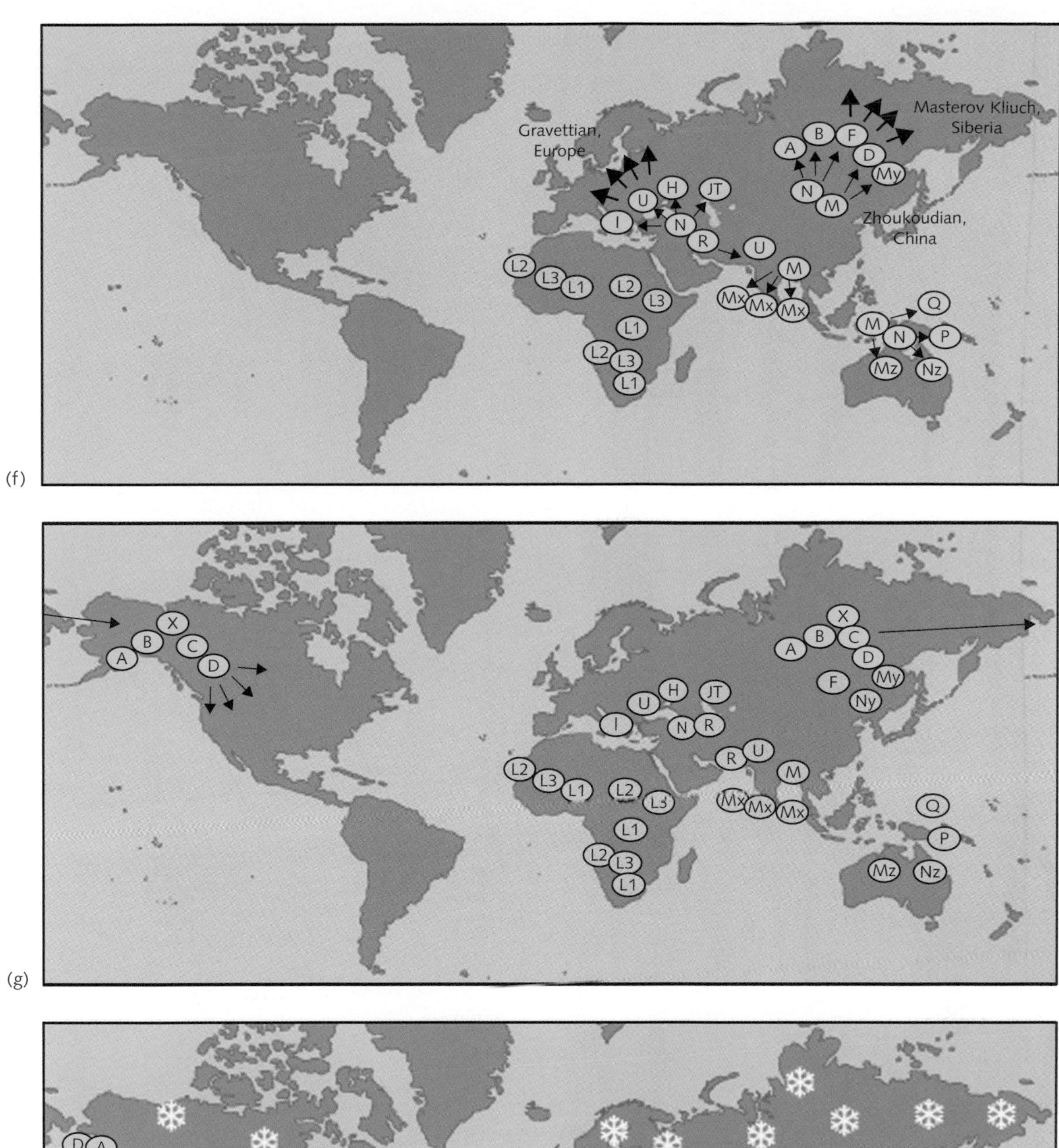

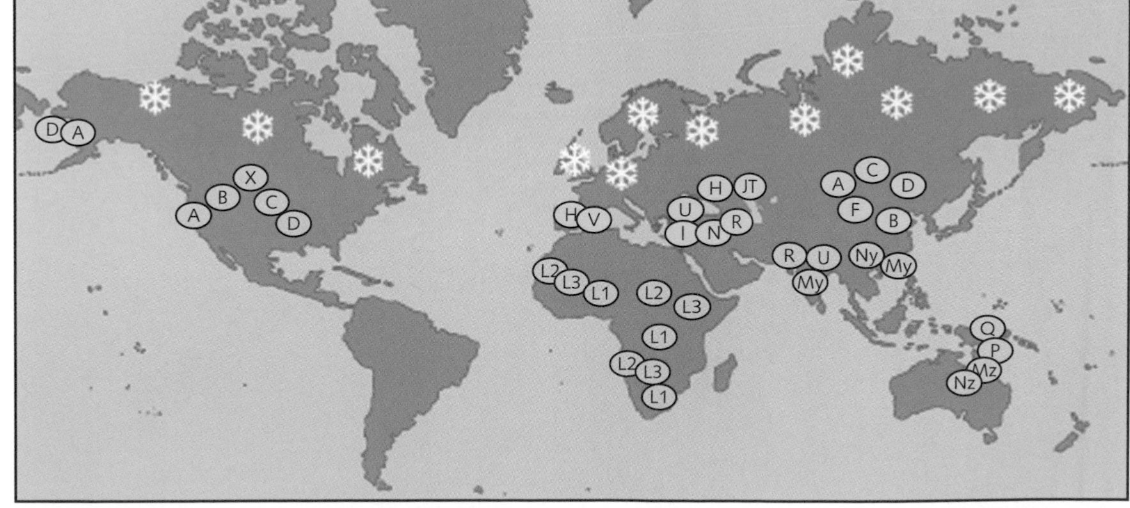

Figure 8.8 (*continued*)

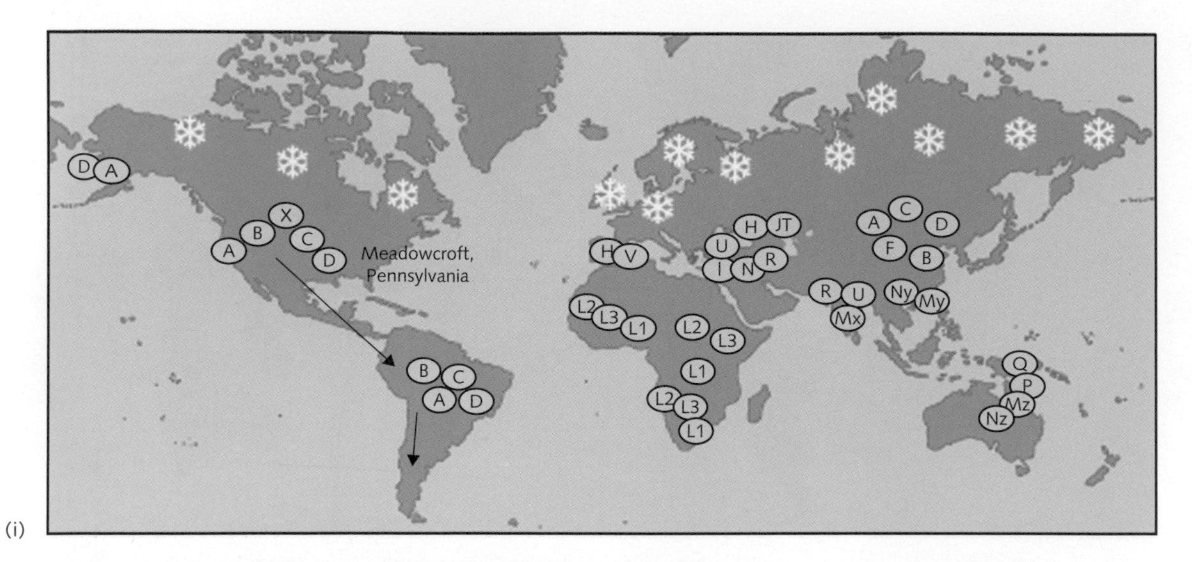

(i)

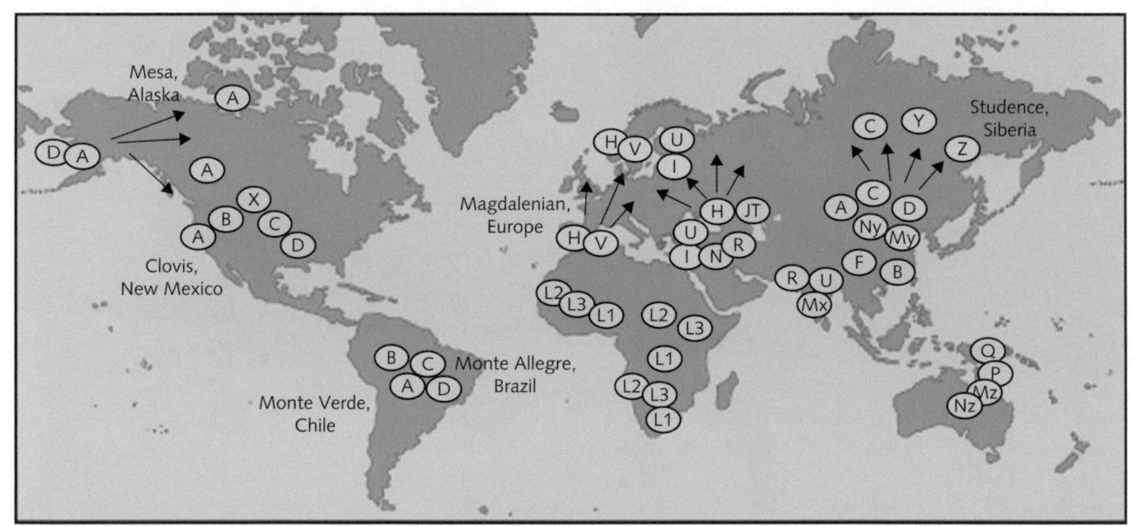

(j)

(k)

Figure 8.8 (*continued*)

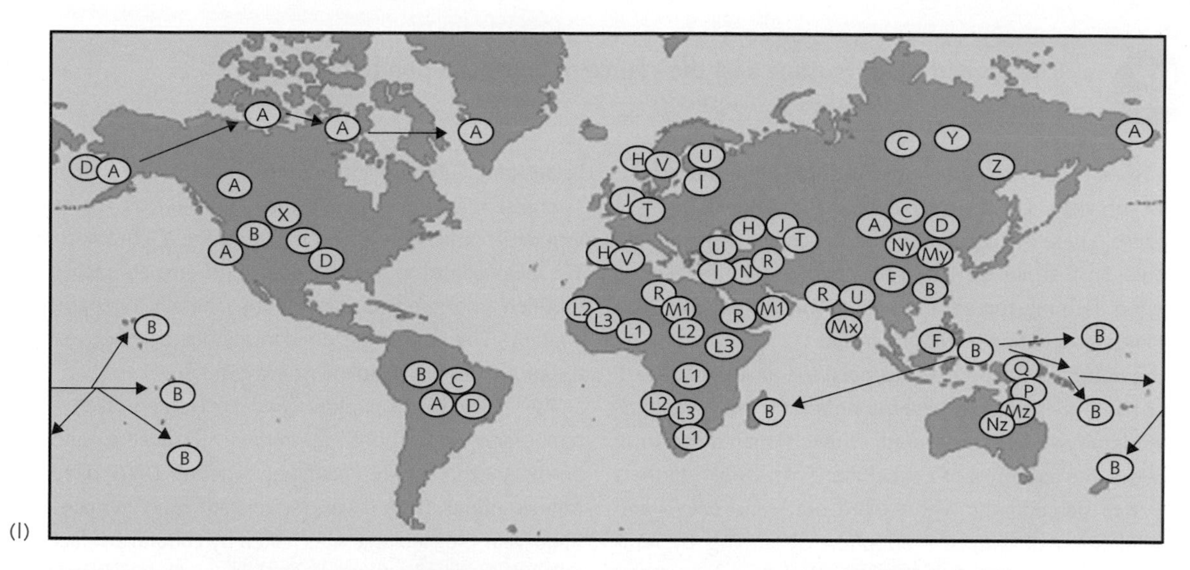

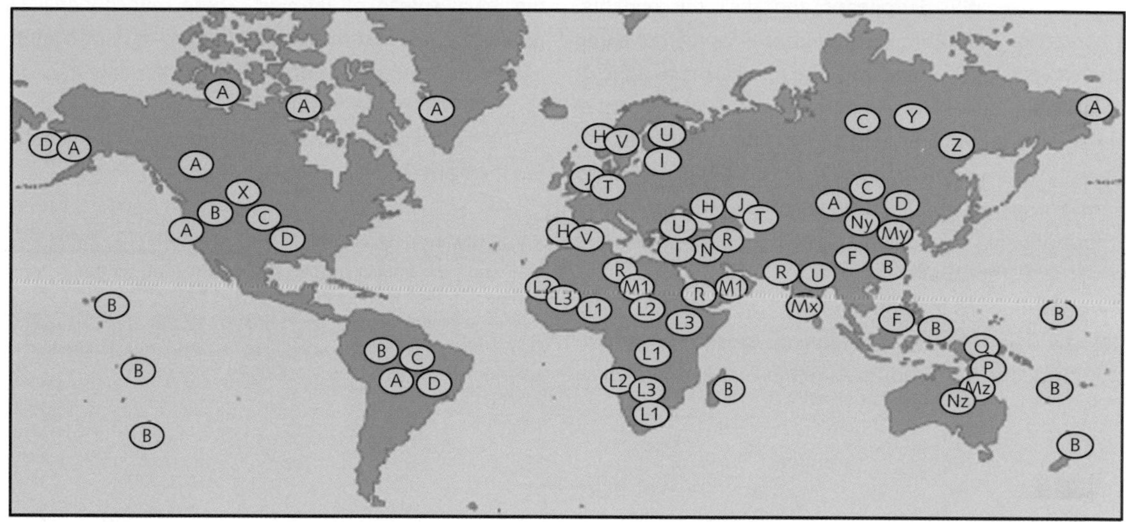

Figure 8.8 *(continued)*

America and to expand southwards (see Figure 8.8g). Human remains found in Alaska have been dated to 9800–9200 years ago. Current evidence suggests that humans first arrived in America 20 000–15 000 years ago.

When glaciers covered northern Europe and arctic America again, humans retreated southwards (see Figure 8.8h and i). Only isolated pockets of the original settlers remained in Europe. One such pocket was ancestral to the Basques, with their now-unique H and V mitochondrial haplogroups. A subsequent warm period, starting 15 000 years ago, saw resettlement of northern Europe (see Figure 8.8j). The genetic diversity in northern Europe is accordingly reduced. The ending of the last ice age, quite abruptly 11 400 years ago, permitted the expansion of agriculture (see Figure 8.8k, and Box 8.6).

- Agriculture reached Britain about 5000 years ago.

Late migrations, during the last 2000 years, populated islands such as Greenland by Inuit from Alaska, Madagascar by people from southeast Asia, and Pacific islands by Austronesians (see Figure 8.8l).

The contemporary distribution of mitochondrial haplogroups contains the records of this history (see Figure 8.8m).

BOX 8.6 The spread of agriculture and the source of European populations

We have discussed crop domestication from the plants' point of view. Let us now examine the human consequences. There is a consensus that agriculture began in the Middle East about 12 000 years ago. After the glaciers receded, farming spread northwards through Europe, supporting a population increasing in size.

The nature of the process has been the subject of debate – a debate that has not become noticeably less heated as more data have been measured. Opinions span a spectrum between the extremes of a movement of *people* – farmers and their descendants who moved north and east – and dissemination of *culture* – adoption of farming by remnants of Palaeolithic Europeans and their descendants. The spread of Indo-European languages along the same northeast vector, correlated with the movement of agriculture, is ambiguous – like agriculture, languages can travel either by migration of people or by cultural transmission. However, surveys of genomes of contemporary and ancient Europeans should be able to detect the relative contributions of Neolithic Middle Eastern farmers and Palaeolithic Europeans.

The contributions to contemporary European DNA gene sequences have been apportioned to several populations including the Basques – representing the original Palaeolithic European inhabitants – and Middle Easterners – representing the originators of agriculture.* The study included markers on autosomes and the Y chromosome. The mean admixture suggested an approximately equal contribution from the two populations. There is a geographic gradient. The Middle-Eastern contribution decreases as the distance from the Middle East increases.

The study by Dupanloup and co-workers used contemporary populations as representatives of ancient ones. Another study examined ancient DNA directly.† Mitochondrial DNA was sequenced from remains at 7500-year-old Neolithic sites linked by cultural artefacts to the initial spread of farming. These mitochondrial DNA results suggest that these early farmers did not contribute a large fraction to the European mitochondrial gene pool. Sequences from the Y chromosome confirm this.

The conclusion is that early farmers did migrate to Europe from the Near East.

* Dupanloup, I., Bertorelle, G., Chikhi, L., & Barbujani, G. (2004). Estimating the impact of prehistoric admixture on the genome of Europeans. Mol. Biol. Evol. **21**, 1361–1372.

† Haak, W., Balanovsky, O., Sanchez, J.J., et al. (2010). Ancient DNA from early European neolithic farmers reveals their near eastern affinities. PLoS Biol. **8**, e1000536.

Genomics and language

Language, as a biological phenomenon, links genomics with several other disciplines, including neurobiology, development, medicine, and anthropology. Language is also a pre-eminent social phenomenon. It underlies interpersonal communication (both face-to-face and worldwide), commerce, and literature. Many important social decisions (e.g. the design of educational systems) depend on appreciating the biological substrates of language. These biological substrates are subtle and our knowledge of them is far from complete.

Language is a feature of all human populations. Of the approximately 6500 languages currently spoken, a few major ones such as Chinese, English, Hindi, and Spanish can each claim over 400 million speakers (see Table 8.1). Others have much smaller communities: Europe retains niche languages such as Basque and Breton. There were estimated to be about 250 languages of indigenous Australians at the time of settlement by Europeans, of which about two-thirds survive. There are over 800 distinct languages spoken on the island of New Guinea.

Human spoken languages have many features in common with biological species. Both languages and species exhibit varying degrees of similarity and diversity. Languages and species can be classified into

Table 8.1 Major languages of the world

Native language	Number of speakers
Mandarin Chinese	>1 000 000 000
English	500 000 000
Hindi	495 000 000
Spanish	425 000 000

taxonomies that, at the highest levels at least, are hierarchical. In both languages and species, ancestor–descendant relationships can be observed. Languages diverge, as species do. The Romance languages, divergent descendants of Latin, are the Galapagos finches of linguistics. Latin itself sits within the larger class of Indo-European languages, of which Sanskrit, the classical language of India, is the extant language putatively closest to their common ancestor. Dialects of languages are analogous to genetic haplotypes.

Just as species have become extinct, so have many languages. Many languages are currently 'endangered', with only a few elderly speakers remaining. Fortunately, like species, languages can be rescued from extinction. Navajo, in decline until the 1940s, is now thriving. Hebrew, which had survived as a written language, re-emerged 50 years ago as the spoken language of the new country of Israel. In some cases, it is possible to reconstruct an extinct common ancestor of surviving descendant languages. This can have interesting implications about the lifestyles of its speakers. For instance, the similarities in some words in Indo-European languages, and dissimilarities in others, suggests that the speakers of the ancestral Indo-European language had domesticated dogs and horses but not cats and camels.

Linguists have developed quantitative methods to measure divergence of languages. Languages develop variation in pronunciation (or spelling, if the language is written) and in structure. An example of variation in pronunciation would be the difference between the English word *ship* and the German cognate *Schiff*. An example of a structural feature of a language is word order: in sentences in English and many other languages, the typical word order is *subject–verb–object*, as in 'The mouse ate the cheese'. In some languages, such as Turkish, typical word order is *subject–object–verb*. Phonological change is much more rapid than structural change – molecular biologists might find a useful analogy to the rapidity of sequence change relative to structural change in proteins!

There are obvious correlations between people's native language, and physical features that are genetically determined; for instance, blue or brown eyes correlate well with native speakers of Swedish or Mandarin, respectively. However, language is clearly transmitted culturally rather than genetically: any child can become a fluent speaker of any language to which he or she is exposed in the cradle. (There is, however, a finite window: it is much more difficult after puberty to become fluent in a new spoken language; some aspect of neural plasticity is shut down. As a result, language has and continues to serve as a litmus test for immigrants. This is, unfortunately, frequently used as a criterion for social discrimination, by persons who may even believe sincerely that there is a correlation between imperfect speech in a non-cradle language and inherent or genetic inferiority.)

Although there is no *direct* genetic link with native language, L.L. Cavalli-Sforza and co-workers found substantial similarities between clustering of genetic variation among human populations and language groups. For instance, just as the Basque language is an isolate, unrelated to any other language spoken in Europe, the Basque people are also genetically distinct. This link between languages and genetics is useful because we can estimate rates of genetic change and thereby infer when languages arose and diverged. One conclusion is that most of the world's language families arose, in a burst, between 6000 and 25 000 years ago. When such inferences from genetics can be compared with datable archaeological evidence or historical records, the results are sometimes, although not always, gratifyingly consistent.

Comparison of genetic and linguistic data can illuminate what happens when two populations collide. Some discrepancies between correlations of genetics and languages appear when conquerors impose a new language on an indigenous population.

In Ivanhoe, Sir Walter Scott contrasted the French word veau for the meat, with calf for the animal, noting that '. . . he is Saxon when he requires tendance, and takes a Norman name when he becomes matter of enjoyment'.

gene families are shared by all four species? (c) How many gene families appear in *A. thaliana*, *P. trichocarpa*, *Glycine max*, and *V. vinifera*, but *not* in *T. cacao?*

Exercise 8.7 How many syntenic groups in the *T. cacao* genome show evidence for triplication of the genome? List the triplets of chromosomes linked by each group.

Problems

Problem 8.1 Jeffreys has recently introduced a newer identification method based on one VNTR locus, D1S8 (or MS32) (see Figure 8.9). This locus contains hundreds of repeats of a 19 bp sequence. One base in the repeat is hypervariable, with the result that each repeat may contain, or lack, a *Hae*III restriction enzyme cut site. The feature identifying any individual's DNA is a binary code specifying the sequence of *presence* or *absence* of *Hae*III restriction sites in successive repeats within the locus. (The sites on homologous chromosomes vary independently, multiplying the variability.)

The identification is read by designing PCR primers: one to a common sequence outside the repeat, and two others to each of the alternative repeat sequences. Amplification using the common primer and one of the others produces a series of fragments: each fragment spans the sequence complementary to the common primer, outside the repeat, to each repeat complementary to the other primer. These fragments can be separated by size on a gel and read off as a bar code. A similar, separate amplification using the other primer can detect fragments that contain the other possible sequence.

Sketch the appearance of a gel containing two lanes, one from each of the amplifications shown in Figure 8.9.

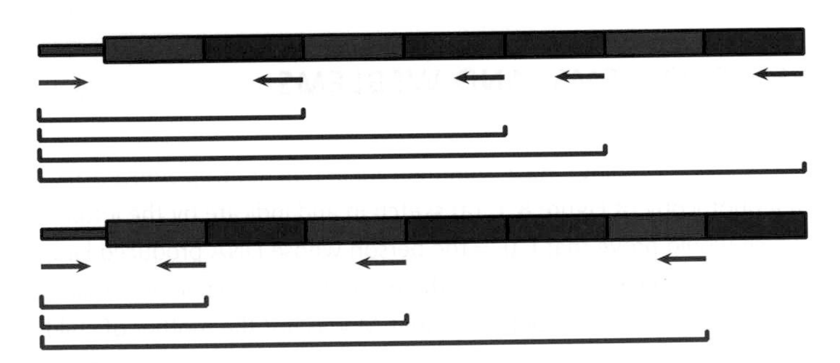

Figure 8.9 Variable repeat of regions containing either of two alternative sequences, one of which contains a cutting site for restriction enzyme *Hae*III. The order of the choices between the alternative repeat sequences provides a 'binary code'. In this diagram, blue represents one repeat sequence and red the other. Top: one PCR amplification is based on primers to one of the alternative sequences (blue arrow) and to a common region flanking the repeat (magenta arrow), which produces fragments starting at each of the occurrences of one of the alternative repeat sequences. The fragments produced are indicated below the arrows. Bottom: a second PCR amplification based on a primer to the other repeat sequence (red arrow) and the common flanking region (magenta arrow) produces fragments starting at each of the occurrences of the other repeat sequence. The fragments produced are indicated below the arrows. (This is a somewhat simplified description of the actual technique.)

Problem 8.2 George Beadle crossed a strain of maize with teosinte and examined the second generation (F_2). Approximately 1/500 plants was identical to the maize grandparent and approximately 1/500 was idential to the teosinte grandparent. Assuming Mendelian inheritance of *n* unlinked genes, such that the teosinte grandparent was homozygous for *n* teosinte alleles and the maize grandparent was homozygous for *n* maize alleles, estimate the value of *n* such that the fraction of F_2 plants genetically identical to the grandparents at all *n* loci is approximately 1/500.

Problem 8.3 Numerous genera of ants cultivate fungi for food. Suggest some similarities and some differences between the domestication of fungi by ants and the domestication of agricultural crops by humans.

Problem 8.4 If humans first entered Alaska from Asia 20 000 years ago and archaeological remains are found in Monte Verde, Chile, dated 14 000 years ago, what mean annual rate of southward migration speed is implied? What two villages or cities in the vicinity of where you live are as far apart as this rate would imply for 10 years of migration?

Problem 8.5 What accounts for the paucity of disease-resistant genes in *T. cacao?* It has been suggested that the genes encoding toll interleukin receptor motif proteins are older than the angiosperm–gymnosperm split, and that they have been lost in lineages including genes in *T. cacao*. (a) What alternative hypothesis is suggested by the observation that *T. cacao* is close to the ancestral eudicot? (b) How might these hypotheses be tested?

Weblems

Weblem 8.1 What is the difference in structure between theobromine and caffeine?

Weblem 8.2 (a) Are there any other species in the genus theobroma? (b) What other genus of plant is most closely related to theobroma?

Weblem 8.3 A region of human mitochondrial DNA sequence, positions 16 047–16 385, was determined to have the following mutations, relative to the Cambridge Reference Sequence (CRS):

	Position (nt)						
Source	**16 147**	**16 172**	**16 189**	**16 223**	**16 248**	**16 320**	**16 355**
CRS	C	T	T	C	C	C	C
Unknown	A	C	C	T	T	T	T

To what mitochondrial haplogroup does the sequence belong?

CHAPTER 9

Microarrays and Transcriptomics

LEARNING GOALS

- To understand the principles underlying microarray technology.
- To be aware of the types of application for which microarrays are suitable.
- To know what a **gene expression table** is and how it is derived from a microarray experiment.
- To be able to distinguish gene-based and sample-based analysis and how they support different types of application.
- To understand how microarrays can be applied to the study of changing gene expression patterns in different physiological states.
- To understand how microarrays can reveal the different time courses of expression of different genes during the yeast diauxic shift.
- To understand the general features of the correlation of gene expression patterns with development in *Drosophila melanogaster*.
- To recognize how microarrays can be used to work out the genes and proteins responsible for specific features of a phenotype.
- To appreciate how microarrays can be used to study the molecular biology of higher mental processes in mammals, including learning and the evolution of human language abilities.

Introduction

Microarrays provide the link between the static genome and the dynamic proteome. We use microarrays: (1) to analyse the mRNAs in a cell, to reveal the expression patterns of proteins; and (2) to detect genomic DNA sequences, to reveal absent or mutated genes.

For an integrated characterization of cellular activity, we want to determine what proteins are present, where and in what amounts.

- The **transcriptome** of a cell is the set of RNA molecules it contains; the **proteome** is its proteins.

We infer protein expression patterns from measurements of the relative amounts of the corresponding mRNAs. Hybridization is an accurate and sensitive way to detect whether any particular nucleic acid sequence is present. Microarrays achieve high-throughput analysis by running many hybridization experiments in parallel (see Box 9.1).

Expression patterns can also help to identify genes that underlie diseases. Some diseases, such as cystic fibrosis, arise from mutations in single genes. For these, isolating a region by genetic mapping can help to pinpoint the lesion. Other diseases, such as asthma, depend on interactions among many genes, with

BOX 9.1 The basic innovation of microarrays is parallel processing

Compare the following types of measurement:

- *'One-to-one'*. To detect whether one oligonucleotide has a particular known sequence, test whether it can hybridize to the oligonucleotide with the complementary sequence.
- *'Many-to-one'*. To detect the presence or absence of a query oligonucleotide in a mixture, spread the mixture out and test each component of the mixture for binding to the oligonucleotide complementary to the query. This is a northern or Southern blot.
- *'Many-to-many'*. To detect the presence or absence of *many* oligonucleotides in a mixture, synthesize a set of oligonucleotides, one complementary to each sequence of the query list, and test each component of the mixture for binding to each member of the set of complementary oligonucleotides. Microarrays provide an efficient, high-throughput way of carrying out these tests in parallel.

To achieve parallel hybridization analysis a large number of DNA oligomers is affixed to known locations on a rigid support, in a regular two-dimensional array. The mixture to be analysed is prepared with fluorescent tags to permit the detection of the hybrids. The array is exposed to the mixture. Some components of the mixture bind to some elements of the array. These elements now show the fluorescent tags. Because we know the sequence of the oligomeric probe in each spot in the array, measurement of the *positions* of the hybridized probes identifies their sequences. This identifies the components present in the sample (see Figure 9.1).

Such a DNA microarray is based on a small wafer of glass or nylon, typically 2 cm^2. Oligonucleotides are attached to the chip in a square array, at densities between 10 000 and 250 000 positions per cm^2. The spot size may be as small as ~150 μm in diameter. The grid is typically a few centimetres across. A **yeast chip** contains over 6000 oligonucleotides, covering all known genes of *Saccharomyces cerevisiae*. A DNA array, or DNA chip, may contain 400 000 probe oligomers. Note that this is larger than the total number of genes, even in higher organisms (excluding immunoglobulin genes). However, the technique requires duplicates and controls, reducing the number of different genes that can be studied simultaneously. Nevertheless, it is possible to buy a single chip containing all known human genes (not all immunoglobulin genes, of course). Also available is a set of 'tiling' chips that cover the entire human genome sequence.

A mixture is analysed by exposing it to the microarray under conditions that promote hybridization, then washing away any unbound oligonucleotides. To compare material from different sources, the samples are tagged with differently coloured fluorophores. Scanning the array collects the data in computer-readable form.

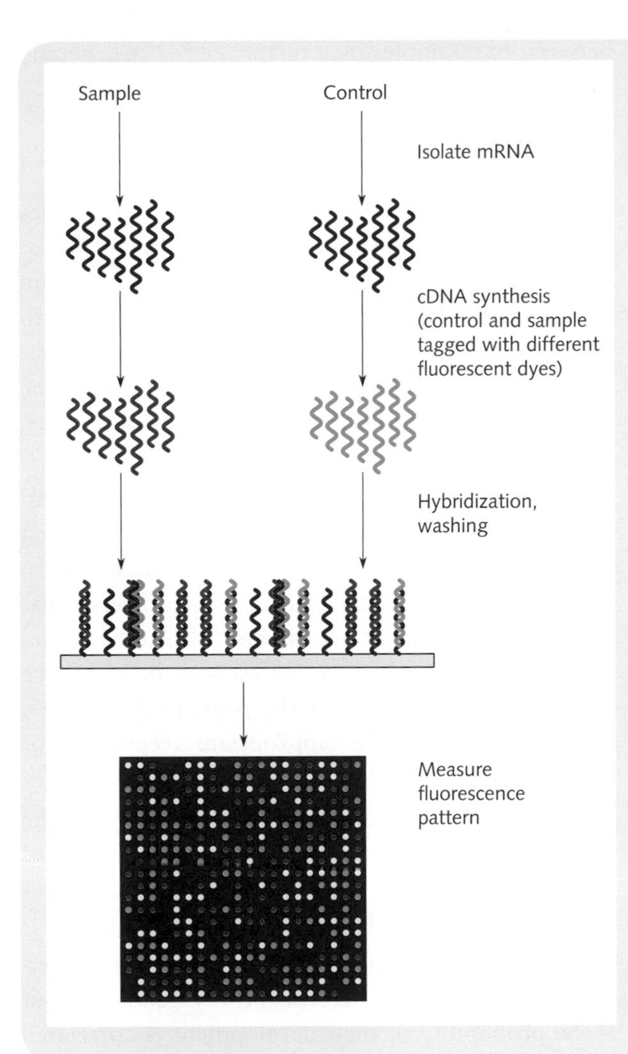

Figure 9.1 Schematic diagram of a microarray experiment. A sample to be tested is compared with a control of known properties. From each source, mRNA is isolated and converted to cDNA, using reagents bearing a fluorescent tag, with different colours for the control and sample. After hybridizing to the microarrays and washing away unbound material, the bound target oligonucleotides appear at specific positions. A red spot indicates binding of oligonucleotides from the sample. A green spot indicates binding of oligonucleotides from the control. A yellow spot indicates binding of both. Each probe, represented here by a wavy black line affixed to the support, really contains *many copies of a single oligonucleotide*. Indeed, for accurate measurement, the concentration of the target must greatly exceed the concentration of the probe. If both red- and green-tagged targets are complementary to the oligonucleotide probe at one spot, both can bind to *different* probe molecules within the same spot.

environmental factors as complications. To understand the aetiology of multifactorial diseases requires the ability to determine and analyse expression patterns of many genes, which may be distributed around different chromosomes.

- The immobilized material on the chip is the **probe**. The sample tested is the **target**.

Microarrays are also used to screen for mutations and polymorphisms. Microarrays containing many sequence variants of a single gene can detect differences from a standard reference sequence.

Different types of chip support different investigations.

- In an **expression chip**, the immobilized oligonucleotides are cDNA samples, typically 20–80 bp long, derived from mRNAs of known genes. This is by far the most common type of microarray. The target sample might contain mRNAs from normal or diseased tissue, for comparison.

  Typically, one position on the chip contains an oligonucleotide with the exact sequence we want to test for and, as a control, another position contains a corresponding **mismatched oligonucleotide**, differing by one base near the centre of the sequence. These form a **probe pair**. To detect a single mRNA, a chip may contain 16–20 probe pairs, spread over the mRNA sequence.

- In **genomic hybridization**, one looks for gains or losses of genes, or changes in copy number.

The probe sequences, fixed on the chip, are large pieces of genomic DNA from known chromosomal locations, typically 500–5000 bp long. The target mixtures contain genomic DNA from normal or disease states. For instance, some types of cancer arise from chromosome deletions, which can be identified by microarrays.

- **Mutation or polymorphism microarray analysis** is the search for patterns of single-nucleotide polymorphisms (SNPs). The oligonucleotides on the chip are selected from reference genomic data. They correspond to many known variants of individual genes.
- **Protein microarrays** are arrays of protein detectors – usually antibodies – that detect protein–protein interactions.
- **Tissue microarrays** collect and assemble microscopic samples of tissue. They permit comparative analysis of the molecular biology and immunohistochemistry of the samples.

> - DNA microarrays analyse the RNAs of a cell, to reveal expression patterns of proteins and of non-protein-coding RNAs; or genomic DNAs, to reveal absent or mutant genes. One of the reasons that microarrays are such a versatile technology is that there are many different kinds of chips. Commercially available chips can target all or at least most known genes for individual species, for example, or even tile an entire genome.

Microarray data are semiquantitative

Microarrays are capable of comparing concentrations of target oligonucleotides. This allows investigation of responses to changed conditions. Unfortunately, the precision is low. Moreover, mRNA levels, detected by the array, do not always quantitatively reflect protein levels. Indeed, usually mRNAs are reverse transcribed into more stable cDNAs for microarray analysis; the yields in this step may also be non-uniform. Microarray data are, therefore, semiquantitative: although the distinction between presence and absence is possible, determination of relative levels of expression in a controlled experiment is more difficult, and measurement of absolute expression levels is beyond the capability of current microarray techniques. A change in expression levels of a gene between two samples by a factor $\geq$1.5–2 is generally considered a significant difference.

Applications of DNA microarrays

- *Investigating cellular states and processes.* Profiles of gene expression that change with cellular state or growth conditions can give clues to the mechanism of sporulation, or to the change from aerobic to anaerobic metabolism. In higher organisms, variations in expression patterns among different tissues, or different physiological or developmental states, illuminate the underlying biological processes.
- *Comparison of related species.* The very great similarity in genome sequence between humans and chimpanzees suggested that the profound phenotypic differences must arise at the level of regulation and patterns of protein and RNA expression, rather than in the few differences between the amino acid sequences of the proteins themselves. Microarrays are the appropriate technique for following up this idea.
- *Diagnosis of genetic disease.* Testing for the presence of mutations can confirm the diagnosis of a suspected genetic disease. Detection of carriers can help in counselling prospective parents.
- *Genetic warning signs.* Some diseases are not determined entirely and irrevocably by genotype, but the probability of their development is correlated with genes or their expression patterns. Microarray profiling can warn of enhanced risk.
- *Precise diagnosis of disease.* Different related types of leukaemia can be distinguished by signature patterns of gene expression. Knowing the exact type of the disease is important for prognosis and for selecting optimal treatment.
- *Drug selection.* Genetic factors can be detected that govern responses to drugs, which in some patients render treatment ineffective and in others cause unusual or serious adverse reactions.
- *Determination of gene function.* A gene with an expression pattern similar to genes in a metabolic pathway is also likely to participate in the pathway.
- *Target selection for drug design.* Proteins showing enhanced transcription in particular disease states might be candidates for attempts at pharmacological intervention.

- *Pathogen resistance.* Comparisons of genotypes or expression patterns between bacterial strains susceptible and resistant to an antibiotic point to the possible involvement of the proteins in the mechanism of resistance.
- *Following temporal variations in protein expression.* This permits timing the course of (1) responses to pathogen infection, (2) responses to environmental change, (3) changes during the cell cycle, and (4) developmental shifts in expression patterns.

Analysis of microarray data

The raw data of a microarray experiment is an image in which the colour and intensity of the fluorescence reflect the extent of hybridization to alternative probes (see Figure 9.1). The two sets of targets are tagged with red and green fluorophores. If only one target hybridizes, the spot appears green; if only the other target hybridizes, the spot appears red. If both hybridize, the colour of the corresponding spot appears yellow.

Extraction of reliable biological information from a microarray experiment is not straightforward. Despite extensive internal controls, there is considerable noise in the experimental technique. In many cases, variability is inherent within the samples themselves. Microorganisms can be cloned; animals can be inbred to a comparable degree of homogeneity. However, experiments using RNA from human sources – for example, a set of patients suffering from a disease and a corresponding set of healthy controls – are at the mercy of the large individual variations that unrelated humans present. Indeed, inbred animals, and even apparently identical eukaryotic tissue-culture samples, show extensive variability.

Data reduction involves many technical details of image processing, checking of internal controls, dealing with missing data, selecting reliable measurements, and putting the results of different arrays on consistent scales. There is extensive redundancy in a microarray – each sequence may be represented by several spots, and in addition to straight duplicates, they may correspond to different regions of a gene. Probe pairs – one perfectly matching oligonucleotide and the other containing a deliberate mismatch – allow data verification. Different oligonucleotides cover different segments of each region of interest in the sequence. Typically, one gene may correspond to ~30–40 spots.

The initial goal of data processing is a **gene expression table.** This is a matrix containing *relative* expression levels, *derived* from the raw data. The rows of the matrix correspond to different genes and the columns to different sources of material. Of course, the gene expression table is not a simple 'replica plate' of the microarray itself. The microarray fluorescence pattern contains the raw data from which the gene expression table must be extracted. Data from many spots on the microarray will contribute to the calculation of the relative expression level of each gene.

A typical experiment *compares* expression patterns in material from two sources – perhaps a control of known properties and a sample to be tested. We may wish to compare organisms growing under different experimental conditions and/or physiological states, or DNA from different individuals or different tissues, or a series of developmental stages.

Two general approaches to the analysis of a gene expression matrix involve (1) comparisons focused on the genes, i.e. comparing distributions of expression patterns of different genes by comparing *rows* in the expression matrix; or (2) comparisons focused on samples, i.e. comparing expression profiles of different samples by comparing *columns* of the expression matrix.

- *Comparisons focused on genes: how do gene expression patterns vary among the different samples?* Suppose a gene is known to be involved in a disease, or linked to a change in physiological state in response to changed conditions. Other genes co-expressed with the known gene may participate in related processes contributing to the disease or change in state. More generally, if two rows (two genes) of the gene expression matrix show similar expression patterns across the samples, this suggests a common pattern of regulation and some relationship between their functions, possibly including (but not limited to) a direct physical interaction.

- *Comparisons focused on samples: how do samples differ in their gene expression patterns?* A consistent set of differences among the samples may distinguish and characterize the classes from which the samples originate. If the samples are from different controlled sources (for instance, diseased and healthy animals), do samples from different groups show consistently different expression patterns? If so, given a novel sample, we could assign it to its proper class on the basis of its observed gene expression pattern.

How can we measure the similarity of different rows or columns? Each row or column of the expression matrix can be considered as a vector in a space of many dimensions. In the row vectors, or **gene vectors** (a row corresponds to a gene), each position refers to the same gene in different samples. The gene vector has as many elements as there are samples. In the column vectors, or **sample vectors** (a column corresponds to a sample), each entry refers to a different gene in a single sample. The sample vector has as many elements as there are genes reported. It is possible to calculate the 'angle' between different gene vectors, or between different sample vectors, to provide a measure of their similarities. The smaller the angle, the more similar the pattern.

The gene vectors and sample vectors correspond, separately, to points in spaces of many dimensions. The number of dimensions is either the number of genes or the number of samples. We may not be able to visualize easily points in a space with more than three dimensions, but all of our intuition about geometry works fine. For instance, it is natural to ask whether subsets of the points form natural clusters – points with high mutual similarity – characterizing either sets of genes or sets of samples.

We have already encountered clustering in relation to phylogenetic trees. Similarly, in analysis of gene expression arrays, after finding clusters we can bring similar genes and samples together. This amounts to reordering the rows and columns of the gene expression matrix. The results are often displayed as a chart, coloured according to the difference in expression pattern. Figure 9.2 contains an example.

Depending on the origin of the samples, what is already known about them, and what we want to learn, data analysis can proceed in different directions.

1. The simplest case is a carefully controlled study, using two different sets of samples of *known characteristics*. For instance, the samples might be taken from bacteria grown in the presence or absence of a drug, from juvenile or adult fruit flies, or from healthy humans and patients with a disease. We can focus on the question: what differences in gene expression pattern characterize the two states? Can we design a classification rule such that, given another sample, we can assign it to its proper class? This would be useful in diagnosis of disease. Subject to the availability of adequate data, such an approach can be extended to systems of more than two classes.

   In computer science, training such a classification algorithm is called 'supervised learning'. The expression pattern of each sample is given by a vector corresponding to a single column of the matrix. This corresponds to a point in a many-dimensional space – as many dimensions as there are genes. In favourable cases, the points may fall in separated regions of space. Then a scientist, or a computer program, will be able to draw a boundary between them. In other cases, separation of classes may be more difficult.

2. In a different experimental situation, we might not be able to *pre-assign* different samples to different categories. Instead, we hope to extract the classification of samples from the analysis. The goal is to cluster the data to *identify* classes of samples and then to investigate the differences among the genes that characterize them.

An intrinsic problem – and a severe one – in interpreting gene expression data is the fact that the number of genes is much larger than the number of samples. We are trying to understand the relationship of one space of very many variables (the genes) to another one (the phenotype) from only a few measured points (the samples). The sparsity of the observations does not give us anywhere near adequate coverage. Statistical methods bear a heavy burden in the analysis to give us confidence in the significance of our conclusions.

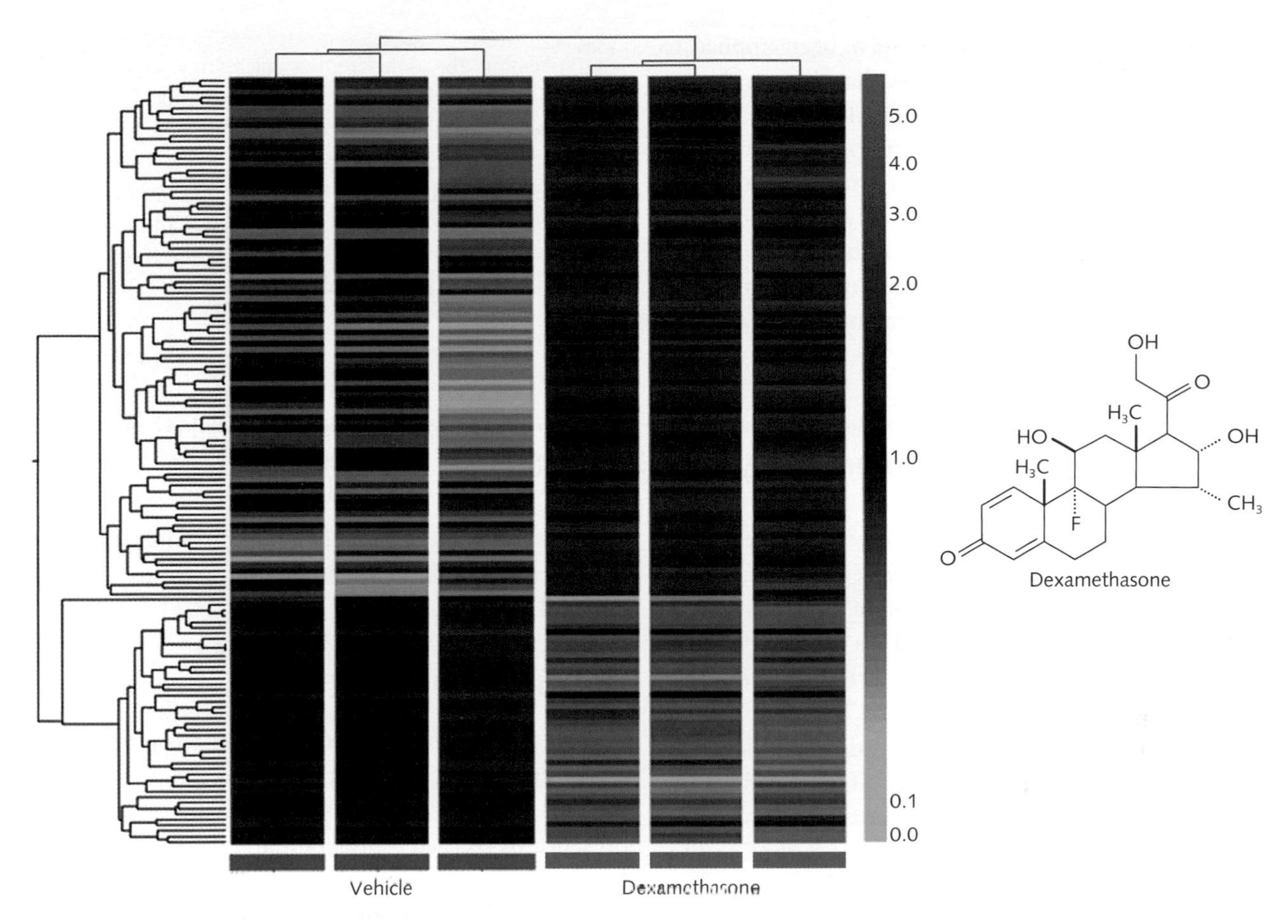

Figure 9.2 Glucocorticoids are regulators of immune system physiology, widely used as anti-inflammatory drugs, for instance in asthma and arthritis. However, cataract formation is a common side effect of prolonged use.

Glucocorticoids are well known to affect gene expression patterns. The steroid binds to a glucocorticoid receptor bound on a cell surface. Like other steroid hormone receptors, the glucocorticoid receptor is a zinc-finger transcription factor. The ligated receptor dimerizes and translocates to the nucleus where it binds a DNA sequence called the glucocorticoid response element, activating other transcription factors to modulate expression of target genes.

Gupta and co-workers studied the effect of dexamethasone, a hydrocortisone analogue, on the protein expression pattern in cultured epithelial cells of the human lens.* Three samples were treated with 1 μM dexamethasone for 4 hours. Three corresponding controls were given only 'vehicle'. (In pharmacology, a **vehicle** is the medium of delivery of the drug.) RNA was extracted, purified, and converted to cDNA. The microarray used was an Affymetrix chip that contained 22 283 known human transcripts and expressed sequence tags (ESTs) for about 15 000 genes.

The figure shows data for these six samples. Levels of expression enhanced or reduced by >1.5-fold were considered significant. Data in red correspond to genes upregulated by dexamethasone; data in green correspond to downregulated genes. The scale on the right relates colour to expression change. The results identified 93 upregulated transcripts and 43 downregulated ones.

In the figure, the data are clustered based on overall expression level (gene vector, according to rows) and *also* based on expression on different chips (sample vector, according to columns). Note that the clusterings by gene and by sample are independent: it would be possible to change the arrangement of the columns without altering the arrangement of the rows and vice versa.

The trees at the top and at the left indicate the similarities among the results, according to sample vector and gene vector, respectively. The sample-vector tree at the top cleanly separates the control triple replicates (vehicle) and the dexamethasone-treated triple replicates. For both control and dexamethasone-treated replicates, the observed expression changes appeared in at least two of the three measurements. The gene-vector tree at the left has a major breakpoint between downregulated and upregulated genes (for one gene the behaviour of different replicates is inconsistent).

To discover the implications of these data for cataract induction, the next step would be to examine the biological functions of the glucocorticoid-sensitive genes. Modern bioinformatics software makes the transition to this computation a facile one. In this case, Gupta et al. found that the modulated genes involved a wide range of functions. It appears that glucocorticoid elicits a network of responses that must be pursued downstream to make direct contact with the mechanism of cataract formation.

* Gupta, V., Galante, A., Soteropoulos, P., Guo, S., & Wagner, B.J. (2005). Global gene profiling reveals novel glucocorticoid induced changes in gene expression of human lens epithelial cells. Mol. Vis. **11**, 1018–1040.

Many clustering algorithms have been applied to microarray data, including those that try to work out simultaneously *both* the number of clusters and the boundaries between them. All algorithms must face the difficulty arising from the sparsity of sampling. Sometimes it is possible to simplify the problem by identifying a small number of combinations of genes that account for a large portion of the variability. This is called **reduction of dimensionality**.

- Processing the data from a microarray experiment produces a gene expression table, or matrix. The rows index the genes and the columns index the samples. We can either focus on the genes, and ask: how do patterns of expression of different genes vary among the different samples? Or we can focus on the samples, and ask: how do the samples differ in their gene expression patterns?

Expression patterns in different physiological states

A fundamental question in biology is how different components of cells smoothly integrate their activities. Measurements of expression patterns tell us part of the story. They provide an inventory of the components but suggest only inferentially how they interact.

Comparisons of alternative physiological states of an organism offer the possibility of extracting, from an entire genome, a subset of genes that underlie a particular life process. An example of shift in physiological state in microorganisms is **diauxy**. Diauxy, or double growth, is the switch in metabolic state of a microorganism when, having exhausted a preferred nutrient, it 'retools' itself for growth on an alternative. The organism may show a biphasic growth curve, with a lag period while the changed complement of proteins is synthesized.

- Jacques Monod discovered diauxy approximately 70 years ago, during his predoctoral work in Paris. He described his observations in his 1941 thesis. Resuming his research career after the war, he and his colleagues at the *Institut Pasteur* made their fundamental discoveries about the mechanism of gene regulation.

The diauxic shift in yeast is the transition from fermentative to oxidative metabolism upon exhaustion of glucose as an energy source. In this chapter, we shall examine the *effects* of the reconfiguration of the expression patterns, comparing the different complements of genes active in the two states. In Chapter 11, we shall return to the yeast diauxic shift to examine the control mechanisms that regulate the transcriptional reprogramming.

As an example from higher organisms, we shall look at changes in gene expression patterns in sleep and wakefulness.

The diauxic shift in *Saccharomyces cerevisiae*

Yeast is capable of adapting its metabolism to a variety of environmental conditions. In the presence of glucose, *Saccharomyces cerevisiae* will – even in the presence of oxygen – preferentially use the Embden–Meyerhof fermentative pathway, reducing glucose to ethanol. Exhaustion of available glucose produces the 'diauxic shift' to oxidative metabolism, sending the products of fermentation through the Krebs (tricarboxylic acid) cycle and mitochondrial oxidative phosphorylation (see Figure 9.3).

The shift is not merely a redirection of metabolic flux through alternative pathways using pre-existing proteins but involves protein synthesis, with a substantial change in expression pattern. Many genes are involved, not only enzymes in the awakened metabolic pathways. Another consequence of switching to respiratory metabolism is the danger of oxidative damage, and the oxidative stress response requires enhanced expression of many other genes.

- Oxygen is essential for aerobic life, yet its reduced forms include some of the most toxic substances with which cells must cope.

The cells are effectively sensing the level of glucose. As glucose itself acts as a repressor of expression of many genes, depletion of glucose releases this repression – 'turning on' a variety of genes.

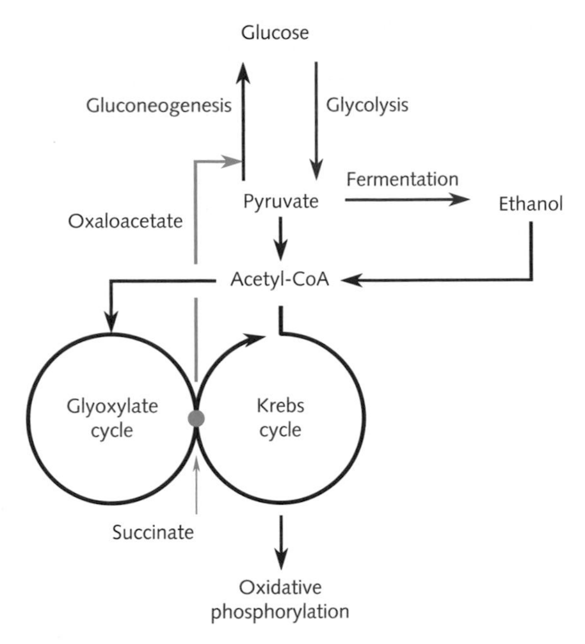

Figure 9.3 Some metabolic pathways in yeast affected by the diauxic shift.

In the presence of ample glucose, yeast will adopt an anaerobic metabolic regimen, converting glucose to ethanol via glycolysis and fermentation (Embden–Meyerhof pathway; red arrows). Upon running out of glucose, it will shift to an aerobic metabolic state in which ethanol is converted to CO_2 and H_2O via the Krebs cycle and oxidative phosphorylation (blue arrows). For the alternative pathways of energy release, pyruvate is the branch compound.

However, the shift to a utilization of a different energy source creates two concomitant problems.

1. The oxidation of ethanol does not provide precursors for essential biosynthetic pathways. Oxidation of ethanol converts all carbon to CO_2. Some must be retained and converted to three- and four-carbon compounds, and even glucose, via the glyoxylate cycle and gluconeogenesis. Acetyl-CoA is the branch compound for this shift – it enters *both* the Krebs cycle and the glyoxylate cycle.

   The glyoxylate cycle shares several intermediates with the Krebs cycle but, to conserve carbon, leaves out the decarboxylations that produce CO_2. Succinate (green dot), one of the metabolites common to the Krebs and glyoxylate cycles, is converted to oxaloacetate in mitochondria. It then feeds into numerous biosynthetic pathways (green arrow).
2. In addition to the metabolic pathways shown in this figure, yeast must also activate pathways of defence against oxidative stress. The danger is from potential chemical attack by **reactive oxygen species**, produced by partial reduction of oxygen: the superoxide radical ($O_2^{-\bullet}$), hydrogen peroxide (H_2O_2), and the hydroxyl radical ($OH^\bullet$). In its defence, yeast increases expression of genes involved in detoxification of reactive oxygen species.

   The mechanism of induction of these genes depends on the transcription factor Yap1p. An increase in the concentration of reactive oxygen species leads to formation of disulphide bridges in Yap1p. This causes a conformational change that masks a nuclear export signal. The result is redistribution of the transcription factor to the nucleus, the site of its activity.

An important component of the mechanism by which high glucose levels lead to repression is the state of phosphorylation of a transcriptional regulatory protein called Mig1. The phosphorylated form of Mig1 stays in the cytoplasm. Upon dephosphorylation, it enters the nucleus and binds to promotors of various genes, repressing their expression. The activity of kinase Snf1–Snf4, which phosphorylates Mig1, depends on the AMP/ATP concentration ratio in the cell. Growth on high levels of glucose generates high levels of ATP via glycolysis, keeping the Snf1–Snf4 kinase inactive. This leaves Mig1 in its active, dephosphorylated form, repressing glucose-sensitive genes.

In a seminal paper in 1977, DeRisi, Iyer, and Brown reported on the changes in gene expression in the diauxic shift in *S. cerevisiae*.* They observed that the patterns of gene expression were stable during exponential growth in a glucose-rich medium. This justifies considering the initial anaerobically growing population as being in a **state** that can be characterized by a set of expression levels of genes. (If, alternatively, there were fluctuations in expression levels that were large compared with the differences between anaerobic and aerobic regimes, it would not be possible to analyse the diauxic shift in terms of expression patterns. In fact, if there are fluctuations, they average out over the population.)

As glucose is depleted, the expression pattern changes (see Figure 9.4, from work by Brauer and co-workers). The changes affect a large fraction, almost 30%, of the genes (Table 9.1).

The genes differentially expressed are associated with proteins in several functional classes. Upon converting from anaerobic to aerobic metabolism several things occur.

- *Metabolic pathways are rerouted.* Synthesis of pyruvate decarboxylase is shut down; synthesis of

Table 9.1 Genes affected when glucose is depleted

Expression ratio in aerobic/anaerobic states	>2	<$\frac{1}{2}$	>4	<$\frac{1}{4}$
Number of genes	710	1030	183	203

* DeRisi, J.L., Iyer, V.R., & Brown, P.O. (1997). Exploring the metabolic and genetic control of gene expression on a genomic scale. Science **278**, 680–686.

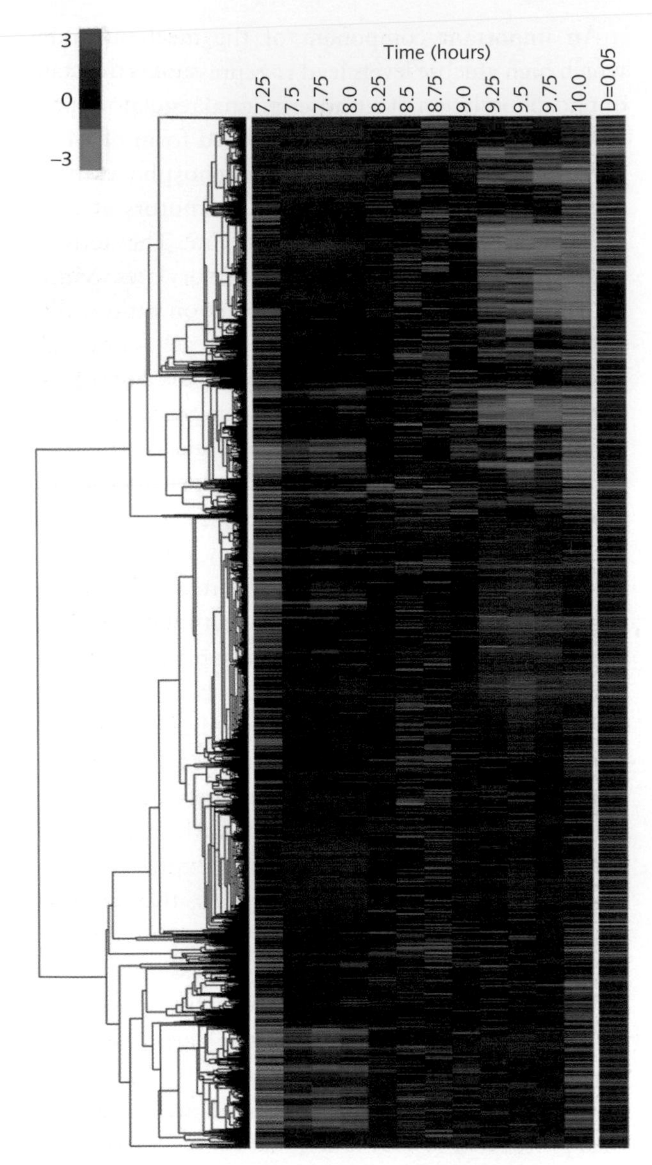

Figure 9.4 Expression patterns of yeast growing initially on glucose, showing the yeast undergoing a diauxic shift. Column headings indicate time in hours after exhaustion of glucose in the samples undergoing diauxic shift.

From: Brauer, M.J., Saldanha, A.J., Dolinski, K., & Botstein, D. (2005). Homeostatic adjustment and metabolic remodeling in glucose-limited yeast cultures. Mol. Biol. Cell **16**, 2503–2517.

pyruvate carboxylase is enhanced. This switches the product of the reaction of pyruvate from acetaldehyde to oxaloacetate. Enhanced synthesis of genes for fructose-1,6-bisphosphatase and phosphoenolpyruvate carboxylase changes the direction of two steps in glycolysis. Expression is enhanced for genes encoding the enzymes that carry out the new Krebs and glyoxylate cycle reactions, and oxidative phosphorylation (blue arrows in Figure 9.3).

- *Genes related to protein synthesis show a decrease in expression level.* These include genes for ribosomal proteins, tRNA synthetases, and initiation and elongation factors. An exception is that genes encoding *mitochondrial* ribosomal molecules have generally enhanced expression.
- *Genes related to a number of biosynthetic pathways show reduced expression.* These include genes encoding enzymes of amino acid and nucleotide metabolism.
- *Genes involved in defence against oxidative stress from reactive oxygen species show enhanced expression.* Proteins encoded include catalases, peroxidases, superoxide dismutases, and glutathione *S*-transferases.

- The diauxic shift in yeast is the transition from fermentative to oxidative metabolism when the yeast runs out of glucose in the medium. The shift is effected by a retooling in which the expression patterns of relevant genes are altered.

Sleep in rats and fruit flies

All humans sleep. The overt characteristics of sleep are familiar: approximately cyclic periods of reduced consciousness, relaxation, and quiescence; a raised arousal threshold; and dreaming (highly correlated to periods of rapid eye movements or REM). Neurophysiologists distinguish different stages of sleep, with different characteristic patterns in electroencephalograms. The consequences of sleep deprivation are also familiar: reduced vigilance and performance and general stroppiness. These demonstrate that sleep has a necessary restorative function.

Other animals sleep. Even *Caenorhabditis elegans* can enter a state of torpor, akin to sleep. Fruit flies sleep, and their sleep shares many features of ours. Like us, they sleep deeper and longer after sleep deprivation. Caffeine keeps them awake. Their sleep correlates with changes in brain electrical activity. We would not expect fruit flies to show all of the higher neurophysiological correlates of sleep, such as dreaming, but it is not unreasonable to hope to find some analogues at the biochemical level.

It is always useful to study a biological phenomenon in the simplest organism that exhibits it, as well as in humans. Comparisons of flies and mammals may reveal fundamental common features disguised within the individual complexities of different species. The results may also illuminate the functions associated with homologous genes. This is not to deny that, in the realm of cognitive phenomena, humans have pushed things farther than other species and present unique features.

Sleep disorders present common and serious medical problems. Sleep deprivation is a major cause of accidents, leading to loss of life and damage to health, property, and productivity. Indeed in rats and flies, prolonged sleep deprivation is itself fatal, after slightly more than 2 weeks.

C. Cirelli and co-workers studied gene expression patterns of rats and fruit flies in three physiological states: spontaneously awake, spontaneously asleep, and sleep deprived. Protocols were similar but not identical in the experiments on the two species.*

Flies were prepared by accustoming them to an alternating regimen: 12 hours with the lights on, awake (8 a.m. to 8 p.m.) and 12 hours in the dark, asleep (8 p.m. to 8 a.m.). A group of flies was then sleep-deprived for 8 hours after the normal end of the waking period, i.e. from 8 p.m. to 4 a.m.

A complication in studying the effect of sleep and wakefulness arises from the circadian rhythms that the expression patterns of many genes are known to obey. The use of the sleep-deprived animals allowed the effects of waking and sleep states to be distinguished from time-of-day effects. To this end, samples of spontaneously asleep and sleep-deprived flies were collected at 4 a.m. Samples from spontaneously awake flies were collected at 4 p.m. (Table 9.2).

Table 9.2 Sleep and wakefulness states of flies

State	Time of sample collection	
	4 p.m.	4 a.m.
Awake	Spontaneously awake	Sleep deprived
Asleep		Spontaneously asleep

* Cirelli, C., LaVaute, T.M., & Tononi, G. (2005). Sleep and wakefulness modulate gene expression in *Drosophila*. J. Neurochem. **94**, 1411–1419; C. Cirelli (2005). A molecular window on sleep: changes in gene expression between sleep and wakefulness. Neuroscientist **11**, 63–74.

The logic is that if the expression pattern of a gene differs between the sleep-deprived and spontaneously awake flies, both in a waking state, the controlling factor must be time of day.

A gene was classified as *sleep related* if its expression was elevated by a factor >1.5 in spontaneously asleep flies relative to *both* spontaneously awake *and* sleep-deprived flies. A gene was classified as *wakefulness related* if its expression was elevated by a factor >1.5 in *both* spontaneously awake *and* sleep-deprived flies relative to spontaneously asleep flies. Some genes showed the influence of *both* physiological state and time of day.

Cirelli and co-workers studied expression patterns of ~10 000 genes of the fly. Of these, 121 were wakefulness related and 12 were sleep related. The expression of a partly overlapping set of 130 genes was moderated by time of day: 87 were more highly expressed at 4 p.m. and 43 were more highly expressed at 4 a.m.

The overlap of sleep/wakefulness-related genes with those modulated by time of day demonstrated a relationship between homeostatic and circadian regulation. Two-thirds of sleep-related genes and one-fifth of wakefulness-related genes were modulated by time of day. This is consistent with the observation that flies with mutations in circadian genes can show abnormal homeostatic regulation of sleep.

In both flies and rats, the genes preferentially expressed in wakefulness and sleep fell into several *different* functional categories.

In flies, genes preferentially expressed in waking states included those encoding proteins involved in detoxification, including cytochrome P450s and glutathione *S*-transferases; genes involved in defence against immune challenge and in lipid, carbohydrate, and protein metabolism; a transcription factor; a nuclear receptor; and the circadian gene *cryptochrome*. Genes preferentially expressed in sleep included the glial gene *anachronism*, the gene encoding the catalytic subunit of glutamate–cysteine ligase, and other genes involved in lipid metabolism.

In rat brains, a much larger number of genes than in flies had raised expression levels during wakefulness. (For rats, a less strict criterion of significant change in expression level was applied: a ratio of >1.2 compared with a ratio of >1.5 in flies.) In contrast to flies, which show fewer sleep-related than

Table 9.3 Enhanced expression of genes in the rat in sleep and wakefulness

	Wakefulness related	Sleep related
Learning and memory	Synaptic plasticity: acquisition, potentiation	Synaptic plasticity: consolidation, depression
Transport	–	Membrane trafficking and maintenance
Metabolism	Energy metabolism Transcription (positive regulation) Translation (negative regulation) Translation	Cholesterol biosynthesis Transcription (negative regulation) Translation (positive regulation) Translation
Stress response	General stress response and unfolded protein response	
Cell signalling	Depolarization-sensitive Glutamatergic neurotransmission	Hyperpolarization-promoting (leakage) GABAergic neurotransmission

From: Cirelli, C. (2005). A molecular window on sleep: changes in gene expression between sleep and wakefulness. Neuroscientist **11**, 63–74.

wakefulness-related genes, in the rat approximately the same number of genes showed enhanced expression in sleep as in wakefulness, as shown in Table 9.3. The genes with enhanced expression are associated with different biological functions.

In the rat, wakefulness-related genes are associated with memory acquisition, energy metabolism, transcription activation, cellular stress, and excitatory neurotransmission. Sleep-related genes are associated with a potassium channel, translation machinery, long-term memory, and membrane trafficking and maintenance, including synthesis and transport of glia-derived cholesterol. Cholesterol is a major component of myelin and other membranes. (Membrane maintenance suggests an analogue, at the molecular level, of Shakespeare's metaphor that sleep '. . . knits up the ravelled sleeve of care . . .'.)

Similarities between rats and flies in the functional categories of sleep- and wakefulness-associated genes are interesting. Wakefulness-associated genes in both species include those for members of the Egr family of transcription factors, the mammalian NGF1-B nuclear receptor and the fly orthologue, homologous MAP kinase phosphatases, and for proteins involved in cholesterol synthesis.

Another window on to the molecular biology of sleep is the effect of mutation. A fruit fly mutant, Shaker, sleeps approximately one-third as long as wild-type flies. The gene involved encodes a potassium channel that affects neural electrical activity. Some humans can get by regularly on only 3–4 hours' sleep per 24 hour interval, but it is not known whether this trait is under the control of the homologous gene. However, what does suggest a link to the fly mutant is a rare human disorder, Morvan's syndrome, the symptoms of which include insomnia. At least one case of Morvan's syndrome has appeared to be of immune origin, involving an autoantibody against a potassium channel.

- If the physiology of sleep is not well understood, the molecular biology of sleep is even more obscure. Measurements of changes in gene expression during sleep and wakefulness give clues as to what distinguishes the states, at the molecular level.

Expression pattern changes in development

Variation of expression patterns during the life cycle of *Drosophila melanogaster*

During their lifetime, insects undergo macroscopic changes in body plan that are more profound than any post-embryonic development in humans and other mammals, and even in amphibians. This allows juvenile and adult flies to occupy different ecological niches. The major stages of a fly's life are embryonic, larval, pupal, and adult. Metamorphosis occurs during the pupal stage: flies spend their 'adolescence' sequestered within a pupa (to the envy of many a parent of a human teenager).

Fly development has been intensively studied at the molecular level. An impressive understanding has been achieved of the mechanism of translation of molecular signals into macroscopic anatomy. The genesis of specific organs – eyes, legs – has been carefully analysed. We have already encountered *HOX* genes and their relationship to body plan.

Arbeitman and co-workers examined changes in transcription patterns in *Drosophila melanogaster* during different stages of its life. When they took up the problem, it was known from earlier work that large-scale changes in gene expression occurred. Microarrays made possible a more systematic and thorough study.

cDNAs containing representatives of 4028 genes (about one-third of the total estimated number in *D. melanogaster*) revealed expression patterns for 66 selected time periods from embryo through to adulthood (see Figure 9.5). Expression levels were compared with pooled mRNA from all life stages, to represent a (weighted) average expression level. The interval between measurements varied from 1 hour (for embryos) up to several days (for adults) until a total age after fertilized egg of up to 40 days.

Stage	Approximate interval between measurements
Embryo	1 hour
Larva	5 hours
Pupa	8 hours
Adult	3–4 days

Most of the genes tested – 3483 out of 4028, or 86% – changed expression levels significantly at *some* stage(s) of life. Of these, 3219 varied by a factor of >4 between their maximum and minimum values.

The data show that in *Drosophila*, as in other species, genes participating in a common process often exhibit parallel expression patterns and similar perturbations of these patterns in mutants. For instance, the expression pattern of the *eyes absent* mutant, which produces an eyeless or at least a reduced eye phenotype, forms a cluster, in the analysis of expression patterns, with 33 genes. Of these, 11 are

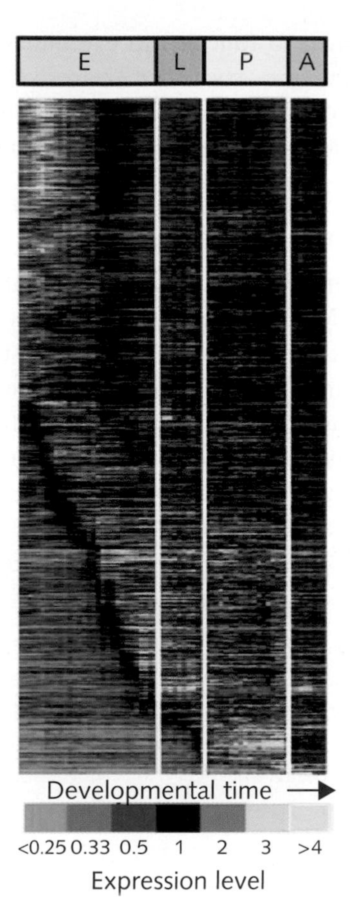

Figure 9.5 Gene expression profiles at different life stages of *D. melanogaster*, ordered by the time of the first rise in transcription levels. E, embryo; L, larva; P, pupa; A, adult. The scale of expression level, relative to that of pooled mRNA samples from all developmental stages, is shown by intensity of colour: black, small change in expression level; dark blue→light blue, increasing downregulation relative to the control; dark yellow→light yellow, increasing upregulation relative to the control. Because of the variation in measurement intervals, the developmental time scale governing the horizontal axis does not correspond to calendar time. The embryonic stage lasts ~1 day, the larval stage ~4 days, the pupal stage 5 days, and the adult stage until cessation of data collection, 30 days.

Looking at the distribution of yellow, it can be seen that some genes are expressed at high levels at single specific stages, but others are expressed at high levels at more than one stage. A gene expressed throughout the life of the fly, at no less than the level of the pooled sample, would appear here as a row containing only black and yellow regions, with no blue.

From: Arbeitman, M.N., et al. (2002). Gene expression during the life cycle of *Drosophila melanogaster*. Science **297**, 2270–2275.

already known to function in eye differentiation or phototransduction. The other 22 are likely to as well; the data provide at least hypotheses, and at best reliable clues, to the function of these genes.

Different life stages make different demands on different genes

Different genes exhibit different temporal patterns of expression. Most of the developmentally modulated genes are expressed in the embryonic stage, as the whole system is getting started. Genes expressed in the early embryo include transcription factors, proteins involved in signalling and signal transduction, cell-adhesion molecules, channel and transport proteins, and biosynthetic enzymes. A third of these are maternally deposited genes; many of these fall off in expression level within 6–7 hours.

- Stathopoulos and Levine* comment that '. . . the genesis of a complex organism from a fertilized egg is the most elaborate process known in biology, and thereby depends on "every trick in the book"'.

The genes studied include one large stage-specific class (36.3%) that shows a single major peak of expression. Some of these remain constitutively expressed (at lower levels) subsequently. Others show sharp peaks in expression level.

Another group of genes (40.3%) shows two peaks in expression. Two patterns are common: genes with their first onset of enhanced expression early in embryogenesis generally have their second at pupation, with elevated expression levels continuing into the pupal stage. Genes with their first onset of enhanced expression late in embryogenesis generally have their second at the late pupal stage, with elevated expression levels continuing into the adult stage. The remaining 23.4% of genes show multiple peaks in expression level.

The observation of similarities between expression patterns in embryonic stages and pupal stages, and between larval and adult stages, is interesting. Certain analogies suggest themselves. In both embryonic and pupal stages, body structures are forming and physical activity is minimal. In contrast, larval and adult stages have a more active lifestyle but stasis of anatomical form, although larvae but not adults grow substantially in size. Consistent with the notion of a 'go back and get it right this time' aspect to metamorphosis is the occurrence of some dedifferentiation in the pupa.

Ideas of this sort can be examined in light of the nature of genes that show different lifelong expression patterns. For example, maximal expression levels of most metabolic genes occur during larval and adult stages. Another set of genes involved in larval and adult muscle development has a similar two-peak expression pattern. More precise analysis is possible and shows that steps in the regulatory hierarchy for muscle development show peaks at different times, with genes expressed later being downstream in the regulatory hierarchy. Similar time-of-onset sequences appear in both larvae and pupae. The two stages at which body plans are formed – embryo and pupa – re-utilize not only the same materials but also some of the same mechanisms.

- Measurement of expression patterns at different stages reveal which batteries of genes are active in development. Particularly striking, in *Drosophila*, is the alternation of time-of-onset of the expression patterns of some genes: embryo and pupa show similarities, and larva and adult.

Flower formation in roses

Roses have long been favourites of gardeners and lovers: for their appearance, for their scent, and, perhaps less commonly, for their purported medicinal qualities – rose hips (fruits) are rich in vitamin C. Breeders have responded by creating tens of thousands of varieties.

- Roses have been known in Europe since antiquity and appear frequently in literature, famously in works by Dante and Shakespeare. Roses symbolized the two armies fighting for control of England at Bosworth in 1485.

Roses have been cultivated for 5000 years. Rose varieties brought to Europe from China at the end of the 18th century introduced the very desirable property of recurrent blooming throughout the season,

* Stathopoulos, A. & Levine, M. (2002). Whole-genome expression profiles identify gene batteries in *Drosophila*. Dev. Cell **3**, 464–465.

rather than flowering only once a year. Some arrived in 1794 with Lord Macartney on his return from the famous embassy to the court of the Qianlong emperor. It took only a single gene change to achieve recurrent flowering. The Chinese strains also enlarged the palette to include yellow and scarlet.

Flower formation involves tissue differentiation and formation of an organ with specialized structure and function. Flowers are the reproductive organs of plants. Their beautiful appearance and scent attract insect pollinators. In addition to homeotic genes that generate the structure of the flower, novel metabolic pathways are activated to produce the small molecules responsible for the colours and scents. Colour and scent are properties of the petals, which are the focus of this section.

Plant developmental biologists distinguish six stages in the development of a rose flower (see Figure 9.6). In formation of petals, an initial stage of cell division may stop while the flower is less than half its final size. Subsequent development occurs by differential cell elongation.

Regulation of pathways that produce scents in roses

Compounds from roses are an important source of fragrances (see Table 9.4). Their production rises to a peak in the mature flower, in consequence of raised expression levels of the enzymes that synthesize them.

- Terpenes are compounds formed from isoprene units (2-methyl-1,3-butadiene). Sesquiterpenes are 15-carbon compounds formed from three isoprene units.

In order to identify enzymes involved in rose scent production, Guterman and co-workers created an expressed sequence tag (EST) database and compared expression patterns in different stages of flower development.* They contrasted two tetraploid cultivars,

- Why an EST library? The rose genome has not yet been fully sequenced. It is about 500–600 Mb long, distributed on seven chromosomes. Most species are diploid or tetraploid; a few are triploid, hexaploid, or octaploid.

* Guterman, I. et al. (2002). Rose scent: genomics approach to discovering novel floral fragrance-related genes. Plant Cell **14**, 2325–2328.

Figure 9.6 Three of the six stages in the development of a rose flower: stages 1 (left), 4 (centre), and 6 (right). At stage 1, petals are beginning to emerge. At stage 4, cells are actively elongated and show increasing pigment concentrations. Finally, at stage 6, the flower is fully open. This figure illustrates the cultivar Fragrant Cloud.

From: Dafny-Yelin, M., et al. (2005). Flower proteome: changes in protein spectrum during the advanced stages of rose petal development. Planta **222**, 37–46.

Table 9.4 Major classes of molecule in the scent of Fragrant Cloud roses

Compounds	Amount emitted (μg/flower per day)
Esters	61
Aromatic and aliphatic alcohols	37
Monoterpenes	18
Sesquiterpenes	10

called Fragrant Cloud and Golden Gate. Fragrant Cloud gives large, red flowers with a strong scent (this is the cultivar illustrated in Figure 9.6). Golden Gate has yellow flowers and a less distinct odour to humans. Two experiments might identify the proteins involved in scent production: (1) a comparison of mature flowers between Fragrant Cloud (highly scent producing) and Golden Gate (poorly scent producing); and (2) a comparison of Fragrant Cloud flowers at an early developmental stage, before scent production ramps up, with the mature Fragrant Cloud flower, which is rich in scent.

The cDNA libraries created from both flowers in stage 4 contained 2139 unique sequences. Of these, 1288 were found only in Fragrant Cloud and 746 only in Golden Gate. Expression patterns were studied of 350 Fragrant Cloud genes associated with:

- primary and secondary metabolism;
- development;
- transcription;

- cell growth;
- cell biogenesis and organization;
- cell rescue;
- signal transduction; and
- unknown functions.

Taking, as a threshold of significance, a twofold difference in expression level, 77 genes (about one-fifth) had higher expression levels in Fragrant Cloud than in Golden Gate, and only three had higher levels in Golden Gate; the rest had similar expression levels in both varieties. Comparing Fragrant Cloud flowers in stage 1 and stage 4, 65 genes had higher expression levels in stage 4 and 14 had lower levels. Common to both sets were 40 genes that were more highly expressed in both Fragrant Cloud relative to Golden Gate and in Fragrant Cloud stage 4 relative to Fragrant Cloud stage 1 flower development (see Figure 9.7).

What are these 40 genes? Fifteen are involved in metabolism and seven appear to have roles in secondary metabolism, i.e. reactions outside of the core metabolic pathways responsible for normal growth, development, and reproduction. Two very strongly upregulated genes encode proteins similar in sequence to known enzymes: glutamate decarboxylase, and (+)-δ-cadinene synthase. The enzyme (+)-δ-cadinene synthase is involved in the synthesis of sesquiterpenes, one of the classes of floral scent molecule.

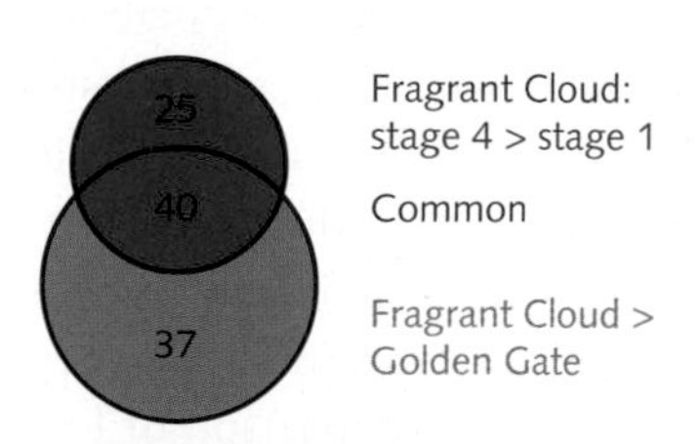

Figure 9.7 Two rose cultivars, Fragrant Cloud and Golden Gate, differ in their maximal odour production in stage 4 of flower development: Fragrant Cloud is rich in scent and Golden Gate is poor. Scent development in Fragrant Cloud is fully developed in stage 4 relative to the immature flowers in stage 1.

In comparisons of expression patterns, 25 genes were expressed more highly in stage 4 Fragrant Cloud rose flowers than in stage 4 Golden Gate rose flowers; 37 genes were expressed more highly in stage 4 Fragrant Cloud rose flowers than in stage 1 Fragrant Cloud rose flowers; and 40 were expressed more highly in stage 4 Fragrant Cloud flowers than in both stage 4 Golden Gate flowers and stage 1 Fragrant Cloud flowers.

These experiments can help to identify proteins preferentially expressed in stage 4 Fragrant Cloud flowers that are involved in scent biosynthesis.

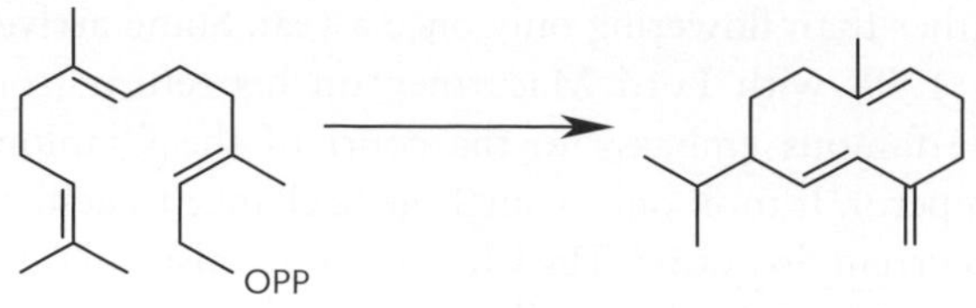

Figure 9.8 Conversion of precursor farnesyl diphosphate to germacrene D, a common scent molecule produced by rose petals.

Cloning of the (+)-δ-cadinene synthase gene to produce recombinant enzyme permitted direct functional studies. It catalyses the reaction of farnesyl diphosphate to produce the sesquiterpene germacrene D, the major sesquiterpene component of the scent of Fragrant Cloud (see Figure 9.8).

Colour: the elusive blue rose

A walk through many neighbourhoods in temperate climates will reveal the many colours that roses can display, notably red, pink, peach, and yellow. Conspicuous by its absence is blue. It has not been possible to produce a blue rose by classical selective breeding. This is not for lack of trying: in 1840 the horticultural societies of Great Britain and Belgium offered a prize of 500 000 francs for one.

The major pigments of flower petals are anthocyanins: cyanidin, pelargonidin, and delphinidin. All are synthesized from a common precursor, dihydrokaempferol (see Figure 9.9). It is delphinidin that is blue, but roses lack the gene for the enzyme that produces dihydromyricetin from dihydrokaempferol (marked by a green * in the figure).

Scientists at the Australian Commonwealth Scientific and Industrial Research Organisation (CSIRO), in collaboration with a Melbourne company, Florigene, and the Suntory Corporation of Japan, have created a blue rose by genetic engineering. There were two challenges: to produce the blue pigment and to turn off synthesis of red and orange pigments. To suppress the red and orange pigments, the target was the enzyme dihydroflavonol reductase (DFR). DFR modifies precursor molecules in all three branches of the pathway; a *DFR*$^{-}$ plant would have white flowers, providing a 'clean slate' for production and display of blue pigments. However, *DFR* is also needed for synthesis of delphinidin.

It was necessary to manipulate the pathways with precision. An interfering RNA (RNAi) was engineered

Dihydroquercetin ← Dihydrokaempferol —* F3′,5′H→ Dihydromyricetin

DFR ↓ DFR ↓ DFR ↓

Cyanidin 3-glucoside | Pelargonidin 3-glucoside | Delphinidin 3-glucoside

Figure 9.9 Simplified scheme showing structures and metabolic relationships of anthocyanin flower pigments. The difficulty in breeding a blue rose is that roses lack the gene that encodes the enzyme flavonoid 3′,5′-hydroxylase (F3′,5′H), which produces the precursor of the blue pigment delphinidin. In other plants, this enzyme acts at the point marked by a green *. The steps corresponding to the vertical arrows are all catalysed by a single enzyme, dihydroflavonol reductase (DFR).

to knock out the rose *DFR*. Then, introduction of a *DFR* from iris, together with the gene for flavonoid 3′,5′-hydroxylase from pansy to supply the enzyme missing in roses, produced a plant with high levels of petal delphinidin and only small amounts of cyanidin. The rose and iris *DFR* genes are quite similar but have two important differences: (1) the natural specificity profile of the iris DFR produces delphinidin predominantly; and (2) the RNAi can be sufficiently specific that it selectively inactivates the rose *DFR* and not the iris homologue.

The new rose is a shade of blue (see Figure 9.10). The reason it is not a purer blue has to do with the relatively acidic pH of rose petals compared with other flowers. (Indeed, anthocyanins can be used as pH indicators.) Finding ways to modify the intracellular pH is the subject of current research.

- Blue roses do not exist in nature, and attempts to breed them have been unsuccessful. Understanding of the underlying metabolic pathways of the pigments allows rational attempts at genetic engineering. It is interesting that the pH of the petals has an effect – the pigments are 'indicators'. If this effect helped to stimulate Sydney Brenner's fascination for molecular biology,* that may be its greatest significance for the field.

Figure 9.10 Picture of delphinidin-rich rose flower, produced by methods similar to those described in the text.
Photograph from Suntory Ltd.

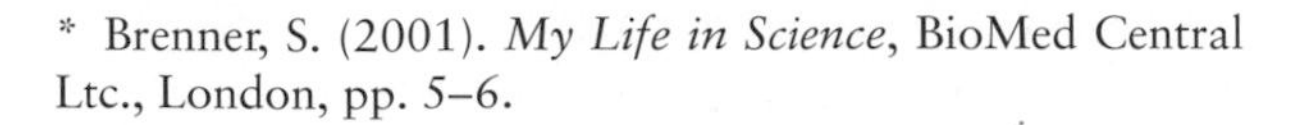

* Brenner, S. (2001). *My Life in Science*, BioMed Central Ltc., London, pp. 5–6.

Expression patterns in learning and memory: long-term potentiation

Learning and memory involve changes in the structure and biochemistry of nerve cells. Nervous systems are dynamic networks, passing signals among cells. Synapses are the sensitive points at which neurons interact. As each neuron integrates inputs from several others, its output depends on the 'weighting' of its inputs – the distribution of the strengths of the input synapses. Increasing or reducing the strengths of individual synaptic connections modulates the dynamics of the network.

Learning and memory must involve some permanent structural change. The observation that memory survives periods of coma, during which neural activity ceases, proves this. The change in the network cannot, therefore, be purely a change in the dynamic state.

Long-term potentiation (LTP) is a neural phenomenon underlying learning and memory. LTP is a persistent increase in strength of a synaptic connection as a result of stimulation of the upstream cell. The original observation, first described by Bliss and Lomo in 1973, was that high-frequency stimulation of a synapse during a finite time interval produced a persistent subsequent enhancement of the postsynaptic response. It is believed that transient effects, lasting <1 hour, require modifications of pre-existing proteins at the synapse. Longer-lasting effects involve protein synthesis and gene transcription, and resulting structural remodelling of synapses.

- Learning must be regarded as a specialized form of development.

In addition to neural plasticity, learning also involves generation of new neurons. LTP stimulates both neurogenesis and enhanced survival of new cells.

Park and co-workers studied genes that change their expression pattern in response to LTP.* They studied cells from the mouse dentate gyrus, a structure within the hippocampus. The hippocampus is known to be involved in memory formation, especially spatial memory (see Box 9.2). Following high-frequency stimulation for four separated 1-second intervals, total RNA was extracted after several time intervals – 30, 60, 90, and 120 minutes after stimulation – and then reverse transcribed into cDNA and hybridized onto Affymetrix GeneChip arrays. The chip reported 12 000 genes and ESTs. Samples of unstimulated tissue provided controls.

Park and colleagues found, within all time points, 1664 genes with statistically significant changed expression patterns. Of these, 39% were upregulated and 61% downregulated. The genes identified suggest that LTP produces changes in a variety of processes affecting cell morphology and affects interactions among cells and between cells and the extracellular matrix.

Specific functional assignment showed several categories of genes, shown in Table 9.5 (this table shows a composite containing genes identified as having changed expression at *any* of the time points; see also Figure 9.11). Most but not all of the categories contain examples of both up- and downregulated genes.

* Park, C.S., Gong, R., Stuart, J., & Tang, S.-J. (2006). Molecular network and chromosomal clustering of genes involved in synaptic plasticity in the hippocampus. J. Biol. Chem. **281**, 30 195–30 211.

London taxi drivers: spatial memory and the hippocampus

London taxi drivers must have an exhaustive knowledge of the metropolitan geography, of optimal routes between points both famous and obscure, and of variations in traffic patterns. As portrayed in the famous 1979 film *The Knowledge*, drivers must pass a strict test to earn their licence before taking control of a classic black London cab. Consistent with the involvement of the hippocampus in spatial memory, brain scans show that London taxi drivers have a larger hippocampus than a control group, and that the hippocampus enlarges with time spent behind the wheel.*

* Maguire, E.A., Gadian, D.G., Johnsrude, I.S., et al. (2000). Navigation-related structural change in the hippocampi of taxi drivers. Proc. Natl. Acad. Sci. USA **97**, 4398–4403; Maguire, E.A., Spiers, H.J., Good, C.D., Hartley, T., Frackowiak, R.S., & Burgess, N. (2003). Navigation expertise and the human hippocampus: a structural brain imaging analysis. Hippocampus **13**, 250–259.

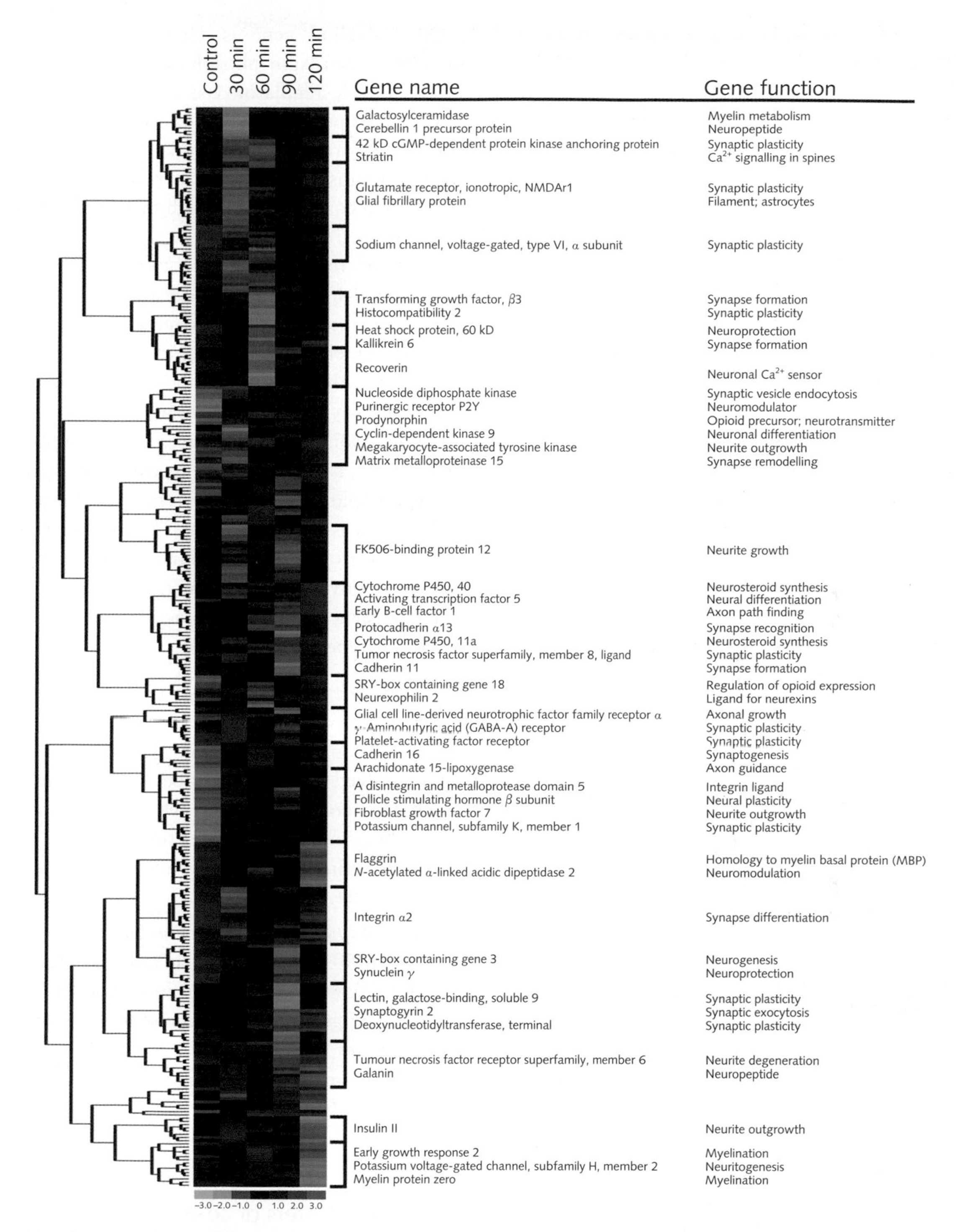

Figure 9.11 Clustering of genes differentially expressed after induction of LTP. The scale at the bottom indicates the expression level relative to the control: green = enhanced expression; red = reduced expression. Brackets indicate clusters of genes with known neural or synaptic functions.

From Park, C.S., et al. (2006). Molecular network and chromosomal clustering of genes involved in synaptic plasticity in the hippocampus. J. Biol. Chem. **281**, 30 195–30 211.

Table 9.5 Genes that change expression pattern in response to long-term potentiation (LTP)

Functional category	Upregulated	Downregulated
The extracellular matrix and its regulation	+	+
Membrane protein/cell surface/adhesion molecule	+	+
Neurosteroid hormone metabolism	–	+
Cytokine/growth factor/ receptor	+	+
Other receptors/signalling	+	+
Ion channel	+	+
Transcription factor/regulation	+	+
Translation	+	+
Neurotransmitter receptor/ neuromodulator	+	+
Regulation of cytoskeleton	+	+
Mitochondrial/energy production	+	+
Proteases/protease inhibitors	+	+
Immunoresponsive proteins/oxidative stress/ neuroprotection/cell death	+	+
Myelin-related proteins	–	+
Chromatin structure	+	–

Expression patterns can identify particular genes involved in neural plasticity. Many of the genes identified by changed expression patterns were already known to play roles in synaptogenesis, synapse differentiation, neurite outgrowth, and synaptic plasticity. Others were not previously known to be involved in LTP, but their altered expression pattern makes them candidates to be tested for their effects on LTP.

For example, transglutaminase is known to be expressed in neural tissue and appears in synapses. However, it was not known to be implicated in LTP. A connection was confirmed by showing that cystamine – a specific antagonist of transglutaminase – impairs LTP. (Cystamine is an inhibitor of transglutaminase and also causes disulphide exchange producing unfolding. Transglutaminase also helps to produce protein aggregates in Huntington's disease. Cystamine does ameliorate Huntington's disease in the mouse, but the mechanism is unclear.)

One puzzling gene is *CDC25B*, an oncogene encoding a tyrosine phosphatase, which functions as a cell-cycle regulator. Use of a specific inhibitor of CDC25B protein product blocked LTP. *CDC25B* must, therefore, have an essential role, but the mechanism remains obscure. This is precisely the kind of unexpected connection that high-throughput methods can turn up.

Some of the differentially expressed genes are coordinated into coherent pathways. These include enhanced expression of genes in the MAPK signalling cascade (which was already known to be important in LTP) and the Wnt signalling pathway (which had not previously been connected with LTP).

Comparison of expression patterns at different times after LTP induction revealed the temporal expression patterns of different genes. An interesting observation is that many genes in the same general functional groups have similar temporal expression profiles. Conversely, different time points are associated with a 'schedule' of activity of genes with particular types of function.

- *Genes with common time profiles*. Genes involved in responses to external stimuli are upregulated at 30 and 60 minutes after LTP induction. These may be involved in interactions between pre- and postsynaptic components. Genes involved in signal transduction and transcription regulation provide less-clear time profiles. It is likely that their effects are indirect.
- *Events happening at particular times*. Many genes active at 30 minutes are involved in cell–cell interaction, synapse formation and remodelling, and neurite outgrowth. These represent a relatively early component of the response. Genes related to the cytoskeleton are downregulated at 30 minutes, but upregulated at 60, 90, and 120 minutes. These may be involved in structural changes at the synapse.

Conserved clusters of co-expressing genes

Mapping the loci of the differentially expressed genes showed that they are concentrated in specific chromosomal regions (tandem duplicates were removed). These clusters tended preferentially to contain genes with similar functions. Comparison of the distributions of homologues in the genomes of rats, humans,

Drosophila, and *C. elegans* suggest that the clustering is conserved in evolution. The clustering may contribute to a mechanism for common regulation of expression, reminiscent of a bacterial operon.

Evolutionary changes in expression patterns

The very high similarity in genome sequence between humans and chimpanzees suggests that the evolutionary differences between such closely related species would not lie primarily in the relatively small changes in sequences of individual proteins, but in expression patterns. Microarrays permit a test of this idea. Nevertheless, amino acid sequence changes in proteins may be significant, even though they may be small. One example is the *FOXP2* gene, discussed in Chapter 4. Another is a contributor to the control of overall cerebral cortical size (a crude but not entirely irrelevant feature of our mental evolution), the gene *ASPM* (abnormal spindle-like microcephaly-associated), which has undergone positive selection in the lineage leading to humans.

- The idea of the importance of changes in expression pattern to evolution appeared in a seminal paper by M.C. King and A.C. Wilson in 1975.

The design of the experiments presents a number of difficulties, however.

- There is a high background of variation in expression pattern among different individuals of any species and among different tissues within any individual. This makes it difficult to identify changes unambiguously attributable to species differences.
- Use of a microarray containing oligomer sequences derived from human genes to measure mRNA levels in chimpanzee tissue underestimates the expression levels in the chimpanzee because of less-effective hybridization resulting from sequence changes.

Nevertheless, carefully controlled experiments show that there are significant differences between human and chimpanzee expression patterns that arose during evolution.

- *The set of genes that show different expression patterns between humans and chimpanzees is rich in transcription factors*. This is in accord with King and Wilson's hypothesis.
- *The differences in expression pattern are not uniform in different tissues*. As far as expression pattern is concerned, our hearts and livers have diverged from chimpanzees in expression pattern more than our brains, both in terms of the numbers of differentially expressed genes and the amount of the differences in transcription levels. (Would you have guessed this?) However, looking at the course of evolution using the macaque, for instance, as an outgroup, shows that the human brain is particularly rich in genes with *increased* expression levels relative to the chimpanzee, consistent with the distinct differences in cognitive abilities. In other tissues, there is a more even distribution of genes expressed more highly in humans or more highly in the chimpanzee.
- *Changes in expression patterns tend to be lower in X and higher in Y chromosomes than in autosomes*. For brain tissue, the average human/chimpanzee ratio of expression level is about 1.51 for autosomes, 1.43 for the X chromosome, and 2.14 for the Y chromosome.
- *Duplicated genes tend to show a higher divergence in expression pattern than non-duplicated genes*. One possible consequence of gene duplication is divergence and specialization of function. This is consistent with the requirement for differential control of gene expression. (Recalling Chapter 1, vertebrate α- and β-globins are exceptions to this paradigm: it is necessary to *calibrate* their levels of expression. It is not yet clear what mechanism achieves this. Synthesis of different amounts of α- and β-globin causes thalassaemias.)
- *The differences in expression patterns are not uniform, even across different autosomes*. During the 6–7 million year period of divergence, there has been substantial chromosome rearrangement between humans and chimpanzees (see Figure 3.6).

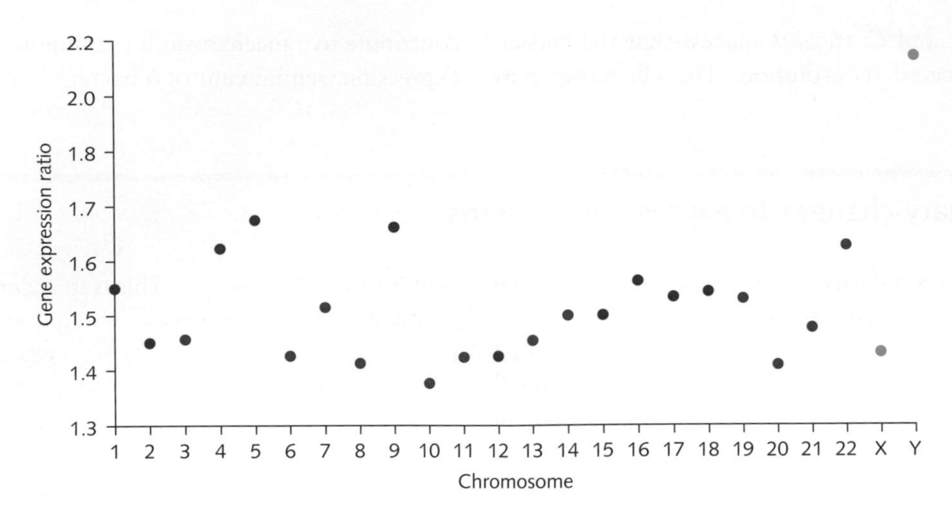

Figure 9.12 Average ratio of gene expression levels between human and chimp cortical tissue in collinear (red) and rearranged (blue) chromosomes. X and Y chromosomes shown in green.

After: Marquès-Bonet, T., Cáceves, M., Bertranpetit, J., Preuss, T.M., Thomas, J.W., & Nawarro, A. (2006). Chromosomal rearrangements and the genomic distribution of gene-expression divergence in humans and chimpanzees. Trends Genet. **20**, 524–529.

Table 9.6 Areas of brain for which human/chimpanzee expression pattern differences were measured

Area of brain	Function in human	Function in chimpanzee if known to differ from human
Dorsolateral prefrontal cortex	Important for higher brain functions: working memory, conscious control of behaviour	
Anterior cingulate cortex	Autonomic functions: heart rate and blood pressure, and cognitive functions such as reward anticipation, decision making, empathy, and emotion	
Broca's area	Mainly language, plus action	Gesture, especially control over orofacial action, including communicative acts
Central part of cerebellum	Coordinating complex movements such as walking	
Caudate nucleus	Regulation and organization of information sent to frontal lobes	
Pre-motor cortex	Sensory guidance of movement, activating proximal and trunk muscles	
Area homologous to Broca's in right hemisphere	Not entirely clear; some involvement in communication without syntactic contribution to language use	

In cortical tissue, changes in expression patterns are larger among genes in rearranged chromosomes than syntenic ones (see Figure 9.12).

The comparative analysis of expression patterns in human and chimpanzee brains has been pursued to high resolution. Khaitovich and colleagues used an array containing ~10 000 human genes and arrays containing ~40 000 human transcripts to test several areas of human and chimpanzee brains* (see Table 9.6).

Differences between the species must be extracted from the variation among individuals. Within each species, typically a few hundred genes (out of ~10 000 tested) vary in expression in different brain regions. There is relatively low variation within the

* Khaitovich, P., et al. (2004). Regional patterns of gene expression in human and chimpanzee brains. Genome Res. **14**, 1462–1473.

cortex itself, but over 1000 genes show differences in expression pattern between the cerebellum and other regions.

It is surprising to observe a *similarity* of expression pattern in humans between Broca's area, associated with speech, and the homologous right-hemisphere area, which is not. This observation suggests that the achievement of language in humans did *not* depend on localized changes in transcription patterns.

Functional analysis showed that genes encoding proteins involved in signal transduction, cell–cell communication, differentiation, and development show a greater-than-random tendency to vary in expression between regions, within both species. Genes encoding proteins involved in protein synthesis and turnover tend to show conserved expression patterns.

Approximately 10% of the genes studied differ in expression pattern between humans and chimpanzees. Most of the differences appear in two or more regions of the brain. The cerebellum contains several genes showing species-specific differences in expression *not* shared by other regions. An analysis of functional categories of the genes showing enhanced or reduced interspecies expression differences does not reveal an enrichment in specific families.

If our goal is to understand at the molecular level the phenotypic differences between humans and chimpanzees involving higher mental functions such as cognition and language, our data must be very accurate and detailed, for as the traits grow more subtle, the molecular signal grows correspondingly fainter. At some point, it will be necessary to trace the origin of the changes in expression patterns to protein and genomic sequences. This does not contradict the King and Wilson hypothesis; after all, amino acid sequence changes modulate the functions of regulatory proteins. We may also be required to address different levels of complexity: the different traits may depend on very complicated patterns of interactions, difficult to infer from properties of individual genes.

Applications of microarrays in medicine

Development of antibiotic resistance in bacteria

The growth in bacterial resistance to antibiotics has created a crisis in disease control.

One of the most powerful antibiotics available for use in humans is vancomycin, a 1.5 kDa glycopeptide antibiotic isolated from a soil bacterium in Borneo, *Amycolatopsis orientalis* (see Figure 9.13). Vancomycin was first used clinically in 1958 when infectious strains of staphylococci developed penicillin resistance (see Box 9.3). It became the antibiotic of choice for many infections and the drug of last resort for some.

Development of drug resistance by pathogenic microorganisms threatens to deprive us of the ability to control infections disease. Widespread use of antibiotics, not only in human clinical medicine, but in raising animals, has contributed to the severity of the problem.

Vancomycin acts by interfering with cell wall synthesis. The cell wall in Gram-positive bacteria is a combination of polysaccharides and peptides. Linear polysaccharides, formed from alternating *N*-acetylglucosamine and *N*-acetylmuramic acid units, are cross-linked by short peptides: L-ala–D-gln–L-lys–D-ala–D-ala. Vancomycin acts by binding to the oligopeptide, preventing the cross-linking. Without a robust cell wall, bacteria cannot stand up to their internal osmotic pressure.

The development of resistant *Staphylococcus aureus* strains occurred gradually. Vancomycin-resistant enterococci appeared in 1977. Twenty years later, *S. aureus* developed resistance. The strains were already methicillin resistant. Vancomycin-resistant *S. aureus* (VRSA) strains appeared in 2002. They have been found in Europe and the USA (see Box 9.3).

Resistance is measured by an increase in the minimum inhibitory concentration (MIC), which is related to the clinically effective dose (Table 9.7).

One contribution to the spread of vancomycin resistance may be the practice in Europe of widespread feeding of avoparcin (see Figure 9.13) to animals. Homologues of the resistance genes are present in the source organism for vancomycin, *A. orientalis*. The finding of bacterial DNA contamination in animal feed-grade avoparcin containing sequences related to the resistance gene cluster strongly suggests that the use of avoparcin has led to gene transfer to bacteria which could be taken up by the animals.

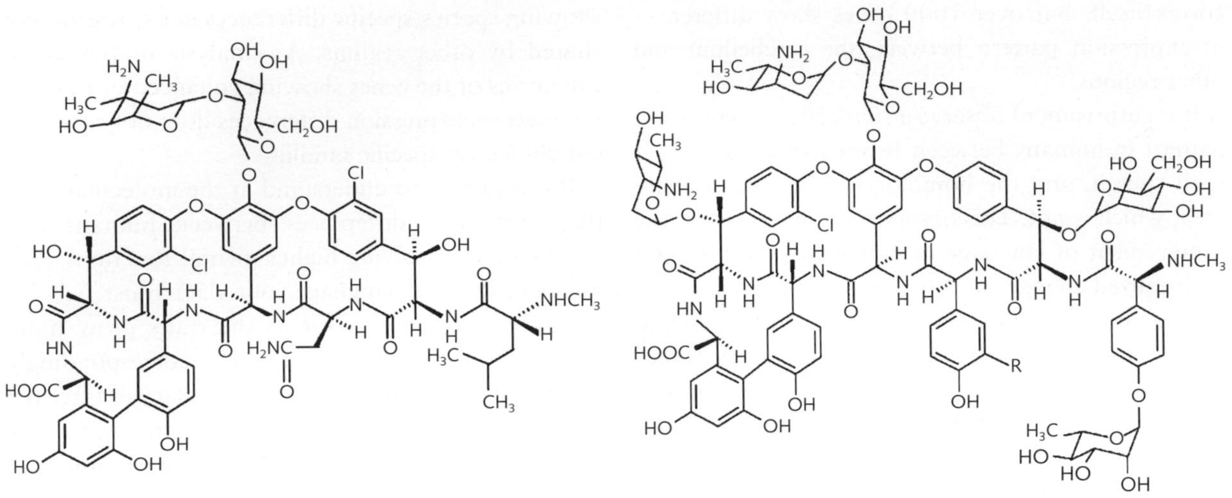

Vancomycin

α-Avoparcin R=H
β-Avoparcin R=Cl

Figure 9.13 (left) Vancomycin, a glycopeptide antibiotic produced by the bacterium *A. orientalis.* (right) Antibiotics α- and β-avoparcin, related to vancomycin and produced by *Streptomyces candidus*.

From: Lu, K., Asano, R., & Davies, J. (2004). Antimicrobial resistance gene delivery in animal feeds. Emerg. Infect. Dis. **10**, 679–683.

BOX 9.3

Development of vancomycin resistance – a chronology*

1941	First clinical use of penicillin G
1942	Appearance of penicillin-resistant *Staphylococcus aureus*
1950s	Multidrug-resistant *S. aureus* widespread
1956	Vancomycin described
1958	First clinical use of vancomycin
1960	First clinical use of methicillin
1961	Appearance of methicillin-resistant *S. aureus*
1960s	Spread of methicillin-resistant *S. aureus*
1970s	Methicillin-resistant *S. aureus* widespread
1988	Appearance of vancomycin-resistant enterococci
1992	Laboratory transfer of high-level vancomycin resistance from enterococci to methicillin-resistant *S. aureus*
1997	Appearance of vancomycin-intermediate *S. aureus* in clinical setting
2002	Appearance of vancomycin-resistant *S. aureus* in clinical setting

* Pfeltz, R.F. & Wilkinson, B.J. (2004). The escalating challenge of vancomycin resistance in *Staphylococcus aureus*. Curr. Drug Targets Infect. Disord. **4**, 273–294.

Table 9.7 Effect of minimum inhibitory concentration (MIC) on resistance

Resistance of *S. aureus* strain	Minimum inhibitory concentration (MIC)	Year first appeared
Sensitive	1 μg/ml	–
Intermediate (VISA)	8–16 μg/ml	1997
Resistant (VRSA)	>32 μg/ml	2002

Eventually the genes found their way to bacteria that infect humans.

- The toxicity of vancomycin in its early days was caused by impurities – the brown preparations were nicknamed 'Mississippi mud'. Since the mid-1980s, purification procedures and the safety of the preparations have improved.

S. aureus has adopted two basic strategies for achieving vancomycin resistance. These approaches can be thought of as defence and attack. Both are effective, if success – for the bacterium – can be defined as attaining a level of resistance that survives doses of vancomycin that would be intolerably toxic to the patient.

Acting defensively, *S. aureus* achieves the *intermediate* stage of vancomycin resistance (VISA) by a

Table 9.8 Genes in vancomycin resistance cluster

Gene	Action of gene product
VanH	Reduces pyruvate to D-lactate
VanA	Esterifies D-ala–D-lactate
VanX	Hydrolyses D-ala–D-ala, leaving D-ala–D-lactate to build the cell wall
VanS	A kinase that senses vancomycin and initiates transcription of the other genes. In the absence of vancomycin, they are not expressed.

number of structural changes, including reduced growth rate, reduction in cell wall cross-links, increased cell-wall thickness, and the appearance of D-glutamic acid instead of D-glutamine in the peptide. The genomic changes responsible for the VISA phenotype can be magnified by vancomycin challenge and selection to produce VRSA strains with MIC = 32 μg/ml.

The alternative, for the bacterium, is a counter-attack on vancomycin, to 'pull its sting'. *S. aureus* has achieved high vancomycin resistance by picking up a specific plasmid from a resistant *Enterococcus*. The plasmid contains a cluster of genes, leading to changing the D-ala–D-ala at the C terminus of the cross-linking pentapeptide to D-ala–D-lactate (Table 9.8). The modified peptide can enter the cell wall but has a lower binding affinity for vancomycin by a factor of ~1000.

- Microorganisms also develop resistance by evolving enzymes that destroy an antibiotic or pump it out of cells. *S. aureus* followed this route to gain resistance to penicillin, which initially led clinicians to turn to vancomycin.

Mongodin and co-workers compared expression patterns of genes in VISA strains (MIC ~8 μg/ml) with VRSA strains produced by selection, not containing the resistance plasmid.* The array contained 2688 oligonucleotides. The experiments were run in parallel, starting with two different clinical VISA isolates. Upon increased vancomycin resistance, 35 genes consistently showed increased expression, some as high as 30-fold, and 16 consistently showed decreased expression.

Genes upregulated with increased vancomycin resistance are associated with the following:

- purine biosynthesis, which is a large component of the change in expression: 15 of the 35 upregulated genes involved purine biosynthesis or transport, and there was a mutation in the regulator of the purine biosynthesis operon;
- cell envelope synthesis, remodelling, and degradation;
- proteins involved in transport and binding of amino acids, peptides and amines, and nucleic acid components (including purines);
- synthesis of staphyloxanthin, an orange carotenoid that gives *S. aureus* its golden colour;
- folic acid synthesis; and
- unknown functions.

Genes downregulated with increased vancomycin resistance are associated with:

- energy metabolism;
- cell envelope biosynthesis;
- proteins involved in transport and binding of carbohydrates, organic alcohols, and acids;
- salvage of nucleic acid components;
- regulatory functions; and
- tetracycline resistance.

It is not always easy to put together the details of a change in expression pattern involving many metabolic subsystems in order to grasp the salient message (see Figure 9.14). However, it is reasonable to think that the goal of the changes is to defend the cell wall, as that is the target of the antibiotic. As Mongodin and colleagues suggested, many of the changes in expression levels combine to funnel metabolites to the formation of ATP. These changes include downregulation of the genes that encode proteins for conversion of ATP to the corresponding deoxynucleoside triphosphate for DNA synthesis (*nrdD*) and for the degradation of AMP (*deoD*). Key enzymes in glycolysis and fermentation are downregulated, diverting glucose 6-phosphate through the pentose phosphate pathway to form the ribose component of ATP.

* Mongodin, E., Finan, J., Climo, M.W., Rosato, A., Gill, S., & Archer, G.L. (2003). Microarray transcription analysis of clinical *Staphylococcus aureus* isolates resistant to vancomycin. J. Bacteriol. **185**, 4638–4643.

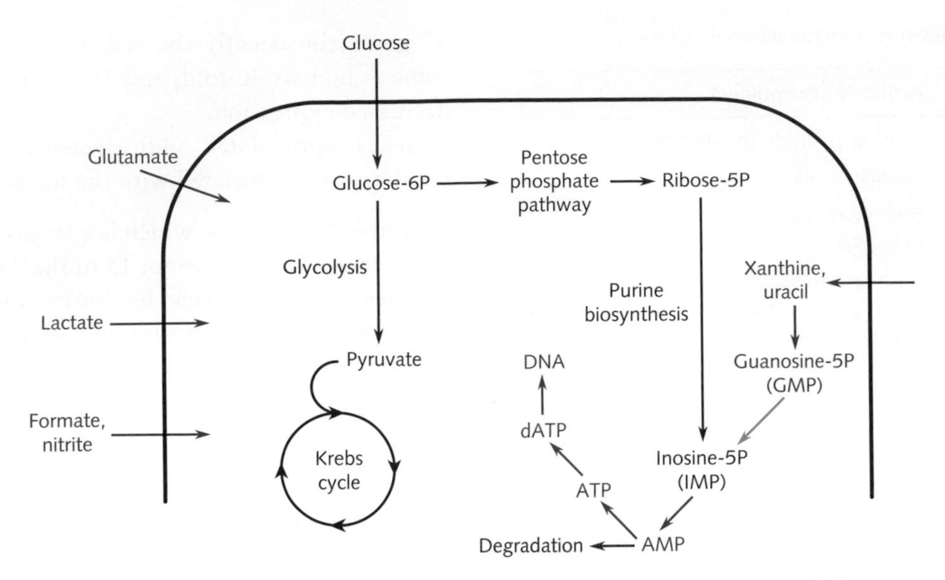

Figure 9.14 With enhancement of vancomycin resistance in the laboratory by selection after vancomycin challenge, expression patterns of genes associated with some processes are upregulated (blue arrows) and others are downregulated (red arrows). A major target of upregulation is purine synthesis, aimed at enhanced production of ATP for energy requirements.

After: Mongodin, E., Finan, J., Climo, M.W., Rosato, A., Gill, S., & Archer, G.L. (2003). Microarray transcription analysis of clinical *Staphylococcus aureus* isolates resistant to vancomycin. J. Bacteriol. **185**, 4638–4643.

Synthesis of the thickened cell wall is a very energy-intensive process. The ratio of cell-wall volume to total cell volume increased by 41% in the vancomycin-resistant cells. Perhaps reduced cellular growth rate is a price that must be paid if a larger fraction of the cell's energy budget goes into cell-wall synthesis.

- Drug resistance in pathogens is a crucial problem in contemporary medicine. Learning, in detail, how it comes about, will be essential for developing ways to prevent it.

Childhood leukaemias

Haemopoietic stem cells are the undifferentiated precursors of all types of blood cell. They mature by differentiating along one of two pathways (see Figure 9.15). B and T cells of the immune system have followed the lymphoid path; red blood cells have followed the myeloid path. An abnormal genetic transformation leading to unregulated proliferation of any blood cell, at any stage of differentiation, gives rise to leukaemia. Leukaemias can be classified according to the type of cell that is proliferating.

Acute lymphoblastic leukaemia is the most common cancer of children, representing almost one-third of childhood cancers. The main treatment is chemotherapy, the success rate for which has greatly improved in the past quarter of a century. Nevertheless, conventional therapy is unsuccessful in about 25% of patients.

Measurements of gene expression patterns have permitted molecular classification of disease subtypes, and correlations with response to chemotherapy and likelihoods of rapid or delayed recurrence or long-term survival.

- From the expression pattern of a group of 50 genes, it is possible to distinguish almost perfectly between lymphoblastic and myeloid leukaemias and, for lymphoblastic leukaemias, to distinguish B- and T-cell lineages. The results are calibrated against established methods based on flow cytometry. The variability in expression pattern is unusually high in acute lymphoblastic leukaemia relative to other types of cancer. It is thereby feasible to create a molecular taxonomy of childhood leukaemias.
- Expression patterns can predict the likelihood of a favourable outcome. This combines questions of the success of therapy, the likelihood of spontaneous relapse after remission, and the development of secondary tumours. A complicating factor of studies of this type in humans is the fact that the samples are taken from patients under a variety of treatments.

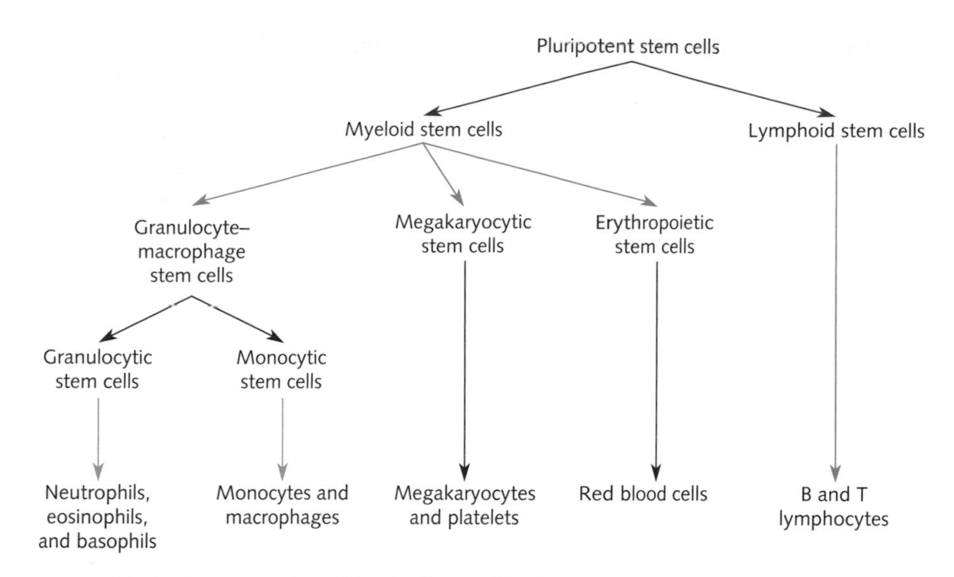

Figure 9.15 Haematopoiesis is the formation of new blood cells. Our blood contains many types of cell:

Cell type	Function
Neutrophils	Respond to bacterial infection
Eosinophils	Respond to allergens and to infections by parasites
Basophils	Respond to allergens
Monocytes and macrophages	Remove dead tissue and respond to infections by bacteria and fungi
Megakaryocytes	Precursors of platelets
Platelets	Involved in blood clotting
Red blood cells (erythrocytes)	Contain haemoglobin and transport O_2 and CO_2
B lymphocytes	Produce specific antibodies
T lymphocytes	Destroy infected cells and regulate immune responses

These cells arise by different developmental pathways from a common stem-cell precursor, which can potentially differentiate into any of the mature cell types, a property called **totipotency**. As maturation proceeds, cells first become **pluripotent** (able to mature into some but not all cell types) and then finally committed to a single ultimate form.

Normally, haematopoiesis produces approximately 175 billion red cells, 70 billion granulocytes (neutrophils, eosinophils, basophils), and 175 billion platelets every day. [One billion is 10^9.] When we are challenged by infection, production can be stepped up by an order of magnitude.

Leukaemia is the uncontrolled proliferation of any of these types of blood cell, either mature forms or their precursors. In mammals, mature erythrocytes, being enucleate, cannot themselves proliferate. However, mutations can occur in *precursor* cells. For example, a mutation affecting the JAK2 signalling pathway in the stem-cell precursor can result in overproduction of erythrocytes, a disease called primary polycythaemia vera. (Secondary polycythaemia vera, also characterized by overproduction of red cells, is a response to lack of oxygen; possible causes include heavy smoking, emphysema, or moving without acclimation to a high altitude.)

Clinical experience shows that the time interval of first remission is a good predictor of long-term survival. A group of genes was found with an expression pattern correlated with length of remission, i.e. these genes are differentially expressed in patients with early and late relapse. Testing the expression levels of these genes can improve the precision of prognosis. Identifying the pathways in which the genes are involved can illuminate the underlying biology of the disease progression. Genes involved in cell proliferation and DNA repair were upregulated in the early-relapse group.

Development of a second cancer, not related to leukaemia in any obvious way, is a common and very serious complication of acute lymphoblastic leukaemia. Brain tumours are one of the most common secondary malignancies. Several genes have been identified, the expression patterns of which correlate with the risk of secondary brain tumours.

- Expression pattern can predict effectiveness of treatment and guide the choice of therapy. In a study using 14 500 probe sets and samples from 173 patients, sets of 20–40 genes were identified that distinguish resistance and sensitivity to four different drugs: prednisolone, vincristine, asparaginase, and daunorubicin. These results, even taken as purely empirical correlations, have clinical utility in guiding treatment. Their interpretation at the genetic level reveals that the activities of the drugs involve some different as well as some common pathways. A set of 45 genes was found to be correlated with resistance to *all four* of the drugs. The majority of these genes involve transcription, DNA repair, cell-cycle maintenance, and nucleic acid metabolism.
- Identification of specific genes involved in diseases can suggest targets for drug development. For instance, one type of acute lymphoblastic leukaemia is associated with overexpression of the gene FLT3 for a receptor tyrosine kinase. Patients with mutations that produce constitutively active receptors have a poor prognosis. FLT3 inhibitors are now in clinical trials.

Thus, expression profiles can permit tailoring of drug therapy, both to the specific disease and subtype, and to the patient.

- Expression profiling can (a) permit precise diagnosis of the subtype of the disease, (b) predict the likely course of the disease, and (c) guide choice of therapy.

Whole transcriptome shotgun sequencing: RNA-seq

Gene expression profiling has been the most common application of microarrays. The goal is to measure the identities and relative amounts of different RNA transcripts in populations of cells. Comparison of transcript profiles between healthy and disease states, or under different external conditions, or as a function of time, reveal the changes in gene expression patterns. Examples of all of these appear in this chapter.

RNA-seq – or whole transcriptome shotgun sequencing – is an alternative method of measuring RNA levels, applying high-throughput sequencing techniques. RNA isolated from cells is fragmented, reverse-transcribed to cDNA, and sequenced. Assembly is easiest by aligning with a reference genome. This will also automatically pick up post-transcriptional edits.

The RNA-seq approach has some advantages over microarrays:

- To construct a microarray, one must make a choice of what probe sequences to include. RNAseq will report whatever is there, with no prior commitment to any set of possible sequences.
- In principle, sequencing methods give more precise measurements of RNA concentrations, by recording how frequently each sequence appears in the pooled results. Not only is sequencing potentially more precise, it has a higher dynamic range.
- If a population of cells includes both a host and a pathogen, the transcriptomes of both are simulaneously measurable. In contrast, without a specially designed chip, a microarray would most likely contain probes only from the host.

Nevertheless, despite many claims, reports of the death of microarrays are greatly exaggerated. Technically, although the sequencing platforms accurately report the relative amounts of different cDNAs presented to them, there is potential bias in the yields of reverse transcription from different RNA molecules. For instance, internal secondary structure of the RNA may interfere with primer binding. (This may be a problem with microarray experiments also.) In addition, it is currently true that microarray measurements are less expensive than the RNAseq approach to collecting equivalent data. Of course, the cost of sequencing is changing rapidly.

- Methods for expression profiling include microarrays and whole transcriptome shotgun sequencing, or RNA-seq. Both methods are in widespread use at present. By asking in detail what one wants from the results, one can make an intelligent choice between them for any particular experiment on any particular system.

RECOMMENDED READING

- General discussions of microarrays:

Butte, A. (2002). The use and analysis of microarray data. Nat. Rev. Drug Discov. **1**, 951–960.

Penkett, C.J. & Bähler, J. (2004). Getting the most from public microarray data. Eur. Pharm. Rev. **1**, 8–17.

- The problem of bacterial drug resistance and the development of novel antibiotics:

Amábile-Cuevas, C.F. (2003). New antibiotics and new resistances. Am. Sci. **91**, 138–149.

Projan, S.J. & Shlaes, D.M. (2004). Antibacterial drug discovery: is it all downhill from here? Clin. Microbiol. Infect. Suppl. **4**, 18–22.

Projan, S.J., Gill, D., Lu, Z., & Herrmann, S.H. (2004). Small molecules for small minds? The case for biologic pharmaceuticals. Expert Opin. Biol. Ther. **4**, 1345–1350.

Thomson, C.J., Power, E., Ruebsamen-Waigmann, H., & Labischinski, H. (2004). Antibacterial research and development in the 21st century – an industry perspective of the challenges. Curr. Opin. Microbiol. **7**, 445–450.

Overbye, K.M. & Barrett, J.F. (2005). Antibiotics: where did we go wrong? Drug Discov. Today **10**, 45–52.

Barrett, J.F. (2005). Can biotech deliver new antibiotics? Curr. Opin. Microbiol. **8**, 498–503.

Talbot, G.H. Bradley, J., Edwards, J.E., et al. (2006). Bad bugs need drugs: an update on the development pipeline from the Antimicrobial Availability Task Force of the Infectious Diseases Society of America. Clin. Infect. Dis. **42**, 657–668.

- Evolution of language:

Fisher, S.E. & Marcus, G.F. (2006). The eloquent ape: genes, brains and the evolution of language. Nat. Rev. Genet. **7**, 9–20.

- A collection of papers describing the emerging field combining genomics and neuroscience:

Jones, B.C. & Mormède, P. (eds.) (2006). *Neurobehavioural Genetics/Methods and Applications*. CRC Press, Boca Raton.

- Presentation of whole transcriptome shotgun sequencing, or RNA-seq:

Wang, Z., Gerstein, M., & Snyder, M. (2009). RNA-seq: a revolutionary tool for transcriptomics. Nat. Rev. Genet. **10**, 57–63.

EXERCISES, PROBLEMS, AND WEBLEMS

Exercises

Exercise 9.1 In the third level (under Hybridization, washing), Figure 9.1 shows, schematically, a row of 14 probe oligomers, corresponding to the leftmost 14 elements in one of the rows of the 19×19 square array of fluorescent dots in the fourth level. Which row?

Exercise 9.2 A professor of molecular biology wanted to design a microarray experiment to detect tRNA sequences. He suggested using the sequences of the anticodon stem–loop as target oligonucleotides (see Figure 1.4). A student pointed out that this was not likely to be successful. For what reason?

Exercise 9.3 On a photocopy of Figure 9.2, (a) indicate the position of a gene that is highly downregulated in one of the vehicle samples but not in the other two; (b) indicate the position of a gene that is more highly upregulated in one of the dexamethasone samples than in the other two.

Exercise 9.4 On a photocopy of Figure 9.4, indicate the position of a gene that is first upregulated and subsequently downregulated during the yeast diauxic shift.

Exercise 9.5 (a) Citrate synthase and (b) pyruvate decarboxylase are two of the enzymes that change expression level in the diauxic shift in yeast. On a photocopy of Figure 9.3, mark the approximate positions of the reactions that they catalyse.

Exercise 9.6 Transcriptome profiling is the measurement of patterns of mRNA concentrations. However, 90% of the RNA in a cell is ribosomal. How could you apply the fact that mRNAs carry a 3′ poly A tail to avoid interference from the high background levels of ribosomal RNA?

Problems

Problem 9.1 The average ratio of gene expression levels between human and chimpanzee is ~1.5 for collinear chromosomes and can be as high as ~1.6–1.7 for rearranged chromosomes (Figure 9.12). (a) Qualitatively describe the differences in banding pattern between human and chimpanzee for human chromosomes 5, 6, 9, and 10 (see Figure 4.18). (b) Are your results consistent with the data in Figure 9.12? (c) Which would you expect to have an average ratio of expression levels closer to 1.0, genes on human chromosome 3 and their gorilla homologues, or genes on human chromosome 8 and their gorilla homologues?

Problem 9.2 Describe the reasons for and against including antibiotics routinely in animal feed. Decide on a conclusion to be drawn from these arguments and formulate a paragraph of recommendations.

Problem 9.3 A recent study of 1737 patients treated with vancomycin for *S. aureus* infections concluded that many patients received less than an adequate dose to maintain serum concentrations above the MIC.* For patients infected by vancomycin-sensitive *S. aureus*, this characterized 7.9% of patients given continuous infusions and 19% of patients to whom the drug was administered by periodic intravenous injections. For patients infected by *S. aureus* strains with intermediate-level resistance to vancomycin, this characterized 79.1% of patients given continuous infusions and 87.8% of patients given periodic intravenous injections. What are the expected effects of this situation on (a) the individual patients involved, and (b) the spread of vancomycin-resistant *S. aureus* strains?

Problem 9.4 You have a sample of mRNA which you convert to cDNA. From this material, what information can you derive from a microarray that would not be available from a high-throughput sequencing run with a Roche 454 Life Sciences Genome Sequencer?

Weblems

Weblem 9.1 An EST for rose germacrene D synthase appears in dbEST (GenBank) as entry BQ105086. Blast this sequence against general sequence databases. What properties of rose germacrene synthase can you infer from the results?

Weblem 9.2 (a) What is the chromosome location of the *ASPM* gene? (b) What mutations are known? (c) What is their clinical effect?

Weblem 9.3 Search for articles on discovery of novel antibiotics and study a number of them to learn about the current situation. (Suggested search terms: future antibiotic development. Alternatively, consult articles cited as recommended reading in this chapter.) (a) Provide facts and figures that describe trends in funding for research by large pharmaceutical companies devoted to novel antibiotic development. (b) Assuming that these figures confirm that large pharmaceutical

* Kitzis, M.D. & Goldstein, F.W. (2006). Monitoring of vancomycin serum levels for the treatment of staphylococcal infections. Clin. Microbiol. Infect. **12**, 92–95.

companies are curtailing the relevant research programmes, what are the reasons for this, in the face of high demand for novel antibiotics? (c) Can you recommend appropriate actions by governments, industry, and the academic community? Give credible reasons that justify the conclusion that if your actions were adopted they would accelerate the development and approval of novel antibiotics. What do you see as the most serious obstacles to acceptance of your ideas? How would you suggest overcoming them? (Warning: consider the possibility that the less ambitious your suggestions, the more credibly you can justify them.)

CHAPTER 10

Proteomics

LEARNING GOALS

- To understand the fundamental chemical structure of proteins: the mainchain and sidechains, types of sidechain, and common post-translational modifications.
- To understand the basic description of protein conformation.
- To be able to distinguish between primary, secondary, tertiary, and quaternary structures.
- To understand the use of polyacrylamide gel electrophoresis (PAGE) to separate proteins.
- To understand the technique and uses of mass spectrometry.
- To appreciate the principles of classifications of protein-folding patterns.
- To understand the possibilities and difficulties of protein structure prediction.
- To understand the goals of structural genomics projects.

Introduction

The proteome is the complete set of proteins associated with a sample of living matter. Proteomics deals with the proteins that form the structures of living things, are active in living things, or are produced by living things. This includes their nature, distribution, activities, interactions, and evolution. Many fields contribute to proteomics.

- *Chemistry and biochemistry*. These include physical methods, such as spectroscopy, kinetics, and techniques of structure determination, and organic and biochemical methods for working out mechanisms of enzymatic catalysis. Techniques for separation and analysis of proteins have their sources in chemistry and molecular biology.
- *Molecular and cellular biology*. These disciplines help to coordinate our knowledge of individual proteins into an understanding of the biological context, and of how protein activities are integrated.
- *Evolutionary biology*. Proteins evolve. Evolution explores variations in amino acid sequences, protein structures, interactions and functions, and patterns of protein expression.
- *Structural genomics*. This activity applies advances in X-ray crystallography and nuclear magnetic resonance (NMR) to high-throughput delivery of coordinate sets of proteins.
- *Bioinformatics*. This brings together the many data streams of genomics, expression patterns, and proteomics, to assemble databases and create links among them. This enables their coordinated application to problems of biology, clinical medicine, agriculture, and technology.

Bioinformatics co-ordinates its efforts with structural genomics to guide and supplement experimental structure determinations by prediction of protein structures from amino acid sequences. Prediction of protein structure is an important technique, given the great disparity between the very large number of experimentally determined sequences and the relatively few structures. Methods for prediction of protein structure from sequence, most of which take advantage of the known protein structures, can provide libraries of three-dimensional models of proteins encoded in genomes.

Protein nature and types

Proteins are where the action is.

- *Proteins have a great variety of functions*. There are structural proteins (molecules of the cytoskeleton, epidermal keratin, viral coat proteins); catalytic proteins (enzymes); transport and storage proteins (haemoglobin, retinol-binding protein, ferritin); regulatory proteins (including hormones, many kinases and phosphatases, and proteins that control gene expression); and proteins of the immune system and the immunoglobulin superfamily (including antibodies and proteins involved in cell–cell recognition and signalling). How can proteins accomplish so many different things? By coming in a great variety of structures, specialized to carry out different functions.
- *The amino acid sequences of proteins dictate their three-dimensional structures and their folding pathways*. Under physiological conditions of solvent and temperature, proteins fold spontaneously to an active native state. The amino acid sequence of a protein must not only preferentially stabilize the native state it must also contain a 'road map' telling the protein how to get there, starting from the many diverse conformations that comprise the unfolded state. This is called the **folding pathway**.
- *Advances in protein science have spawned the biotechnology industry*. It is now possible to design and test modifications of known proteins and to design novel ones with desired functions.

Protein structure

The chemical structure of proteins

Chemically, protein molecules are long polymers typically containing several thousand atoms, composed of a uniform repetitive **backbone** (or **mainchain**) with a particular **sidechain** attached to each residue (see Figure 10.1). The amino acid sequence of a protein specifies the order of the sidechains.

- A protein is a message written in a twenty-letter alphabet.

The sidechains in proteins show a variety of physicochemical features: some are charged, some are uncharged but polar, and others are **hydrophobic** (see Box 10.1). Different types of residue make different types of interaction.

- *Hydrogen bonding*. The hydrogen bond is an interaction between two polar atoms (oxygen or nitrogen; occasionally sulphur) mediated by a hydrogen atom. Several types of hydrogen bond are *extremely* important to biology:
  - Water is an extensively hydrogen-bonded liquid. This accounts for its physicochemical properties, for instance its high boiling point. The structure of water determines the solubility of different substances. Water is not merely the medium of most biochemical processes, it is an active participant in many.
  - Hydrogen bonds between nucleic acid bases mediate the complementarity between adenine and thymine, and between guanine and cytosine.
  - Hydrogen bonds between C=O and H–N groups stabilize the structure of proteins.

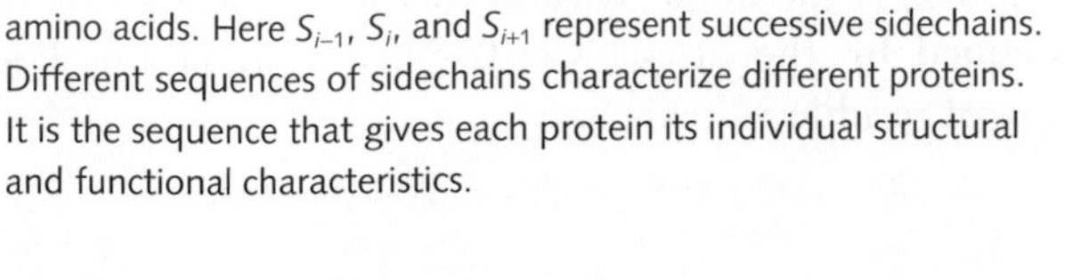

Residue $i-1$ Residue i Residue $i+1$

S_{i-1} S_i S_{i+1} } Sidechains variable

··· N—Cα—C—N—Cα—C—N—Cα—C— ··· } Mainchain constant

Figure 10.1 Proteins contain a mainchain of constant structure. Attached at regular intervals are sidechains of variable structure, each chosen (with few exceptions) from the canonical set of 20 amino acids. Here S_{i-1}, S_i, and S_{i+1} represent successive sidechains. Different sequences of sidechains characterize different proteins. It is the sequence that gives each protein its individual structural and functional characteristics.

Hydrogen bonds are about 20 times weaker than covalent chemical bonds. In aqueous solution, the solvent water can form hydrogen bonds to polar groups in nucleic acids and proteins. There is a competition between intramolecular hydrogen bonds and solute–solvent hydrogen bonds. Hydrogen bonds in solution can be easily broken and reformed.

BOX 10.1 The amino acids

Glycine	$^+NH_3$–Cα–COO^-
Alanine	–Cα–CH_3
Serine	–Cα–CH_2–OH
Cysteine	–Cα–CH_2–SH
Threonine	–Cα–CH(OH)–CH_3
Proline	(five-membered ring) N–Cα
Valine	–Cα–CH–$(CH_3)_2$
Leucine	–Cα–CH_2–CH–$(CH_3)_2$
Isoleucine	–Cα–CH(CH_3)–CH_2–CH_3
Methionine	–Cα–CH_2–CH_2–S–CH_3
Phenylalanine	–Cα–CH_2–(benzene ring)
Tyrosine	–Cα–CH_2–(benzene ring)–OH
Aspartic acid	–Cα–CH_2–COO^-
Glutamic acid	–Cα–CH_2–CH_2–COO^-
Histidine	–Cα–CH_2–C (imidazole ring: N, NH)
Asparagine	–Cα–CH_2–$CONH_2$
Glutamine	–Cα–CH_2–CH_2–$CONH_2$
Lysine	–Cα–CH_2–CH_2–CH_2–CH_2–NH_3^+
Arginine	–Cα–CH_2–CH_2–CH_2–NH–C(NH_2)=NH_2^+
Tryptophan	–Cα–CH_2–(indole ring, N–H)

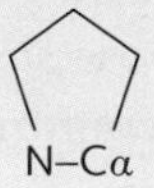

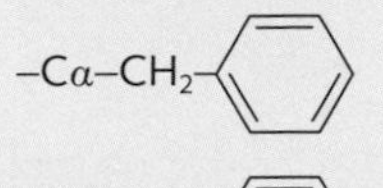

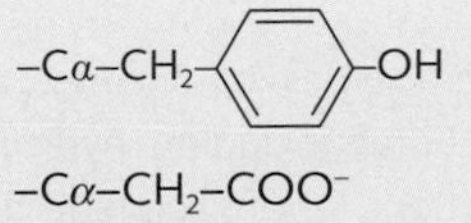

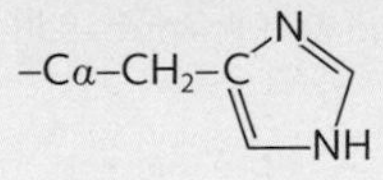

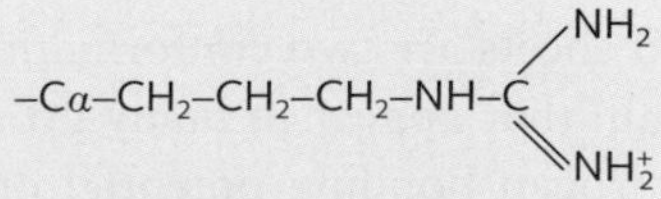

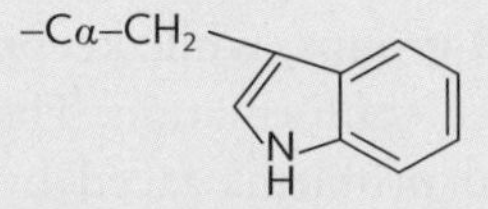

- *Hydrophobic interactions*. Hydrophobic residues have sidechains that are primarily hydrocarbon in nature. They have thermodynamically unfavourable interactions with water. Salad dressing is an everyday example of the hydrophobic effect: the thermodynamic unfavourability of dissolving oil in water causes a phase separation. It is energetically favourable to bury hydrophobic sidechains in the interior of a protein, where they are not exposed to the solvent. This is a general feature of the structures of globular proteins.
- *Disulphide bridges*. In addition to the primary chemical bonds in the individual residues and the peptide bonds joining the residues into a polymer, cysteine residues in proteins, with sidechain $-CH_2SH$, can form disulphide bonds: $-CH_2S-SCH_2-$. Disulphide bonds contribute to the stability of native states. In order to denature proteins fully, it is necessary to break any disulphide bonds.

Different possible conformations of the backbone of a protein bring different types of residue into spatial proximity and expose some but not all residues to the solvent. Every conformation therefore has a different associated energy that depends on the distribution of favourable and unfavourable interactions. The native state of a soluble globular protein is the conformation that optimizes the set of interactions among the residues and between the residues and the solvent.

- Different types of residues make different types of interactions, including hydrogen bonds, hydrophobic interactions, and disulphide bridges. Formation of the native structure allows optimal formation of favourable inter-residue and residue–solvent interactions.

Helices and sheets

Underlying the great variety of protein folding patterns are some recurrent structural themes. Helices and sheets are two conformations of the polypeptide chain that appear in many proteins. They satisfy the hydrogen bonding potential of the mainchain N–H and C=O groups, while keeping the mainchain in an unstrained conformation. They thereby solve certain structural problems faced by *all* globular proteins. They present general solutions – where in this context 'general' means 'compatible with all (or at least almost all) amino acid sequences'.

Helices and sheets are like Lego® pieces, standard units of structure of which many proteins are built and which can be put together in different ways. Helices are formed from a single consecutive set of residues in the amino acid sequence. They are therefore a **local structure** of the polypeptide chain, i.e. they form from a set of residues consecutive in the sequence. The mainchain hydrogen-bonding pattern of an α-helix, the most common type of helix, links the C=O group of residue i to the H–N group of residue $i + 4$.

Sheets form by lateral interactions of several independent sets of residues to create a hydrogen-bonded network that is often nearly flat, but sometimes cylindrical (forming a **barrel structure**). Unlike helices, sheets need not form from consecutive regions of the chain but may bring together sections of the chain separated widely in the sequence.

- Helices and sheets are recurrent structures, stabilized by mainchain hydrogen bonding, that appear in many protein structures.

Conformation of the polypeptide chain

The conformation of a polypeptide chain can be described in terms of angles of internal rotation around the bonds in the mainchain (see Figure 10.2). The bonds between the N and Cα, and between the Cα and C, are single bonds. Internal rotation around these bonds is not restricted by the electronic structure of the bond, only by possible steric collisions in the conformations produced (see Box 10.2).

The entire conformation of the protein can be described by these angles of internal rotation. Each set of four successive atoms in the mainchain defines an angle. In each residue i (except for the N and C termini), the angle ϕ_i is the angle defined by atoms C(of residue $i - 1$)–N–Cα–C, and the angle ψ_i is the angle defined by atoms N–Cα–C–N(of residue $i + 1$). Then ω_i is the angle around the peptide bond itself, defined by the atoms Cα–C–N(of residue $i + 1$)–Cα(of residue $i + 1$).

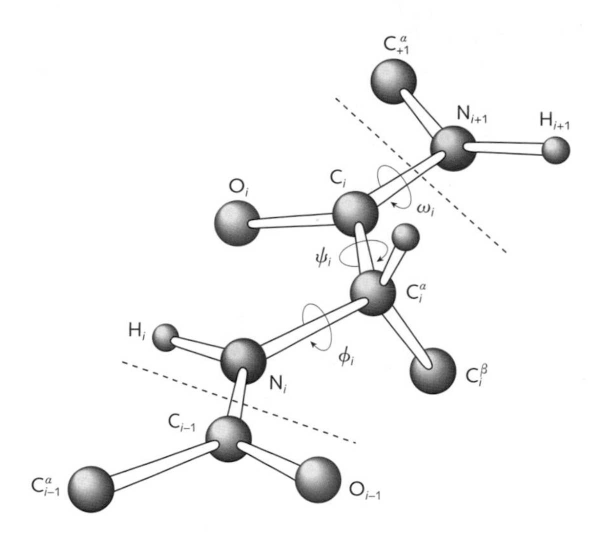

Figure 10.2 Conformational angles describing the folding of the polypeptide chain.

The mainchain of each residue (except the C-terminal residue) contains three chemical bonds: N–Cα, Cα–C, and the peptide bond C–N linking the residue to its successor. The conformation of the mainchain is described by the angles of rotation around these three bonds:

Rotation around:	N–Cα bond	Cα–C bond	Peptide bond (C–N)
Name of angle:	ϕ	ψ	ω

ω is restricted to be close to 180° (*trans*) or, infrequently, close to 0° (*cis*).

The peptide bond has a partial double-bond character and adopts two possible conformations: *trans* (by far the more common) and *cis* (rare). Angle ω is restricted to be close to 180° (*trans*) or 0° (*cis*).

Proline is an exception: the sidechain is linked back to the N of the mainchain to form a pyrrolidine ring. This restricts the mainchain conformation of proline residues. It disqualifies the N atom as a hydrogen-bond donor, for instance in helices or sheets. Also, the energy difference between *cis* and *trans* conformations is less for proline residues than for others. Most *cis* peptides in proteins appear before prolines.

Protein folding patterns

Focusing on the backbone, in the native state the polypeptide chain follows a curve in space. The general spatial layout of this curve defines a **folding pattern.** We now know over 70 000 protein structures. There is great but not infinite variety: many proteins have similar folding patterns. The native states are selected from a large but finite repertoire.

In describing protein structures, the Danish protein chemist K.U. Linderstrøm-Lang described a hierarchy of levels of protein structure. The amino acid sequence – roughly the set of chemical bonds – is

The Sasisekharan–Ramakrishnan–Ramachandran diagram

The mainchain conformation of each residue is determined primarily by the two angles ϕ and ψ, assuming the common *trans* conformation of the peptide bond, $\omega = 180°$.

For some combinations of ϕ and ψ, atoms would collide, a physical impossibility. V. Sasisekharan, C. Ramakrishnan, and G.N. Ramachandran first plotted the sterically allowed regions (see Figure 10.3). There are two main allowed regions, one around $\phi = -57°$, $\psi = -47°$ (denoted α_R) and the other around $\phi = -125°$, $\psi = +125°$ (denoted β) with a 'neck' between them. The mirror image of the α_R conformation, denoted α_L, is allowed for glycine residues only. (As glycine is achiral – identical to its mirror image – a Ramachandran plot specialized to glycine must be right–left symmetric. For non-glycine residues, collisions of the Cβ atom forbid the α_L conformation.)

The two major allowed conformations of the mainchain, α_R and β, correspond to the two major types of secondary structure: α-helix and β-sheet. The α-helix is right-handed, like the threads of an ordinary bolt. In the β region, the chain is nearly fully extended.

A graph showing the ϕ and ψ angles for the residues of a protein against the background of the allowed regions is called a Sasisekharan–Ramakrishnan–Ramachandran plot, often called a Ramachandran plot for short.

It is no coincidence that the same conformations that correspond to low-energy states of individual residues also permit the formation of structures with extensive mainchain hydrogen bonding. The two effects thereby cooperate to lower the energy of the native state.

→

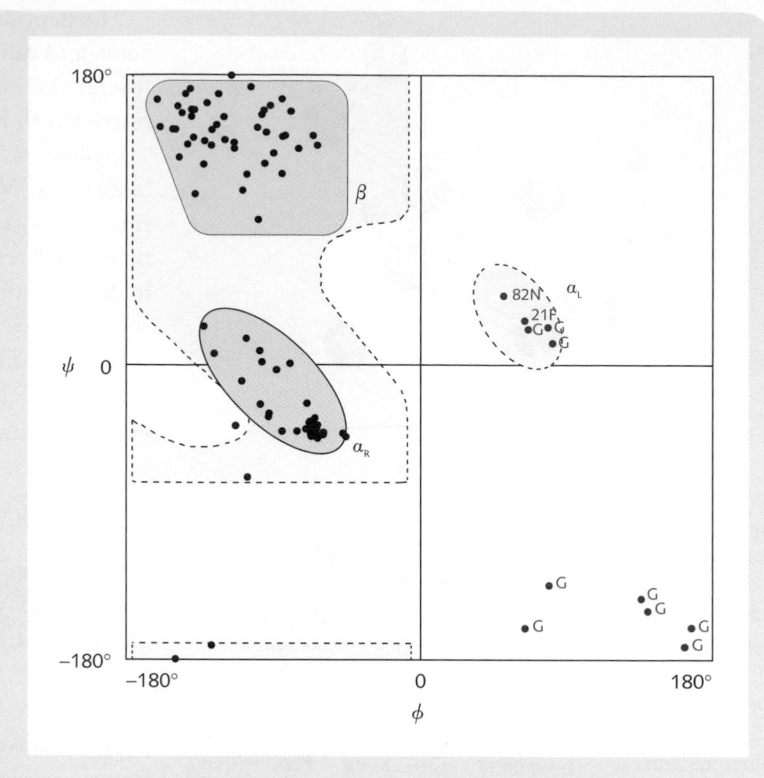

Figure 10.3 A Sasisekharan–Ramakrishnan–Ramachandran plot of bovine acylphosphatase [2ACY]. Sterically most-favourable regions are shown in green and sterically allowed regions in yellow. Residues with $\phi > 0$, mostly glycines, appear in red.

called the **primary structure**. The assignment of helices and sheets – the hydrogen-bonding pattern of the mainchain – is called the **secondary structure**. The assembly and interactions of the helices and sheets is called the **tertiary structure**. For proteins composed of more than one subunit, J.D. Bernal called the assembly of the monomers the **quaternary structure** (see Figure 10.4 and Box 10.3).

Some proteins change their quaternary structure as part of a regulatory process. Cyclic AMP activates protein kinase A by a mechanism involving subunit dissociation. The resting, inactive form of protein kinase A is a tetramer of two catalytic subunits and two regulatory subunits. In this resting state, the regulatory subunits inhibit the activity of the catalytic subunits. Binding of cyclic AMP to protein kinase A dissociates the tetramer, releasing individual catalytic subunits in active form.

In some cases, evolution can merge proteins – changing quaternary to tertiary structure. For example, five separate enzymes in *Escherichia coli* that catalyse successive steps in the pathway of biosynthesis of aromatic amino acids correspond to five regions of a single protein in the fungus *Aspergillus nidulans*.

- We describe protein folding patterns according to a hierarchy of primary, secondary, tertiary, and quaternary structures. See Box 10.3 and Figure 10.4.

Domains

One way that proteins have evolved increasing complexity is by assembling a large protein from a set of smaller quasi-independent subunits, either by forming stable oligomers, as in haemoglobin (see Figure 10.4), or by concatenating units within a *single* polypeptide chain. **Domains** are compact units within the folding pattern of a single chain. Justifications for regarding them as quasi-independent include the observation that domains can be 'mixed and matched' in different proteins and, in many cases, the similarities of their

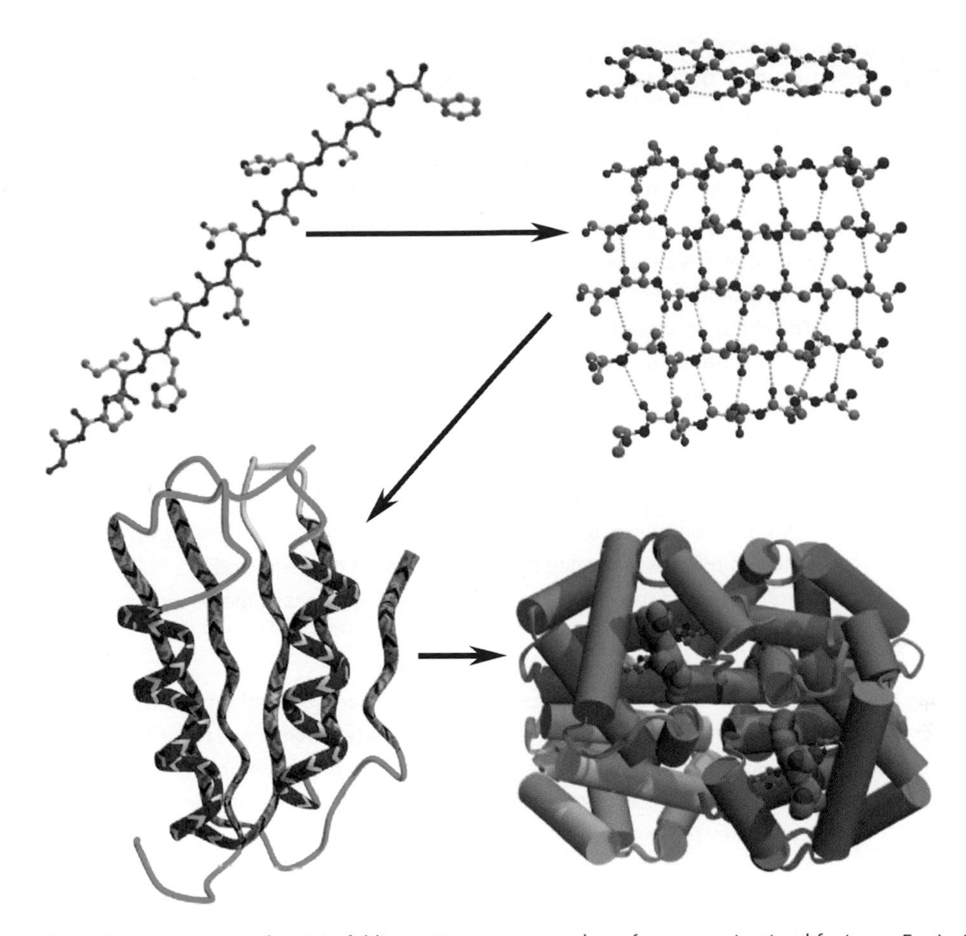

Figure 10.4 Underlying the great variety of protein folding patterns are a number of common structural features. For instance, α-helices and β-sheets are standard elements of the 'parts list' of many protein structures. α-helices and β-sheets were modelled by L. Pauling before their experimental observation. Pauling recognized that helices and sheets provide convenient ways for the residues to achieve comfortable steric relationships and satisfy the requirements for backbone hydrogen bonding in an (almost) sequence-independent manner.

This figure shows, at the upper left, the primary structure in terms of a simple extended chain. The standard secondary structures, the α-helix and β-sheet, are shown at the upper right, with hydrogen bonds indicated by broken lines. Tertiary structure is represented, at the lower left, by acylphosphatase, which contains two α-helices packed against a five-stranded β-sheet. Human haemoglobin, a tetramer containing two copies of two types of chain, illustrates quaternary structure, at the lower right. (Acylphosphatase is *not* a subunit of haemoglobin.)

folding patterns to those of homologous monomeric proteins.

Domains form the basis of the higher-level protein structural organization typical of eukaryotic proteins. **Modular proteins** are multidomain proteins that often contain many copies of closely related domains. For example, fibronectin, a large extracellular protein involved in cell adhesion and migration, contains 29 domains including multiple tandem repeats of three types of domain, F1, F2, and F3 (see Figure 4.16). It is a linear array of the form: $(F1)_6(F2)_2(F1)_3(F3)_{15}(F1)_3$. Fibronectin domains also appear in other modular proteins. (See http://www.bork.embl-heidelberg.de/Modules/ for pictures and nomenclature.)

To create new proteins, inventing new domains is an unusual event. It is far more common to create different combinations of existing domains in increasingly complex ways. These processes can occur independently, and take different courses, in different phyla.

BOX 10.3

Protein structure – basic vocabulary

Polypeptide chain	Linear polymer of amino acids.
Mainchain	Atoms of the repetitive concatenation of peptide groups . . . N–Cα–(C=O)N–Cα–(C=O) . . .
Sidechains	Sets of atoms attached to each Cα of the mainchain. Most sidechains in proteins are chosen from a canonical set of 20.
Primary structure	The chemical bonds linking atoms in the amino acid sequence in a protein.
Hydrogen bond	A weak interaction between two neighbouring polar atoms, mediated by a hydrogen atom.
Secondary structure	Substructures common to many proteins, compatible with mainchain conformations, free of interatomic collisions, and stabilized by hydrogen bonds between mainchain atoms. Secondary structures are compatible with all amino acids, except that a proline necessarily disrupts the hydrogen-bonding pattern.
α-Helix	Type of secondary structure in which the chain winds into a helix, with hydrogen bonds between residues separated by four positions in the sequence.
β-Sheet	Another type of secondary structure, in which sections of mainchain interact by lateral hydrogen bonding.
Folding pattern	Layout of the chain as a curve through space.
Tertiary structure	The spatial assembly of the helices and sheets, and the pattern of interactions between them. (Folding pattern and tertiary structure are nearly synonymous terms.)
Quaternary structure	The assembly of multisubunit proteins from two or more monomers.
Native state	The biologically active form of a protein, which is compact and low energy. Under suitable conditions, proteins form native states spontaneously.
Denaturant	A chemical that tends to disrupt the native state of a protein; for instance, urea.
Denatured state	Non-compact, structurally heterogeneous state formed by proteins under conditions of high temperature, or high concentrations of denaturant.
Post-translational modification	Chemical change in a protein after its creation by the normal protein-synthesizing machinery.
Disulphide bridge	Sulphur–sulphur bond between two cysteine sidechains. A simple example of a post-translational modification.

Post-translational modifications

How much does genomics actually tell us about the proteome? Even if we could identify coding regions of genomes with complete accuracy, we would not know about:

- levels of transcription – or even absence of transcription;
- formation of different splice variants (in eukaryotes);
- mRNA editing – exclusive of splicing – before translation, which alters the amino acid sequence;
- the nature and binding sites of ligands integral to the final structure; and
- post-translational modifications, the subject of this section.

The ribosome synthesizes proteins by using the genetic code to direct the incorporation of a sequence of amino acids chosen from the canonical 20. Selenomethionine and pyrrolysine are two natural rare extensions of the standard genetic code.

However, the protein world is richer than the standard genetic code suggests. Many proteins contain ligands, such as metal ions or small organic molecules, as intrinsic and permanent parts of the structures. The nature of the binding between protein and ligand depends on the protein as well as the ligand. For instance, the haem group is bound covalently to cytochrome *c* but non-covalently to (almost all) globins. (Of course, proteins bind many molecules *transiently*. Enzyme–substrate complexes provide many examples.)

Post-translational modifications can take several forms (see also Box 10.4).

- Attaching various groups to sidechains, including but not limited to acetate, phosphate, lipids, and carbohydrates. Disulphide bridge formation is a related example. Some additions, such as sulphation, are permanent modifications; others, notably phosphorylations, are in many cases reversible.
- Conversions, for instance deamidation of asparagine (or glutamine) to aspartic acid (glutamic acid), or deimination of arginine to citrulline.
- Removing peptides, either from a terminus or from the middle of the chain, and in a few cases even making cyclic permutations.
- Addition of other peptides or proteins, not always by extension of the mainchain, through peptide linkages.

Major types of post-translational modification

- Attachments of groups to termini and sidechains. Although acetylation of protein N termini is not uncommon, most modifications involve sidechains. Many possible derivatives are observed.
  - **Reversible phosphorylation** of serine, threonine, or tyrosine sidechains is a very common means of regulating protein activity. However, **irreversible phosphorylation** of tau protein contributes to the development of Alzheimer's disease. The neurotoxicity of some organophosphorus compounds arises from their irreversible phosphorylation of acetylcholinesterase.
  - Attachment of sugars or oligosaccharides to proteins to make **glycoproteins**. In mammals, many glycoproteins appear on cell surfaces, to mediate cell–cell recognition and communication and immune system recognition. The difference between O, A, and B blood groups resides in the carbohydrate attached to serum glycoproteins and to (non-protein) glycolipids on cell surfaces. Many viruses, including influenza and HIV-1, gain entry to cells via cell-surface glycoprotein receptors. Lectins are carbohydrate recognition proteins that mediate cell–cell recognition and communication and sugar transport. In vertebrates, recognition of sugars on the surfaces of bacteria is a component of the immune response to infection.

    Deficiencies in turnover lead to **glycoprotein storage diseases** involving accumulation of incompletely degraded glycoproteins or oligosaccharides. Examples include α- and β-mannosidosis and aspartylglucosaminuria. Tay–Sachs disease is a related condition, part of a larger family of diseases called the **lysosomal storage diseases**. Tay–Sachs disease arises from a mutation in the α subunit of the hexosaminidase A gene. In Tay–Sachs disease, the dysfunction of the mutant protein impedes the degradation of a ganglioside, rather than the degradation of a glycoprotein.
  - Addition of oligomers of the small protein ubiquitin to lysine residues targets proteins for degradation by the proteasome. Conversely, the methylation of lysine is believed to 'protect' proteins against ubiquitinylation and thereby against degradation.
- Post-translational modification by proteolytic cleavage implies that the protein is synthesized with extra amino acids beyond those required to form the native state. What is the purpose of the additional residues?
  - Some proteins are synthesized with **N-terminal signal peptides** that direct their transport to particular subcellular compartments or organelles, or mark them for secretion.
  - It would be dangerous to turn rogue proteases loose on the cells in which they are synthesized. Many proteases are synthesized in inactive forms and then activated by cleavage.

→

- Facilitation of folding. Insulin contains two polypeptide chains, one of 21 residues and the other of 30 residues. The precursor proinsulin is a single 81-residue polypeptide chain from which *excision of an internal peptide* produces the mature protein. Insulin contains one intrachain and two interchain disulphide bridges. Attempts to renature mature insulin – after unfolding and breaking the disulphide bridges – give poor yields. Many incorrectly paired disulphide bridges form. *In vivo*, the precursor proinsulin folds into a three-dimensional structure with the cysteines in proper relative positions to form the correct disulphide bridges. Excision of a central region by endopeptidases then produces the mature dimer. Unfolded proinsulin *can* spontaneously refold correctly.

- Most post-translational cleavage reactions are carried out by proteases. Alternatively, **inteins** are proteins that have a 'self-splicing' activity. They autocatalytically excise internal peptides and join the ends. (In contrast, peptide excision from proinsulin leaves two chains that are *not* joined by a peptide bond.)
- The lectin concanavalin A is synthesized in a precursor form that is a cyclic permutation of the final structure. Thus, during maturation of the protein, there is cleavage of an internal peptide bond and formation of a new peptide bond between the original N and C termini. For concanavalin A, the DNA sequence of the gene is not co-linear with the amino acid sequence of the mature protein.

Why is there a common genetic code with 20 canonical amino acids?

Almost all organisms synthesize proteins containing a canonical set of 20 amino acids.

However, both nature and the laboratory show that 20 amino acids are not a fundamental limitation. Selenomethionine and pyrrolysine are natural exceptions. P. Schultz and co-workers have extended the genetic code by introducing modified tRNAs and synthetases into *E. coli*, yeast, and even mammalian cells in tissue culture.* Approximately 70 novel amino acids are now available to be introduced into proteins at specific sites. Some of the novel amino acids show designed steric or electronic properties; others contain chromophores as fluorescent reporters or are susceptible to photocross-linking; there are glycosylated amino acids; iodine derivatives to facilitate X-ray structure determination; and sidechains containing other types of reactive group.

In understanding the contents and layout of the common genetic code, can we go beyond F. Crick's comment that the code is a 'frozen accident'?

There is now consensus that prokaryotes were and are engaged in widespread horizontal gene transfer (see p. 120). This suggests – leaving aside the question of what the optimal genetic code should be – that it would be to the advantage of any participating species to conform to *some* standard, as that would give it access to all of the other genes. Analogously, anyone can run any operating system on a computer that they want, but the obvious advantages of running the same system as many other people exert pressure to conform to *some* standard. Perhaps that is at least a partial explanation of why almost all species have the same genetic code.

It is true that if different species adopted different genetic codes, this might protect them against viruses jumping from other species.

But why not a code with many more than 20 amino acids? Certainly, one perfectly feasible way to introduce greater versatility into the components of proteins is by expanding the genetic code. However, keeping within the general framework of a triplet code, introducing more amino acids at the expense of the redundancy of the code threatens to reduce robustness. An alternative approach to greater versatility without this cost is to effect post-translational modifications of individual amino acids. Whether or not this reasoning is the correct explanation, post-translational modification is the choice that nature seems largely to have made.

* See http://schultz.scripps.edu/research.html and Wang, L. & Schultz, P.G. (2004). Expanding the genetic code. Angew. Chem. Int. Ed. Engl. **44**, 34–66.

Separation and analysis of proteins

The complete complement of a cell's proteins is a large and complex set of molecules. Metazoa contain tens of thousands of protein-encoding genes. Different splice variants multiply the number of possible proteins. Vertebrate immune systems generate billions of molecules by specialized techniques of combinatorial gene assembly.

To give some idea of the 'dynamic range' required of detection techniques, the protein inventory of a yeast cell varies from 1 copy per cell to 1 million copies per cell.

Examples of techniques for separating mixtures of proteins include gel filtration, chromatography, and electrophoresis. All methods of separating molecules require two things:

1. A difference in some physical property, between the molecules to be separated; and
2. a mechanism, taking advantage of that property, to set the molecules in motion; the speed differing according to the value of the property selected. This moves apart molecules with different properties.

In some separation methods, one component can stand still and the other(s) move away from it. Affinity chromatography is an example. With others, different species can all move, at different rates, and spread themselves out.

- To measure an inventory of the proteins in a sample, the proteins must be: (1) separated, (2) identified, (3) counted.

Polyacrylamide gel electrophoresis (PAGE)

In electrophoresis, an electric field exerts force on a molecule. The force is proportional to the molecule's total or net charge. In a vacuum, the corresponding acceleration would be inversely proportional to the mass. However, counteracting the acceleration from the electric field are retarding forces from the medium through which the proteins move. **Polyacrylamide gels** contain networks of tunnels, with a distribution of sizes. Smaller proteins can enter smaller tunnels as well as larger ones, and therefore move faster through the gel than larger proteins. Proteins with different mobilities move different distances during a run, spreading them out on the gel.

The mobility of a native protein depends on its mass and its shape. Higher mass tends to reduce mobility; more compact shape tends to increase it. In particular the mobility of denatured proteins is lower than that of the corresponding native states. To achieve a separation that depends solely on molecular weight, denature the proteins. Common denaturing media include urea (which competes for hydrogen bonds), and the reducing agent dithiothreitol to break S–S bridges (and iodoacetamide to prevent their reformation).

Sodium dodecyl sulphate (SDS) is a negatively charged detergent that helps to denature proteins. Multiple detergent molecules bind all along the polypeptide chain. The result is a protein–detergent complex that has an extended shape, with a uniform charge density along its length.

Carrying out **SDS-PAGE** in one dimension spreads out a mixture of proteins or nucleic acids into bands. Running several samples on the same gel in parallel lanes is a familiar procedure if only from sequencing gels. The results of protein gels can be made visible ('developed') by staining with **Coomassie Blue**, or, if the samples are radioactively labelled, by **autoradiography**. Often markers of known molecular weight are run in a separate lane for calibration.

Two-dimensional polyacrylamide gel electrophoresis (2D-PAGE)

One-dimensional PAGE will not adequately separate a very complicated mixture of proteins. The bands in a lane on a gel will overlap, and contain mixtures of proteins with similar sizes. To achieve better resolution, a two-stage procedure first separates proteins according to charge; then an SDS-PAGE step, run in a direction 90° from the original direction, separates according to size.

The charge on a protein depends on the charged residues it contains, and the pH of the medium. At different values of pH, ionizable groups on proteins have different charges. For instance, a free histidine

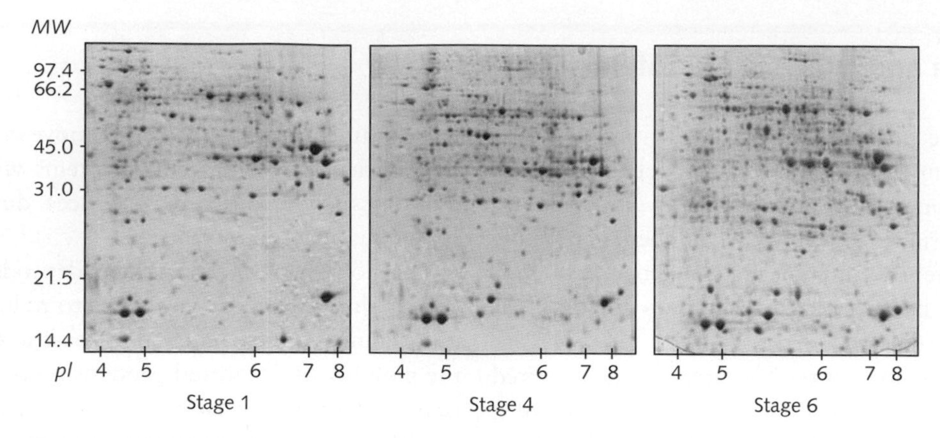

Figure 10.5 Two-dimensional PAGE gels of rose petal proteins at developmental stages 1, 4, and 6. Each gel contains over 600 proteins, of which 421 are common to all three stages. About 12% of the proteins are stage specific.

From: Dafny-Yelin, M., et al. (2005). Flower proteome: changes in protein spectrum during the advanced stages of rose petal development. Planta 222, 37–46.

sidechain is uncharged below pH ~5, and positively charged above pH ~7. For any protein, there is a pH at which it has a net charge of 0. This is called its **isoelectric point.**

A protein at its isoelectric point will feel no force in an electric field. It will not migrate in electrophoresis. To separate proteins according to their isoelectric points, establish a pH gradient in a medium and apply an electrophoretic field. The proteins will migrate, changing their charge as they pass through regions of different pH, until they reach their isoelectric points and then they will stop. The result, called **isoelectric focusing**, spreads proteins out according to their charged sidechains.

After the proteins are spread out along a lane by isoelectric focusing, running PAGE at 90° spreads them out in two dimensions (Figure 10.5). It is possible to compare the resulting patterns. Spots of interest can be eluted and identified by mass spectrometry (see next section).

- Polyacrylamide gel electrophoresis (PAGE) is a common method for protein separation. Proteins migrate through a gel with different mobilities depending on their mass, shape, and charge. Proteins separated by SDS-PAGE are denatured by the detergent sodium dodecyl sulphate, creating a protein–detergent complex with a uniform layer of negative charge. Protein mobility in SDS-PAGE depends only on relative molecular mass.

Mass spectrometry

Mass spectrometry is a physical technique that characterizes molecules by measurement of the masses of their ions, or of ions formed from their fragments. Applications to molecular biology include:

- rapid identification of the components of a complex mixture of proteins;
- sequencing of proteins and nucleic acids (see p. 309);
- analysis of post-translational modifications or substitutions relative to an expected sequence; and
- measuring extents of hydrogen–deuterium exchange to reveal the solvent exposure of individual sites (providing information about static conformation, dynamics, and interactions).

Identification of components of a complex mixture

First, the components are separated by electrophoresis, then the isolated proteins are digested by trypsin to produce peptide fragments with relative molecular masses of about 800–4000. Trypsin cleaves proteins after Lys and Arg residues. Given a typical amino acid composition, a protein of 500 residues yields about 50 tryptic fragments. The mass spectrometer measures the masses of the fragments with very high accuracy (see Figure 10.6). The list of fragment masses, called the **peptide mass fingerprint**, characterizes the protein (Figures 10.7 and 10.8). Searching a database of fragment masses identifies the unknown sample.

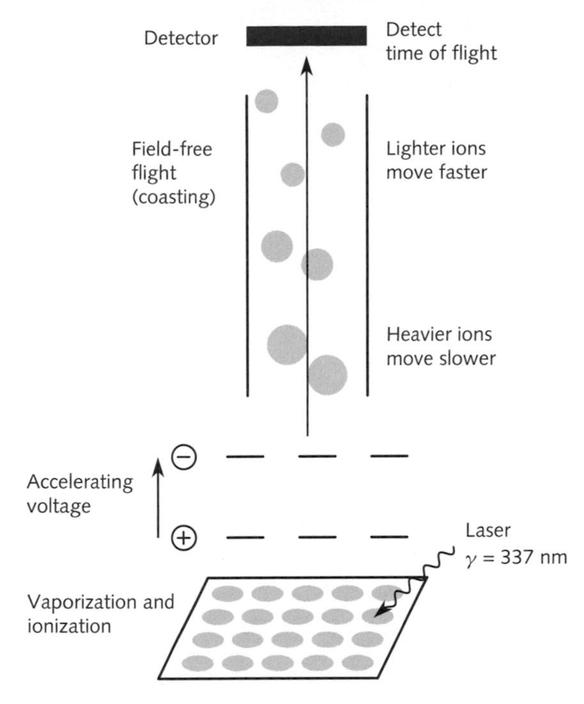

Figure 10.6 Schematic diagram of a mass spectrometry experiment.

Mass spectrometry is sensitive and fast. Peptide mass fingerprinting can identify proteins in subpicomole quantities. Measurement of fragment masses to better than 0.1 mass units is quite good enough to resolve isotopic mixtures. It is a high throughput method, capable of processing 100 spots/day (although sample preparation time is longer). However, there are limitations. Only proteins of known sequence can be identified from peptide mass fingerprints, because only their predicted fragment masses are included in the databases. (As with other fingerprinting methods, it would be possible to show that two proteins from different samples are likely to be the same, even if no identification is possible.) Post-translational modifications interfere because they alter the masses of the fragments.

Protein sequencing by mass spectrometry

Fragmentation of a peptide produces a mixture of ions. Conditions under which cleavage occurs primarily at

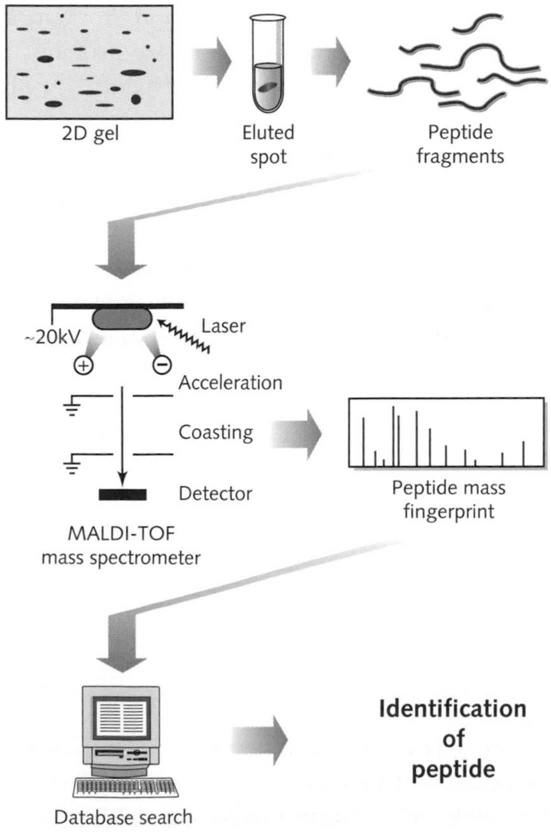

2D gel
Eluted spot
Peptide fragments
~20kV
Laser
Acceleration
Coasting
Detector
MALDI-TOF mass spectrometer
Peptide mass fingerprint
Database search
Identification of peptide

Figure 10.7 (*left*) Identification of components of a mixture of proteins by elution of individual spots, digestion, and fingerprinting of the peptide fragments by MALDI–TOF (matrix-assisted laser desorption ionization–time of flight) mass spectrometry, followed by looking up the set of fragment masses in a database.

The operation of the spectrometer involves the following steps.

1. Production of the sample in an ionized form in the vapour phase. Proteins being fairly delicate objects, it has been challenging to vaporize and ionize them without damage. Two 'soft-ionization' methods that solve this problem are:
   - **matrix-assisted laser desorption ionization** (MALDI): the protein sample is mixed with a substrate or matrix that moderates the delivery of energy; a laser pulse absorbed initially by the matrix vaporizes and ionizes the protein;
   - **electrospray ionization** (ESI): the sample in liquid form is sprayed through a small capillary with an electric field at the tip to create an aerosol of highly charged droplets, which fragment upon evaporation, ultimately producing ions, which may be multiply charged, devoid of solvent; these ions are transferred into the high vacuum region of the mass spectrometer.
2. Acceleration of the ions in an electric field. Each ion emerges with a velocity proportional to its charge/mass ratio.
3. Passage of the ions into a field-free region, where they 'coast'.
4. Detection of the times of arrival of the ions. The time of flight (TOF) indicates the mass-to-charge ratio of the ions.
5. The result of the measurements is a trace showing the flux as a function of the mass-to-charge ratio of the ions detected (see Figure 10.8).

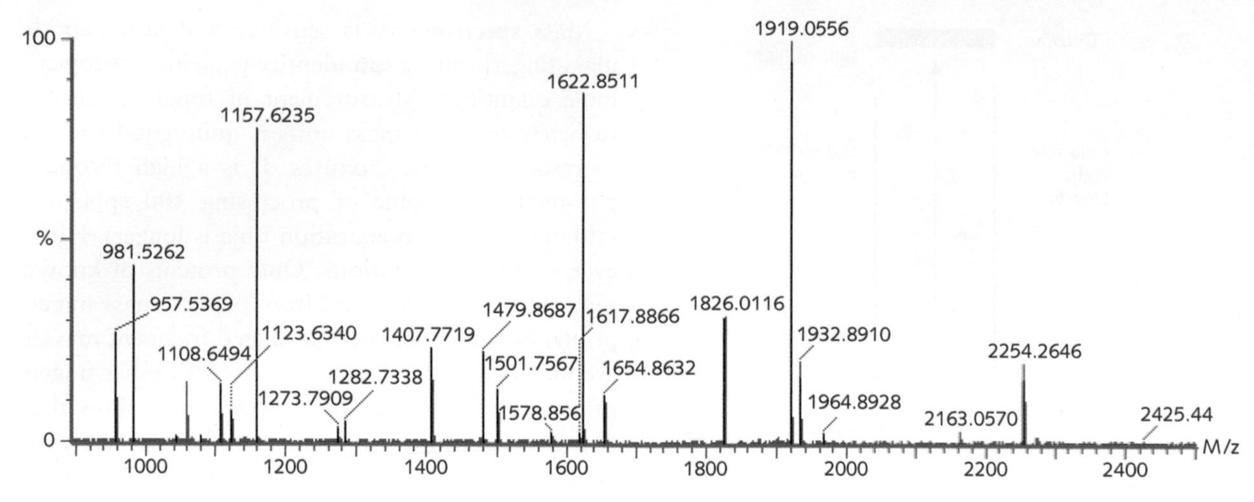

Figure 10.8 Mass spectrum of a tryptic digest. Of the 21 highest peaks (shown in black), 15 match expected tryptic peptides of the 39 kDa subunit of cow mitochondrial complex I. This easily suffices for a positive identification.

Figure courtesy of Dr I.M. Fearnley, MRC Dunn Human Nutrition Unit, Cambridge, UK.

peptide bonds yield a series of ions differing by the masses of single amino acids. The y ions are a set of nested fragments containing the C terminus (see Figure 10.9a) (b ions are nested fragments containing the N terminus). The difference in mass between successive y ions is the mass of a single residue. The amino acid sequence of the peptide is, therefore, deducible from analysis of the mass spectrum (see Figure 10.9b).

Two ambiguities remain: Leu and Ile have the same mass and cannot be distinguished, and Lys and Gln have almost the same mass and usually cannot be distinguished. Discrepancies from the masses of standard amino acids signal post-translational modifications. In practice, the sequence of about 5–10 amino acids can be determined from a peptide of length <20–30 residues.

Measuring deuterium exchange in proteins

If a protein is exposed to heavy water (D_2O), mobile hydrogen atoms will exchange with deuterium at rates dependent on the protein conformation. By exposing proteins to D_2O for variable amounts of time, mass spectrometry can give a conformational map of the protein. Applied to native proteins, the results give information about the structure. Applied to initially denatured proteins brought to renaturing conditions using pulses of exposure, the method can give information about intermediates in folding.

- Mass spectrometry is often used to characterize proteins isolated from mixtures. The peptide mass fingerprint – the list of fragment masses – is usually sufficient to identify a protein.

Classification of protein structures

Several web sites offer hierarchical classifications of the entire Protein Data Bank (wwPDB; see p. 107) according to the folding patterns of the proteins. These include:

- SCOP: Structural Classification of Proteins
  http://scop.mrc-lmb.cam.ac.uk/scop/
- CATH: Class/Architecture/Topology/Homologous superfamily
  http://cathwww.biochem.ucl.ac.uk/latest/
- DALI: based on extraction of similar structures from distance matrices
  http://ekhidna.biocenter.helsinki.fi/dali/start

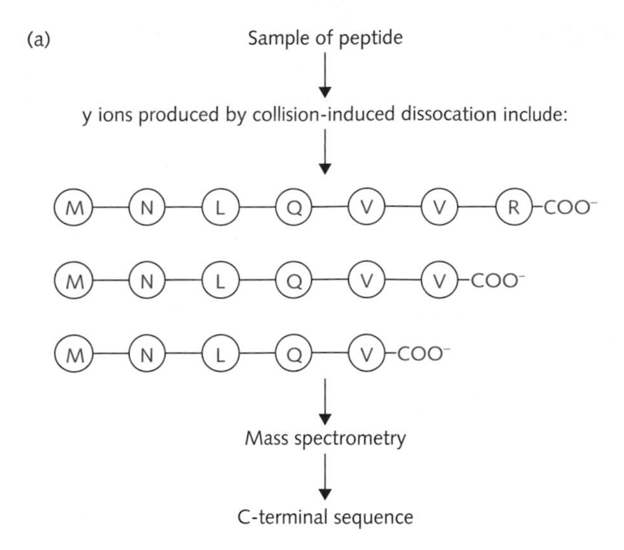

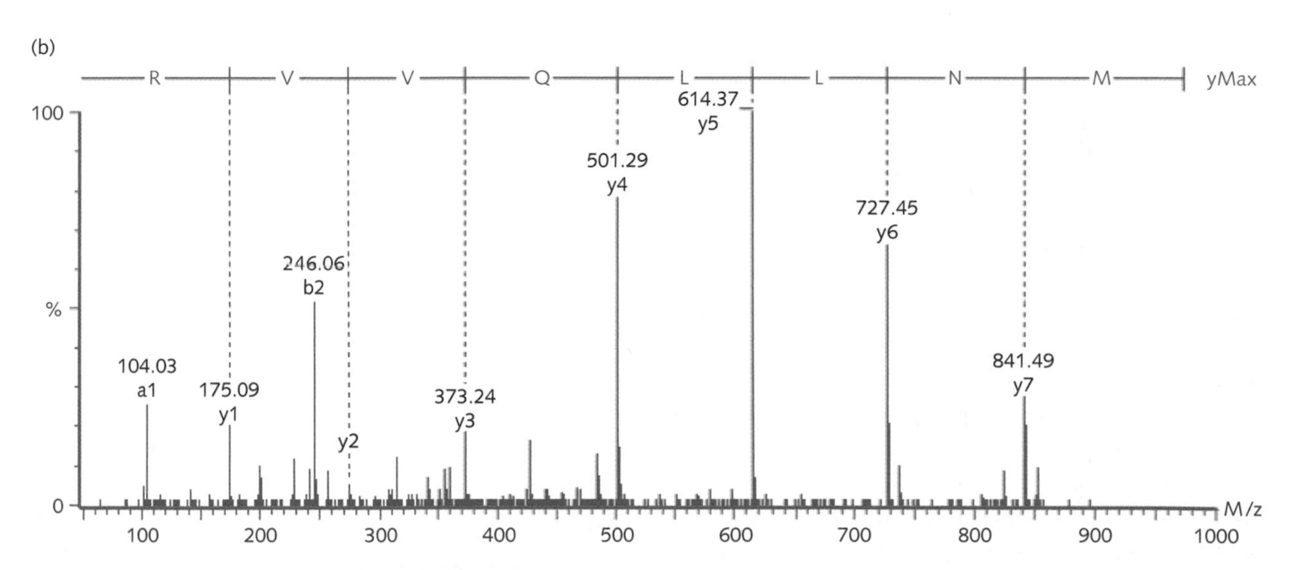

Figure 10.9 Peptide sequencing by mass spectrometry. Collision-induced dissociation produces a mixture of ions. (a) The mixture contains a series of ions, differing by the masses of successive amino acids in the sequence. The ions are not *produced* in sequence as suggested by this list, but the mass-spectral measurement automatically sorts them in order of their mass/charge ratio. (b) Mass spectrum of fragments suitable for C-terminal sequence determination. The greater stability of y ions over b ions in fragments produced from tryptic digests simplifies the interpretation of the spectrum. The mass differences between successive y-ion peaks are equal to the individual residue masses of successive amino acids in the sequence. Because y ions contain the C terminus, the y-ion peak of smallest mass contains the C-terminal residue etc., and therefore the sequence comes out 'in reverse'. The two leucine residues in this sequence could not be distinguished from isoleucine in this experiment.

From: Carroll, J., Fearnley, I.M., Shannon, R.J., Hirst, J., & Walker, J.E. (2003). Analysis of the subunit composition of complex I from bovine heart mitochondria. Mol. Cell Proteomics **2**, 117–126 (supplementary figure S138).

- CE: a database of structural alignments
  http://cl.sdsc.edu/

These sites describe projects *derived* from the primary archival databases of macromolecular coordinates. They are useful general entry points to protein structural data.

- It is in the tertiary structure of domains that proteins show their individuality and variety. Classifying proteins according to their tertiary structure indicates evolutionary relationships (or, at the very least, interesting structural similarities) between proteins that might have diverged so far that the relationship is not detectable by comparing their amino acid sequences.

SCOP

SCOP, by A.G. Murzin, L. Lo Conte, B.G. Ailey, S.E. Brenner, T.J.P. Hubbard, and C. Chothia, organizes protein structures in a hierarchy according to evolutionary origin and structural similarity. At the lowest level of the hierarchy are individual domains. SCOP groups sets of domains into **families** of homologues, for which the similarities in structure and sequence (and sometimes function) imply a common evolutionary origin. Families, containing proteins of similar structure and function but for which the evidence for evolutionary relationship is suggestive but not compelling, form **superfamilies**. Superfamilies that share a common folding topology, for at least a large central portion of the structure, are grouped as **folds**. Finally, each fold group falls into one of the general **classes**. The major classes in SCOP are α, β, $\alpha + \beta$, α/β, and miscellaneous 'small proteins', many of which have little secondary structure and have structures stabilized by disulphide bridges or ligands. Box 10.5 shows the SCOP classification of *E. coli* CheY (see Figure 10.10).

- SCOP (Structural Classification of Proteins) offers facilities for searching on keywords to identify structures, navigation up and down the hierarchy, generation of pictures, access to the annotation records in the PDB entries, and links to related databases.

The latest SCOP release contains 38 221 PDB entries split into 110 800 domains. The distribution of entries at different levels of the hierarchy is shown in Table 10.1.

To locate a protein of interest in SCOP, the user can traverse the structural hierarchy, or search via

BOX 10.5 SCOP classification of CheY protein of *Escherichia coli*

1. *Root*: SCOP.
2. *Class*: Alpha and beta proteins (α/β).
   Mainly parallel β-sheets (β–α–β units).
3. *Fold*: Flavodoxin-like.
   Three layers, $\alpha/\beta/\alpha$; parallel β-sheet of five strands, order 21345.
4. *Superfamily*: CheY-like.
5. *Family*: CheY-related.
6. *Protein*: CheY protein.
7. *Species*: *Escherichia coli*.

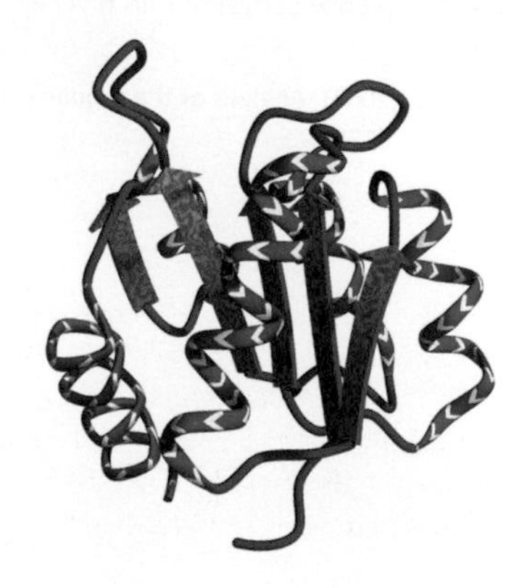

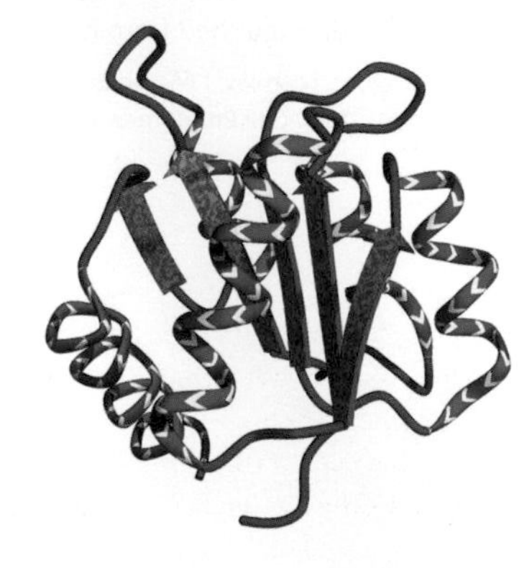

Figure 10.10 *E. coli* CheY protein. CheY is a bacterial signal-transduction protein involved in regulation of flagellar dynamics in chemotaxis. Activation of a receptor causes phosphorylation of CheY. Phosphorylated CheY interacts with flagellum protein FliM to induce tumbling [3CHY].

Table 10.1 The distribution of SCOP entries at different levels of the hierarchy

Class	Number of folds	Number of superfamilies	Number of families
All α proteins	284	507	871
All β proteins	174	354	742
Alpha and beta proteins (α/β)	147	244	803
α and β proteins ($\alpha + \beta$)	376	552	1055
Multi-domain proteins	66	66	89
Membrane and cell surface proteins	58	110	123
Small proteins	90	129	219
Total	1195	1962	3902

keywords, such as protein name, PDB code, function (including Enzyme Commission number), and name of fold (for instance, barrel). For each structure, SCOP provides textual information, pictures, and links to other databases.

Numerous other web sites offering classifications of protein structures are indexed at: http://www.bioscience.org/urllists/protdb.htm.

Changes in folding patterns in protein evolution

Proteins identified by SCOP as related by evolution show recognizably similar but not identical folding patterns. Figure 10.11 compares spinach plastocyanin and cucumber stellacyanin. For illustrations of the degree of similarities of proteins grouped together at

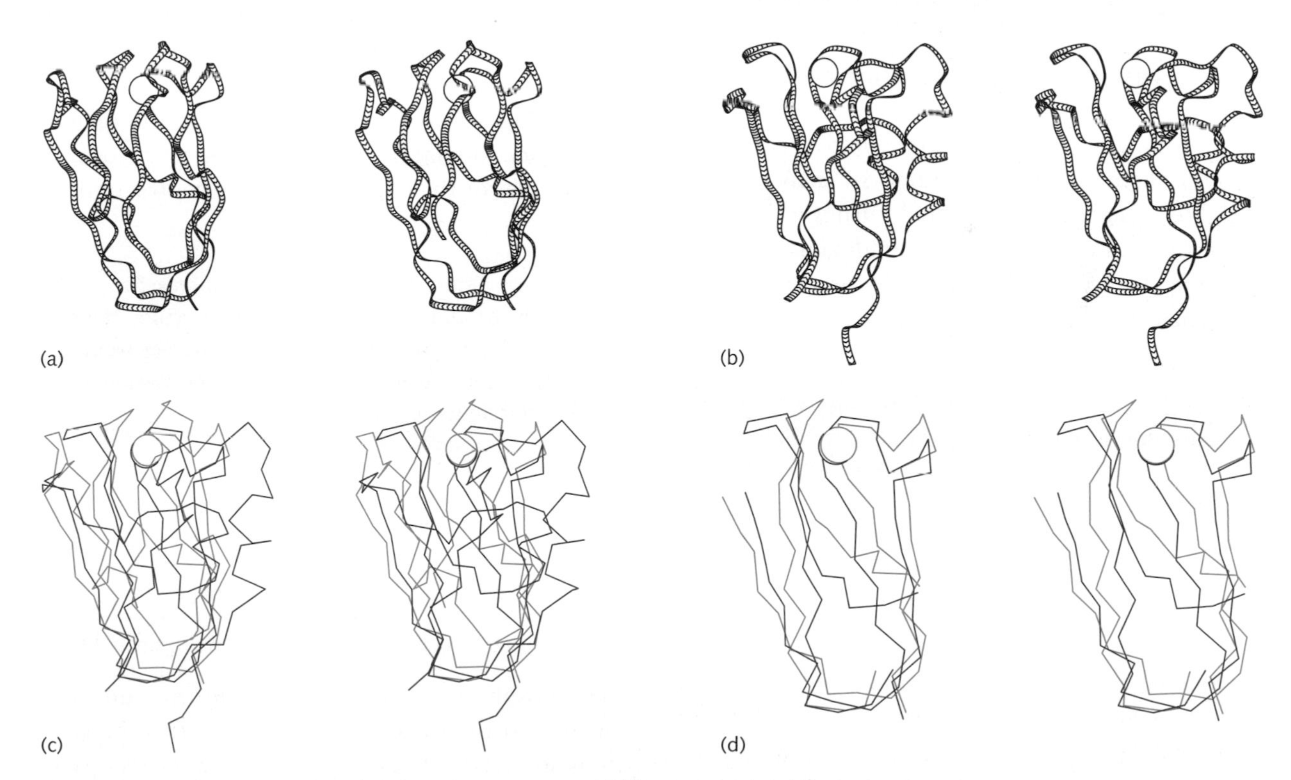

Figure 10.11 Two related proteins that share the same general folding pattern, but differ in detail. Circles represent copper ions. (a) Spinach plastocyanin [1AG6], (b) cucumber stellacyanin [1JER]. Superposition showing (c) the entire structures and (d) only the well-fitting core (plastocyanin, green; stellacyanin, magenta). The main secondary structural elements of these proteins are two β-sheets packed face-to-face. It is seen in the superposition that several strands of β-sheet are conserved but displaced, and that the helix at the right of the cucumber stellacyanin structure has no counterpart in the spinach plastocyanin structure. Even the (relatively) well-fitting core shows the conservation of folding topology but nevertheless reveals considerable distortion.

different levels of the hierarchy, and discussion of other classification schemes, see Chapter 4 of *Introduction to Protein Architecture* (Lesk, A.M. Oxford University Press, Oxford), which contains a large number of pictures of protein structures suitable for browsing by any reader interested in exploring the stunning variety of folding patterns seen in nature.

Many proteins change conformation as part of the mechanism of their function

The fundamental principle is that proteins fold to unique native structures. However, the mechanism of action of many proteins requires flexibility, or conformational change, during the active cycle. Most protein conformational changes are responses to binding.

- An enzyme may change structure after binding a substrate and/or a cofactor.
- Conformational changes arising from interactions with one or more other proteins, or nucleic acids, are a common component of regulatory mechanisms.
- Some proteins are microscopic motors, interconverting chemical and mechanical energy.
- Some serpins (serine protease inhibitors) are synthesized in a metastable active state and convert spontaneously to an inactive native state. This gives them a limited lifetime of activity, under tighter control than normal turnover processes. (Serpins also undergo an analogous structural change when they are cleaved by proteases, as part of their mechanism of inhibition.)

> - Many proteins are microscopic machines, with internal parts moving in precise ways to support their function.

Conformational change during enzymatic catalysis

A general scheme for enzyme catalysis is:

1. An enzyme (E) reversibly binds a substrate (S) to form a Michaelis complex (ES).
2. The ES complex changes to a transition state ($ES^{\ddagger}$), the peak of the thermodynamic barrier between substrate and product.
3. The transition state converts to the enzyme–product complex (EP).
4. The EP complex breaks down to release product (P) and re-form the original enzyme.

$$E + S = ES = ES^{\ddagger} \rightarrow EP \rightarrow E + P$$

Many enzymes have different structures in the unligated state, in the Michaelis complex, in the transition state, and/or in the enzyme–product complex.

Many binding sites occur in clefts between protein domains. Binding often induces conformational changes involving reorientation of domains, to close the structure around the ligand. Some of these changes can be described as a 'hinge motion', in which the two domains remain individually rigid but change their relative orientation by means of structural changes in only a few residues in the regions linking the domains. Hinge motion in myosin is responsible for the impulse in muscle contraction.

> - By their nature, transition states are reactive and difficult to trap long enough for structure determination. Possible solutions include enzymes binding transition-state analogues or inhibitors, or lowering the temperature to slow down the reaction.

Arginine kinase

Arginine kinase catalyses the reaction:

L-arginine + ATP = *N*-phospho-L-arginine + ADP

The phosphate group is added to the nitrogen attached to the $C\alpha$ of the arginine, not to the sidechain. In invertebrates, phosphoarginine serves as an energy store from which ATP can be regenerated. In vertebrates, phosphocreatine plays an analogous role. (Some athletes take creatine orally to increase their energy storage capacity.)

- One enzyme family – creatine kinase, arginine kinase, glycocyamine kinase, taurocyamine kinase, hypotaurocyamine kinase, lombricine kinase, opheline kinase, and thalassamine kinase – maintains ATP concentrations during bursts of muscular activity.

The structure of arginine kinase from the horseshoe crab (*Limulus polyphemus*) differs between the unligated state and the state binding ADP, arginine, and nitrate, a simulation of the transition state in which the nitrate mimics the phosphate group being transferred (see Figure 10.12).

Closure of interdomain clefts also occurs in transport proteins. Transport proteins act as carriers for their ligands, without catalysing reactions.

Ribose-binding protein is one of many periplasmic proteins in bacteria that are involved in chemotaxis and transport (see Figure 10.13). These proteins scavenge for nutrients in the cell's environment by coupling ligation to interaction with transporters or chemotaxis receptors in the inner membrane.

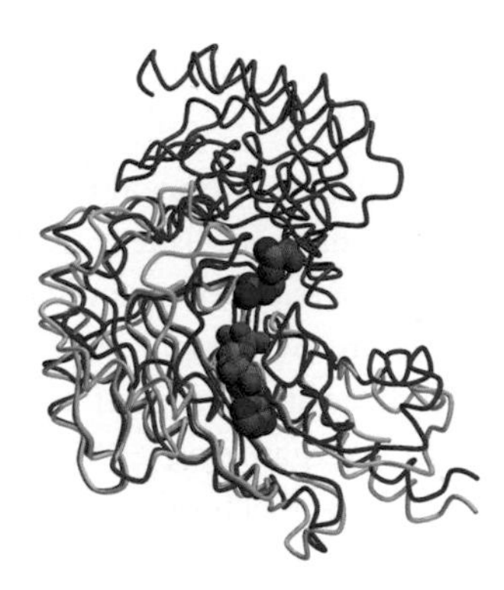

Figure 10.12 Superposition of two conformational states of horseshoe crab arginine kinase [1M15, 1M80]. The unligated state is shown in pink and purple; the ligated state in dark green and cyan. The ligands, arginine and ADP, appear in the ligated structure only. There are steric clashes between the ligands and the unligated structure in its position in the picture.

The nature of the conformational change has reminded many people of the Venus flytrap. Regions of the structure have come together around the ligands. The motion of the small domain at the top of the picture is primarily a 'hinge' motion – the mobile domain moves almost rigidly around an axis through the interdomain interface. The axis is approximately perpendicular to the page.

The parts of the structure at the lower right also deform. This protein is showing 'induced fit' – in response to ligation.

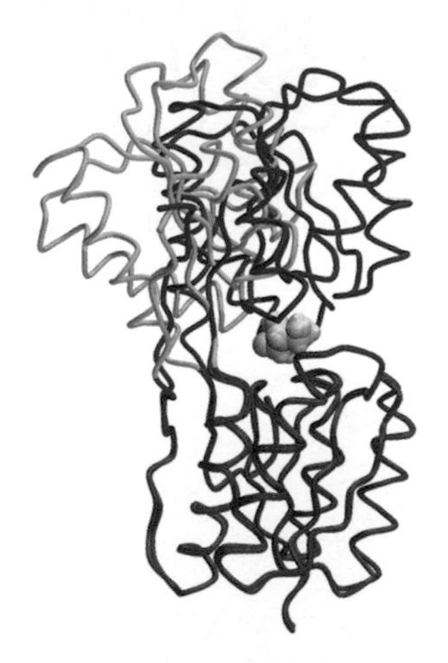
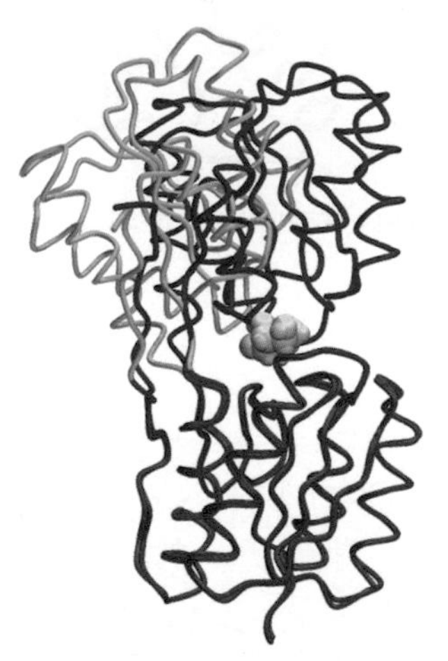

Figure 10.13 Superposition of two structures of ribose-binding protein [2DRI, 1URP]. The unligated structure is shown in pink and cyan; the ligated structure is shown in dark green and purple. The ribose, in yellow, appears only in the ligated structure.

Compared with arginine kinase (Figure 10.12), this is a more pure 'hinge' motion: the individual domains remain nearly rigid. The conformational change is achieved by rotations about bonds in only a few residues in the hinge region itself.

Ribose-binding protein, like many other members of this family, undergoes conformational changes upon ligation such that domains of the protein close around the ligand. The structural changes increase the protein–ligand interactions. They also create a new surface recognized by transport complexes.

Regulation of G protein activity

GTP-binding proteins (or G proteins) are an important class of signal transducer. One of them, p21 Ras, is a molecular switch in pathways controlling cell growth and differentiation. Ras has two conformational states, which differ in the structure of a local mobile region (see Figure 10.14). The resting, inactive state binds GDP. Membrane-bound G-protein-coupled receptors (see p. 362) trigger a GDP–GTP exchange transition, associated with a conformational change (see Figure 10.15). Activated Ras binds Raf-1, a serine/threonine kinase. The Ras–Raf-1 complex initiates the MAP kinase phosphorylation cascade. Ultimately, the signal enters the nucleus, where it activates transcription factors regulating gene expression.

Ras has a GTPase activity to reset it to the inactive state. Mutations that abrogate the GTPase activity are oncogenic. Mutants that are trapped in the active state continuously trigger proliferation. Mutations in Ras appear in 30% of human tumours.

- G proteins are a large family of signal transducers. Their proper function requires that they alternate between two states of different structure and activity. Mutations that leave the G protein Ras trapped in an active state are oncogenic.

Motor proteins

Motor proteins use chemical energy to set molecules in *controlled* motion. (Heat is *random* molecular motion – conversion of chemical energy to heat is *easy*!) There are two requirements: (1) coupling ATP hydrolysis to conformational change, to generate a force; and (2) organizing a cycle of attachment and detachment to a mechanical substrate, to allow the force to generate movement.

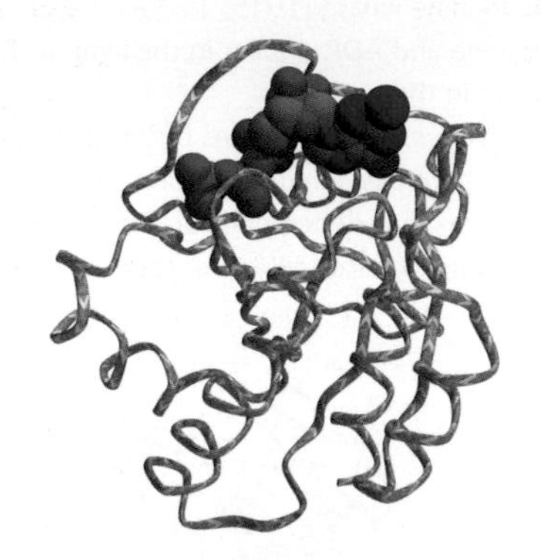
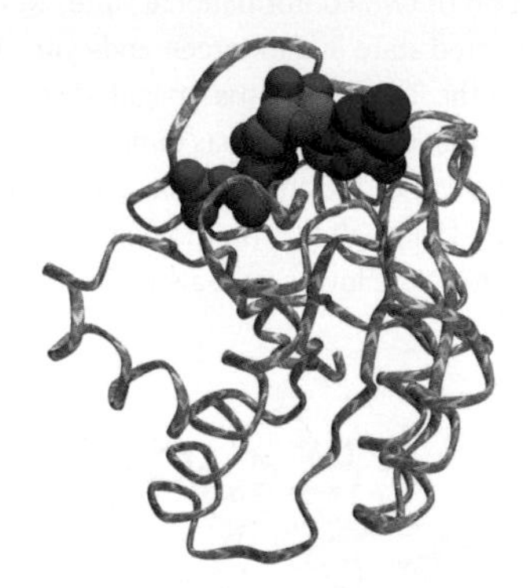

Figure 10.14 p21 Ras bound to GTP. Although an active GTPase, the system was stabilized for crystal-structure analysis by cooling to 100 K [1QRA].

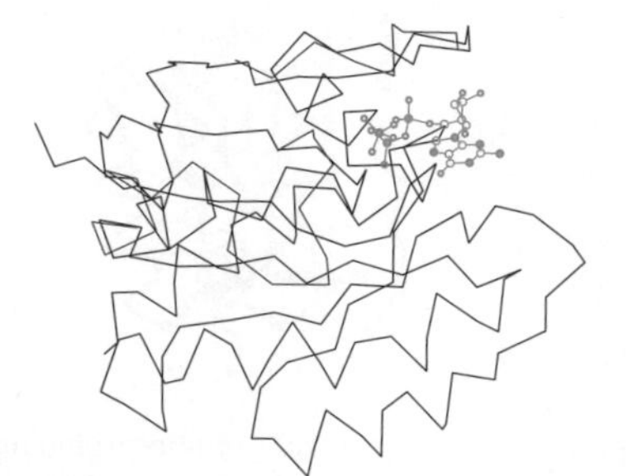

Figure 10.15 The conformational change in p21 Ras from the inactive GDP-binding conformation to the active GTP-binding conformation primarily involves two regions (shown here in red) that form a patch on the molecular surface [1QRA, 1Q21].

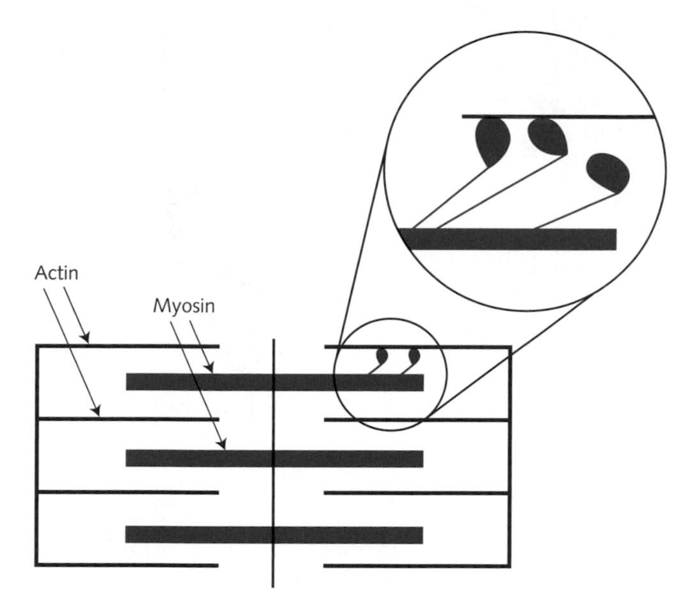

Figure 10.16 Schematic diagram of a sarcomere. Thick myosin filaments (red) overlap thin actin filaments (black). In the main diagram, it is cursorily indicated that multiple myosin molecules from thick filaments interact with adjacent thin filaments. In fact, each thick filament contains several hundred myosin molecules. The inset shows different stages of the power stroke. From left to right: attachment, conformational change propelling the thin filament inwards by ~10 nm, detachment (followed by recovery of original conformation of the myosin head.)

Some motor proteins propel themselves – and their cargo – by exerting force against a stationary object, such as a cytoskeletal filament. Others remain stationary and propel movable objects.

- **Myosins** interact with actin during muscle contraction.
- **Kinesins** and **dyneins** interact with microtubules, mediating organelle transport, chromosome separation in mitosis, and movements of cilia and flagella.

Myosins, kinesins, and dyneins are primarily **linear motors.** In contrast ATPase is a **rotary motor.**

- ATPase rotates during its action. Oxidative phosphorylation and photosynthesis create pH gradients across the membranes of mitochondria and chloroplasts, respectively. The mechanical step of ATPase activity is part of the mechanism for converting the osmotic energy of the potential gradient across a membrane to the high-energy phosphate bond of ATP.

> - Motor proteins are energy transducers. Some involve conversion of chemical energy – via ATP hydrolysis – to mechanical energy. ATP synthase converts chemiosmotic energy to chemical energy by ATP formation, with a rotary motor as part of its mechanism.

The sliding filament mechanism of muscle contraction

The structural and mechanical unit of vertebrate skeletal muscle is an intracellular organelle called the **sarcomere.** Sarcomeres contain interdigitating filaments of actin and myosin (Figure 10.16). The actin filaments are fixed to structures called the Z-disks at the ends of the sarcomere. The motor protein myosin pulls the actin filaments inwards towards the centre of the sarcomere. During contraction, the actin and myosin filaments do not themselves shorten but slide past one another, shortening the sarcomere by increasing the region of overlap. Think of the shortening of a bicycle pump during its compression stroke.

A large muscle may contain ~10^4–10^5 sarcomeres, laid end to end. Each sarcomere has a resting length of ~2.5 μm and can contract by ~0.3 μm. Therefore, the entire muscle can contract by about ~1–2 cm.

Individual myosin molecules are large fibrous proteins of relative molecular mass ~5×10^5. They contain a fibrous section ~1.6–1.7 μm long, and a globular head. Each thick filament contains ~200–300 myosin molecules. The mechanical coupling between actin and myosin occurs through the myosin head, as shown in Figure 10.16.

During the power stroke, the myosin head undergoes a cycle of attachment–detachment and conformational change. From left to right in the inset in Figure 10.16, attachment of the myosin head is followed by conformational change that propels the

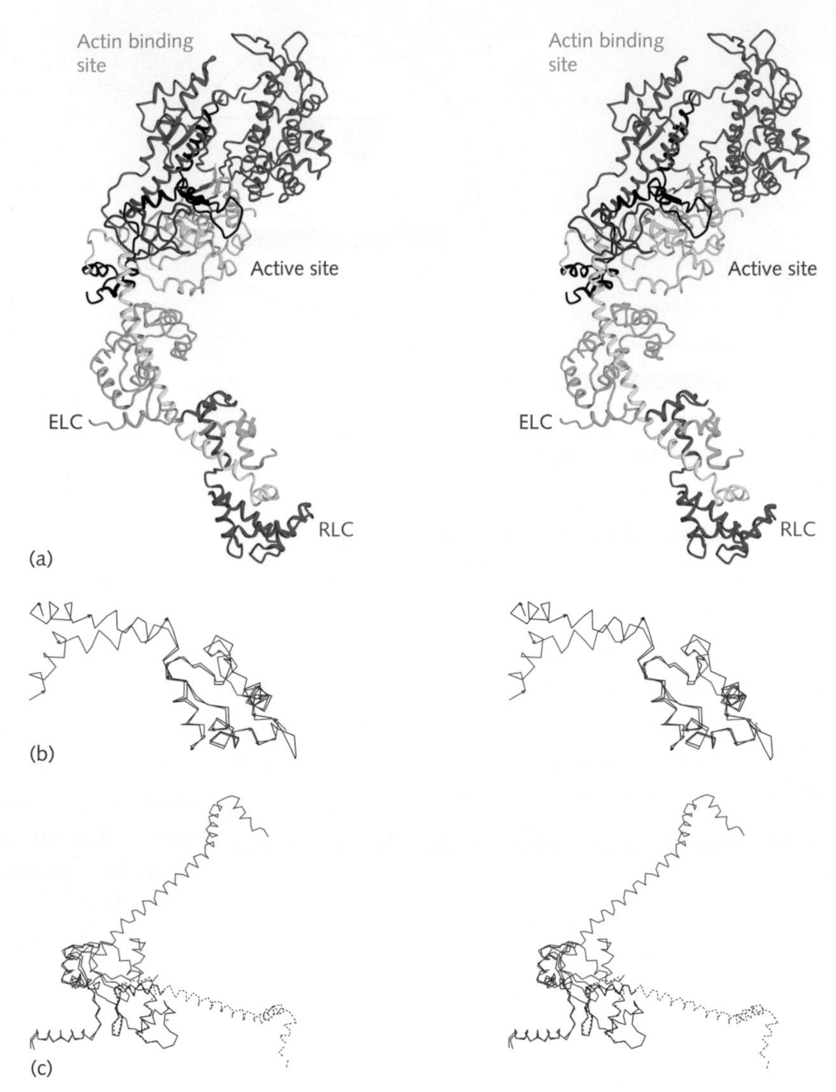

Figure 10.17 The contraction of muscle is a transformation of chemical energy to mechanical energy. It is carried out at the molecular level by a hinge motion in myosin, while myosin is attached to an actin filament. The cycle of *attach to actin–change conformation–release from actin* in a large number of individual myosin molecules creates a macroscopic force within the muscle fibre. (a) The structure of myosin subfragment 1 from chicken. The active site binds and hydrolyses ATP. ELC and RLC are the essential and regulatory light chains [2MYS]. (b) Hinge motion in myosin. Comparison of parts of chicken myosin open form [2MYS] (no nucleotide bound) and closed form binding the ATP analogue ADP·AlF_4^- [1BR2]. This shows the segments of the structure that surround the hinge region. (c) Model of the swinging of the long helical region in myosin as a result of the hinge motion. The dashed line shows a model of the position that the complete long helix would occupy in the closed form [2MYS] and [1BR1]. This conformational change is coupled to hydrolysis of ATP. It takes place while myosin is bound to actin, providing the power stroke for muscle contraction. In the context of the assembly and mechanism of function of a muscle filament, it is arguable that one should regard the helix as fixed and the head as swinging. However, this would not show the magnitude of the conformational change as dramatically.

actin towards the centre of the sarcomere. Detachment is followed by restoration of the initial conformation of the myosin. The myosin heads are like oars that 'row' the actin filaments towards the centre of the sarcomere. The displacement of the actin is ~10 nm per myosin molecule per cycle. Hydrolysis of one molecule of ATP during each cycle of each myosin molecule provides the energy.

Structures of fragments of myosin containing the globular head have defined the mechanism of the conformational change (see Figure 10.17).

Allosteric regulation of protein function

Allostery is modulation of the activity of a protein at one site by structural changes caused by binding a molecule at a distant site.

Allosteric proteins show 'action at a distance': ligand binding at one site affects activity at another. An impulse at the first site must transmit a conformational change affecting the second. In contrast, GTP-activated p21 Ras (Figures 10.14 and 10.15) shows ligand-induced regulation of activity, but the structural change is adjacent to the ligand. It is more challenging to explain the properties of haemoglobin, in which the shortest distance between binding sites is over 20 Å (2 nm).

- An allosteric protein with multiple binding sites for the same ligand may show cooperative binding.

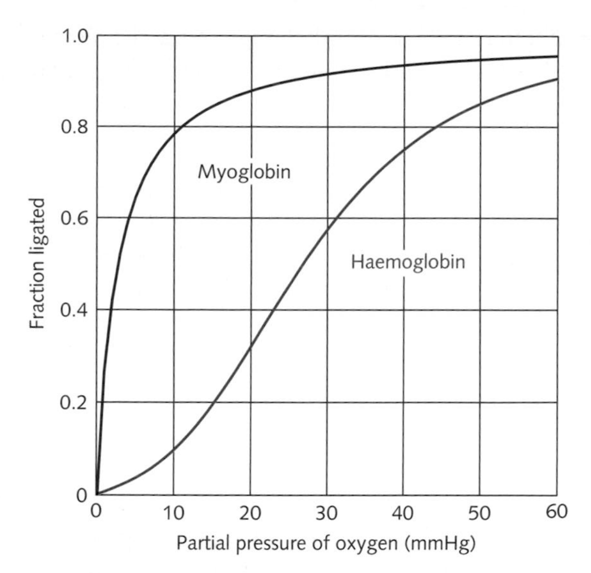

Figure 10.18 Oxygen-dissociation curves for myoglobin and haemoglobin. Myoglobin shows a simple equilibrium, with a binding constant independent of oxygen concentration. Haemoglobin shows positive cooperativity, the binding constant for the first oxygen being several orders of magnitude smaller than the binding constant for the fourth oxygen. The units for partial pressure are traditional in the literature about this topic: 760 mmHg = 1 atmosphere = 101 325 Pa.

Allosteric changes in haemoglobin

To play its physiological role in oxygen distribution effectively, haemoglobin must capture oxygen in the lungs as efficiently as possible and release as much as possible to other tissues. To achieve this 'take from the rich, give to the poor' effect, haemoglobin has a high oxygen affinity at high oxygen partial pressure (pO_2) and a low affinity at low pO_2. Haemoglobin shows **positive cooperativity:** binding of oxygen *increases* the affinity for additional oxygen (Figure 10.18). Some proteins show negative cooperativity: binding *reduces* the affinity for additional ligand.

In contrast to a ligand that induces cooperative binding of the *same* ligand, an **effector** alters the activity of a protein towards a *different* ligand. Bisphosphoglycerate (BPG) is an effector for haemoglobin. It binds preferentially to the deoxy form, decreasing the oxygen affinity and enhancing oxygen release. People and animals living at high altitude have higher concentrations of BPG than their sea-level relatives.

- Allosteric proteins deviate from the Michaelis–Menten curve in ligand binding or, in the cases of allosteric enzymes, in reaction velocity as a function of substrate concentration. The cooperativity is achieved by ligation-induced conformational change.

A mammalian foetus depends on its mother for oxygen. It is perhaps surprising that the oxygen affinity of isolated human foetal haemoglobin is *lower* than that of adult haemoglobin. However, foetal haemoglobin has a lower affinity for the effector BPG. This difference in the interaction with the effector gives foetal haemoglobin a higher oxygen affinity than the maternal haemoglobin.

The vertebrate haemoglobin molecule is a tetramer containing two identical α chains (α_1 and α_2) and two identical β chains (β_1 and β_2) (see Figure 1.16). It can adopt two structures: deoxyhaemoglobin (unligated) and oxyhaemoglobin (four oxygen molecules bound).

- The difference in colour between arterial and venous blood reveals the different state of the iron in ligated and unligated haemoglobin.

The oxygen affinity of the oxy form of haemoglobin is similar in magnitude to that of isolated α and β subunits and to that of myoglobin, a monomeric globin. The oxygen affinity of the deoxy form is much less: the ratio of binding constants for the first and fourth oxygens is 1:150–300, depending on conditions. Therefore, it is the deoxy form that is special, as it has had its oxygen affinity 'artificially' reduced. J. Monod, J. Wyman, and J.-P. Changeux proposed that the reduced oxygen affinity of the subunits in the deoxy state of haemoglobin arises from structural constraints that hold the subunits in a 'tense' (T), internally inhibited form, whereas the oxy form is in a 'relaxed' (R) form, as free to bind oxygen as the isolated monomer.

A general model for cooperativity is that the subunits of a protein are in equilibrium between the T and R forms. This model rationalizes the properties of haemoglobin.

1. At low partial pressures of oxygen, all of the subunits of haemoglobin are in the T form and are unligated (i.e. not binding oxygen). The binding constant for oxygen is low, because binding to the T state is inhibited.
2. At high partial pressures of oxygen, all of the subunits of haemoglobin are in the R form and each subunit binds an oxygen. The binding constant for oxygen is high, because binding of oxygen to the R state is unconstrained.
3. In the erythrocyte, haemoglobin is an equilibrium mixture of deoxy and oxy forms; the concentration of partially ligated forms is tiny. Binding of between two and three oxygen molecules shifts the subunits *concertedly* from all being T state to all being R state.
4. Effector molecules such as BPG modify oxygen affinity by shifting the $T \rightleftharpoons R$ equilibrium, by preferentially stabilizing one of the two forms.

The interpretation of this scheme in structural terms was one of the early triumphs of protein crystallography. The structures of haemoglobin in different states of ligation have been studied with intense interest, because of their physiological and medical importance and because they were thought to offer a paradigm of the mechanism of allosteric change. The two crucial questions to ask of the haemoglobin structures are:

1. What is the mechanism by which the oxygen affinity of the deoxy form is reduced?
2. How is the equilibrium between low- and high-affinity states altered by oxygen binding and release?

Comparison of the oxy and deoxy structures has defined the changes in tertiary structures of individual subunits, and in the quaternary structure. The allosteric change involves an interplay between changes in tertiary and quaternary structure (see Figure 10.19).

- The details of the quaternary structure – the relative geometry of the subunits and the interactions at their interfaces – is determined by the way the subunits fit together.
- The fit of the subunits depends on the shapes of their surfaces.
- The tertiary structural changes alter the shapes of the surfaces of the subunits, changing the way they fit together.

The haemoglobin tetramer can be thought of as a pair of dimers: $\alpha_1\beta_1$ and $\alpha_2\beta_2$. The allosteric change involves a rotation of 15° of the $\alpha_1\beta_1$ dimer with respect to the $\alpha_2\beta_2$ around an axis approximately perpendicular to their interface. (The motion is like that of a pair of shears with α_1 and α_2 as the blades and β_1 and β_2 as the handles.)

Starting from the deoxy structure, ligation of oxygen creates strain at the haem group, arising from a change in the position of the iron and the histidine sidechain linked to it. To relieve this strain, there are shifts in the F helix and the FG corner (the region of the chain between the F and G helices). To accommodate these shifts, a set of tertiary structural changes alter the overall shape of the $\alpha_1\beta_1$ and $\alpha_2\beta_2$ dimers, notably the shifting of the relative positions of the FG corners. In consequence, the deoxy quaternary structure is destabilized because the dimers no longer fit together properly (having changed their shape). Adopting the alternative quaternary structure requires the tertiary structural changes to take place even in subunits not yet liganded. As a result of the quaternary structural change, these unligated subunits have been brought to a state of enhanced oxygen affinity. It is important to emphasize that this is a sequence of steps in a logical process and not a description of a temporal pathway of a conformational change.

Conformational states of serine protease inhibitors (serpins)

Figure 10.20 shows the serpin antithrombin III in two conformational states, native and latent.

Serpins show multiple conformational states with different folding patterns. Under physiological conditions, the native states of inhibitory serpins are metastable, converting spontaneously to the latent state. In the native state (Figure 10.20a), the main β-sheet (green) has five strands (the rightmost much shorter than the others). The reactive-centre loop (red) is exposed, not participating in any secondary structure. It is available to interact with a protease. In the latent state (Figure 10.20b), the reactive-centre loop forms a sixth strand within the main β-sheet. The two

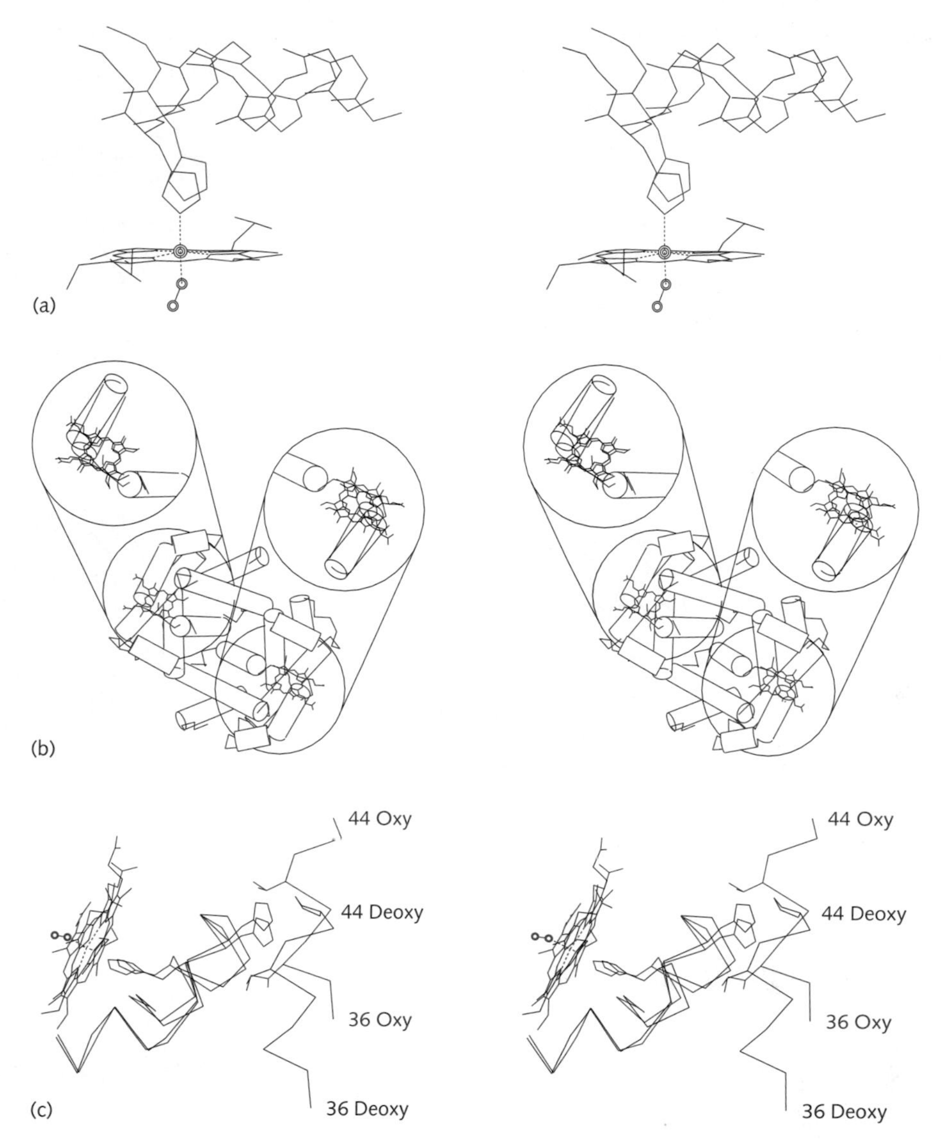

Figure 10.19 Some important structural differences between oxy- and deoxyhaemoglobin [1HHO, 2HHB]. (a) Changes at the haem group in human haemoglobin result in a change in state of ligation. This figure shows the F helix, proximal histidine, and haem group of the β chain in the oxy (black) and deoxy (red) forms; only the oxy haem is shown. The structures were superposed on the haem group. (b) The $\alpha_1\beta_1$ dimer in oxy (red) and deoxy (black, in blown-up regions only) forms. In the blown-up regions, only the F helix, FG corner, and haem group are shown. The oxy and deoxy $\alpha_1\beta_1$ dimers have been superposed on their interface; in this frame of reference, there is a small shift in the haem groups and a shift and conformational change in the FG corners. (c) Alternative packing of α_1 and β_2 subunits in oxyhaemoglobin (red) and deoxyhaemoglobin (black). The oxy and deoxy structures have been superposed on the F and G helices of the α_1 monomer. Although for the purposes of this illustration we have regarded the α_1 subunit as fixed and the β_2 subunit as mobile, only the relative motion is significant.

structures have identical amino acid sequences and chemical bonding patterns, but topologically different secondary and tertiary structures.

The latent state resembles the state produced by cleavage of the reactive-centre loop, as part of the mechanism of inhibitory action of this molecule.

- For a description of the mechanism of inhibition and its dependence on cleavage and conformational rearrangement, see Lesk, A.M. *Introduction to Protein Science*, 2nd ed., p. 219 (Oxford University Press, Oxford).

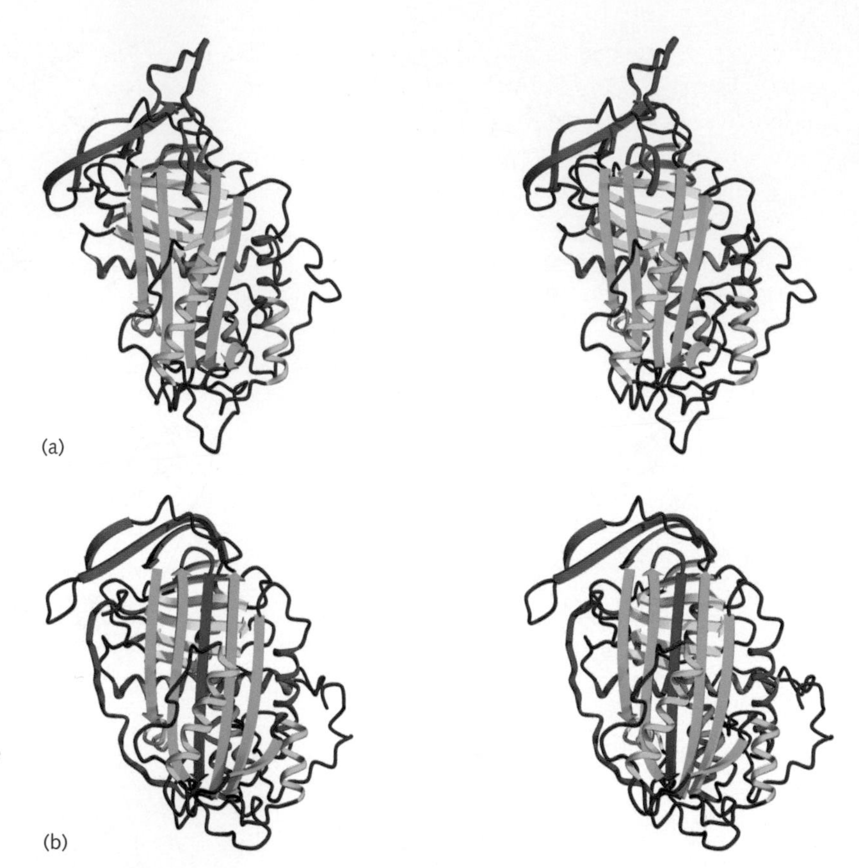

Figure 10.20 Antithrombin III, a serine proteinase inhibitor.
(a) Native conformation;
(b) latent conformation [1ATH].

Protein structure prediction and modelling

The observation that each protein folds spontaneously into a unique three-dimensional native conformation implies that nature has an algorithm for predicting protein structure from amino acid sequence. Some attempts to understand this algorithm are based solely on general physical principles; others are based on observations of known amino acid sequences and protein structures. A proof of our understanding would be the ability to reproduce the algorithm in a computer program that could predict protein structure from amino acid sequence. The Critical Assessment of Structure Prediction (CASP) programmes provide 'blind' tests of the state of the art (see Box 10.6).

Most attempts to predict protein structure from basic physical principles alone try to reproduce the interatomic interactions in proteins, to define a numerical energy associated with any conformation. Computationally, the problem of protein structure

Critical Assessment of Structure Prediction (CASP)

Judging of techniques for predicting protein structures requires blind tests. To this end, J. Moult initiated biennial CASP programmes. Crystallographers and NMR spectroscopists in the process of determining a protein structure are invited to (1) publish the amino acid sequence several months before the expected date of completion of their experiment; and (2) commit themselves to keeping the results secret until an agreed date. Predictors submit models, which are held until the deadline for release of the experimental structure. Then the predictions and experiments are compared.

The results of CASP evaluations record progress in the effectiveness of predictions, which has occurred partly because of the growth of the databanks but also because of improvements in the methods.

prediction then becomes a task of finding the global minimum of the conformational energy function over all possible backbone and sidechain conformations. So far this approach has not generally succeeded, partly because of the imprecision of the energy function and partly because the minimization algorithms tend to get trapped in local minima.

The alternative to *a priori* methods are approaches based on assembling clues to the structure of a target sequence by finding similarities to known structures. These empirical or 'knowledge-based' techniques have become very powerful and are currently the most successful methods known.

- *Homology modelling*. Suppose a target protein of known amino acid sequence but unknown structure is related to one or more proteins of known structure. Then we expect that much of the structure of the target protein will resemble that of the known protein. The related protein of known structure can therefore serve as a basis for a model of the target protein. The challenge is to predict how the differences between the sequences are reflected in differences between the structures. This can be thought of as the 'differential' rather than the 'integral' form of the folding problem.
- *Attempts to predict secondary structure* without attempting to assemble these regions in three dimensions. The results are lists of regions of the sequence predicted to form α-helices and regions predicted to form strands of β-sheet.
- *Fold recognition*. Given a library of known structures, determine which of them shares a folding pattern with a query protein of known sequence but unknown structure. If the folding pattern of the target protein does not occur in the library, such a method should recognize this. The results are a nomination of a known structure that has the same fold as the query protein, or a statement that no protein in the library has the same fold as the query protein.
- *Prediction of novel folds*, either by *a priori* or knowledge-based methods. The results are a complete coordinate set for at least the mainchain and sometimes the sidechains also. The model is intended to have the correct folding pattern, but would not be expected to be comparable in quality to an experimental structure.

D. Jones has likened the distinction between fold recognition and *a priori* modelling to the difference between a multiple-choice question on an examination and an essay question.

Homology modelling

Model building by homology is a useful technique when one wants to predict the structure of a target protein of known sequence, when the target protein is related to at least one other protein of known sequence *and* structure. If the proteins are closely related, the known protein structures – called the parents – can serve as the basis for a model of the target. It is on homology modelling that we depend to extend the results of structural genomics to the entire protein world.

The completeness and quality of the results depend crucially on how similar the sequences are. As a rule of thumb, if the sequences of two homologous proteins have 50% or more identical residues in an optimal alignment, the structures are likely to have similar conformations over more than 90% of the model. This is a conservative estimate, as Figure 10.21 shows.

Although the quality of the model will depend on the degree of similarity of the sequences, it is possible to specify this quality before experimental testing. Therefore, knowing how good a model is *necessary for the intended application* permits intelligent prediction of the probable success of the exercise.

Steps in homology modelling are as follows.

1. Align the amino acid sequences of the target and the protein or proteins of known structure. Usually, insertions and deletions will lie in the loop regions between helices and sheets.
2. Determine mainchain segments to represent the regions containing insertions or deletions. Stitching these regions into the mainchain of the known protein creates a model for the complete mainchain of the target protein.
3. Replace the sidechains of residues that have been mutated. For residues that have not mutated, retain the sidechain conformation. Residues that have mutated tend to keep the same sidechain conformational angles and could be modelled on this basis. However, computational methods are

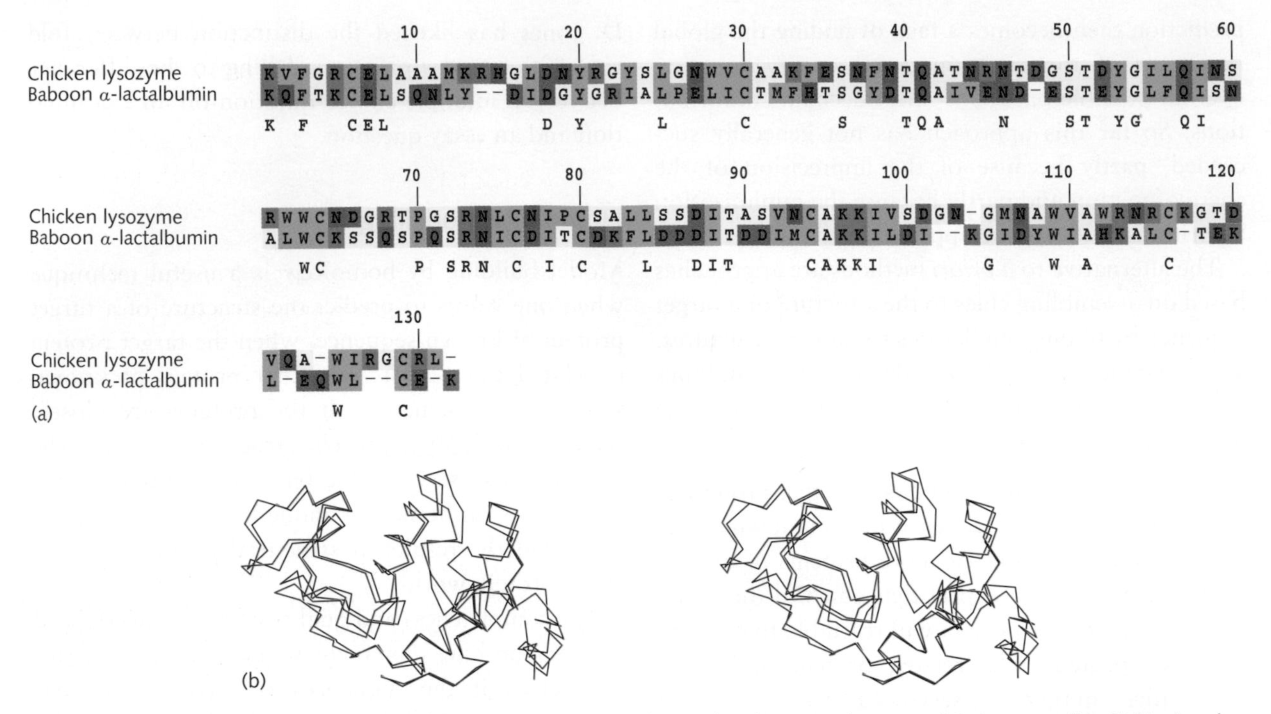

Figure 10.21 (a) Aligned sequences and (b) superposed structures of two related proteins, hen egg white lysozyme (black) [1AKI] and baboon α-lactalbumin (red) [1ALC]. The sequences are related (37% identical residues in the aligned sequences) and the structures are very similar. Each protein could serve as a good model for the other, at least as far as the course of the mainchain is concerned.

now available to search over possible combinations of sidechain conformations.

4. Examine the model – both by eye and using programs – to detect any serious collisions between atoms. Relieve these collisions, as far as possible, by manual manipulations.
5. Refine the model by limited energy minimization. The role of this step is to fix up the exact geometrical relationships at places where regions of the mainchain have been joined together and to allow the sidechains to wriggle around a bit to place themselves in comfortable positions. The effect is really only cosmetic – energy refinement will not correct serious errors in such a model.

In most families of proteins, the structures contain relatively constant regions and more variable ones. The core of the structure of the family retains the folding topology, although it may be distorted, but the periphery can entirely refold (see Figure 10.11); in contrast, hen egg white lysozyme and baboon α-lactalbumin, shown in Figure 10.21, are closely related and quite similar in structure.

A single parent structure will permit reasonable modelling of the conserved portion of the target protein but will fail to produce a satisfactory model of the variable portion. A more favourable situation occurs when several related proteins of known structure can serve as parents for modelling a target protein. These reveal the regions of constant and variable structure in the family. The observed distribution of structural variability among the parents dictates an appropriate distribution of constraints to be applied to the model.

Mature software for homology modelling is available. SWISS-MODEL is a web site that will accept the amino acid sequence of a target protein, determine whether a suitable parent or parents for homology modelling exist and, if so, deliver a set of coordinates for the target. SWISS-MODEL (http://www.expasy.org/swissmodel/SWISS-MODEL.html) was developed by T. Schwede, M.C. Peitsch, and N. Guex, now at the Geneva Biomedical Research Institute. Another program in widespread use, MODELLER, was developed by A. Šali. Each of these is associated with a library of models corresponding to amino acid sequences in databanks. MODBASE (http://salilab.org/modbase) and 3DCrunch (http://swissmodel.expasy.org/SM_3DCrunch.html) collect homology models of proteins of known sequence.

- Homology modelling is one of the most useful techniques for protein structure prediction – when it is applicable.

Secondary structure prediction

The goal of secondary structure prediction is to identify those residues within the sequence that will form helices and strands of β-sheet in the native structure, independent of their spatial arrangement in the tertiary structure.

The original motivations for secondary structure prediction included:

- a belief that prediction of secondary structure should be substantially easier than prediction of tertiary structure;
- a belief that prediction of secondary structure would be a step towards prediction of tertiary structure; and
- both experimental evidence from co-polymers of amino acids and statistical evidence from the observed residue compositions of helices and β-sheets in solved protein structures implying that there are preferences among the residues for forming (or breaking) helices.

Early work on secondary structure prediction made *a priori* predictions based on tables of residue preferences. Methods were tested according to a three-state model in which each residue in the prediction and in the experimental structure was assigned to the classes 'helix', 'sheet', and 'other', and the percentage of residues assigned to the correct class was defined as a measure of success called Q_3. Such methods achieved typical accuracies corresponding to Q_3 ~55%.

Progress depended on the recognition that tables of many aligned sequences contained consensus information that could improve the prediction accuracy. The idea is to apply pattern-recognition algorithms to a set of aligned sequences homologous to the sequence of an unknown structure for which the secondary structure is to be predicted. The patterns are based on the distribution of residues at the positions in the alignment table.

The most powerful pattern recognition algorithms now being applied to secondary structure prediction include neural networks and hidden Markov models. The basic idea is to develop a network of implications containing a large set of adjustable parameters governing the assignment of each residue to the three classes: helix, sheet, or other. The systems are quite general and the parameters must be adjusted by 'training' on known sequences and structures.

The best current methods claim average accuracies of Q_3 ~75%.

- CASP categories change as the field progresses. Secondary structure prediction and fold recognition have been discontinued. Prediction of residue – residue contacts, of disordered regions, and the ability to refine models have been added.

Prediction of novel folds: ROSETTA

ROSETTA is a program developed by D. Baker and colleagues that predicts protein structure from amino acid sequence by assimilating information from known structures. At several recent CASP programmes, ROSETTA showed the most consistent success on targets in both the 'novel fold' and 'fold recognition' categories.

ROSETTA predicts a protein structure by first generating structures of fragments using known structures and then combining them. For each contiguous region of three and nine residues, instances of that sequence and related sequences are identified in proteins of known structure. For fragments this small, there is no assumption of homology to the target protein. The distribution of conformations of the fragments in the proteins of known structure models the distribution of possible conformations of the corresponding fragments of the target structure.

ROSETTA explores the possible combinations of fragment conformations, evaluating compactness, paired β-sheets, and burial of hydrophobic residues. The procedure carries out 1000 independent simulations, with starting structures chosen from the fragment conformation distribution patterns evaluated favourably. The structures that result from these simulations are clustered and the centres of the largest clusters presented as predictions of the target structure. The idea is that a structure that emerges many times from independent simulations is likely to have favourable features.

There is a general belief that the work of Baker and colleagues represents a major breakthrough in the field of protein structure prediction.

Robetta (http://robetta.bakerlab.org) is a web server designed to integrate and implement the best of the protein structure prediction tools. The central pipeline of the software involves first the parsing of a submitted amino acid sequence of a protein of unknown structure into putative domains. Then homology modelling techniques are applied to those domains for which suitable parents of known structure exist and the *de novo* methods developed by Baker and co-workers are applied to other domains. In addition, the user will receive the results of other prediction methods based on software developed outside the Robetta group. These include, for example, predictions of secondary structure, coiled coils, and transmembrane helices.

Some methods are specialized to particular types of structure.

Prediction of transmembrane proteins and signal peptides

Many proteins are designed to sit within membranes. Membrane proteins mediate the exchange of matter, energy, and information between cell interiors and surroundings. Examples of membrane protein functions include energy transduction via the generation or release of concentration gradients across cell or organelle membranes, and signal reception and transmission.

It is estimated that, in the human genome, approximately 30% of genes encode membrane proteins. Approximately 70% of known targets of drugs are membrane proteins. Given that membrane proteins are so common, it is important to have reliable tools for their identification. Relatively few membrane protein structures have been determined experimentally. This places a greater burden on computational tools for sequence analysis, to identify and characterize them.

Among their adaptations, membrane proteins contain regions of mostly non-polar residues that interact with the organic layer. Many membrane proteins contain a set of seven consecutive α-helices that traverse the membrane, oriented approximately perpendicular to the plane of the membrane (see Figure 4.17). These helices are connected by loops that protrude into the aqueous surroundings. A second class of membrane protein structures contains a β-barrel. Transmembrane helices are typically 15–30 residues long. Although enriched in hydrophobic residues, they contain some polar sidechains, usually in interfaces between α-helices packed together in the structure.

A useful clue to the orientation of the helices across the membrane is the '+ inside rule'. The loops between helices lie either entirely inside or entirely outside the cell or organelle. Those inside contain a preponderance of positively charged residues.

A simple approach to prediction of membrane proteins involves looking for amino acid segments of 15–30 residues in length that are rich in hydrophobic residues. However, signal peptides also contain hydrophobic helices: the signal sequence typically comprises a positively charged n-region, followed by a helical hydrophobic h-region, followed by a polar c-region. Methods for recognizing transmembrane helices in amino acid sequences tend to pick up the h-regions of signal peptides as false positives. Methods for recognizing signal peptides in amino acid sequences tend to pick up transmembrane helices as false positives.

L. Käll, A. Krogh, and E.L.L. Sonnhammer trained hidden Markov models to test simultaneously for transmembrane helices and signal peptides. The goals are to find both at the same time, to discriminate between them in the results, and to predict not only the positions of the transmembrane helices but also the locations – cytoplasmic or interior – of the loops. The method, called Phobius, is available at http://phobius.cgb.ki.se/.

Phobius is the most successful algorithm currently available for recognizing signal peptides and helical transmembrane proteins, and for predicting the orientation of the transmembrane segments. Phobius is capable of distinguishing h-domains of signal peptides from transmembrane helices: the number of false classifications of signal peptides was 3.9%, and

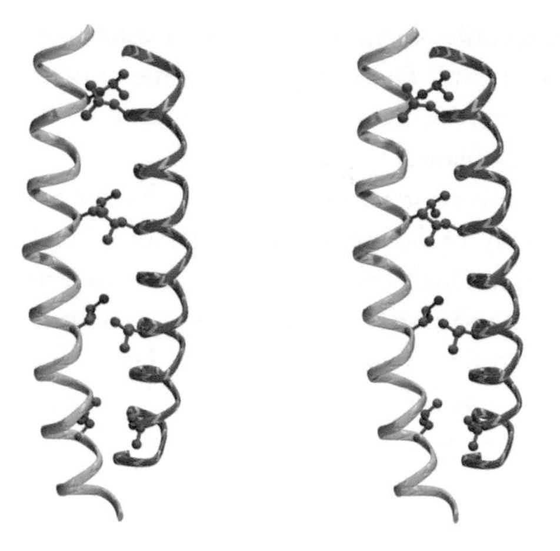

Figure 10.22 Coiled-coil BZIP domain encoded by proto-oncogene c-*jun* [1JNM].

the number of false classifications of transmembrane helices was 7.7%. These results represent a great improvement over previous methods. It is interesting that addressing the two problems at once proved to be more successful than treating them separately.

Coiled-coil regions

Proteins containing coiled coils are known among structural proteins such as α-keratin and also occur in a variety of globular proteins associated with a number of functions, prominently including transcription regulation. Figure 10.22, showing a leucine zipper, is a typical example.

Such coiled-coil domains contain a signature pattern in their amino acid sequences. They show *heptad repeats* – seven-residue patterns – containing positions denoted *a*, *b*, *c*, *d*, *e*, *f*, and *g*, of which the first and fourth positions – *a* and *d* – are usually hydrophobic. Here is the sequence of the leucine zipper protein GCN4, with the heptads demarcated and the hydrophobic positions indicated by asterisks:

```
  abcdefg abcdefg abcdefg abcdefg
  *  *    *  *    *  *    *  *
R|MKQLEDK|VEELLSK|NYHLENE|VARLKKL|VG
```

Programs for predicting coiled coils include Coils, by A. Lupas and J. Lupas (http://www.ch.embnet.org/software/COILS_form.html), and Paircoil, by B. Berger, D.B. Wilson, E. Wolf, T. Tonchev, M. Milla, and P.S. Kim (http://paircoil.lcs.mit.edu/webcoil.html).

Available protocols for protein structure prediction

Here we collect a few of the many web sites that deal with protein structure prediction.

Many sites act as 'dispatchers' for other sites, accepting a sequence and resending it to many other web servers. These allow users to submit a sequence once and have it processed by many other sites to which the query is distributed. These are sometimes called 'metaservers'.

Homology modelling

SWISS-MODEL: http://www.expasy.org/swissmod/SWISS-MODEL.html. Results of the application of SWISS-MODEL to proteins of known sequence are available through 3DCrunch: http://swissmodel.expasy.org/SM_3DCrunch.html.

MODELLER (homology modelling software): http://salilab.org/modeller/modeller.html. Results of the application of MODELLER to proteins of known sequence are available through MODBASE: http://salilab.org/modbase.

Secondary structure prediction

A list of secondary-structure prediction servers can be found at: http://abs.cit.nih.gov/main/otherservers.html. Quite a few sites are listed. Many methods are available through the following site: http://cubic.bioc.columbia.edu.

Prediction of full three-dimensional structure

Available at the Robetta web site: http://robetta.bakerlab.org.

Prediction of antibody structure

Available at: http://www.biocomputing.it/pigs.

Prediction of transmembrane helices and signal peptides

Phobius is available at http://phobius.cgb.ki.se/.

Prediction of coiled coils

See Coils, by A. Lupas and J. Lupas, at: http://www.ch.embnet.org/software/COILS_form.html and Paircoil, by B. Berger and co-workers, at: http://paircoil.lcs.mit.edu/webcoil.html.

- Because proteins fold into native structures at the dictation of the amino acid sequence, it should be possible to write a computer program to predict protein structure from amino acid sequence. Many people have tried to predict different features of protein structures – prediction of the full three-dimensional structure remains the ultimate goal. The CASP (Critical Assessment of Structure Prediction) programme subjects the efforts to objective tests.

Structural genomics

In analogy with full-genome sequencing projects, structural genomics has the commitment to deliver the structures of the complete protein repertoire. X-ray crystallographic and NMR experiments will solve a 'dense set' of proteins, such that all proteins are close enough to one or more experimentally determined structures to model them confidently. More so than genomic sequencing projects, structural genomics projects combine results from different organisms. The human proteome is of course of special interest, as are proteins unique to infectious microorganisms.

The goals of structural genomics have become feasible partly by advances in experimental techniques, which make high-throughput structure determination possible, and partly by advances in our understanding of protein structures, which define reasonable general goals for the experimental work and suggest specific targets.

How many structures are needed? The theory and practice of homology modelling suggests that at least 30% sequence identity between target and some experimental structure is necessary. This means that experimental structure determinations will be required for an exemplar of every sequence family, including many that share the same basic folding pattern. Experiments will have to deliver the structures of something like 10 000 domains. In the year 2010, 7936 structures were deposited in the PDB, so the throughput rate is not far from what is required.

Directed evolution and protein design

One strand of Darwin's thinking that led to the theory of evolution was the observation that farmers could improve the quality of livestock by selective breeding. He drew an analogy between this **artificial selection** and the idea of **natural selection** that he was proposing as the mechanism of evolution. We now recognize that evolution by natural selection takes place at the molecular level. Why not artificial selection also?

Natural proteins do many things, but not everything we would like them to. For applications in technology, it would be useful to have proteins that would:

- have activities unknown in nature;
- show activity towards unnatural substrates, or altered specificity profiles;
- be more robust than natural proteins, retaining their activity at higher temperature or in organic solvents; and
- show different regulatory responses, enhanced expression, or reduced turnover.

One day, we shall be able to design amino acid sequences *a priori* that will fold into proteins with desired functions. As this is not yet possible, scientists have used **directed evolution** – or artificial selection – to generate molecules with novel properties starting from natural proteins.

Evolution requires the generation of variants and differential propagation of those with favourable features. Molecular biologists dealing with microbial evolution have advantages over the farmers that Darwin observed. We can generate large numbers of variants artificially. Screening and selection can, in many cases, be done efficiently, by stringent growth conditions, and there are virtually no limits on the size of the 'flock' or 'litter'. Darwin might well have been envious. He wrote:

> . . . as variations manifestly useful or pleasing to man appear only occasionally, the chance of their appearance will be much increased by a large number of individuals being kept. Hence, number is of the highest importance for success. On this principle Marshall formerly remarked,

with respect to the sheep of parts of Yorkshire, 'as they generally belong to poor people, and are mostly *in small lots*, they never can be improved.'

– *The Origin of Species*, Chapter 1.

The procedure of directed evolution comprises these steps:

1. Create variant genes by mutagenesis or genetic recombination.
2. Create a library of variants by transfecting the genes into individual bacterial cells.
3. Grow colonies from the cells and screen for desirable properties.
4. Isolate the genes from the selected colonies and use them as input to step 1 of the next cycle.

Strategies for generating variants include (a) single and multiple amino acid substitutions, (b) recombination, and (c) formation of chimaeric molecules by mixing and matching segments from several homologous proteins. Each method has its advantages and disadvantages. The smaller the change in sequence, the more likely that the result will be functional. Yet, multiple substitutions or recombinations give a greater chance of generating novel features. The choice depends in part on the nature of the goal. For instance, it is easier to lose a function than to gain one. (Why would you want to *lose* a function? Removal of product inhibition to enhance throughput in an enzymatically catalysed process is an example.)

Directed evolution of subtilisin E

Subtilisins are a family of bacterial proteolytic enzymes. Subtilisin E, from the mesophilic bacterium *Bacillus subtilis*, is a 275-residue monomer. It becomes inactive within minutes at 65°C. Directed evolution has produced interesting variants, with features including enhanced thermal stability and activity in organic solvents.

Enhancement of thermal stability by directed evolution

Thermitase, a subtilisin homologue from *Thermoactinomyces vulgaris*, remains stable up to 80°C. The existence of thermitase is reassuring, because it shows that the evolution of subtilisin to a thermostable protein is possible. However, subtilisin E and thermitase differ in 157 amino acid residues. Do we have to go this far? Are all of the changes essential for thermostability, or has there been considerable neutral drift as well?

A thermostable variant of subtilisin E, produced by directed evolution, differs from the wild type by only eight residue substitutions. The variant is identical to thermitase in its temperature of optimum activity: 76°C (17°C higher than the original molecule) and stability at 83°C (a 200-fold increase relative to the wild type).

The procedure involved successive rounds of generation of variants, and screening and selection of those showing favourable properties. The formation of mutations, via error-prone PCR to produce an average of two to three base changes per gene, was alternated with *in vitro* recombination to find the best combinations of substitutions at individual sites (Figure 10.23). At each step, several thousand clones were screened for activity and thermostability.

The optimal variant differed from the wild type at eight positions: N188S, S161C, P14L, N76D, G166R, N181D, S194P, and N218S. Figure 10.24 shows their distribution in the structure. Most of the substitutions are far from the active site, which is not surprising as the wild type and variant do not

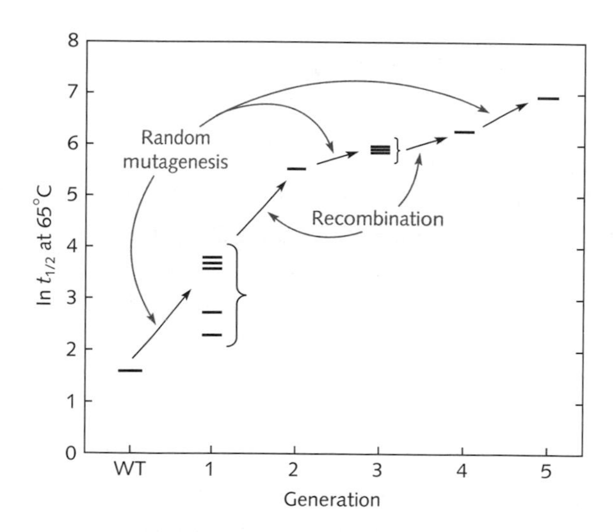

Figure 10.23 Directed evolution of a thermostable subtilisin. The starting wild type (WT) was subtilisin E from the mesophilic bacterium *B. subtilis*. Steps of random mutagenesis were alternated with recombination. At each step, screening for improved properties and artificial selection chose candidates for the next round. ($t_{1/2}$ measured in minutes.)

After: Zhao, H. & Arnold, F.H. (1999). Directed evolution converts subtilisin E into a functional equivalent of thermitase. Prot. Eng. **12**, 47–53.

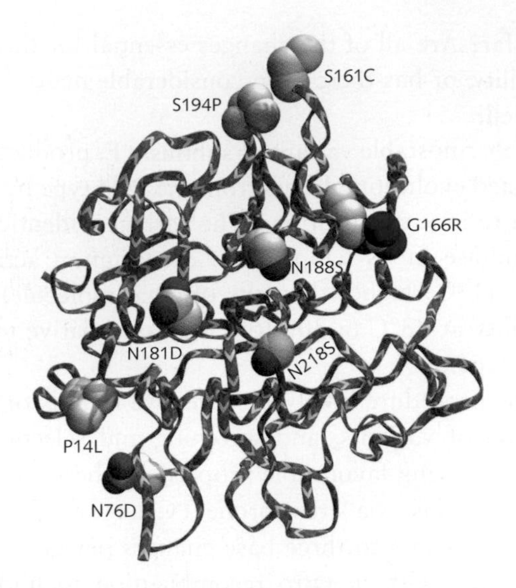

Figure 10.24 The sites of mutation in *B. subtilis* subtilisin E that produced a thermostable variant by directed evolution. The sidechains shown are those of the final product.

differ in function. Most of the sites of substitution are in loops between regions of secondary structure. These regions are the most variable in the natural evolution of the subtilisin family. However, only two of the substitutions produce the amino acids that appear in those positions in thermitase. Two of the substitutions are in α-helices, including P14L. P14L has a certain logic: proline tends to destabilize an α-helix because it costs a hydrogen bond.

Enzyme design

We know that mutations can change the structure and function of proteins, in some cases in radical ways. We observe this in natural protein evolution and can achieve it by artificial selection (see preceding section). We can do a reasonable job at predicting the structural changes arising from mutations. Putting these together suggests that we should be able to design proteins with altered or even novel functions *in silicio*.

Gramicidin S is a cyclic decapeptide antibiotic from *Bacillus brevis*. It contains unusual amino acids, including D-phenylalanine and ornithine. Synthesis of gramicidin S is independent of the normal ribosomal protein-synthesizing machinery. Instead, the enzyme gramicidin synthetase isomerizes the substrate L-phenylalanine to D-phenylalanine, and activates it to the amino acid adenylate. Another component of the enzyme effects the polymerization, in a sequence-specific manner.

C.-Y. Chen, I. Georgiev, A.C. Anderson, and B.R. Donald computationally redesigned gramicidin synthetase to accept other amino acids as substrates. The wild-type enzyme has no activity towards Arg, Glu, Lys, and Asp. Their computational modelling approach successfully predicted the sequences of modified enzymes with activity for each of these four unnatural substrates.

Protein complexes and aggregates

Within cells, life is organized and regulated by a set of protein–protein and protein–nucleic acid interactions.

Interacting proteins and nucleic acids span a range of structures and activities:

- simple dimers or oligomers in which the monomers appear to function independently;
- oligomers with functional 'cross-talk', including ligand-induced dimerization of receptors and allosteric proteins such as haemoglobin (Figure 1.16), phosphofructokinase, and asparate carbamoyltransferase;
- large fibrous proteins such as actin or keratin;
- non-fibrous structural aggregates such as viral capsids;
- large aggregates with dynamic properties such as F1-ATPase, pyruvate dehydrogenase, the GroEL–GroES chaperonin, and the proteasome;
- protein–nucleic acid complexes, including ribosomes, nucleosomes, transcription regulation complexes, splicing and repair particles, and viruses;
- many proteins, whether monomeric or oligomeric, which *function* by interacting with other proteins. These include all enzymes with protein substrates and many antibodies, inhibitors, and regulatory proteins.

BOX 10.7 Diseases associated with protein aggregates

Disease	Aggregating protein	Comment
Sickle-cell anaemia	Deoxyhaemoglobin–S	Mutation creates hydrophobic patch on surface
Classical amyloidoses	Immunoglobulin light chains, transthyretin, and many others	Extracellular fibrillar deposits
Emphysema associated with Z-antitrypsin	Mutant α_1-antitrypsin	Destabilization of structure facilitates aggregation
Huntington's	Altered huntingtin	One of several polyglutamine-repeat diseases
Parkinson's	α-Synuclein	Found in Lewy bodies
Alzheimer's	$A\beta$, τ	$A\beta$ = 40–42 residue fragment
Spongiform encephalopathies	Prion proteins	Infectious, despite containing no nucleic acid

Protein aggregation diseases

Protein interactions are frequently associated with disease, caused by misfolded or mutant proteins that are prone to aggregation. **Amyloidoses** are diseases characterized by extracellular fibrillar deposits. Alzheimer's and Huntington's diseases are also associated with protein aggregation. Many aggregates contain proteins in a common crossed-β-sheet structure, different from their native state. There are a variety of causes of protein aggregation, including overproduction of a protein, destabilizing mutations, and inadequate clearance in renal failure (see Box 10.7).

Alzheimer's disease

Alzheimer's disease is a neurodegenerative disease common in the elderly. It is associated with two types of deposits:

(a) dense insoluble extracellular protein deposits, called **senile plaques**. These contain the $A\beta$ fragment (the N-terminal 40–43 residues) of a cell surface receptor in neurons, the β-protein precursor (βPP) or **amyloid precursor protein (APP)**.

(b) **neurofibrillary tangles**, twisted fibres inside neurons, containing microtubule-associated protein tau. There is some evidence that amyloid deposits promote tangle formation.

Abnormalities in tau appear in other neurodegenerative diseases, the **tauopathies**.

Prion diseases – spongiform encephalopathies

Prion diseases are a set of neurodegenerative conditions of animals and humans, associated with deposition of protein aggregates in the brain, and a characteristic sponge-like appearance of the brains of affected individuals seen in postmortem investigation.

Prion diseases are unusual among protein deposition diseases in that they are transmissible; and unusual among transmissible diseases in that the infectious agent is a protein. Moreover – another unusual feature – some prion diseases are hereditary, for example, **familial Creutzfeld–Jacob disease (CJD)**. (Distinguish between a disease transmitted from mother to baby perinatally by passage of an infectious agent – as in many AIDS cases – with a truly hereditary disease depending on parental genotype.) All hereditary human prion diseases involve mutations in the same gene, one that encodes the protein found in the aggregates deposited in the brains of sufferers of *both* hereditary and infectious prion diseases.

The prion protein can exist in two forms: the normal PrP^{C} and the dangerous PrP^{Sc}. PrP^{Sc} but not PrP^{C} can (a) form aggregates, (b) catalyse the conversion of additional PrP^{C} to PrP^{Sc} within the brain of an individual person or animal, and (c) infect other individuals, by various routes including ingestion of nervous tissue from an affected animal (or person, in the case of kuru).

- PrP^C = prion protein-*Cellular*;
  PrP^{Sc} = prion protein-*Scrapie*.

Prion disease presents widespread health problems for humans and animals. In 2001 a serious epidemic of **bovine spongiform encephalopathy** (colloquially, 'mad-cow disease') devastated the United Kingdom countryside. There was an apparent association with the appearance of human cases of **variant Creutzfeld–Jacob disease (vCJD)**. In the hereditary disease familial CJD, symptoms began to appear in people aged 55–75. Variant CJD affected people in their twenties. It is hypothesized that these outbreaks were associated with transmission of prion protein infections across species barriers: sheep to cows for BSE, and cows to humans for vCJD.

Prion proteins form a family of homologous proteins in many species of animals, and also in yeast, but apparently not in *C. elegans* or *Drosophila*. Normal human prion protein is synthesized as a 253-residue polypeptide. This comprises: an N-terminal signal peptide, followed by a domain containing ~5 tandem repeats of the octapeptide PHGGGWGQ (in mammals), a conserved 140-residue domain, and a C-terminal hydrophobic domain. The signal domain and the C-terminal domain are cleaved off, and the protein is anchored to the extracellular side of the cell membrane of neurons by a GPI (glycosylphosphatidylinositol) group bound to the C-terminal residue Ser231 of the mature protein.

The normal role of PrP^C is not clear. Mice in which PrP^C has been knocked out develop normally for a time, and eventually die of apparently unrelated developmental defects. In fact, PrP^C-knockout mice are not susceptible to infection with PrP^{Sc}, an observation important in proving the mechanism of the disease.

The nature of the conformational change is still not entirely clear. The change from α to β structure shown by circular dichroism is one clue. In principle, prion proteins show multiple structures from one polypeptide sequence. However differences in glycosylation patterns between PrP^C and PrP^{Sc} have been reported; these may play a role in defining the conformation.

The mechanism by which PrP^{Sc} catalyses the transformation of additional PrP^C to PrP^{Sc} is also not clear. Inherited prion diseases are associated with mutants, presumably increasing the tendency for conformational mobility (Table 10.2). A related question concerns the kinetics of the process – what governs the rate of accumulation of aggregates that causes many prion diseases to appear only among the elderly?

- Many diseases arise from formation of protein aggregates, including: sickle-cell anaemia, amyloidoses, Alzheimer's disease, Huntington's disease, familial and variant Creutzfeld–Jacob disease, prion diseases. Most of these are genetic, some are infectious.

Properties of protein–protein complexes

Stoichiometry – what is the composition of the complex?

Protein complexes vary widely in the numbers and variety of molecules they contain. Some contain only

Table 10.2 Some diseases associated with prion proteins

Disease	Species affected	Symptoms
scrapie	sheep	hypersensitivity, unusual gait, tremor
bovine spongiform encephalopathy, or 'mad cow disease'	cow	similar to scrapie
kuru	human	loss of coordination, dementia
Creutzfeld–Jacob disease (CJD)	human, age 55–75	impaired vision and motor control, dementia
variant CJD	human, age 20–30	psychiatric and sensory anomalies preceding dementia
Gerstmann–Straüssler–Scheinker syndrome	human	dementia
fatal familial insomnia	human	sleep disorder

BOX 10.8 Evolution of the proteasome

The archaeal proteasome contains 14 identical α subunits and 14 identical β subunits. They are arranged in four stacked rings, α_7–β_7–β_7–α_7. All β subunits have protease activity. The core of eukaryotic proteasome also contain the α_7–β_7–β_7–α_7 stacked ring structure, but each ring contains seven diverged and non-identical subunits. Thus, the eukaryotic proteasome contains seven homologous but non-identical α subunits and seven homologous but non-identical β subunits. Only three of the eukaryotic β subunits have protease activity. The eukaryotic proteasome also contains large regulatory subunits in addition to the α–β rings, which select ubiquitinylated proteins for degradation.

a few proteins; others are very large. For example, pyruvate dehydrogenase contains hundreds of subunits and some viral capsids contain thousands.

Some prokaryotic proteins containing identical subunits are homologous to eukaryotic proteins containing related but non-identical subunits, arising by gene duplication and divergence. The proteasome is an example (see Box 10.8). Some viruses achieve diversity *without* duplication, by combining proteins with the same sequence but different conformations.

Affinity – how stable is the complex?

The measure of the affinity of a complex is the **dissociation constant**, K_D, the equilibrium constant for the *reverse* of the binding reaction:

$$\text{protein–ligand} = \text{protein} + \text{ligand} \quad K_D = \frac{[P][L]}{[PL]}$$

where [P], [L], and [PL] denote the numerical values of the concentrations of protein (P), ligand (L), and protein–ligand complex (PL), respectively, expressed in mol l^{-1}. The lower the K_D, the tighter the binding. K_D corresponds to the concentration of free ligand at which half the proteins bind ligand and half are free: [P] = [PL].

- The Michaelis constant of an enzyme is the dissociation constant of the enzyme–substrate complex.

Table 10.3 Dissociation constants of some protein–ligand complexes

Biological context	Ligand	Typical K_D
Allosteric activator	Monovalent ion	10^{-4}–10^{-2}
Co-enzyme binding	NAD, for instance	10^{-7}–10^{-4}
Antigen–antibody complexes	Various	10^{-4}–10^{-16}
Thrombin inhibitor	Hirudin	5×10^{-14}
Trypsin inhibitor	Bovine pancreatic trypsin inhibitor	10^{-14}
Streptavidin	Biotin	10^{-15}

Dissociation constants of protein–ligand complexes span a wide range, as shown in Table 10.3.

Structural studies have elucidated several important features of the interactions between soluble proteins, that contribute to affinity.

- *What holds the proteins together?* Burial of hydrophobic sidechains, hydrogen bonds and salt bridges, and Van der Waals forces. A typical protein–protein interface might involve 22 residues and 90 atoms, of which 20% would be mainchain atoms, and an occasional water molecule (Box 10.9). Burial of 1 Å^2 contributes ~100 J mol^{-1} to stability. There is, on average, one intramolecular hydrogen bond per 170 Å^2 of interface area. The average value of the surface area buried in binary protein complexes is ~1600 Å^2. The minimum buried surface for stability of a protein–protein complex is ~1000 Å^2.
- *Do proteins change conformation in complexing?* In some cases the interaction energy has to 'pay for' the conformational change and the interface tends to be correspondingly larger. Complexes that involve conformational changes generally bury >2000 Å^2.
- *What determines specificity?* Complementarity of the occluding surfaces, in shape, hydrogen-bonding potential, and charge distribution. Prediction of protein complexes from the structures of the partners is the **docking problem.** Reliable solution of this problem, together with progress in structural genomics, would permit *in silicio* screening of proteomes for interacting partners.

How are complexes organized in three dimensions?

When two proteins form a complex, each leaves a 'footprint' on the surface of the other, defining the portion of the surface involved in the interaction. If two proteins interact using the *same* surface on both, the complex is *closed*. If two proteins interact through *different* surfaces, the complex is *open*. The significance is that a closed complex does not allow additional proteins to bind with the same interaction.

A protein–protein interface: phage M13 gene III protein and *E. coli* TolA

During infection of *E. coli* by phage M13, a complex forms between the N-terminal domain of the minor coat gene 3 protein of the phage and the C-terminal domain of a receptor protein in the bacterial cell membrane, TolA (see Figure 10.25).

The complex is stabilized by burial of 1765 Å^2 of surface area, by combination of β-sheets from both proteins to form an extended β-sheet (see Figure 10.25a) and by several linkages of sidechains by hydrogen bonds and salt bridges. The area buried in the complex is divided almost evenly between the two partners.

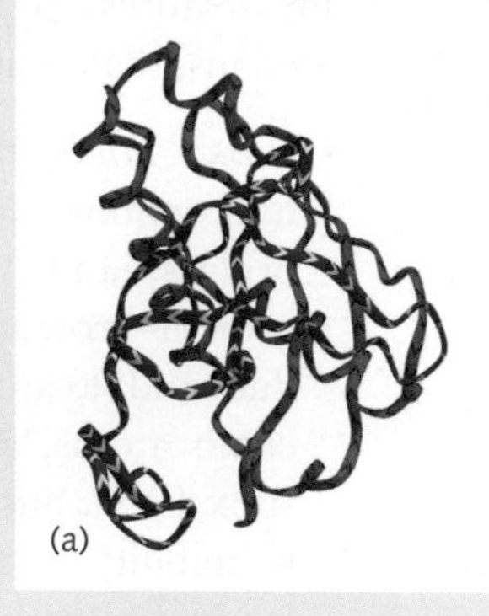

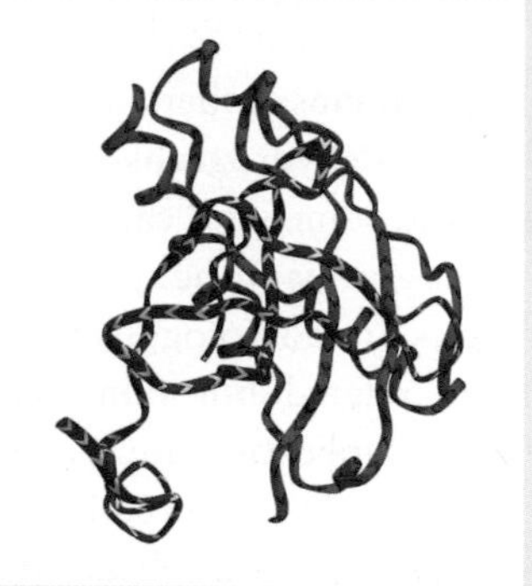
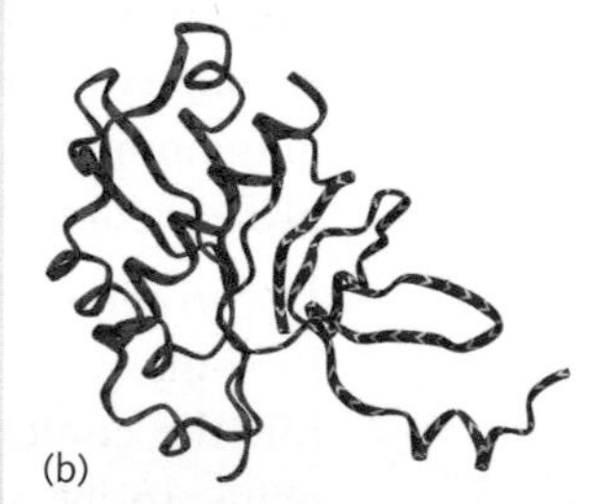

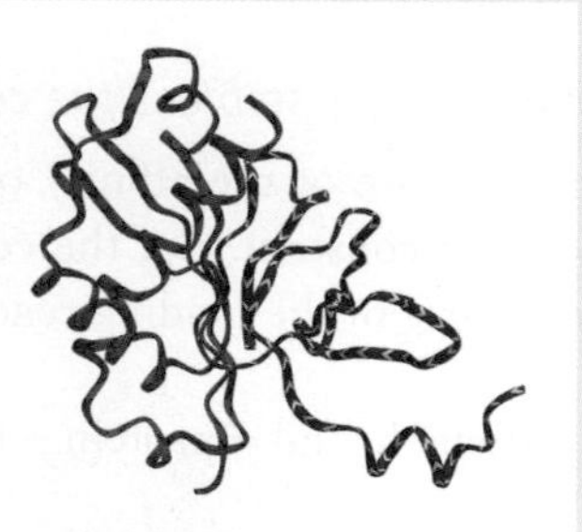
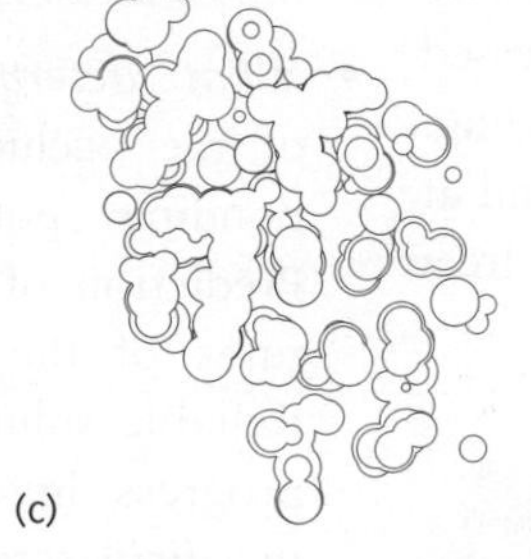

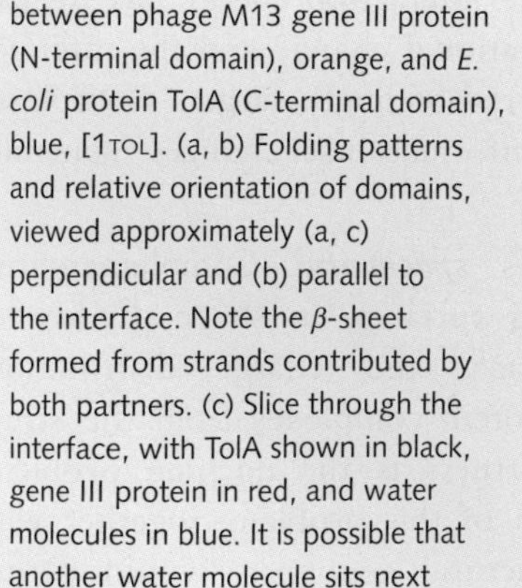
Figure 10.25 The interface between phage M13 gene III protein (N-terminal domain), orange, and *E. coli* protein TolA (C-terminal domain), blue, [1TOL]. (a, b) Folding patterns and relative orientation of domains, viewed approximately (a, c) perpendicular and (b) parallel to the interface. Note the β-sheet formed from strands contributed by both partners. (c) Slice through the interface, with TolA shown in black, gene III protein in red, and water molecules in blue. It is possible that another water molecule sits next to the one inside the structure.

An open complex, in which the surface of potential interaction is not occluded, can grow by accretion of additional subunits. Thus, open but not closed complexes are compatible with the formation of aggregates by continued addition of monomers making the same interaction.

Multisubunit proteins

An important class of protein–protein complexes is oligomeric or multisubunit proteins. We appeal to structural biology to address the following questions.

- *What is the stoichiometry?* How many different types of subunit appear and how many of each are present? Most proteins are homodimers or homotetramers. Monomers and heterooligomers are less common. The ribosome is an extreme example of a heterooligomer. Proteins containing odd numbers of subunits are rarer than those containing even numbers of subunits.
- *What is the relationship between the contributions of different subunits to the interface?* Consider a dimer of two identical subunits: in **isologous** binding, the interface is formed from the same sets of residues from both monomers; in **heterologous** binding, different monomers contribute different sets of residues to the binding site. A handshake is isologous.

- An isologous open structure is not possible. Why?

- *Is the structure open or closed?* In an open structure, at least one of the sites forming the binding surface is exposed in at least one of the subunits, so that additional subunits could be added on. In a closed structure, all binding surfaces are in contact with partners and the assembly is saturated. Domain swapping – exchange of segments between two interacting domains – often but not always produces closed isologous dimers.
- *What is the symmetry of the structure?* Symmetry is the rule, rather than the exception, in structures of oligomeric proteins. The subunits in most dimers are related by an axis of twofold symmetry. Yeast hexokinase is an exception. It forms an asymmetric dimer. In the human growth hormone receptor, a nearly symmetric dimer binds an asymmetric ligand (see Figure 10.26).
- *Do any of the subunits undergo conformational changes on assembly?* Often we don't know. In cases of extensively interlocked interfaces, such as the Trp repressor, the monomers could not adopt the same structure in the absence of their partners. Allosteric proteins can undergo ligand-dependent conformational changes. In ATP synthase, a threefold symmetric complex of $\alpha\beta$ subunits is distorted by interaction with the γ subunit.

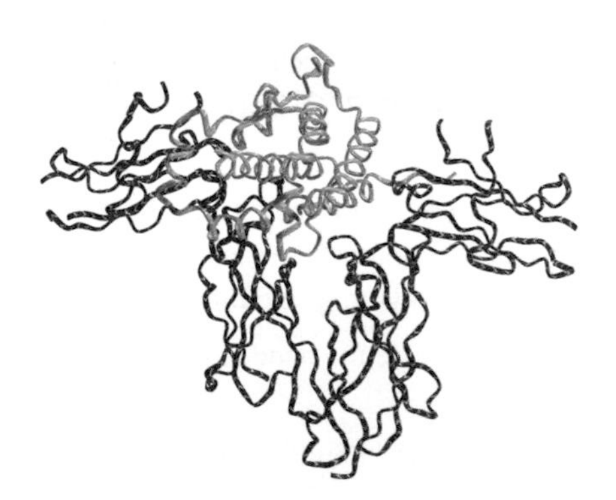
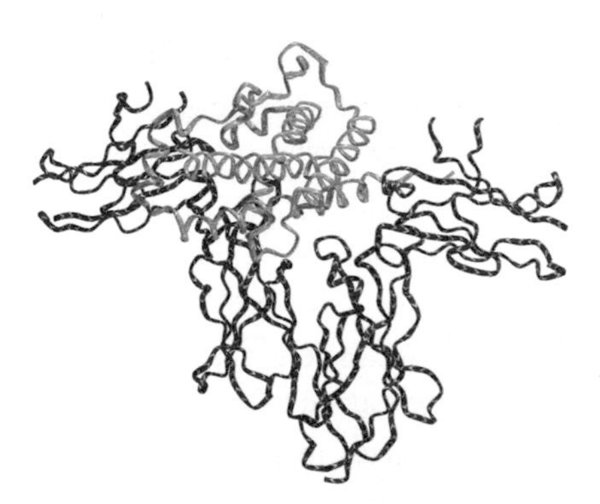

Figure 10.26 Human growth hormone (blue) in complex with two molecules illustrating the dimerized exterior domain of its receptor (green, orange) [3hhr].

RECOMMENDED READING

- Two fine general references:

Branden, C.-I. & Tooze, J. (1999). *Introduction to Protein Structure*, 2nd edn. Garland, New York.

Liljas, A., Liljas, L., Piskur, J., Lindblom, G., Nissen, P., & Kjeldgaard, M. (2009). *Textbook of Structural Biology*. World Scientific, Singapore.

- Reviews of current techniques in proteomics:

de Hoog, C.L. & Mann, M. (2004). Proteomics. Annu. Rev. Genomics Hum. Genet. **5**, 267–293.

Domon, B. & Aebersold, R. (2006). Mass spectrometry and protein analysis. Science **312**, 212–217.

- The current state of the art in protein structure prediction:

Moult, J. (2005). A decade of CASP: progress, bottlenecks, and prognosis in protein structure prediction. Curr. Opin. Struct. Biol. **16**, 285–289.

Janin, J. (2005). Assessing predictions of protein–protein interaction: the CAPRI experiment. Prot. Sci. **14**, 278–283.

Tramontano, A. (2006). *Protein Structure Prediction: Concepts and Applications*. Wiley–VCH, Weinheim.

- Articles about protein complexes:

Russell, R.B., et al. (2004). A structural perspective on protein–protein interactions. Curr. Opin. Struct. Biol. **14**, 313–324.

Sali, A. & Chiu, W. (2005). Macromolecular assemblies highlighted. Structure **13**, 339–341.

- A companion volume to this one, treating the topics of this chapter in more detail:

Lesk, A.M. (2010). *Introduction to Protein Science/Architecture, Function and Genomics*. Oxford University Press, Oxford.

EXERCISES, PROBLEMS, AND WEBLEMS

Exercises

Exercise 10.1 For each of the following amino acids, say whether they are hydrophobic, polar, positively charged at pH 7, or negatively charged at pH 7: (a) leucine; (b) aspartic acid; (c) glutamine; (d) phenylalanine; (e) lysine.

Exercise 10.2 (a) Identify a hydrophobic amino acid that is more bulky than alanine but less bulky than leucine. (b) Identify two amino acids that have almost the same size and shape (differing only in that one has a methyl group and the other has a hydroxyl group).

Exercise 10.3 On a photocopy of Figure 10.2, indicate the bond, a rotation around which would correspond to the conformational angle ψ_{i-1}.

Exercise 10.4 Would it be possible, by rotation around bonds shown in Figure 10.2, to convert residue *i* from the L to the D conformation?

Exercise 10.5 Estimate the values of ϕ and ψ that correspond to the α_L conformation in Figure 10.3.

Exercise 10.6 Figure 10.2 shows the *trans* conformation of the polypeptide chain. (a) If the angle labelled ω_i in that figure is changed from $\omega = 180°$ *trans* to $\omega = 0°$ *cis*, keeping the positions of all atoms in residues $i - 1$ and i fixed, what atom would occupy the position currently occupied by H_{i+1}? (b) In the structure shown in Figure 10.2, $\phi_i = 180°$. The only unlabelled atom is the hydrogen connected to C_i^{α}. Assuming a rotation that keeps the positions of N_i, H_i, and the atoms of residue $i - 1$ fixed, estimate the value of ϕ_i that would place this unlabelled hydrogen atom at the position that C_i^{β} occupies in Figure 10.2.

Exercise 10.7 Describe and compare the nature of the accelerating and retarding forces on the molecules in mass spectrometry and SDS-PAGE.

Exercise 10.8 In a typical protein–protein interface of area 1700 Å^2: (a) how many intermolecular hydrogen bonds would you expect to be formed? (b) How many fixed water molecules would you expect to find in the interface? (c) If the entire buried area were hydrophobic, what contribution to the free energy of stabilization would you estimate it to make?

Exercise 10.9 In the dimer between syntrophin and neuronal nitric oxide synthase (see Figure 10.27), (a) is the dimer structure open or closed? (b) What secondary structure element is shared between the two domains?

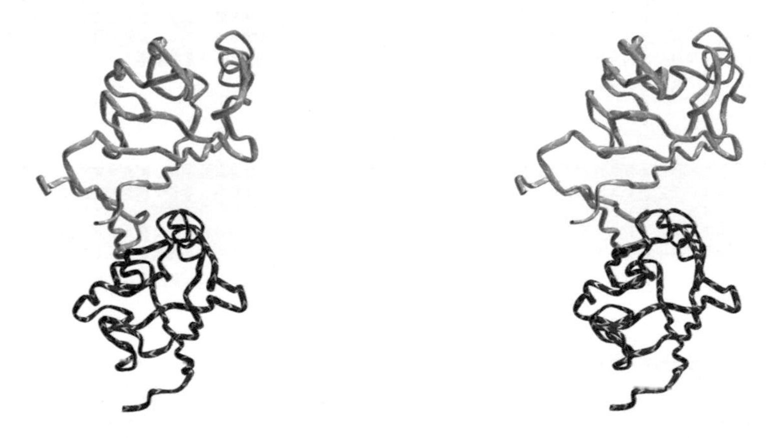

Figure 10.27 Interaction between PDZ domains in syntrophin (cyan) and neuronal nitric oxide synthase (magenta) [1QAV].

Problems

Problem 10.1 P. Schultz has posed the question: would an extended genetic code – perhaps one in which one of the rarely used stop codons coded for a novel amino acid with a somewhat unusual size, shape, or charge distribution – be 'better' than the normal one? The question could be taken to apply to either natural or artificial exensions of the code. How would you design experiments to answer this question? What precautions would you consider necessary?

Problem 10.2 As a general rule, vertebrates use creatine as a phosphogen and invertebrates use arginine. Figure 10.28 shows the sequence alignment of creatine kinases (CK) from rabbit and chicken, and arginine kinases (AK) from sea cucumber, horseshoe crab, and abalone. The numbers of identical residues in pairs of sequences in this alignment are:

	Rabbit CK	Chicken CK	Sea cucumber AK	Horseshoe crab AK	Giant abalone AK
Rabbit CK	378	252	226	147	132
Chicken CK	252	381	219	129	129
Sea cucumber AK	226	219	370	154	137
Horseshoe crab AK	147	129	154	357	191
Giant abalone AK	132	129	137	191	358

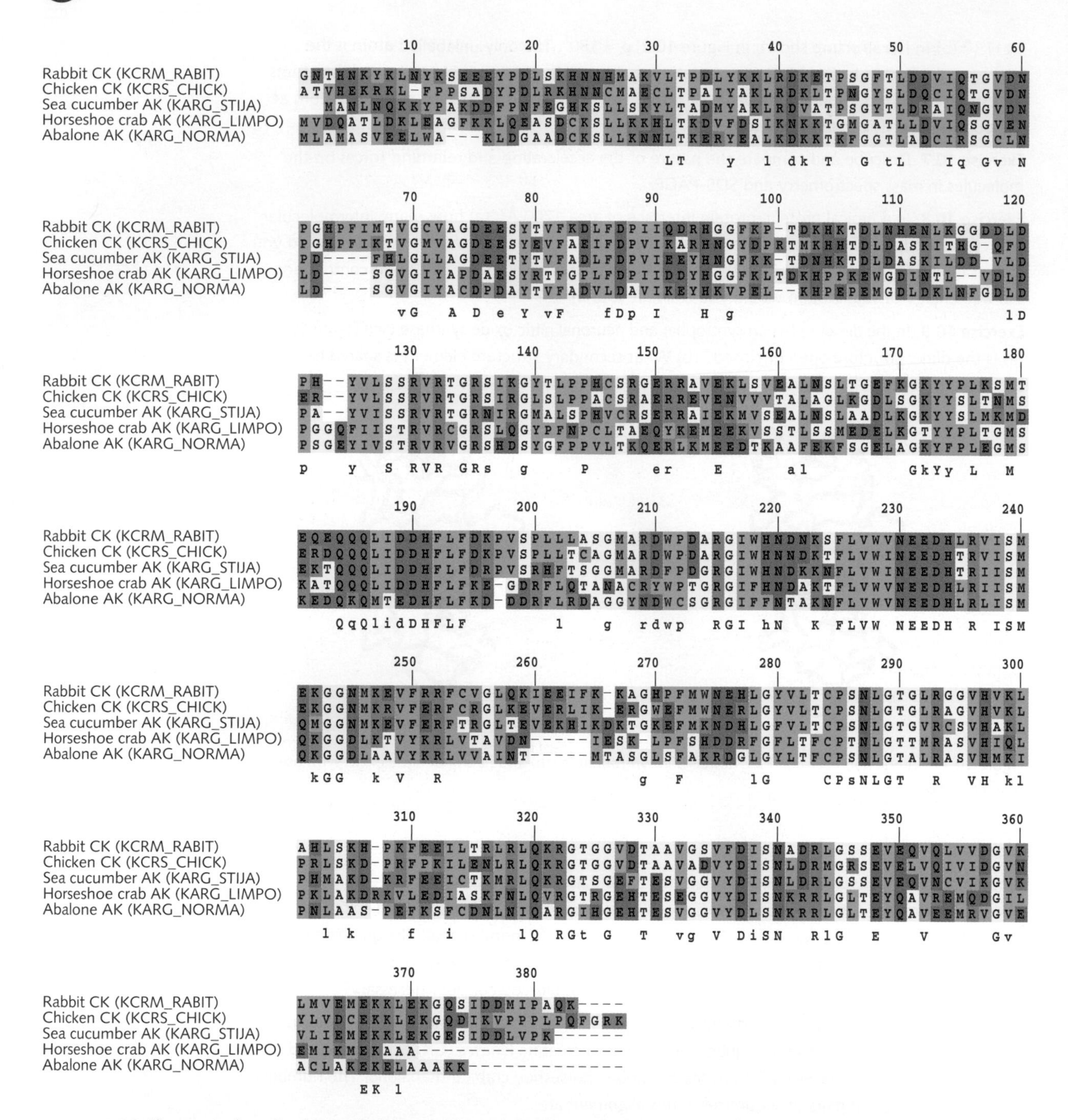

Figure 10.28 Alignment of creatine kinase (CK) from rabbit and chicken, and arginine kinase (AK) from sea cucumber, horseshoe crab, and abalone.

(a) Does sea cucumber arginine kinase appear to be more related to vertebrate creatine kinases or other invertebrate arginine kinases? (b) On a photocopy of Figure 10.28, circle (at least two) regions, each at least four residues long, in which sea cucumber arginine kinase resembles vertebrate creatine kinases more closely than it resembles other invertebrate arginine kinases, and circle (at least two) regions, each at least four residues long, in which sea cucumber arginine kinase resembles other invertebrate arginine kinases more closely than it resembles vertebrate creatine kinases. (c) Can you identify any residues that might conceivably be responsible for the difference in substrate specificity between arginine and creatine? (d) Outline how you could test the hypothesis you presented as your answer to part (c), using only computational and not wet laboratory methods. (e) Which is more likely: that arginine kinase activity evolved once and that no protein in any ancestor of sea cucumber had creatine kinase activity, or that sea cucumber arginine kinase evolved from a precursor it shared with present vertebrate creatine kinases? Explain your reasoning.

Weblems

Weblem 10.1 Choose one of the following glycoprotein storage diseases: (1) aspartylglucosaminuria, (2) α-mannosidosis, (3) β-mannosidosis, (4) Sandhoff–Jatzkewitz disease, or (5) sialidosis. (a) What is the inheritance pattern of this disease? (b) What is the incidence of this disease in the USA; i.e. what fraction of the population is affected? (c) What is the biochemical defect that causes this disease?

Weblem 10.2 Some athletes take creatine in order to build up their ability to store energy as phosphocreatine. Athletes find that creatine certainly improves performance in 'burst' events – all-out effort for up to 10 seconds – but its efficacy for 'endurance events' is debated. Examples of burst events include the 100 metre dash, a point in tennis, or a play in football. In all of these cases, there is a pause between energy bursts to allow replenishment of phosphocreatine by oxidative metabolism. This takes ~30–60 seconds. (The International Tennis Federation rules allow no more than 20 seconds between points.) (a) How does the US Food and Drug Administration classify creatine? (b) Is creatine banned by the Olympics or major commercial sports enterprises? (c) Why would it be difficult to enforce a ban on creatine supplements?

Weblem 10.3 The bacterium *Pseudomonas fluorescens* and the fungus *Curvularia inaequalis* each possesses a chloroperoxidase, an enzyme that catalyses halogenation reactions. Do these enzymes have the same folding pattern?

CHAPTER 11

Systems Biology

LEARNING GOALS

- To gain a sense of the discipline of systems biology as an integrative approach to all the 'omics' disciplines.
- To understand the idea of networks and their representation as graphs.
- To appreciate that many aspects of metabolic networks are shared by different organisms, and that they evolve.
- To know the databases dealing with metabolic pathways, including EcoCyc, KEGG, WIT, and BRENDA, and be fairly fluent at using them and the links they contain.
- To distinguish between static and dynamic aspects of biological networks.
- To understand the ideas of stability and robustness and the mechanisms by which life achieves them.
- To appreciate how computational concepts important for systems biology, such as randomness and complexity, have been made precise and quantitative.
- To understand the structure, dynamics, and evolution of metabolic networks.
- To know the different ways of experimentally determining protein–protein and protein–nucleic acid interactions.
- To be familiar with some of the basic types of DNA-binding protein.
- To understand the structures, dynamics, and evolution of regulatory networks.
- To appreciate the adaptability of the yeast regulatory network.

Introduction to systems biology

The goal of systems biology is the synthesis of all biological data into a unified picture of the structure, dynamics, logistics, and ultimately the logic of living things. Systems biology focuses on the integration of gene, RNA, and protein activity.

Molecules are social animals and life depends on their interactions. As individual molecules have specialized functions, control mechanisms are required to organize and coordinate their activities. Failure of control mechanisms can lead to disease and even death.

Two parallel networks: physical and logical

Systems biology deals with networks. Networks consist of sets of molecules and the interactions among them. There are networks of genes, of RNAs, of proteins, and of metabolites. The same set of molecules may be connected by different types of interaction or relationship, to form different networks (see Box 11.1).

In cells, two interaction networks are in operation: (a) a *physical network* of protein–protein and protein–nucleic acid complexes, and (b) a *logical network* of control cascades. Interactions may be physical or logical. Often, they are both. *Physical and logical networks operate in parallel.* A macromolecular complex such as the ribosome is a network of proteins and RNAs, interacting through the *physical* contacts in their assembly. A transcription regulatory network is a network of genes, exerting *logical* control over expression patterns via the synthesis of specific DNA-binding proteins. A transcription factor that acts by binding to DNA may never interact physically with the proteins the expression of which it controls. Metabolic pathways have a similar duality: many but not all metabolic pathways are mediated by physical protein–protein interactions and regulated by logical ones.

Examples of purely physical interactions include macromolecular complexes – both multiprotein complexes and protein–nucleic acid complexes. Examples of logical interactions *not* mediated entirely by direct physical interaction between proteins include feedback loops in which the increase in concentration of a product of a metabolic pathway inhibits an enzyme catalysing one of the early steps in the pathway, or the secretion of a small molecule as a signal to other cells, in 'fire and forget' mode (see Box 11.2). In these cases, the logical interaction is transmitted by diffusion of a small molecule, rather than physical contact between source and recipient of the signal.

The allosteric change in haemoglobin is an example of simultaneous physical and logical interaction: the subunits of haemoglobin respond to changes in oxygen levels by a conformational change that alters oxygen affinity. Another example is the transmission of a signal from the surface of a cell across the membrane to the interior by dimerization of a receptor. This can be the initial trigger of a process that ultimately affects gene expression. Not all links of this process need involve protein–protein interactions; some may be mediated by diffusion of small molecules such as cyclic AMP.

Even though particular complexes may participate in both physical and logical networks, the two networks remain distinct in terms of their organization and their biological function, and it is useful to keep the distinction between them in mind, *especially* when they overlap.

BOX 11.1 Some networks in systems biology

Network	Element of network	Connection between elements
Genomes	Gene	Homology *or* shared expression pattern *or* linkage
Protein	Protein	Homology *or* regulatory relationship *or* shared expression pattern *or* physical complex formation
Metabolite	Chemical compound e.g. glucose	Substrate and product of an enzymatic reaction *or* similarity in structure *or* similarity in reactivity

- Cells contain both physical and logical networks. Many interactions are common to both.

Cell–cell communication in microorganisms: quorum sensing

Control mechanisms *not* involving direct protein–protein interactions mediate intercellular signalling in microorganisms. *Vibrio fischeri* is a marine bacterium that can adopt alternative physiological states in which bioluminescence is active or inactive. (Literally a 'light switch'.) The organism can live free in seawater or colonize the light organs of certain species of fish or squid. It is bioluminescent only when growing within the animal.

What controls the switch? The bacteria respond to the local density of bacterial cells, a form of communication called **quorum sensing**. In *V. fischeri*, quorum sensing is mediated by secretion and detection of a small signalling molecule, *N*-(3-oxohexanoyl)-homoserine lactone. Related species use other *N*-acyl homoserine lactones, abbreviated to AHL. AHL can diffuse freely out of the cells in which it is synthesized. Within the light organs, culture densities can reach 10^{10}–10^{11} cells ml^{-1}, and the AHL concentration can exceed the threshold of about 5–10 nM for flipping the physiological switch.

Bacterial genes *luxI* and *luxR* govern the regulation. The product of *luxI* is involved in the synthesis of AHL. The *luxR* gene product, LuxR, contains a membrane-bound domain, which detects the AHL signal, and a transcriptional activator domain. LuxR activates an operon that includes (1) genes for synthesis of luciferase (the enzyme responsible for the bioluminescence); and (2) *luxI*, expression of which synthesizes additional AHL, amplifying the signal and sharpening the transition.

The host also senses the bacteria: the light organs of squid grown in sterile salt water do not develop properly. This appears to be a reaction to the luminescence, rather than to the AHL. For the animal, the luminescence contributes to camouflage: disguise from predators at lower depths, by blending with illumination from the sky. The masking of shadows is a natural form of 'make-up'. (The bioluminescence also regularly surprises diners in seafood restaurants, who jump to the conclusion that their glowing dinner is of extraterrestrial origin. However, most bioluminescent bacteria are harmless, although some strains of the related *Vibrio* species, *V. cholerae*, the causative agent of cholera, are weakly bioluminescent. In fact, the virulence of *V. cholerae* is also under the control of quorum sensing, by a related mechanism.)

Statics and dynamics of networks

Networks have both **static** and **dynamic** aspects. The reaction sequences and enzyme names memorized by generations of biochemistry students, and appearing on wall charts, mugs, etc., reflect the **static structure** of the metabolic pathway network. The patterns of flow of metabolites through the network, and the control of expression patterns of metabolic enzymes in response to changing nutrient availability, are **dynamic** features.

Stability is an important goal of regulatory dynamics. Concentrations of metabolites in cells are carefully controlled. Sometimes they remain roughly constant (steady-state conditions). Sometimes they vary cyclically. Sources of metabolic stability include:

- constant rate of input (for instance, rhythmic breathing ensures a regular supply of oxygen);
- feedback inhibition of enzymes;
- allosteric control of activity of enzymes;
- turning proteins 'on' and 'off' by phosphorylation and dephosphorylation; and
- control of amounts of proteins by regulation of expression.

Robustness is another crucial feature of the dynamics of biological networks. Biological systems need to be robust, both for survival of individuals under stress and for the plasticity required for evolution.

- Under unchanging environmental conditions, an organism's biochemical systems must be stable.
- Under rapidly changing conditions, the system must accommodate both neutral and stressful perturbations. Internally driven short-term adjustments dampen out fluctuations and choreograph programmes such as the cell cycle. Responses to external stimuli adjust to changes in the composition or levels of nutrients or oxygen.
- Longer-term regulatory controls in an individual organism include changes of physiological state, such as sporulation in bacteria, the course of our response to and recovery from viral infections such as a cold, and the unfolding of developmental stages during a lifetime. Populations respond to long-term changes in conditions by evolving.

Pictures of networks as graphs

We think of networks in terms of graphs (see Boxes 11.3 and 11.4). Any network contains a set of elements called nodes, and edges connecting the nodes that stand for interactions. The familiar map of the London Underground is a network, taking the stations as the nodes and the tracks connecting the stations as edges. Two stations interact if they are connected by tracks on one or more lines. Note that the modern London Underground map shows the **topology** of the network but does not quantitatively represent the geography of the city. (An early map, from 1925, did maintain geographical accuracy. This was possible when the system was simpler than it is now.) Some of the maps now posted in the Paris Métro are fairly accurate geographically. *Considered as graphs, a geographically accurate map and a simplified map with the same edges, or connections, correspond to the same network.*

BOX 11.3 The idea of a graph

- A graph consists of a set of vertices, V, and a set of edges, E. The vertices correspond to the nodes of the network.
- Each edge is specified by a pair of vertices.
- In a **directed graph**, the edges are **ordered** pairs of vertices.
- In a **labelled graph**, there is a value associated with each edge. (A directed graph is a special case of a labelled graph: consider the arrowheads as labels.)

An undirected unlabelled graph specifies the connectivity of a network but *not* the distances between vertices (the topology but not the geometry, as in the modern London Underground map). Optionally, labels on the edges can indicate distances. For example, some phylogenetic trees indicate only the topology of the ancestor–descendant relationships. Others indicate quantitatively the amount of divergence between species. Phylogenetic trees are often drawn with the lengths of the branches indicating the time since the last common ancestor. This is a pictorial device for labelling the edges.

Many graphs do not correspond to physical structures, and in any event edge labels need not reflect geometry in the usual sense. For example, the links in a network of metabolic pathways might be labelled to reflect flow patterns.

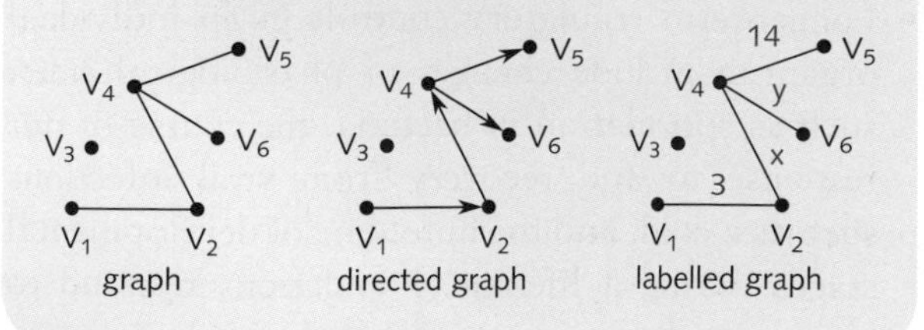

BOX 11.4 Examples of graphs

- Sets of people who have met each other
- Electricity distribution systems
- Phylogenetic trees
- Metabolic pathways
- Chemical bonding patterns in molecules
- Citation patterns in scientific literature
- The World Wide Web

- Graphs are abstract representations of networks. They show the connectivity of the network. Labelled graphs can show physical distances beween nodes, or other properties of edges such as throughput capacity.

A fundamental property of a network is its **connectivity.** If V_A and V_Z are vertices in a graph:

- a **path** from V_A to V_Z is a series of vertices: V_A, V_B, $V_C \ldots V_Z$, such that an edge in the graph connects each successive pair of vertices; the edges of a path in a directed graph must be traversed in the proper direction (in city traffic, a path must obey designations of 'one-way' streets);
- the number of edges in the chain is called the **length** of the path; and

- a **cycle** is a path of length >2 for which the initial and final end points are the same, but in which no intermediate link is repeated.

Sequences of consecutive metabolic reactions are pathways in a graph of metabolites. An irreversible reaction corresponds to a directed edge. A concatenation of signal-transduction events is a pathway in a regulatory network.

- A vitamin is a compound that we must eat because we cannot synthesize it. Therefore, there can be no path in the metabolic network leading to a vitamin.

For some networks, such as metabolic pathways or patterns of traffic in cities, the **dynamics** of the system depend on the transmission capacities of the individual links. These capacities can be indicated as labels of the edges of the graph. This allows modelling of patterns of flow through the network. Examples include route planning, in travel or deliveries. Note that the shortest path may well not give optimal throughput. In many cities, taxi drivers are exquisitely sensitive – and insensitively voluble – about currently optimal traffic paths.

A graph that contains a path between any two vertices is said to be **connected**. Alternatively, a graph may split into several connected components. The graph on page 344 has two connected components, one containing five vertices and one containing only one vertex. (In the extreme case, a graph could contain many vertices but no edges at all.) It is often useful to determine the *shortest path* between any two nodes, and to characterize a network by the distribution of shortest path lengths. The phrase 'six degrees of separation' – the title of a play by John Guare, made into a film – refers to the assertion (attributed originally to Marconi) that if the people in the world are vertices of a graph and the graph contains an edge whenever two people know each other, then the graph is connected and there is a path between any two vertices with length ≤6.

The London Underground network is connected in that there is (usually) a route between any two stations. Many questions familiar to commuters are shared in the analysis of biological networks; for example: what are the paths connecting station A and station B? Regarding different lines as subnetworks, how easy is it to transfer from one to another, i.e. what is the nature of the patterns of connectivity? In case of failure of one or more links, is the network robust, i.e. does it remain connected?

Trees

A **tree** is a special form of graph (see p. 182). *A tree is a connected graph containing only one path between each pair of vertices.* A hierarchy is a tree: examples include military chains of command and the Linnaean taxonomy. A tree cannot contain a cycle: if it did, there would be two paths from the initial point (= the final point) to each intermediate point. In the undirected graph on page 344, the subgraph consisting of vertices V_1, V_2, V_4, V_5, and V_6 is a tree. Adding an edge from V_1 to V_5 would create an alternative path from V_1 to V_5, and the cycle $V_1 \rightarrow V_2 \rightarrow V_4 \rightarrow V_5 \rightarrow V_1$; the graph would no longer be a tree.

The **density of connections** is the mean number of edges per vertex and characterizes the structure of a graph. A **fully connected** graph contains an edge between every pair of nodes. A fully connected graph of N vertices has $N - 1$ connections per vertex. A graph with no edges has 0 connections per node. Nervous systems of higher animals achieve their power not only by containing large number of neurons but also by high degrees of connectivity.

Sometimes there are limits on numbers of connections. For many human societies, in the graph in which individuals are the vertices and edges link people married to each other, each node has connectivity 0 or 1. Hydrocarbon structures can be represented as graphs, with the hydrogen and carbon atoms as vertices and the chemical bonds as the edges. The rules of valence require that each node corresponding to a carbon atom has ≤4 connections.

In other networks, connectivities follow statistical regularities. For instance, the World Wide Web can be considered to be a directed graph: individual documents are the nodes and hyperlinks are the edges. The distribution of incoming and outgoing links follows **power laws:** $P(k)$ = probability of k edges = k^{-q}, where $q = 2.1$ for incoming links and $q = 2.45$ for outgoing links (see Box 11.5).

The density of connections is very important in defining the properties of a network. For instance,

BOX 11.5

'Small-world' networks

Many observed networks, including biological networks, the World Wide Web, and electric power distribution grids, have the characteristics of high clustering and short path lengths. They include relatively few nodes with very large numbers of connections, called 'hubs', and many nodes with few connections. These combine to produce short path lengths between all nodes. From this feature, they are called 'small-world networks'. Such networks tend to be fairly robust, staying connected after failure of random nodes. Failure of a hub would be disastrous but is unlikely, because there are few hubs.

Many networks, notably the World Wide Web, are continuously adding nodes. The connectivity distribution tends to remain fairly constant as the network grows. These are called 'scale-free' networks.

the interactions that spread disease among humans and/or animals form a network. Whether a disease will cause an epidemic depends not only on the ease of transmission in any particular interaction but also on the density of connections. As the density of connections – the rate of interactions – increases, the system can exhibit a *qualitative* change in behaviour, analogous to a phase change in physical chemistry, from a situation in which the disease remains under control to an epidemic spreading through an entire population. The classic approach of 'quarantine' – isolating people for 40 days – works by cutting down the degree of connectivity of the disease-transmission network. Note that a carrier who shows no symptoms – 'Typhoid Mary'* was a classic case – serves as a hub of the disease transmission network.

Two historical epidemics associated with wars demonstrate the distinction between topology and geometry in network connectivity.

In the early years of The Peloponnesian War, Athens suffered a severe epidemic. (From Thucydides' detailed description of the symptoms, the disease was probably bubonic plague.) A factor contributing to its transmission was the crowding of people into the city from the surrounding countryside, out of fear of greater vulnerability to military invasion.

After World War I, an epidemic of influenza killed ~50–100 million people, more than died in the war itself. Long-distance travel by soldiers returning from the war helped spread the disease. Any epidemic needs an infectious agent, and a high density of routes of transmission.

These examples show that the controlling factor is the density of the *connections* and not the density of the people.

Sources of ideas for systems biology

Just as molecular biology calls upon chemistry, systems biology calls upon mathematics for help in making quantitative the general ideas about network properties described in the preceding sections.

Several related ideas are important in coping with the static and dynamic aspects of systems biology. These include complexity, entropy, randomness, redundancy, robustness, predictablility, and chaos. We deal with these in our daily lives, but without the need to define them precisely and quantitatively. How well do we really understand these concepts? What are the relationships among them? And how can they be used to illuminate biology in general and systems biology in particular?

Complexity of sequences

The simplest complex object in biology is a sequence. We have all heard of random sequences and probably agree that the more random the sequence, the more complex it is. For example, genomic sequences contain 'low-complexity' regions. In the human genome, such regions include simple repeats, or microsatellites, or regions of highly skewed nucleotide composition such as AT-rich or GC-rich regions, or polypurine or

* Mary Mallon (1869–1938) presented the following unfortunate combination of features: (1) she was infected with typhoid; (2) she did not show symptoms; and (3) she worked for many families as a cook.

polypyrimidine stretches. Are these regions more or less random than a region containing a gene that encodes a specific protein? How can such properties of sequences be measured?

Here is an approach to such questions. Take a sequence of characters:

AGTCTCTA . . . , or AATAAAAATAAA . . . ,
or ABZXUVJFLT. . . .

What determines the amount of information needed to specify the next character in each sequence? Less information is required if the set of possible characters – A, T, G, C – is very small or if the distribution is very skewed – AATAAAAATAAA – than if the set is very large and the ratio of different characters is more even.

How can we make this quantitative? Genomic sequences are limited to characters A, T, G, and C. To identify each symbol, it is enough to ask two 'yes-or-no' questions. For instance:

1. Is it a purine (or a pyrimidine)? (Purine implies it is A or G.)
2. Is it 6-amino (or 6-keto)? (6-Amino implies it is A or C.)

Knowing the answers to these two questions is enough for us to identify one of the four bases uniquely.

Representing yes = 1 and no = 0, each 'yes-or-no' question provides 1 binary digit, or **one bit** of information. We could encode each nucleotide of a genome sequence as a **two-bit** binary string.

To identify a character of the ordinary alphabet – *abcd . . . z* – requires *more* than two yes-or-no questions. It is therefore reasonable to think that a character string of full text is more complex than a genomic sequence of the same length containing only characters A, T, G, and C.

Questions of how much information is needed to specify an amino acid appear in the genetic code itself. How many nucleotides are required to encode 20 amino acids? If each position in a gene can contain one of four nucleotides, then there are only 16 possible dinucleotides – not enough. So, if the same number of nucleotides is to be required for each amino acid, there must be at least three nucleotides per codon, as observed. As there are only 20 amino acids, the triplet code contains redundancy.

Why not make do with fewer amino acids? If 15 amino acids (plus a stop signal) – not unreasonable – would suffice, then from the information point of view, a doublet code would be possible. However, a two-base codon/two-base anticodon interaction would probably not have adequate stability.

Shannon's definition of entropy

In 1948, C.E. Shannon introduced the concept of **entropy** into information theory, as part of his analysis of signal transmission. Suppose a text contains symbols with relative probability p_i. Shannon's measure of entropy is:

$$H = -\sum_i p_i \log_2 p_i$$

The Shannon entropy, H, can be interpreted as the minimum average number of bits per symbol required to transmit the sequence.

For example, for a genomic sequence with equimolar base composition ($p_A = p_T = p_G = p_C = 0.25$),

$$\begin{aligned} H &= -\sum_i p_i \log_2 p_i \\ &= -[0.25 \log_2 0.25 + 0.25 \log_2 0.25 \\ &\quad + 0.25 \log_2 0.25 + 0.25 \log_2 0.25] \\ &= 2 \end{aligned}$$

($\log_2 0.25 = -2$.)

The result $H = 2$ for the gene sequence with equimolar base composition recovers our informal result that two bits, or two 'yes-or-no' questions, are required. For a sequence limited to *two* equiprobable characters A and T: $p_A = p_T = 0.5$, $H = -[0.5 \log_2 0.5 + 0.5 \log_2 0.5] = 1$. This also makes sense, because, knowing that the only choices are A and T, we can decide which it is with *one* 'yes-or-no' question, or one bit.

Suppose that a sequence is known to have the skewed nucleotide composition $p_A = p_T = 0.42$, and $p_G = p_C = 0.08$. Then:

$$\begin{aligned} H = -[&0.42 \log_2 0.42 + 0.42 \log_2 0.42 \\ &+ 0.08 \log_2 0.08 + 0.08 \log_2 0.08] = 1.63 \end{aligned}$$

What is the significance of the fact that the value $H = 1.63$ is less than 2? The uncertainty in each transmitted symbol is not complete – it is more likely to be A or T than G or C. In principle, we can use this knowledge to improve the coding efficiency.

The Morse code for telegraphy took such advantage of unequal letter distribution frequencies to encode common letters with short sequences and uncommon letters with longer ones. For instance E = dot (length one) and J = dot-dash-dash-dash (length four).

Depending on the content of the message, the efficiency of transmission depends on *how* it is encoded. A message containing repetitions – GACGACGACGACGAC... – would not have particularly low entropy if encoded one nucleotide at a time, but an entropy of 0 if encoded one triplet at a time. Compression algorithms are sensitive to such nuances and optimize the encoding.

Conversely, looking at distributions of oligonucleotides (dinucleotides, triplets, etc.) is a useful way of detecting biologically significant patterns. Codon-usage patterns in protein-coding regions are examples. Some algorithms for gene identification make use of biases in coding regions of frequencies of hexanucleotides.

Although the actual genetic code does not achieve the theoretical efficiency that entropy calculations suggest, and indeed there does not even seem to be selection for reduction in the size of non-viral genomes, it is clear that the redundancy in the genetic code has biological significance. Many single-base mutations are silent. Conservative mutations allow proteins to evolve with small non-lethal changes that, cumulatively, can achieve large changes in structure and function. And of course the redundancy in having two copies of the genetic information in two strands of DNA is used to detect and correct errors in replication, and to repair DNA damage.

Shannon entropy is linked with thermodynamic entropy through the general notion of disorder or randomness. The relationship has been explored by physicists, including J.C. Maxwell and L. Szilard, in their discussions of 'Maxwell's demon', and by E.T. Jaynes.

Randomness of sequences

The Shannon entropy of sequences is related to the idea of randomness, another concept that we know from everyday life without worrying too much about exactly what it means. A.N. Kolmogorov defined the complexity of a sequence of numbers as *the length of the shortest computer program that can reproduce the sequence*. Thus the sequence 0,0,0,0,0,0,0 ... is far from random, as it is the output of the very short program:

Step 1: print 0.
Step 2: go back to step 1.

Periodic sequences, such as:

Monday, Tuesday, Wednesday, Thursday, Friday, Saturday, Sunday, Monday, ...

are also of low complexity. In contrast, *a truly random sequence has no description shorter than the sequence itself*.

The relationship between complexity, randomness, and compressibility

One way to shorten the specification of a non-random sequence is to compress it. We all use compression algorithms on our files to save disk space. If a sequence is truly random, in the sense of Kolmogorov, it cannot be compressed. By definition, non-random sequences *can* be compressed.

One basic principle of compression is that: *if you can predict what is coming next, you can compress effectively*.

The reason that sequences such as 0,0,0,0, ... and Monday, Tuesday, Wednesday, Thursday, Friday, Saturday, Sunday, Monday, ... are so effectively compressible is that it is simple to decide what the successor of any element is. Even sequences for which it is not possible to decide unambiguously what the next element is can be compressed if some indications are available. It is not even necessary that the rules be supplied 'up front' as they can be for sequences such as 0,0,0,0, ... and Monday, Tuesday, Wednesday, Thursday, Friday, Saturday, Sunday, Monday, ... The rules and statistics of prediction of a successor can be generated on the fly from the incoming data. The rule, 'the weather on some day is likely to be the same as the weather the day before' would – in most places – be good enough for effective compression of a series of weather reports.

Putting together these considerations suggests a general idea that *the harder it is to predict the contents of a data set from a subset of the data, the more complex the data set is*.

The relationships among complexity, predictability, and compressibility, which we have so far described for character strings, apply to the static structures of other types of object, including images, three-dimensional structures, and – especially – networks. Indeed, most types of biological information can be regarded as networks. For instance, a nucleotide sequence is equivalent to a network in which the individual bases are the nodes, and each base is connected by a directed edge pointing to the next base. That's a perfectly proper graph! Conversely, recognizing that sequences are networks can usefully lead us to ask – can we define analogues of sequence alignment for more general networks? (Yes, we can.)

In biology, we are also interested in the complexity of **processes**.

Static and dynamic complexity

One dimension of complexity is time. Is it possible to distinguish static from dynamic complexity? If we could define and measure the static complexity of a system, this would provide an approach to dynamic complexity: we could ask how the static complexity of a system changes with time. Such changes appear to be governed by at least some general rules. If you stop someone on the street, they might well say that in closed systems the laws of thermodynamics *require* that complexity always increases in natural processes. Other passers-by might say that the solar system is structurally complex but, ignoring tidal effects, dynamically simple. Will these statements hold up to rigorous analysis?

Within classical Newtonian mechanics, we could base an analysis of dynamic complexity on the definition and description of the trajectories of a system of particles. From the Kolmogorov point of view, the initial positions and velocities of the particles, knowledge of the forces between them, and Newton's laws of motion together provide a concise description of the dynamics of such a system.

However, even within the framework of classical dynamics, this concise description can break down in the case of chaotic states. In chaotic states, very small changes in the initial conditions can lead to very large changes in the ensuing trajectories. Prediction of the dynamics requires very precise statement of the initial conditions and very precise knowledge of the forces. Specification of the information required to describe the dynamics cannot in these cases be concise. Chaos is an extreme form of dynamic complexity.

Another way to look at this is directly relevant to systems biology: the dynamics of non-chaotic systems are robust to small changes in initial conditions, but the dynamics of chaotic systems are not robust to small changes in initial conditions.

Chaos and predictability

The discovery of the laws of mechanics in the 17th century – Newton's *Principia* was published in 1687 – gave rise to the hope that the dynamics of the solar system in particular (and much, if not all, of the universe in general) was predictable. Laplace expressed the view that:

> If we can imagine a consciousness great enough to know the exact locations and velocities of all the objects in the universe at the present instant, as well as all forces, then there could be no secrets from this consciousness. It could calculate anything about the past or future from the laws of cause and effect.

Leaving aside philosophical questions of the implications about free will and responsibility, there are also issues of computability. How much information do we really need, and how accurately do we need it, to predict the dynamics of the solar system? The weather? The universe? In chaotic systems, accurate prediction of the dynamic development requires unachievably accurate knowledge of the initial conditions (to the point where Heisenberg's uncertainty principle killed off Laplace's hope of perfect determinism.)

It is true that in classical mechanics even chaotic systems are subject to Poincaré's recurrence principle: any system of particles held at fixed total energy will eventually return arbitrarily closely to any set of initial positions and velocities. (What rescues the second law of thermodynamics is that the closer the reapproach demanded, the longer the time required, i.e. the rarer the fluctuations that achieve the recurrence.) However, knowing that the configuration will recur does not simplify the calculation of the trajectories of the particles.

Through unpredictability, chaotic dynamics is associated with complexity. However, chaotic dynamics is not entirely incompatible with order and even the 'spontaneous' generation of order. In governing the time course of evolution of a system, chaotic dynamics does sometimes produce stable states or approximations to stable states – these are called **attractors**.

Sometimes these are unique points; in other cases they are periodic and/or localized states. There have been examples of apparent generation of order in model systems evolving 'at the edge of chaos'.

There are even examples of static or structural order in chaotic systems. Many sequences associated with chaotic behaviour have a **fractal** structure. This means that if an object is dissected into parts, the parts have a structure similar to that of the whole (as well as to one another). B. Mandelbrot has produced many beautiful images. This self-similarity at different scales implies that if we know part of such a structure we can predict a larger segment of it. This should trigger the idea that predictability should permit compressibility and effectively reduce complexity. Indeed, such internal structural relationships have been applied to compression. Fractal image compression is an effective tool for reducing the size of images to a form from which the recovered image is not exactly the same as the starting image but perceptually equivalent.

Fractal structures in biology include branching patterns of plants and of the circulatory systems of vertebrates. At the molecular level, the storage polysaccharide glycogen has features of a fractal structure.

Computational complexity

Perhaps the best-developed area of analysis of the complexity of processes comes from studies of the complexities of computational problems.

An algorithm in computer science defines a process for solving a computational problem. For some problems, the execution time required to solve it is directly proportional to the size of the problem. These problems are said to be of order $\mathcal{O}(N)$ (read 'Oh-N'). For instance, searching for a number in an *unsorted* table requires an execution time proportional to the length N of the table, $\mathcal{O}(N)$. For some problems, the execution time increases only as $N \log N$. Sorting a list is an $\mathcal{O}(N \log N)$ problem. For some problems, the execution time increases as N^2, N^3, . . . The alignment of two sequences by dynamic programming (see Chapter 3) is an $\mathcal{O}(N^2)$ problem. These are said to be polynomial-time problems. Still other problems have even greater time demands. Enumerating all subsets of a set containing N members is $\mathcal{O}(2^N)$.

Computer scientists define the complexity of a problem in terms of the dependence of execution time on problem size (see Box 11.6).

Classes P and NP

A problem that can be solved in polynomial time is said to be in **class P**. $\mathcal{O}(N \log N)$ algorithms are faster than $\mathcal{O}(N^2)$ and are, therefore, in class P.

Suppose, however, that the optimal algorithm to solve a problem has order worse than polynomial – for instance, it might have exponential order $\mathcal{O}(2^N)$ – but that if you *propose* a solution, it can be *checked* in polynomial time. Such a problem is said to be of **class NP**. (NP does *not* stand for non-polynomial, but for non-deterministic polynomial, referring to a different model for the computation. Don't worry about this technical distinction.)

Consider the problem of sorting a list of numbers into order. That is, given a series of N numbers: 2,1,7,5,8,4,3, . . . an algorithm must produce as output the numbers rearranged into order: 1,2,3,4,5,7,8, . . . Whatever the order of the optimal algorithm that *solves* the problem, an algorithm to *verify* that 1,2,3,4,5,7,8, . . . is a solution (or that 1,8,7,2,4,5,3, . . . , is *not* a solution) can run in time *linear* in the length of the list. It is necessary only to check that each number is greater than or equal to its predecessor, which can be done by looking at each element of the list once. Therefore, sorting a list of numbers into order is a problem in class NP. (Sorting also happens to be in class P; sorting algorithms are known with order $\mathcal{O}(N \log N)$.)

NP-complete problems. Does P = NP?

Many NP problems have equivalent complexities, in the sense that if a polynomial algorithm were discovered for one, it could be applied to solve others. The set of NP-complete problems is the set of NP problems such that if we could solve any one of them in polynomial time, we would be able to solve all of them in polynomial time. In other words, the discovery of a polynomial-time algorithm for *any* problem known to be NP-complete would cause the classes P and NP-complete to coalesce. But are there *any* NP problems that are *not* in class P? This is the famous unsolved conjecture of computer science: does P = NP?

What is the a relationship between computational complexity and our notions of the complexity associated with entropy, randomness, and predictability? Think of an algorithm as operating on a set of input data. The algorithm might extract information from the data, as in a program that solves an equation. Or the algorithm might modify the data, as in sorting a list of numbers. The successive steps of the algorithm leave a trace as a sequence of intermediate steps. We can analyse the complexity of the trace, just as we do any other sequences, including genome sequences.

The theory of computational complexity places general limits on the efficiency of computations, independent of the nature of the hardware. In principle, these limits constrain cells as much as they do human programmers. Cells do lots of computations – for instance, how much of each gene should be transcribed at the moment. Now, the classical theory of computational complexity applies to traditional computer architectures, in which successive operations are executed one at a time. The inherent complexity of many biological computations implies that cells could not use this organization in their calculations. This is an inherent constraint on the design of living regulatory systems. In fact, many computations within cells operate as parallel processes. These are not subject to the constraints derived from the classical theory of computational complexity.

Computer scientists are extending the theory of complexity to alternative computer architectures. The comparison of the constraints imposed by different computer architectures affords insight into how cells organize different calculations.

The metabolome

Classification and assignment of protein function

Proteins have a very wide variety of functions. In systems biology, there are two particular classes of function that form dynamic networks. (a) The enzymes that run the biochemistry of the cell. (b) Regulatory networks exercise control to provide stability and robustness.

The Enzyme Commission

The first detailed classification of protein functions was that of the **Enzyme Commission (EC)**. In 1955, the General Assembly of the International Union of Biochemistry (IUB), in consultation with the International Union of Pure and Applied Chemistry (IUPAC), established an International Commission on Enzymes, to systematize nomenclature. The Enzyme Commission published its classification scheme, first on paper and now on the web: http://www.chem.qmul.ac.uk/iubmb/enzyme/.

EC numbers (looking suspiciously like IP numbers) contain four numeric fields, corresponding to a four-level hierarchy. For example, EC 1.1.1.1 corresponds to the reaction:

an alcohol + NAD = the corresponding aldehyde or ketone + $NADH_2$

Note that several reactions, involving different alcohols, would share this number (whether or not the same enzyme catalysed them); but that the same dehydrogenation of one of these alcohols by an enzyme using the alternative cofactor NADP would not. It would be assigned EC 1.1.1.2.

The first field in an EC number indicates to which of the six main divisions (classes) the enzyme belongs:

Class 1. Oxidoreductases

Class 2. Transferases

Class 3. Hydrolases

Class 4. Lyases

Class 5. Isomerases

Class 6. Ligases

The significance of the second and third numbers depends on the class. For oxidoreductases the second number describes the substrate and the third number the acceptor. For transferases, the second number describes the class of item transferred, and the third number describes either more specifically what they transfer or in some cases the acceptor. For hydrolases, the second number signifies the kind of bond cleaved (e.g. an ester bond) and the third number the molecular context (e.g. a carboxylic ester or a thiolester). (Proteinases, a type of hydrolase, are treated slightly differently, with the third number including the mechanism: serine proteinases, thiol proteinases, and acid proteinases are classified separately.) For lyases the second number signifies the kind of bond formed (e.g. C–C or C–O), and the third number the specific molecular context. For isomerases, the second number indicates the type of reaction and the third number the specific class of reaction. For ligases, the second number indicates the type of bond formed and the third number the type of molecule in which it appears. For example, EC 6.1 for C–O bonds (enzymes acylating tRNA), EC 6.2 for C–S bonds (acyl-CoA derivatives), etc. The fourth number gives the specific enzymatic activity.

The **Enzyme Structures Database** at PDBe links Enzyme Commission numbers to proteins of known structure (http://www.ebi.ac.uk/thornton-srv/databases/enzymes/).

The Gene Ontology™ Consortium protein function classification

In 1999, Michael Ashburner and many co-workers faced the problem of annotating the soon-to-be-completed *Drosophila melanogaster* genome sequence. As a classification of function, the EC classification was unsatisfactory, if only because it was limited to enzymes. Ashburner organized the **Gene Ontology™ Consortium** to produce a standardized scheme for describing function.

- An ontology is a formal set of well-defined terms with well-defined interrelationships; that is, a dictionary and rules of syntax.

The Gene Ontology™ Consortium (http://www.geneontology.org) has produced a systematic classification of gene function, in the form of a dictionary of terms, and their relationships.

Organizing concepts of the Gene Ontology project include three categories:

- **Molecular function:** a function associated with what an individual protein or RNA molecule does in itself; either a general description such as *enzyme*, or a specific one such as *alcohol dehydrogenase*. (specifying a catalytic activity, not a protein). This is function from the biochemist's point of view.
- **Biological process:** a component of the activities of a living system, mediated by a protein or RNA, possibly in concert with other proteins or RNA molecules; either a general term such as *signal transduction*, or a particular one such as *cyclic AMP synthesis*. This is function from the cell's point of view.

Because many processes are dependent on location, Gene Ontology (GO) also tracks:

- **Cellular component:** the assignment of site of activity or partners; this can be a general term such as *nucleus* or a specific one such as *ribosome*.

Figure 11.1 shows an example of the GO classification.

Neither the EC nor the GO classification is an assignment of function to individual proteins. The EC emphasized that: '*It is perhaps worth noting, as it has been a matter of long-standing confusion, that enzyme nomenclature is primarily a matter of naming reactions catalysed, not the structures of the proteins that catalyse them*'. (See http://www.chem.qmul.ac.uk/iubmb/nomenclature/.)

Assigning EC or GO numbers to proteins is a separate task. Such assignments appear in protein databases such as UniProtKB.

Comparison of Enzyme Commission and Gene Ontology classifications

Enzyme Commission identifiers form a strict four-level hierarchy, or tree. For example, isopentenyl-diphosphate D-isomerase is assigned EC number 5.3.3.2. The initial 5 specifies the most general category, 5 = isomerases, 5.3 comprises intramolecular isomerases, 5.3.3 those enzymes that transpose C=C bonds, and the full identifier 5.3.3.2 specifies the particular reaction. In the molecular function ontology, GO assigns the identifier 0004452 to isopentenyl-diphosphate Δ-isomerase. (The numerical GO identifiers themselves have no interpretable significance.)

Figure 11.2 compares the EC and GO classifications of isopentenyl-diphosphate D-isomerase. The figure shows a path from GO:0004452 to the root node of the molecular function graph, GO:0003674. In this case there are four intervening nodes, progressively more general categories as we move up the figure. Note that the GO description of this enzyme as an oxidoreductase is inconsistent with the EC classification, in which a committed choice between oxidoreductase and isomerase must be made at the highest level of the EC hierarchy.

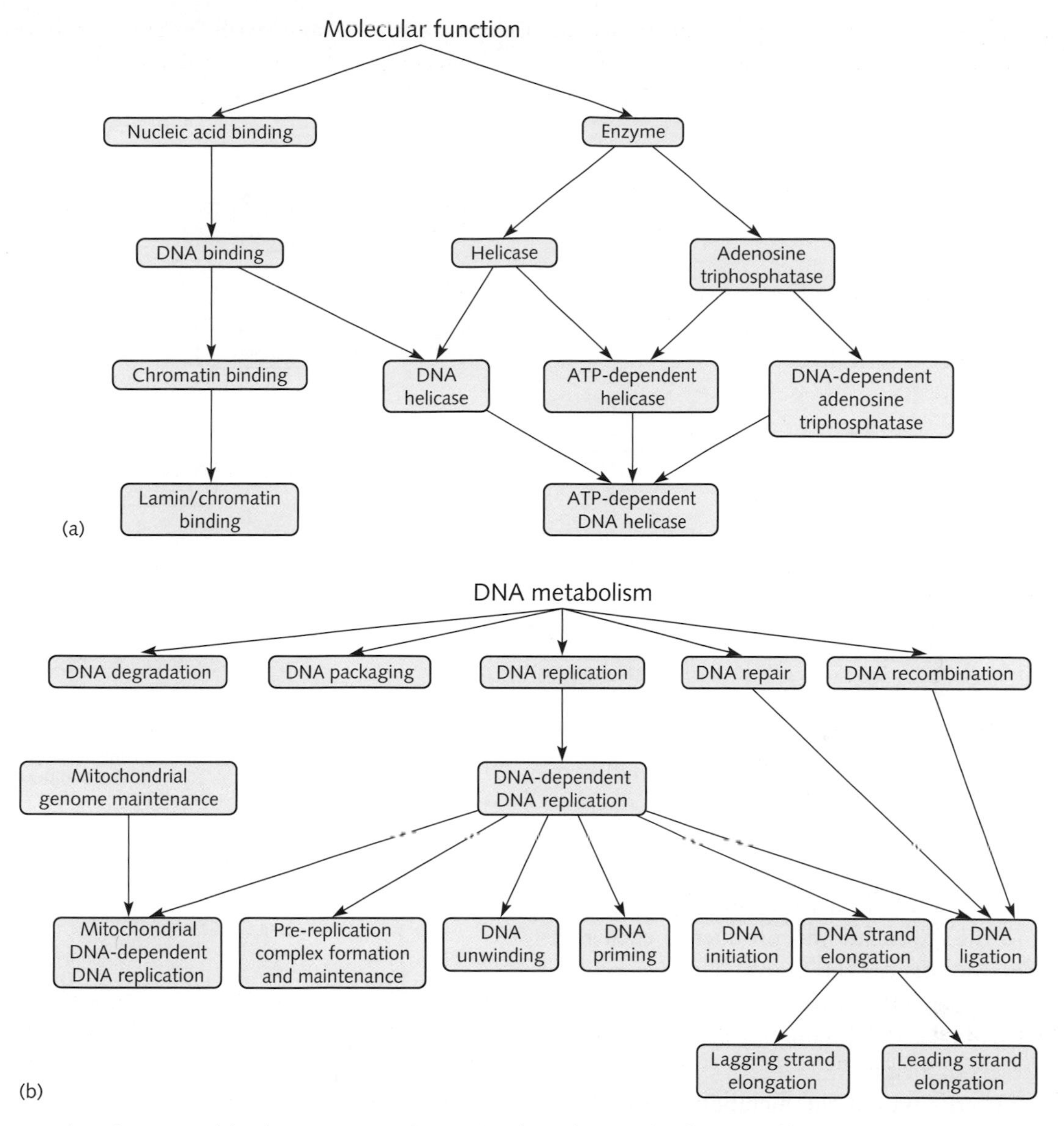

Figure 11.1 Selected portions of the three categories of Gene Ontology, showing classifications of functions of proteins that interact with DNA.

(a) *Biological process*: DNA metabolism.

(b) *Molecular function*: including general DNA binding by proteins, and enzymatic manipulations of DNA.

(c) *Cellular component*: Different places within the cell.

These pictures illustrate the general structure of the Gene Ontology classification. Each term describing a function is a node in a graph. Each node has one or more parents and may have one or more descendants: arrows indicate direct ancestor–descendant relationships. A path in the graph is a succession of nodes, each node the parent of the next. Nodes can have 'grandparents', and more remote ancestors.

Unlike the EC hierarchy, the Gene Ontology graphs are not *trees* in the technical sense, because there can be more than one path from an ancestor to a descendant. For example, there are two paths in (a) from enzyme to ATP-dependent helicase. Along one path helicase is the intermediate node. Along the other path adenosine triphosphatase is the intermediate node.

Although the nodes are shown on discrete levels to clarify the structure of the graph, all the nodes on any given level do not necessarily have a common degree of significance; unlike family, genus, and species levels in the Linnaean taxonomic tree, or the ranks in military, industrial, academic, etc. organizations. GO terms could not have such a common degree of significance, given that there can be multiple paths, of different lengths, between different nodes.

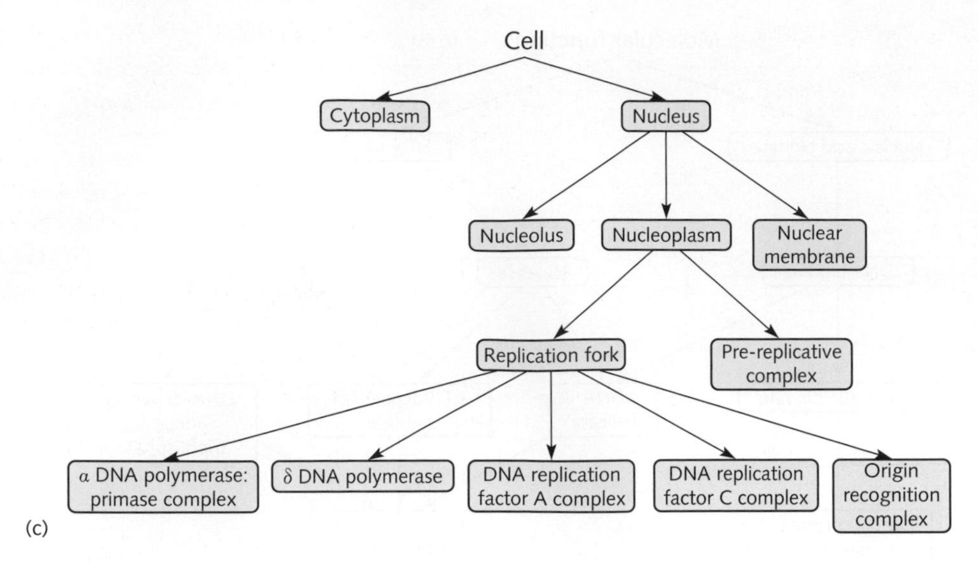

Figure 11.1 (*continued*)

Metabolic networks

Metabolism is the flow of molecules and energy through pathways of chemical reactions. Substrates of metabolic reactions can be macromolecules – proteins and nucleic acids – as well as small compounds such as amino acids and sugars.

The full panoply of metabolic reactions forms a complex network. The *structure* of the network corresponds to a graph in which metabolites are the nodes and the substrate and product of each reaction define an edge in the graph. The *dynamics* of the network depend on the flow capacities of all of the individual links, analogous to traffic patterns on the streets of a city.

Some patterns within the metabolic network are linear pathways. Others form closed loops, such as the tricarboxylic acid (Krebs) cycle. Many pathways are highly branched and interlock densely. However, metabolic networks also contain recognizable clusters or blocks, for instance catabolic and anabolic reactions. There is a relatively high density of internal connections within clusters and relatively few connections between them.

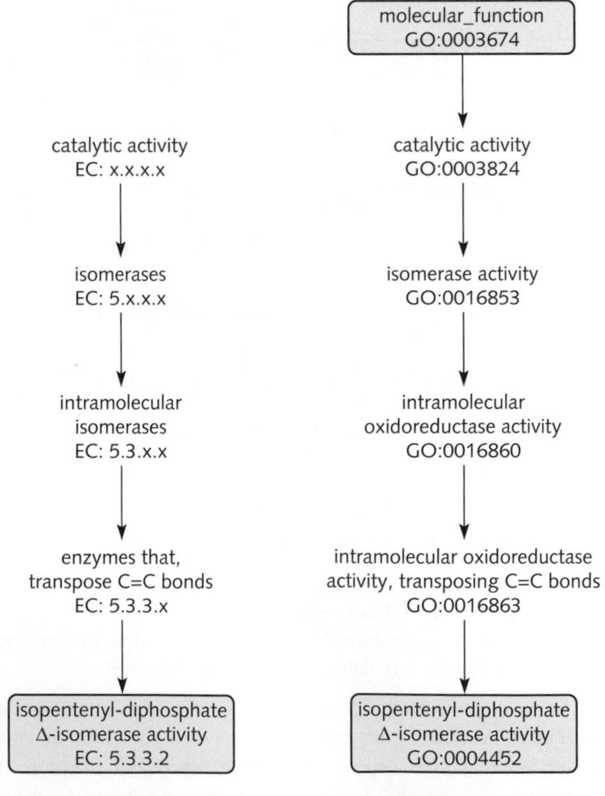

Figure 11.2 Comparison of Enzyme Commission and Gene Ontology classifications of isopentenyl-diphosphate Δ-isomerase.

Databases of metabolic pathways

Biochemists have learned a lot about different enzymes in different species. Approximately an eighth of the sequences in the UniProt database are enzymes (over 2 million sequences in all). They come from ~200 000 species. They represent ~100 000 different enzymatic activities.

Databases organize this information, collecting it within a coherent and logical structure, with links to

Table 11.1 Databases of metabolic pathways

Database	Home page
EcoCyc	http://ecocyc.org
BioCyc	http://www.biocyc.org
KEGG	http://www.genome.jp/kegg/
WIT	www.mcs.anl.gov/compbio

other databases that provide different data selections and different modes of organization. EcoCyc deals with *E. coli*. It is the model for – and linked with – numerous parallel databases, with uniform web interfaces, treating other organisms. BioCyc is the 'umbrella' collection. KEGG, the Kyoto Encyclopedia of Genes and Genomes, contains information from multiple organisms. WIT contains metabolic reconstructions derived from genome sequences (Table 11.1).

EcoCyc

EcoCyc is a database representing what we know about the biology of *E. coli*, strain K-12 MG1655. It contains:

- *the genome:* the complete sequence, and for each gene its position and function if known;
- *transcription regulation:* operons, promoters, and transcription factors and their binding sites;
- *metabolism:* the pathways, including details of the enzymology of individual steps; for each enzyme the reaction, activators, inhibitors, and subunit structure are given;
- *membrane transporters:* transport proteins and their cargo; and
- *links to other databases:* protein and nucleic-acid sequence data, literature references, and comparisons to different *E. coli* strains.

Methionine synthesis in *Escherichia coli*

A tiny subset of the *E. coli* metabolic network is the pathway for synthesis of methionine from aspartate (see Figure 11.3).

To appreciate the logic of the system from diagrams such as Figure 11.3, keep in mind that both the reaction sequence and the control cascades are embedded in much larger networks.

- The first step, phosphorylation of L-aspartate, is common to the biosynthesis of methionine, lysine, and threonine. *E. coli* contains three aspartate kinases, encoded by three separate genes, each specific for one of the end-product amino acids. They catalyse the same reaction but are subject to separate regulation.
- The third step, conversion of L-aspartate-semialdehyde to L-homoserine, is common to the methionine and threonine synthesis pathways. Two homoserine dehydrogenases are separately encoded. Regulation of expression of the aspartate kinases and homoserine dehydrogenases suffices to control all three pathways.
- The piece of the regulatory network is also extracted from a more complex tapestry. For example, CRP (catabolite repressor protein) regulates more than 200 genes!
- Methionine is converted to *S*-adenosylmethionine, a common participant in methyl group transfers. *S*-Adenosylmethionine activates the Met repressor (encoded by *metJ*) (see Figure 11.3). This is a more complicated form of feedback. In classic feedback inhibition, a product interacts directly with an enzyme that produces one of its precursors. In this case, the product interacts with a repressor, which reduces the expression of enzymes that produce its precursors. (See page 47.)

In the EcoCyc web page that contains the information corresponding to this figure, the items are active. Links to other internal pages expand information about metabolites, cofactors, enzymes, genes, and regulators. It is possible to 'zoom' in or out by controlling the level of detail. For instance, asking for less detail than the contents of Figure 11.3(a) would first eliminate the information about the genes and enzymes and then reduce the pathway to an outline showing only critical intermediates:

L-aspartate → → → homoserine → → → L-homocysteine → L-methionine (⊖ feedback from L-methionine to homoserine)

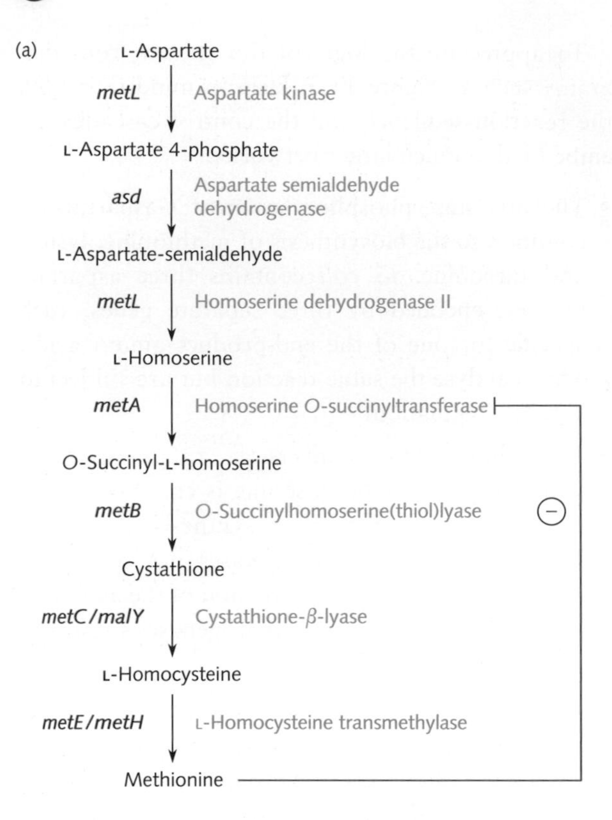

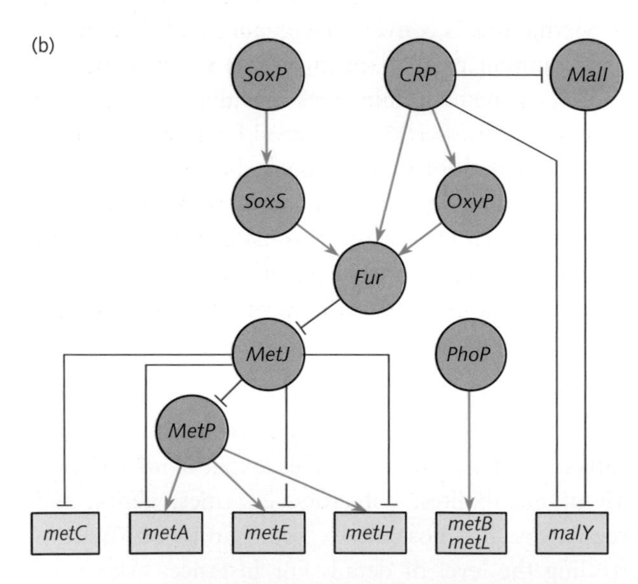

Figure 11.3 (a) The synthesis of methionine from aspartate is a seven-step pathway through a linear sequence of intermediates (black). Different enzymes (green) catalyse different steps. They are encoded by the genes shown in blue. *metC* and *malY* encode alternative cystathione-β-lyases. The final step, conversion of L-homocysteine to L-methionine, is also catalysed by two different L-homocysteine transmethylases, encoded by two genes, *metE* and *metH*. One mechanism of control is at the protein level: there is 'feedback inhibition' by the product, methionine, which inhibits homoserine *O*-succinyltransferase. This is shown by the red line; note that it connects the product, methionine, to an *enzyme*, not to a gene. (b) Expression of the genes in the L-aspartate → L-methionine pathway (blue rectangles) is subject to regulation. Circles contain the genes for the transcription factors that control expression. Regulated genes appear in rectangles. Molecules that tend to *enhance* transcription are connected to their targets by green arrows. Molecules that tend to *repress* transcription are connected to their targets by red lines ending in a 'T'. In addition to the links shown here, most of the regulatory proteins feed back on themselves; in most cases, the self-regulatory signal is a repression.

Control is exerted on every protein of the pathway, from a variety of points of initiation.

- Readers are urged to explore the EcoCyc web site on their own, either deliberately or serendipitously, or guided by weblems in this chapter or at the Online Resource Centre.

It is also possible to explore in other dimensions. The methionine synthesis pathway is embedded in larger networks. One of these involves synthesis of the amino acids lysine and threonine in addition to methionine, all starting with aspartate (see Figure 11.4).

The Kyoto Encyclopedia of Genes and Genomes (KEGG)

The Kyoto Encyclopedia of Genes and Genomes (KEGG) is an extremely comprensive battery of databases for molecular biology and genomics. One of its special strengths is an integration of metabolic and genomic information. KEGG contains pathway maps, which describe potential networks of molecular activities, both metabolic and regulatory. Figure 11.5 shows a pathway from KEGG, the reductive carboxylate cycle in photosynthetic bacteria. This pathway is basically the Krebs cycle, run backwards.

- Several databases present information about metabolic pathways. These include EcoCyc, BioCyc, WIT, and KEGG. By linking the steps in metabolic pathways to individual genes and proteins, it is possible to reconstruct pathways in particular species, and to compare metabolic pathways in different slecies.

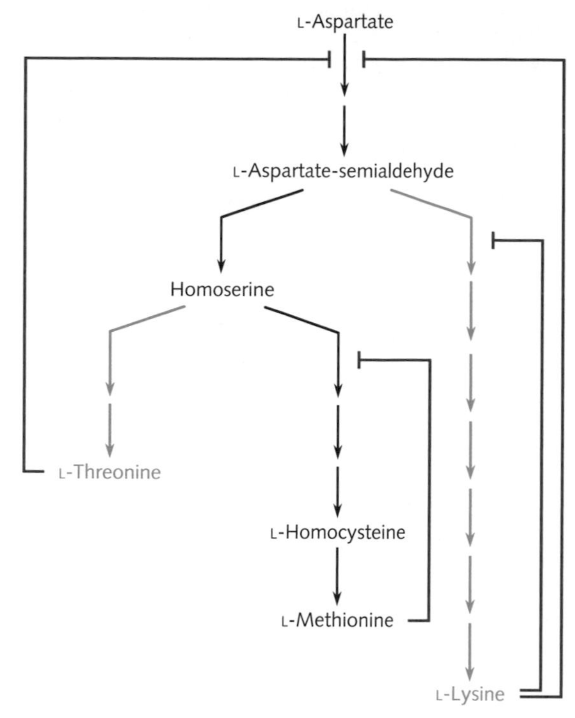

Figure 11.4 The pathway of amino acid biosynthesis from aspartate branches after aspartate semialdehyde. In this figure, the black sequence corresponds to the previous example, and the green pathways are the immediate context. The aspartate → methionine sequence is a subnetwork of the network shown here. Each amino acid plays a regulatory role, exerting feedback inhibition over its own synthesis, without affecting the others. It looks as if threonine and lysine both individually inhibit the first step of the synthesis of all three products, but this step is catalysed by *three* separate aspartate kinases, allowing specialized regulation.

KEGG derives its power from the very dense network of links among these categories of information, and additional links to many other databases to which the system maintains access. Two examples of the kinds of questions that can be treated with KEGG are:

(a) It has been suggested that simple metabolic pathways evolve into more complex ones by gene duplication and subsequent divergence. Searching the pathway catalogue for sets of enzymes that share a folding pattern will reveal clusters of linked paralogues.

(b) KEGG can take the set of known enzymes from some organism and check whether they can be integrated into established metabolic pathways. A gap in a pathway suggests a missing enzyme or an unexpected alternative pathway.

- Several databases assemble biochemical reactions into metabolic pathways. Individual steps are linked to Enzyme Commission and Gene Ontology Consortium classifications of function, and to individual proteins that catalyse the reactions. These databases are useful in organizing the assignment of function to proteins identified in newly sequenced genomes.

Evolution and phylogeny of metabolic pathways

Most organisms share many common metabolic pathways. But there are many individual variations.

Some organisms have metabolic competence completely absent from others. Plants but not humans have enzymes for reactions involved in photosynthesis and cell-wall formation.

Some organisms achieve the same overall metabolic transformation but use alternative pathways; that is, different sets of intermediates. For instance, classical glycolysis and the Entner–Doudoroff pathway are alternative routes from glucose to pyruvate, in which there is a whole succession of reactions possible (Figure 11.6). Often, organisms will share many steps in a metabolic transformation but some will extend or truncate the pathway. Humans have lost activity in the last enzyme in vitamin C synthesis, L-gulonolactone oxidase, and we must include it in our diet. Most mammals have a working copy of the gene for this protein, and can synthesize vitamin C. We have an inactive pseudogene.

Another example is the pathway for nitrogen excretion (see Figure 11.7). Organisms with more water available in their immediate surroundings use more of the reactions.

We can represent the metabolic networks of different species as graphs. The nodes are metabolites. There are edges between pairs of metabolites if the organism has an enzyme that will convert one to the other, or if the interconversion is spontaneous. We can then compare the graphs to get a quantitative measure of the divergence. Intuitively, we expect that the divergence in metabolic network should correspond to the divergence between species as measured from comparing genome sequences.

The procedure outlined here deals with a static and binary picture of the metabolic network. Either a transformation is possible, or it is not. It is entirely

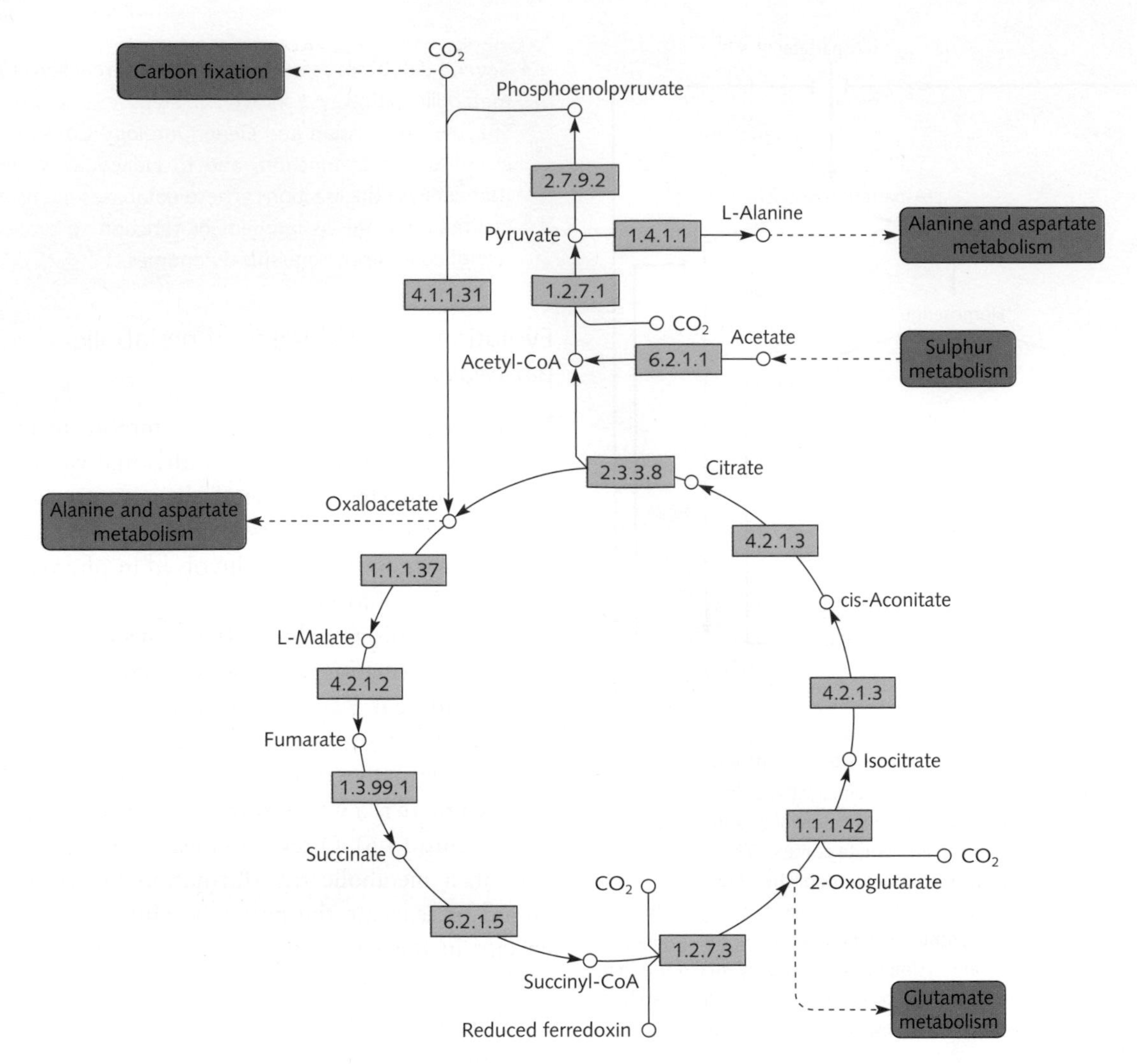

Figure 11.5 Metabolic pathway map from The Kyoto Encyclopedia of Genes and Genomes (KEGG). This figure shows the reductive carboxylate cycle, and its links to other metabolic processes. The numbers in square boxes are EC numbers identifying the reactions at each step.

possible that enzymes that catalyse corresponding steps in the network have very different kinetic constants in two species, or are subject to different kinds of regulation. In this case the dynamic patterns of traffic through the network might be quite different, even if the topology of the network is the same. Think of the difference in traffic flow through a city during rush hour, and at midnight. The roads haven't changed, but the kinetics has.

Carbohydrate metabolism in archaea

The common pathway from glucose to pyruvate in bacteria and eukaryotes is the Embden–Meyerhof glycolytic route (see Figure 11.6).

B. Siebers and P. Schönheit have studied the metabolic pathways of carbohydrate metabolism in archaea. In the initial conversion of glucose to pyruvate, they observed a number of differences in the pathway, from either the standard Embden–Meyerhof glycolytic pathway, or the Entner–Doudoroff alternative.

Pyrococcus furiosus, *Thermococcus celer*, *Archaeoglobus fulgidus* strain 7324, *Desulfurococcus amylolyticus*, and *Pyrobaculum aerophilum* use a modified Embden–Meyerhof pathway (Figure 11.8). *Sulfolobos solfataricus* and *Haloarcula marismortui* use a modified Entner–Doudoroff pathway (Figure 11.9). *Thermoproteus tenax* uses both.

In addition to the differences in the sequence of metabolites, the enzymes that catalyse even the same

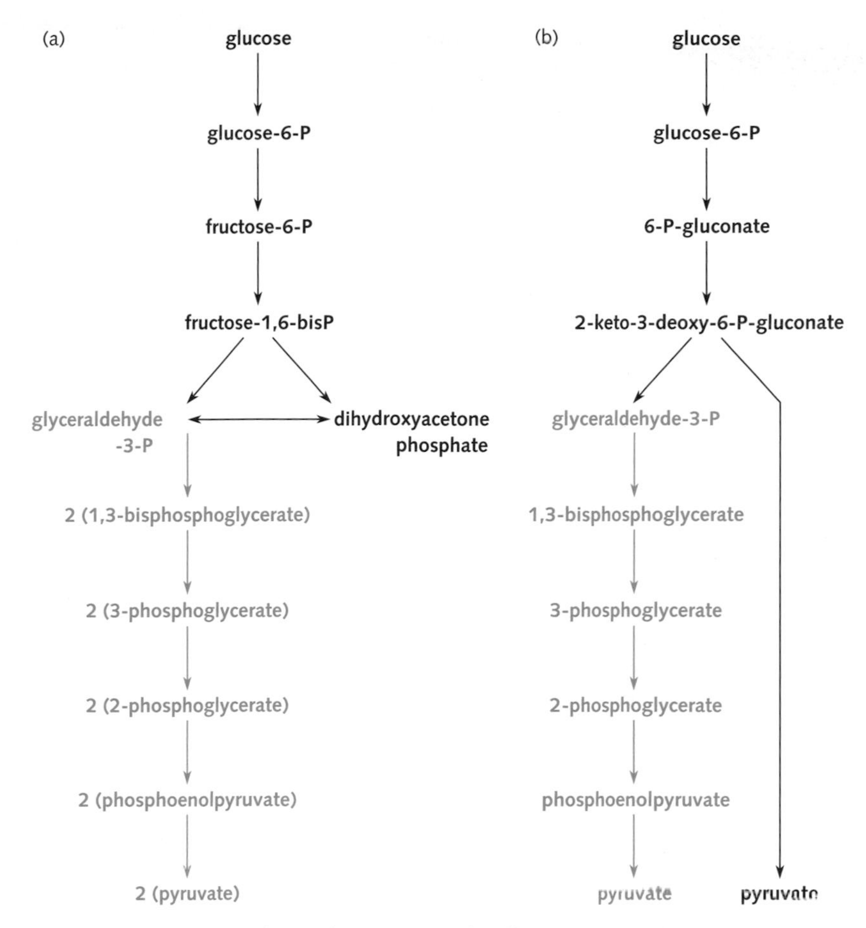

Figure 11.6 (a) Embden–Meyerhof glycolytic pathway, (b) Entner–Doudoroff pathway. Note that the enzymatic conversion of glyceraldehyde-3-phosphate to pyruvate is the same in both pathways (green branch).

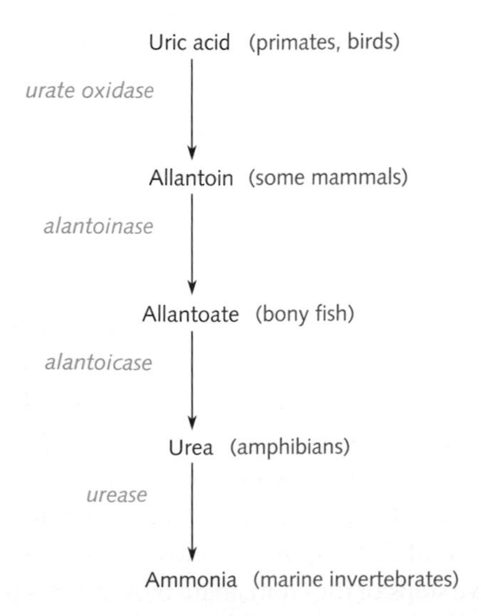

Figure 11.7 Succession of reactions to produce excreted forms of end products of nitrogen metabolism.

reactions are almost always not homologues of bacterial or eukaryotic ones. Many of them use different cofactors. Bacterial and eukaryotic phosphofructokinases (that convert fructose-6-phosphate to fructose-1,6,bisphosphate) use ATP as the phosphoryl donor. The archaeal enzymes that catalyse this reaction can use ATP, ADP, or even inorganic pyrophosphate. In addition, some of the familiar enzymes are under allosteric control. The control relationships are not retained in the corresponding archaeal enzymes.

- Of particular interest for comparative genomics are facilities to compare pathways among different organisms. Alignment and comparison of pathways can expose how pathways have diverged between species. Even if the pathways are the same, in some cases the enzymes are non-homologous.

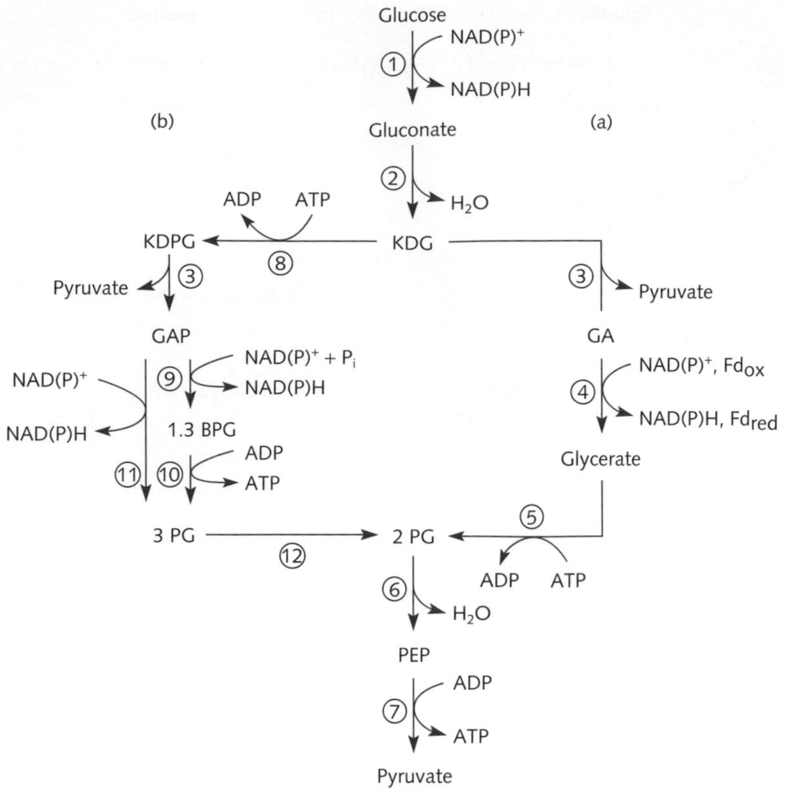

Figure 11.8 Modifications of the Entner–Doudoroff (ED) pathway in Archaea. (a) The non-phosphorylative ED pathway in *Thermoplasma acidophilum*. (b) The semi-phosphorylative ED pathway in halophilic Archaea. A branched ED (combining (a) and (b)) appears in *S. solfataricus* and *T. tenax*. Abbreviations: 1.3 BPG, 1,3-bisphosphoglycerate; Fdox and Fdred, oxidized and reduced ferredoxin; GA, glyceraldehyde; GAP, glyceraldehyde-3-phosphate; KDG, 2-keto-3-deoxygluconate; KDPG, 2-keto-3-deoxy-6-phosphogluconate; PEP, phosphoenolpyruvate; 2 PG, 2-phosphoglycerate; 3 PG, 3-phosphoglycerate. Enzymes are numbered as follows: 1, glucose dehydrogenase; 2, gluconate dehydratase; 3, KD(P)G aldolase; 4, glyceraldehyde dehydrogenase (proposed for *T. acidophilum*), glyceraldehyde:ferredoxin oxidoreductase (proposed for *T. tenax*) or glyceraldehyde oxidoreductase (proposed for *S. acidocaldarius*); 5, glycerate kinase; 6, enolase; 7, pyruvate kinase; 8, KDG kinase; 9, GAPDH; 10, phosphoglycerate kinase; 11, GAPN; 12, phosphoglycerate mutase.

From: Siebers, B. & Schönheit, P. (2005). Unusual pathways and enzymes of central carbohydrate metabolism in Archaea. Curr. Opin. Micro. **8**, 695–705.

Reconstruction of metabolic networks

Pathway comparison can be useful for annotation of genomes. It is often possible to assign function to proteins on the basis of similarity to sequences of proteins of known function in other organisms. However, sometimes there are several weak similarities to other proteins and it is unclear which is the true homologue. Conversely, sometimes an organism has a metabolic pathway but no annotated enzyme for an essential step. Confronting the unannotated proteins with the unassigned functions can sometimes identify the protein that fills the gap in the pathway.

If an enzyme needed for a pathway cannot be identified, even by weak sequence similarity, it may be that the organism has evolved a non-homologous enzyme for the task. For example, the archaeon *Methanococcus jannaschii* has a pathway for biosynthesis of chorismate from 3-dehydroquinate. Enzymes for most of the steps have homologues in bacteria and/or eukaryotes. However, shikimate kinase was not identifiable from sequence similarity. *M. jannaschii* must have *some* protein with this function. How can it be found?

Although in bacteria, genes consecutive in pathways are often consecutive in operons in the genome, this is not true of *M. jannaschii*. However, the genes for successive steps of the chorismate biosynthesis pathway *are* clustered and consecutive in another archaeon, *Aeropyrum pernix*. It was possible to propose a gene

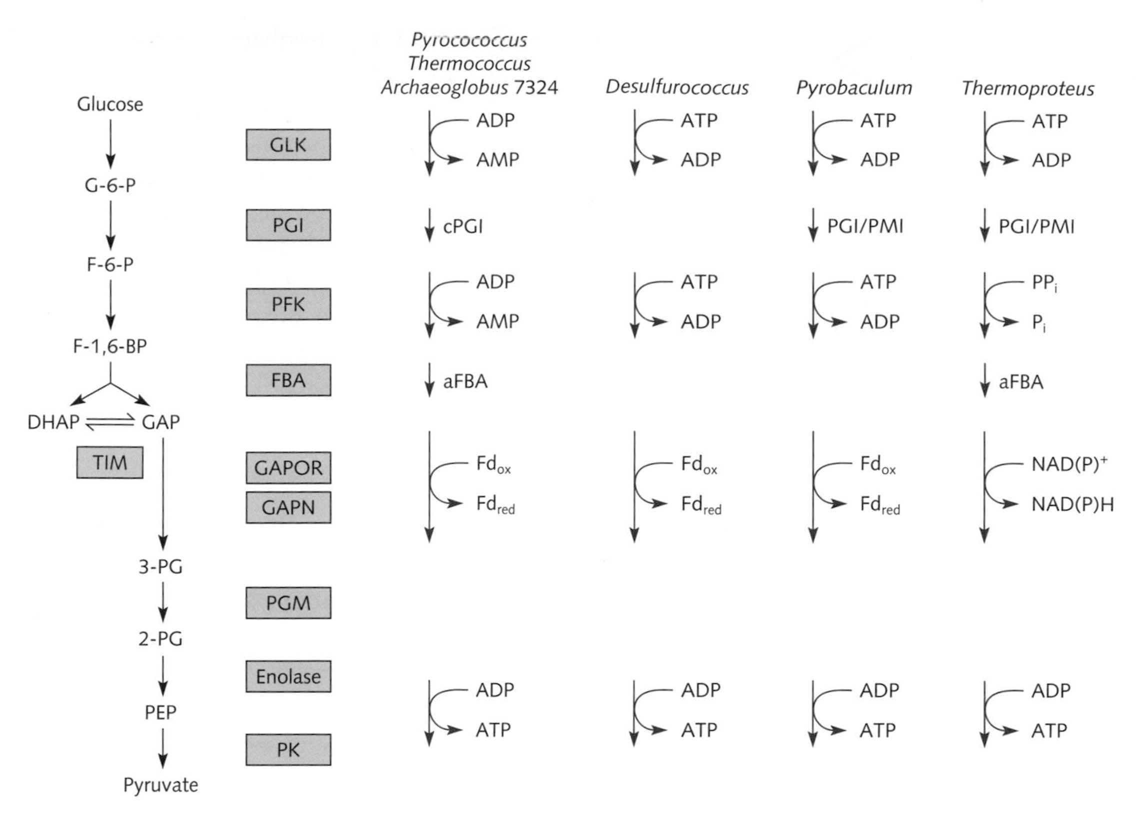

Figure 11.9 Modifications of the Embden–Meyerhof (EM) pathway in Archaea. In this case most of the reactions are the same. The enzymes are not homologous to those that catalyse the corresponding reactions in bacteria and eukarya. Note the differences in cofactors. The steps and mechanisms of regulation also differ. Abbreviations: aFBA, archaeal class I FBA; cPGI, cupin PGI; DHAP, dihydroxyacetone phosphate; FBA, fructose 1,6-bisphosphatae aldolase; F-1,6-BP, fructose 1,6-bisphosphate; Fdox and Fdred, oxidized and reduced ferredoxin; F-6-P, fructose-6-phosphate; GAP, glyceraldehyde-3-phosphate; GAPN, non-phosphorylative glyceraldehyde 3-phosphate dehydrogenase; GAPOR, glyceraldehyde-3 phosphate-ferredoxin oxidoreductase; GLK, glucokinase (ADP- or ATP-dependent); G-6-P, glucose-6-phosphate; PEP, phosphoenolpyruvate; PFK, 6-phosphofructokinase; 2-PG, 2-phosphoglycerate; 3-PG, 3-phosphoglycerate; PGI/PMI, bifunctional phosphoglucose/phosphomannose isomerase); PGI, phosphoglucose isomerase; PGM, phosphoglycerate mutase; PK, pyruvate kinase; TIM, triosephosphate isomerase.

From: Siebers, B. & Schönheit, P. (2005). Unusual pathways and enzymes of central carbohydrate metabolism in Archaea. Curr. Opin. Micro. **8**, 695–705.

for a shikimate kinase in *A. pernix* and to identify a homologue of that gene in *M. jannaschii*.

Experiments confirmed the prediction that the *M. jannaschii* gene thus identified (*MJ1440*) encoded a shikimate kinase. It has no sequence similarity to bacterial or eukaryotic shikimate kinases. A protein from a different family has been recruited for the archaeal pathway.

Regulatory networks

Regulatory networks pervade living processes. Control interactions are organized into linear signal transduction cascades and reticulated into control networks.

Any individual regulatory action requires (1) a stimulus; (2) transmission of a signal to a target; (3) a response; and (4) a 'reset' mechanism to restore the resting state (see Figure 11.10). Many regulatory actions are mediated by protein–protein complexes. Transient complexes are common in regulation, as dissociation provides a natural reset mechanism.

Some stimuli arise from genetic programmes. Some regulatory events are responses to current internal

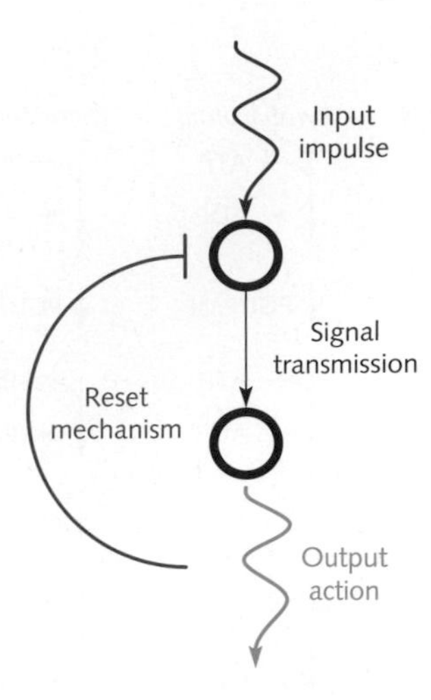

Figure 11.10 The elementary step in a regulatory network. An input impulse is received by a node, which transmits a signal to a downstream node, causing an output action. This is followed by reset of the upstream node to its inactive state. Combination of such elementary diagrams gives rise to the complex regulatory networks in biology.

metabolite concentrations. Others originate outside the cell; the signal is detected by surface receptors and transmitted across the membrane to an intracellular target.

G-protein-coupled receptors (GPCRs) illustrate the components of signal transduction. Recall that GPCRs contain seven transmembrane helices, with a binding site for triggering ligands on the extracellular side and a binding site for the downstream recipient of the signal, a heterotrimeric G protein, on the intracellular side.

G proteins consist of three subunits: G_α, G_β, and G_γ. G_α and G_γ are anchored to the membrane. In the resting inactive state, G_α binds GDP. An activated GPCR binds to a specific G protein and catalyses GTP–GDP exchange in the G_α subunit. This destabilizes the trimer, dissociating G_α:

$$G_\alpha(\mathrm{GDP})G_\beta G_\gamma \rightleftharpoons G_\alpha(\mathrm{GTP}) + G_\beta G_\gamma$$

The separated components, G_α and $G_\beta G_\gamma$, activate downstream targets, such as adenylate cyclase.

A single activated GPCR can interact successively with many G-protein molecules, amplifying the signal. It is therefore essential to turn the signal off, after it has had its effect. Mutations that render a GPCR constitutively active cause a number of diseases, the symptoms emerging from a war between the rogue receptor and the feedback mechanisms that are unequal to the task of restraining its effects. Different GPCRs have different mechanisms for restoring the resting state. Rhodopsin, for example, is inactivated by cleavage of the isomerized chromophore.

The activity of the heterotrimeric G proteins is turned off by the GTPase activity of G_α, converting G_α(GTP) to G_α(GDP). G_α(GDP) does not bind to its receptors – shutting down that pathway of signal transmission. G_α(GDP) rebinds the $G_\beta G_\gamma$ subunits. This resets the system.

Signal transduction and transcriptional control

The signal transduction network exerts control '*in the field*' by a variety of mechanisms including inhibitors, dimerization, ligand-induced conformational changes including but not limited to allosteric effects, GDP–GTP exchange or kinase–phosphorylase switches, and differential turnover rates. This component acts fast, on subsecond timescales. The transcriptional regulatory network exerts control '*at headquarters*' through control over gene expression. This component is slower, acting on a timescale of minutes.

General characteristics of all control pathways include the following:

- A single signal can trigger a single response or many responses.
- A single response can be controlled by a single signal or influenced by many signals.
- Each response may be stimulatory – increasing an activity – or inhibitory – decreasing an activity.
- Transmission of signals may damp out stimuli or amplify them.

There are ample opportunities for complexity, opportunities of which cells have taken extensive advantage.

Structures of regulatory networks

Think of control, or regulatory, networks as assemblies of **activities**. Although mediated in part by physical assemblies of macromolecules (protein–protein and protein–nucleic acid complexes), regulatory networks:

1. *Tend to be unidirectional.* A transcription activator may stimulate the expression of a metabolic enzyme, but the enzyme may not be involved directly in regulating the expression of the transcription factor.
2. *Have a logical dimension.* It is not enough to describe the connectivity of a regulatory network. Any regulatory action may stimulate or repress the activity of its target. If two interactions combine to activate a target, activation may require *both* stimuli (logical 'and'), or *either* stimulus may suffice (logical 'or').
3. *Produce dynamic patterns.* Signals may produce combinations of effects with specified time courses. Cell-cycle regulation is a classic example.

The structure of a regulatory network can be described by a graph in which edges indicate steps in pathways of control. Regulatory networks are directed graphs (see p. 344): the influence of vertex A on vertex B is expressed by a directed edge connecting A and B. An edge directed from vertex A to vertex B is called an **outgoing connection** from A and an **incoming connection** to B. Conventionally, an arrow indicates a stimulatory interaction, and a 'T' symbol indicates an inhibitory interaction. An edge connecting a vertex to itself indicates auto-regulation. A double-headed arrow indicates reciprocal stimulation of two nodes; note that this is *not* the same as an undirected edge.

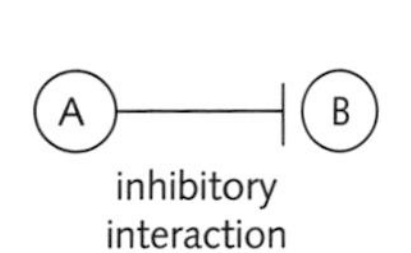

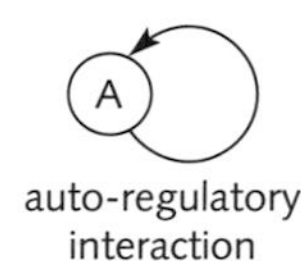

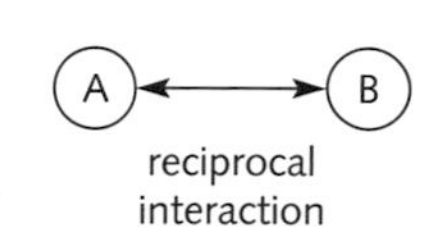

Dynamics, stability, and robustness

An unlabelled, undirected graph gives a *static* picture of the topology of a network. The dynamic states are more complex (see Box 11.7), including:

- equilibrium;
- steady state;
- states that vary periodically;
- unfolding of developmental programmes;
- chaotic states;
- runaway or divergence; and
- shutdown.

Although much is known about the mechanisms of individual elements of control and signalling pathways, understanding their integration is a subject of current research. For instance, the idea that healthy cells and organisms are in stable states is certainly no more than an approximation (in most cases, it is an idealization).

Understanding *how* cells achieve even an apparent approximation to stability is also quite tricky. It is likely that great redundancy of control processes lies at its basis. Regulation is based on the result of many individual control mechanisms – here a short feedback loop, there a multistep cascade. Somehow the independent actions of all of the individual signals combine to achieve an overall, integrated result. It is like the operation of the 'invisible hand' that, according to Adam Smith, coordinates individual behaviour into the regulation of national economies.

- Robustness is more than stability. Stability is keeping your composure in unchanging conditions. Robustness is keeping your composure in changing conditions.

Robustness through redundancy

In principle, networks can achieve robustness through an extension of the mechanism by which redundancy confers stability. The most direct approach is simple **substitutional redundancy**: if two proteins are each capable of doing a job, knock out one and the other takes over. In the London Underground, this would correspond to a second line running over the same route. For instance, when the Circle Line is not running, passengers travelling between Paddington and King's Cross stations can use the Hammersmith & City line that runs on the same tracks. In yeast, for example, single-gene knockouts of over 80% of the ~6200 open reading frames are survivable injuries.

Dynamic states of a network of processes

- At **equilibrium**, one or more forward and reverse processes occur at compensating rates, to leave the amounts of different substances unchanged:

$$A \rightleftharpoons B$$

Chemical equilibria are generally self-adjusting upon changes in conditions or in concentrations of reactants or products.

- A **steady state** will exist if the total rate of processes that produce a substance is the same as the total rate of processes that consume it. For instance, the two-step conversion

$$A \longrightarrow B \longrightarrow C$$

could maintain the amount of B constant, provided that the rate of production of B (the process $A \rightarrow B$) is the same as the rate of its consumption (the process $B \rightarrow C$). The net effect would be to convert A to C.

A cyclic process could maintain a steady state in all of its components:

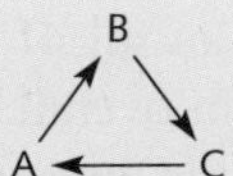

A steady state in such a cyclic process with all reactions proceeding in one direction is very different from an equilibrium state. Nevertheless, in some cases, it is still true that altering external conditions produces a shift to another, neighbouring steady state.

- States that vary **periodically** appear in the regulation of the cell cycle, circadian rhythms, and seasonal changes such as annual patterns of breeding in animals and flowering in plants. Circadian and seasonal cycles have their origins in the regular progressions of the day and year, but have evolved a certain degree of internalization.

- Many equilibrium and some steady-state conditions are **stable**, in the sense that concentrations of most metabolites are changing slowly if at all and the system is robust to small changes in external conditions. The alternative is a **chaotic state**, in which small changes in conditions can cause very large responses. Weather is a chaotic system: the meteorologist E. Lorenz asked, 'Does the flap of a butterfly's wings in Brazil set off a tornado in Texas?' In a carefully regulated system, chaos is usually well worth avoiding, and it is likely that life has evolved to damp down the responses to the kinds of fluctuation that might give rise to it. Chaotic dynamics does sometimes produce approximations to stable states – these are called **strange attractors**. Understanding stability in dynamic systems subject to changing environmental stimuli is important but is beyond the scope of this book.

- **Unfolding of developmental programmes** occurs over the course of the lifetime of the cell or organism. Many developmental events are relatively independent of external conditions and are controlled primarily by regulation of gene expression patterns.

- **Runaway** or **divergence**. Absence of a predator can lead to uncontrolled multiplication of a species. An example is the growth of the rabbit population of Australia from an 'inoculum' of 24 animals in 1859. Breakdown in control over cellular proliferation leads to unconstrained growth in cancer.

- **Shutdown** is part of the picture. Apoptosis is the programmed death of a cell, as part of normal developmental processes or in response to damage that could threaten the organism, such as DNA strand breaks. Breakdown of mechanisms of apoptosis – for instance, mutations in the protein p53 – is an important cause of cancer.

Some duplicated genes contribute to substitutional redundancy. For example, in studying models for diabetes it appears that mice and rats (but not humans) have two similar but non-allelic insulin genes. Substitutional redundancy requires equivalence not only of function but of expression levels. In the mouse, knocking out either insulin gene leads to compensatory increased expression of the other, producing a normal phenotype.

Coordinated expression patterns, providing substitutional redundancy, are more probable among duplicated genes than among unrelated ones. For example, *Escherichia coli* contains two fructose-1,6-bisphosphate aldolases. One, expressed only in the presence of special nutrients, is non-essential under normal growth conditions. However, the other is essential. *In this case, functional redundancy does* not *provide robustness.* These two enzymes are probably homologous, but they are distant relatives, not the product of a recent gene duplication. One is a member of a family of fructose-1,6-bisphosphate aldolases typical of bacteria and eukaryotes, whereas the other is a member of another family that occurs in archaea. *E. coli* is unusual in containing both.

An alternative mechanism of network robustness is **distributed redundancy**: equivalent effects achieved through different routes. In normal *E. coli*, approximately two-thirds of the NADPH produced in metabolism arises via the pentose phosphate shunt, which requires the enzyme glucose-6-phosphate dehydrogenase. Knocking out the gene for this enzyme leads to metabolic shifts, after which increased levels of NADH produced by the tricarboxylic acid cycle are converted to NADPH by a transhydrogenase reaction. The growth rate of the knockout strain is comparable to that of the parent.

Dynamic modelling

Diagrams such as Figure 11.3 give a static picture of the structure of a metabolic pathway and its control.

Can we model the dynamics? What would it mean to do so?

A challenge that might – naively – appear relatively simple would be to predict the effect of knocking out an enzyme. An easy guess would be to expect a build-up of the substrate of the missing enzyme. However, if the metabolic pathways branch in the vicinity of that metabolite, the consequences of a knockout are more complex.

For example, the disease phenylketonuria results most commonly from a specific dysfunctional (i.e. knocked-out) enzyme, phenylalanine hydroxylase. The normal function of phenylalanine hydroxylase is to convert phenylalanine to tyrosine (see page 60). In phenylketonuria, phenylalanine does indeed build up. However, the excess phenylalanine is converted by phenylalanine transaminase to phenylpyruvic acid:

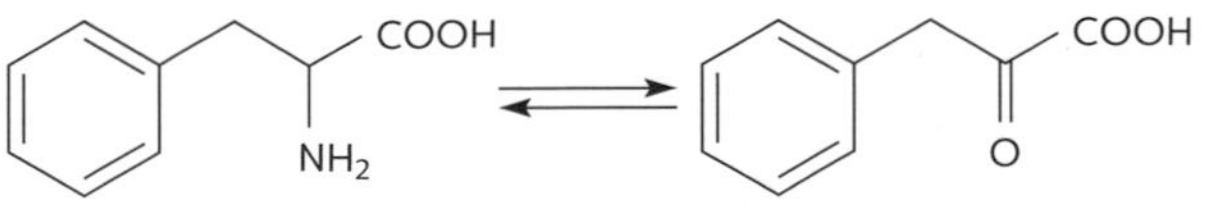

COOH
NH_2
COOH
O

phenylalanine ⇌ phenylpyruvic acid

Both compounds accumulate. As phenylpyruvic acid is less readily absorbed by the kidneys than phenylalanine, it is excreted into the urine, giving the disease its name. (Phenylalanine is not a ketone.) The Guthrie test for phenylketonuria measures the concentration of phenylpyruvic acid in the blood of newborns.

A challenge greater than predicting the effect of a single knockout would be to simulate the entire metabolic network: given an initial set of metabolite concentrations, to predict the concentrations as a function of time. The idea would be to combine predictions of the rates of individual reactions, assuming a simple model such as Michaelis–Menten kinetics, or more complex models of allosteric enzymes. This requires knowing accurately the kinetic constants of all of the enzymes, including effects of inhibitors. It requires being able to give a sensible treatment of the idea of 'substrate concentration' within a cell divided into compartments and to deal with questions of rates of diffusion in a crowded intercellular environment. Longer-term simulation would require knowing the kinetics of transcription regulation, for which no simple model analogous to the Michaelis–Menten equation is available. There are also serious computational issues involving how precisely the kinetic parameters must be known, and the extent to which simplifying assumptions – for instance, the steady-state approximation – are justified.

Accurate simulation of metabolic patterns of entire cells is a clear target for research in the field. However, the problem is a difficult one. Current approaches include:

- *Attempts at detailed numerical analysis of simple networks.* For instance, a simulation of the asparate → threonine pathway (see Figure 11.4) in *E. coli* represented the enzymatic transformations and feedback inhibition as a set of coupled equations.* Changes in expression pattern were not included. Steady-state solutions were compared with experimental measurements on cell extracts. It was possible to:
  - simulate the time course of threonine synthesis and the effects of changes in initial metabolite concentrations;
  - predict the steady-state concentrations of intermediates;

* Chassagnole, C., Raïs, B., Quentin, E., Fell, D.A., & Mazat, J.P. (2001). An integrated study of threonine-pathway enzyme kinetics in *Escherichia coli*. Biochem. J. **356**, 415–423.

- predict the effects of changes in concentrations of individual enzymes on overall throughput, expressed as **flux control coefficients**; such data can help to guide development of microbial factories for increased yield of particular products;
- for different steps, distinguish whether the substrates and products are approximately at equilibrium.

- The flux control coefficient is the percentage change in flux divided by the percentage change in amount of enzyme. It is not a property of the enzyme, but a property of a reaction within a metabolic network. A flux control coefficient equal to 1 would correspond to a rate-limiting step.

- *Focusing not on individual enzymes but on potential sets of flow rates.* Represent the metabolic network as a graph. Metabolites are the nodes. Edges correspond to reactions: an edge connects two compounds if there is a reaction, or possibly several reactions, that interconvert them. The goal is to predict the flow rate through each edge. Recently the models have been generalized to include regulation of expression. There are general constraints on the set of flow rates:
  - under steady-state conditions, the fluxes through each node must add up to zero; i.e. for each compound, the amount that is synthesized or supplied externally must equal the amount used up or secreted;
  - the flux control coefficients of all of the reactions contributing to a single flux must add up to 1;
  - the flux through any edge is limited by the values of the Michaelis–Menten parameter V_{max} for all enzymes contributing to the edge; and
  - the thermodynamic properties of each reaction determine whether or not the reaction is reversible: this is a property of the substrate and product of the reaction, not of the enzyme; the flux of an irreversible reaction must be ≥ 0.

- It is interesting to see whether the space of possible metabolic states is connected or broken up into separated regimens.

In general, many possible flow patterns, or metabolic states, are consistent with the constraints. To determine a single metabolic state to compare with experiments, it is possible to select from the feasible states the one that is optimal for ATP production or for growth rate.

A variety of observable quantities are predictable.

- The effects of changes of medium or gene knockouts: which enzymes are essential for growth on different carbon sources?
- What are limiting factors in growth?
- What are maximal theoretical yields of ATP, or assimilation of carbon, etc.?
- What are the fluxes through individual pathways? This is difficult but not impossible to measure.
- What are the flux control coefficients of different enzymes?
- For optimal growth, how much oxygen and carbon source are taken up?

Such models have been constructed for several organisms, including prokaryotes and eukaryotes. Predictions have generally achieved good agreement with experiments.

Protein interaction networks

The units from which interaction networks are assembled are:

- for physical networks, a protein–protein or protein–nucleic acid complex;
- for logical networks, a dynamic connection in which the activity of a process is affected by a change in external conditions, or by the activity of another process.

Most experiments reveal only pairwise interactions. The challenges are to integrate pairwise interactions into a network and then to study the structure and dynamics of the system.

Many techniques detect physical interactions directly. These include:

- *X-ray and NMR structure determinations* can not only identify the components of the complex, but reveal how they interact, and whether conformational changes occur upon binding.
- *X-ray tomography of cells* produces images by measuring differential absorption of low-energy X-rays (see Figure 11.11). From a series of tilted views, three-dimensional reconstruction is possible.
- *Two-hybrid screening systems.* Transcriptional activators such as Gal4 contain a DNA-binding domain and an activation domain. Suppose these two domains are separated, and one test protein is fused to the DNA-binding domain and a second test protein is fused to the activation domain. Then a reporter protein will be expressed only if the components of the activator are brought together by formation of a complex between two test proteins (Figure 11.12). High-throughput methods allow parallel screening of a 'bait' protein for interaction with a large number of potential 'prey' proteins.
- *Chemical crosslinking* fixes complexes so that they can be isolated. Subsequent proteolytic digestion and mass spectrometry permits identification of the components.
- *Coimmunoprecipitation.* An antibody raised to a 'bait' protein binds the bait together with any other 'prey' proteins that interact with it. The interacting proteins can be purified and analysed, for instance by western blotting, or mass spectrometry.
- *Chromatin immunoprecipitation* identifies DNA sequences that bind proteins (Figure 11.13).
- *Phage display.* Genes for a large number of proteins are individually fused to the gene for a phage coat protein, to create a population of phage each of which carries copies of one of the extra proteins exposed on its surface. Affinity purification against an immobilized 'bait' protein selects phage displaying potential 'prey' proteins. DNA extracted from the interacting phages reveals the amino acid sequences of these proteins.
- *Surface plasmon resonance* analyses the reflection of light from a gold surface to which a protein has been attached. The signal changes if a ligand binds

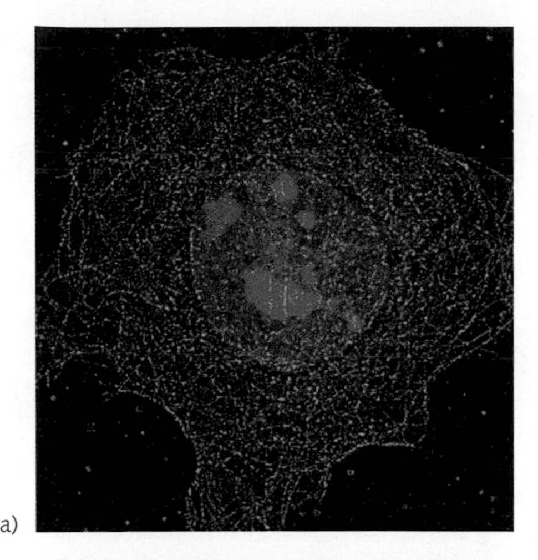

(a)

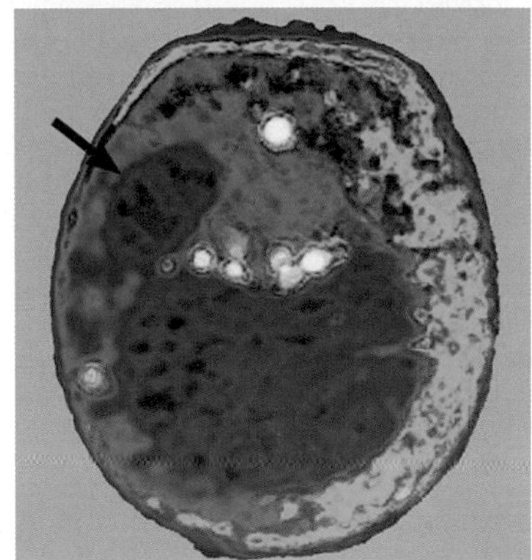

(b)

Figure 11.11 (a) X-ray image of the microtubule network of a mouse epithelial cell labelled using metal-conjugated antibodies (to form individual particles ~50 nm in diameter). Different regions of the cell have measurably different X-ray absorbances. Colouring the image according to X-ray absorbance brings out contrasts; of course, there is no suggestion that the colours correspond to a realistic interaction with visible light. Here the microtubule network appears in blue, and the nucleus and nucleoli in orange. The total width of the field is 120 mm. (b) Cryo X-ray tomography of a yeast cell (*S. cerevisiae*). This image shows a 0.5 mm section at 60 nm resolution. Lipid droplets are coloured white, the vacuole and nucleus are red. The arrow points to the nucleus. Other cytoplasmic structures appear green and orange. Cell diameter, 5 mm.

(a) From: Meyer-Ilse, W., et al. (2001). High resolution protein localization using soft X-ray microscopy. J. Microsc. **201**, 395–403.
(b) From: Larabell, C.A. & Le Gros, M.A. (2004). X-ray tomography generates 3-D reconstructions of the yeast, *Saccharomyces cerevisiae*, at 60-nm resolution. Mol. Biol. Cell **15**, 957–962.

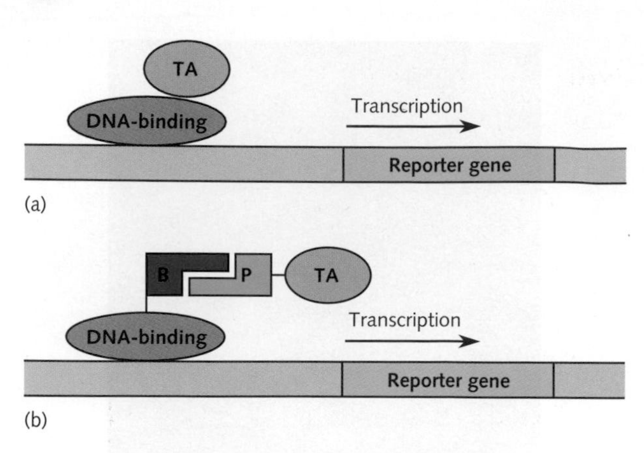

Figure 11.12 (a) A transcription activator contains two domains, a DNA-binding domain (pink) and a transcription-activator domain (TA; blue). Together they induce expression of a reporter gene. The *lacZ* gene encoding β-galactosidase (see p. 380) is a common choice of reporter gene, because chromogenic substrates make β-galactosidase easy to detect. (b) Transcription proceeds if the DNA-binding domain and TA domain are separated in different proteins that can form a complex. A 'bait' protein B (red) is fused to the DNA-binding domain and a 'prey' protein P (cyan) is fused to the TA domain. Formation of a complex between bait and prey brings together the DNA-binding and TA domains, inducing transcription.

This is the basis of a high-throughput approach for detecting pairs of interacting proteins.

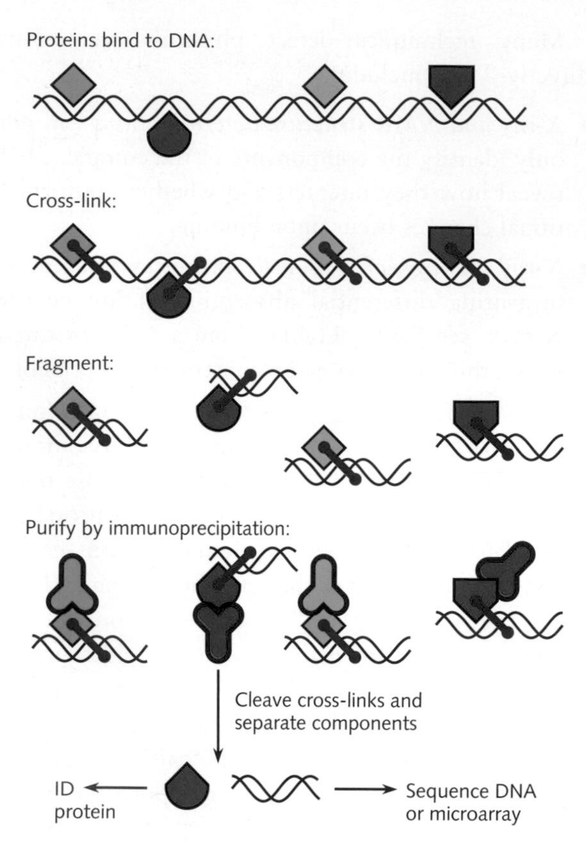

Figure 11.13 Chromatin immunoprecipitation. Treatment with formaldehyde cross-links proteins and DNA, fixing the complexes that exist within a cell. After isolation of chromatin, breaking the DNA into small fragments allows separation of proteins by binding to specific antibodies, carrying the DNA sequences along with them. Reversal of the cross-link followed by sequencing of the DNA identifies the specific DNA sequence to which each protein binds. To identify multiple sites in the genome to which the protein binds, the DNA fragments can be analysed using a microarray. To avoid the requirement for antibodies specific for each protein to be tested, the proteins can be fused to a standard epitope, or to a sequence that can be biotinylated, taking advantage of the very high biotin–streptavidin affinity.

to the immobilized protein. (The method detects localized changes in the refractive index of the medium adjacent to the gold surface. This is related to the mass being immobilized.)

- *Fluorescence resonance energy transfer.* If two proteins are tagged by different chromophores, transfer of excitation energy can be observed over distances up to about 60 Å.
- *Tandem affinity purification (TAP)* allows probing cells *in vivo* for partners that bind to a selected 'bait' protein.

Fusion of the bait protein to *two* affinity tags, separated by a cleavage site, permits high-efficiency in extraction of complexes, using two successive, or tandem, affinity purification steps separated by a cleavage step to expose the second tag. The cleavage steps are specific and require only mild conditions, in order to leave the bait–prey complex intact (Figure 11.14).

The fusion proteins contain an individual bait protein extended at its N or C terminus by a calmodulin-binding peptide, a TEV protease cleavage site (TEV is a cysteine protease from tobacco etch virus), and protein A (which binds to high affinity to an available antibody). The double tag, and the two-step purification, gives superior performance relative to a single-tag technique, in yield of low-concentration complexes.

In vivo, the expressed tagged bait protein binds to a set of prey proteins. Figure 11.14 shows the purification protocol to recover the complexes from cell extracts.

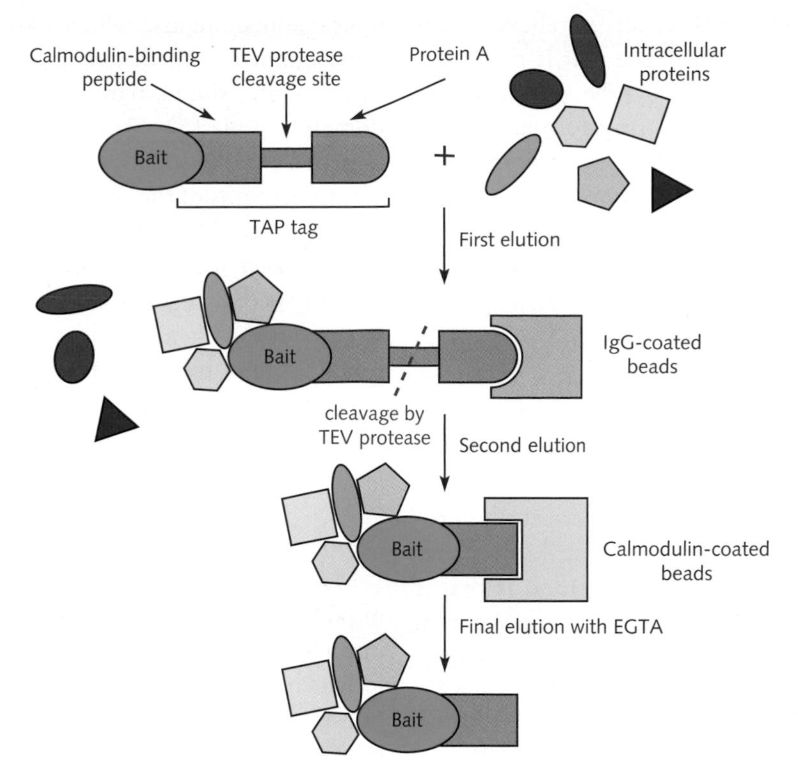

Figure 11.14 The construct and protocol in the tandem affinity purification (TAP) method for purification of complexes with a selected 'bait' protein. The fusion protein containing the bait and the two tags separated by the TEV cleavage site binds *in vivo* to proteins in the cell. A first affinity purification step binds the bait protein to a column containing IgG-coated beads that bind specifically to the first tag, protein A. After thorough washing, cleavage by TEV protease releases the bound complexes, and exposes the second tag. A second affinity purification step binds the bait protein to a column containing calmodulin-coated beads, which bind specifically to the second tag, the calmodulin-binding peptide. After washing, elution with the chelating agent ethylene glycol tetraacetic acid (EGTA) releases the purified complexes.

Determination of a protein interaction network requires measurements of many bait proteins. Because all bait proteins carry the same fusion sequences, the purification and cleavage steps in TAP are unique. There is no need to design different purification protocols for different bait proteins.

Other methods provide complementary information:

- *Domain recombination networks.* Many eukaryotic proteins contain multiple domains. A feature of eukaryotic evolution is that a domain may appear in different proteins with different partners. In some cases proteins in a bacterial operon catalysing successive steps in a metabolic pathway are fused into a single multidomain protein in eukarya. The domains of the eukaryotic protein are individually homologous to the separate bacterial proteins. (Examples of proteins fused in eukarya and separate in prokaryotes are also known.)

  It is possible to create a network by defining an interaction between two protein domains whenever homologues of the two domains appear in the same protein. This is evidence for some functional link between the domains, even in species where the domains appear in separate proteins.

- *Coexpression patterns.* Clustering of microarray data identifies proteins with common expression patterns. They may have the same tissue distribution, or be up- or down-regulated in parallel in different physiological states. This is also suggestive evidence that they share some functional link. In the response of *M. tuberculosis* to the drug isoniazid, genes for the Fatty Acid Synthesis complex are coordinately upregulated. They are on an operon-like gene cluster, and in fact these proteins do form a physical complex. On the other hand, alkyl hydroperoxidase (AHPC) is also upregulated in response to isoniazid. AHPC acts to relieve oxidative stress. There is no evidence that it physically interacts with the Fatty Acid Synthesis complex, or that it mediates a metabolic transformation coupled to fatty acid synthesis. It is a second, independent, component of the response to isoniazid.

- *Phylogenetic distribution patterns.* The *phylogenetic profile* of a protein is the set of organisms in which it and its homologues appear. Proteins in a common structural complex or pathway are functionally linked and expected to co-evolve. Therefore proteins that share a phylogenetic profile are likely to have a functional link, or at least to have a common subcellular origin. There need be no sequence or structural similarity between the proteins that share a phylogenetic distribution pattern. A welcome feature of this method is that it derives information about the function of a protein from its relationship to *nonhomologous* proteins.

Each of these methods provides a basis for a protein interaction network. The networks formed by combining each set of interactions are different, although they overlap, to a greater or lesser extent. They give different views of the kinds of relationships between proteins that exist in cells. It is possible to form a more comprehensive network by combining different types of interactions. For instance, the DIP database is a curated collection of experimentally determined protein–protein interactions (see http://dip.doe-mbi.ucla.edu/). It contains data about 71 276 interactions between 23 201 proteins from 372 organisms.

A limitation that remains is the difficulty of determining structures of transient complexes, or of systems showing substantial conformational changes upon assembly. The situation is shared with much of current molecular biology: we are coming to grips with static structures but are awaiting the development of methods for treating the dynamics.

Structural biology of regulatory networks

Many molecules involved in regulation are multidomain proteins. Each domain in a multidomain protein is relatively free to interact with other molecules. An *interaction domain* is a part of a protein that confers specificity in ligation of a partner. Regulatory proteins contain a limited number of types of interaction domains, which have diverged to form large families with different individual specificities. For instance, the human genome contains 115 SH2 domains, and 253 SH3 domains. (*S*rc-*H*omology domains SH2 and SH3 are named for their homologies to domains of the src family of cytoplasmic tyrosine kinases.) Many individual interaction domains even interact with different partners as they participate in successive steps of a control cascade. Initial interactions may also trigger recruitment of additional proteins to form large regulatory complexes.

Figure 11.15 shows types of interaction domain complexes with ligands, including binding of peptides (which may be attached to proteins), and protein–protein complexes. Protein–nucleic acid complexes will appear next.

Many interaction domains are sensitive to the state of post-translational modification of their ligands, for instance binding preferentially to states of a ligand in which specific tyrosines, serines, or threonines are phosphorylated. These and other post-translational modifications function as switches, turning on or interrupting/resetting a signalling cascade.

Protein–protein complex formation allows a cell to detect a signal molecule in the external medium and report its arrival to the cell interior, without the signal molecule itself ever needing to enter the cell. Many receptors use an ingenious dimerization mechanism. The receptor has external, transmembrane, and internal segments. An external ligand binds to *two* molecules of receptor (see Figure 10.26). The juxtaposition of the external portions also brings the internal portions together, because they are tethered to the external regions by the transmembrane segments. Interaction between the interior segments triggers a conformational change that activates a process such as phosphorylation of a protein. This may initiate a signal transduction cascade that can transmit and amplify the original stimulus (see Box 11.8). Within the cell, ligand-induced dimerization may activate DNA-binding domains (Figure 11.16).

Large-scale protein interaction networks are built up from many individual interactions. Figure 11.17 shows a portion of an interaction network of yeast proteins, based on sets of proteins that have been found together in solved structures.

L-Dopa is used to treat Parkinson's disease.

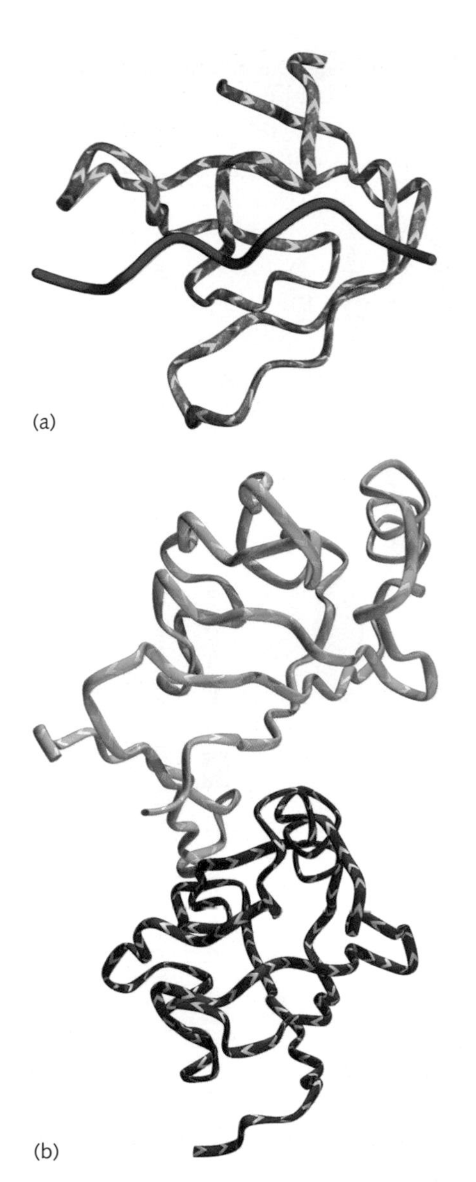

Figure 11.15 Some types of interactions involved in regulatory signalling. (a) Binding of a peptide (magenta) by an SH3 domain [1CKA]. SH3 domains are common constituents of regulatory proteins. Functions of SH3 domains include signal transduction, protein and vesicle trafficking, cytoskeletal organization, cell polarization, and organelle biosynthesis. (b) Domain–domain interaction. PDZ domains in syntrophin (magenta) and neuronal nitric oxide synthase (cyan) [1QAV].

Regulation of tyrosine hydroxylase illustrates several control mechanisms

Tyrosine hydroxylase catalyses the conversion of L-tyrosine to L-3,4-dihydroxyphenylalanine (L-dopa) in neurons, a step in the synthesis of the neurotransmitters dopamine and adrenaline. Tyrosine hydroxylase is the focus of many diverse forms of regulation, including control over transcription and RNA processing and turnover. One regulatory pathway is triggered by the arrival of a neurotransmitter at the external cell surface.

- External binding to a receptor activates adenylate cyclase inside the cell.
- Binding of cyclic AMP activates protein kinase A. Protein kinase A forms inactive tetramers containing two catalytic and two regulatory subunits, or dissociates into active monomers. Binding of cyclic AMP to protein kinase A breaks up the tetramer, releasing monomeric catalytic subunits in active form.
- Active protein kinase A phosphorylates tyrosine hydroxylase at Ser40, upregulating its activity.
- The mechanism balancing this stimulation is the specific dephosphorylation of Ser40 by phosphatase 2A.

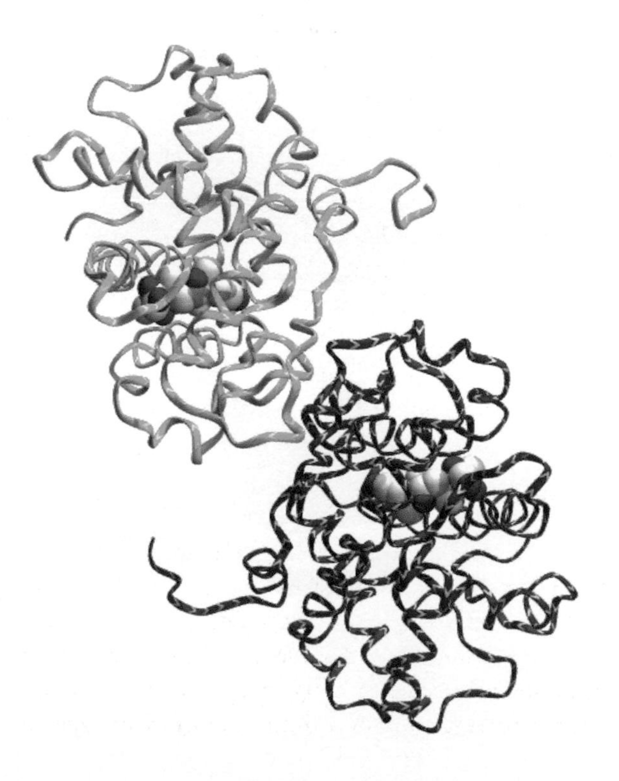

Figure 11.16 Glucocorticoid receptor binding domains with dexamethasone, a cortisol analogue that has anti-inflammatory and immunosupressant activities [1M2Z] (see page 271). Ligation-induced dimerization can lead to translocation from the cytoplasm into the nucleus, and activation of gene transcription by DNA-binding domains (not shown).

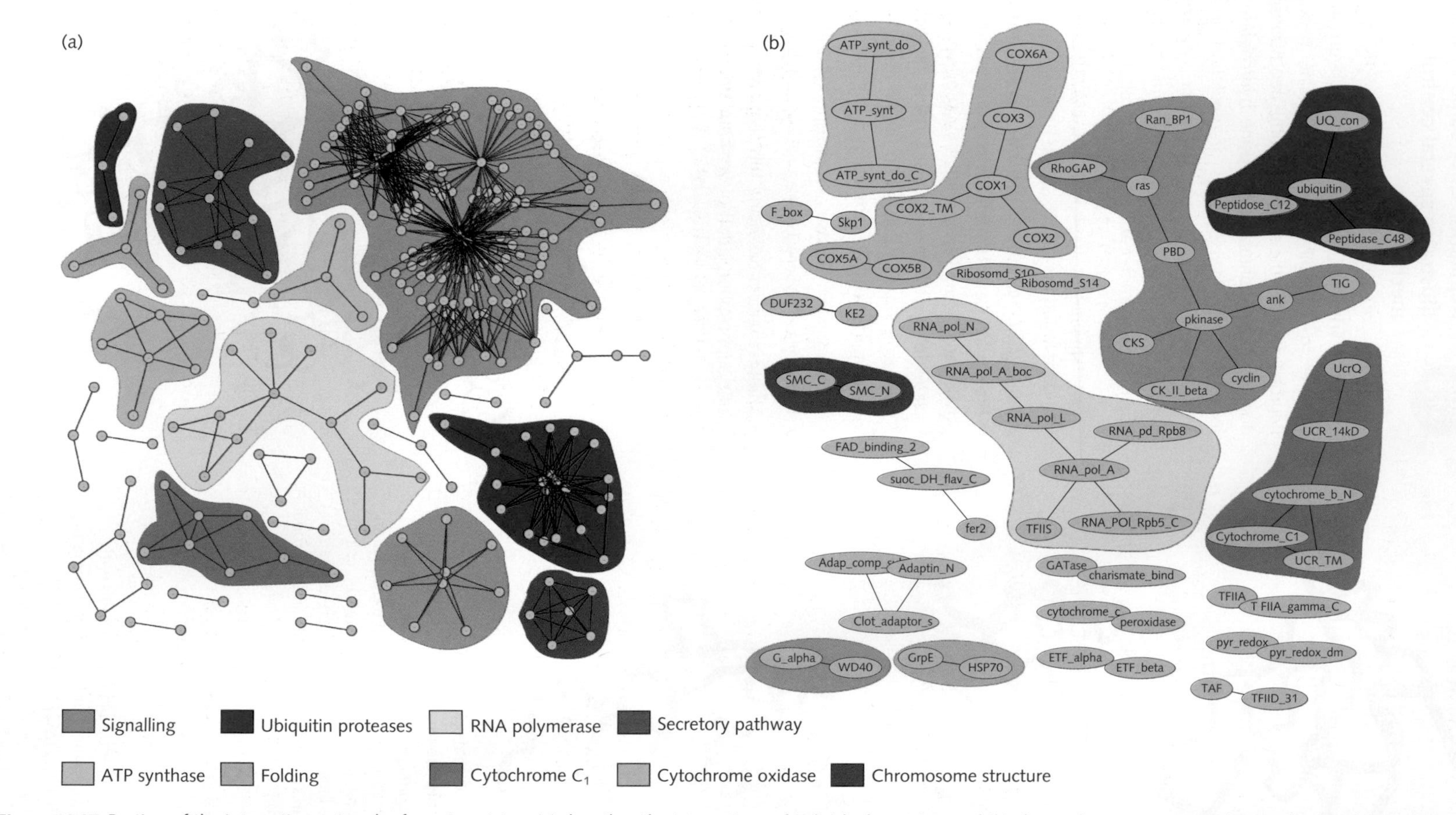

Figure 11.17 Portion of the interaction network of yeast proteins: (a) describes the interactions of individual proteins, and (b) shows the interactions within a subnetwork based on representations of different protein families, in different functional categories, linked in (a). This figure is based on structural data and modelling. Each relationship implies a physical interaction between the proteins. Some of the interactions involve stable complexes (for instance, RNA polymerase II); others involve transient complexes.

From: Aloy, P. & Russell, R. (2005). Structure-based systems biology: a zoom lens for the cell. FEBS Lett. **579**, 1854–1858.

Protein–DNA interactions

DNA–protein complexes mediate several types of process:

- replication, including repair and recombination;
- transcription;
- regulation of gene expression; and
- DNA packaging, including nucleosomes and viral capsids.

Different processes require different degrees of DNA-sequence specificity (see Box 11.9).

Structural themes in protein–DNA binding and sequence recognition

What does a protein looking at a stretch of DNA in the standard B conformation see? (See Figure 3.12.) What could it hope to grab hold of? Prominent general features are the sugar–phosphate backbone, including charged phosphates suitable for salt bridges and potential hydrogen-bond partners in the sugar hydroxyl groups. Contact with the bases is accessible through the major and minor grooves, although unless the DNA is distorted the bases are visible only 'edge on'. Hydrogen-bonding patterns between bases in the grooves and particular amino acids account for some of the DNA-sequence specificity in binding. However, many protein–DNA hydrogen bonds are mediated by intervening water molecules, an effect that tends to *reduce* the specificity.

The idea that an α-helix has the right size and shape to fit into the major groove of DNA was noted in the 1950s. The structures of the first protein–DNA complexes confirmed this prediction. It became the paradigm for protein–DNA interactions. Indeed, when a student solving the structure of the Met repressor–DNA complex told his supervisor that, in the electron-density map he was interpreting, it looked as if a β-sheet were binding in the major groove, he was advised, with patience strongly tinged with condescension, to go back and look for the helix.

We now recognize great structural variety in DNA–protein interactions. A few examples include:

- Helix-turn-helix domains. These appear in prokaryotic proteins that regulate gene expression,

Specificities of DNA-binding proteins

DNA-binding proteins show varying degrees of DNA-sequence specificity.

- Some DNA-binding proteins are relatively non-specific with respect to nucleotide sequence, including DNA replication enzymes and histones.
- Some, for instance *Eco*RV, *bind* to DNA with low specificity but cleave only at GATATC. This combination permits a mechanism of finding the target sequence by initial non-specific binding followed by diffusion in one dimension along the DNA.
- Some recognize specific nucleotide sequences. For example, the *Eco*R1 restriction endonuclease binds specifically to GAATCC sequences with almost absolute specificity. It is a homodimer that recognizes palindromic sequences.
- Some DNA-binding proteins recognize consensus sequences. For example, the phage Mu transposase and repressor proteins bind 11 bp sequences of the form CTTT[A/T]PyNPu[A/T]A[A/T] (where [A/T] = A or T, Py = either pyrimidine (C or T), Pu = either purine (A or G), and N = any of the four bases).
- Some recognize nucleotide sequences indirectly, via modulations of local DNA structure. For example, the TATA box-binding protein takes advantage of the greater flexibility of AT-rich sequences to form complexes in which the DNA is very strongly bent (see Figure 11.23). The distinction between sequence specificity achieved through direct interaction with bases and specificity through recognition of local structure has been termed 'digital versus analogue readout'.
- Some recognize general structural features of DNA, such as mismatched bases or supercoiling.
- Some DNA-binding proteins form an initial complex with high DNA-sequence specificity, followed by recruitment of other proteins of low specificity to enhance overall binding affinity or create a functional complex.

eukaryotic homeodomains involved in developmental control, and histones that package DNA in chromosomes.

- Zinc fingers, including eukaryotic transcription factors, and steroid and hormone receptors.
- Proteins with β-sheets that interact with DNA, for instance the gene regulatory proteins, Met and Arc repressors, and the TATA box-binding protein.
- Leucine zippers that act as eukaryotic transcriptional regulators.
- The 'high mobility group' proteins in eukaryotes and the prokaryotic protein HU, which bind sequences non-specifically and bend DNA.
- Enzymes that interact with DNA and are involved in replication, translation, repair, and uncoiling. Some are relatively small; others are large multiprotein complexes. They show many different types of folding pattern. Many distort the DNA structure in order to get access to the bases that are the target of their activity.
- Viral capsid proteins and histones, which package DNA into compact forms.

These examples are an anecdotal list, not a classification.

RNA-binding proteins have a separate variety. Some resemble DNA-binding proteins. Others bind to RNA molecules of defined structure; for instance, enzymes that interact with tRNA, including but not limited to amino acid tRNA synthetases, and the ribosome itself.

From the point of view of systems biology, a very important class of DNA-binding proteins is transcriptional regulators. These proteins and their complexes with DNA have been a focus of structural biology. There is a great variety of structures and a few recurrent structural themes.

An album of transcription regulators

λ Cro

Bacteriophage λ is a virus containing a double-stranded DNA genome of 48 502 bp. A λ phage infecting an *E. coli* cell chooses – depending on which genes are active – between **lysis** or **lysogeny**. Phage λ can replicate and **lyse** the cell, releasing ~100 progeny.

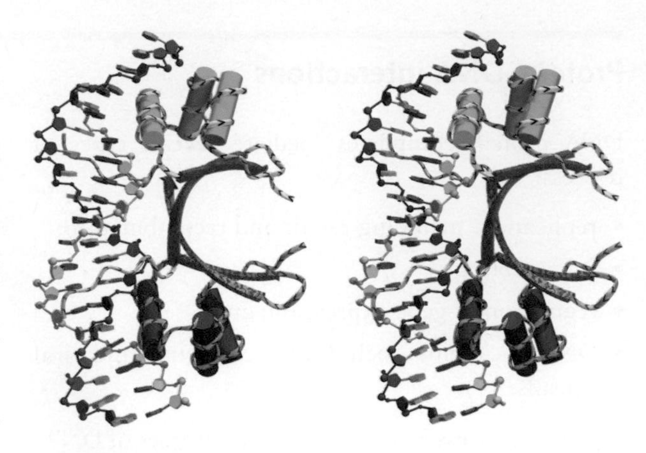

Figure 11.18 Bacteriophage λ Cro is an example of the 'helix-turn-helix' structural motif. Following along the chain, the first secondary structure is a helix, followed by two more helices that frame the motif. The second of these two helices (the third helix in the molecule) – called the **recognition helix** – lies in the major groove and makes extensive contacts with the DNA. A long C-terminal tail wraps around the DNA, following the minor groove [1CRO].

Alternatively, it can integrate its DNA into the host genome. The phage in such a **lysogenized** cell is dormant and can be released by stimuli that switch it from the lysogenic to the lytic state.

λ Cro binds to DNA as a symmetrical dimer (see Figure 11.18). Its target sequence is approximately palindromic, in the sense that two complementary strands contain approximately the same sequences in reverse order (see Exercise 11.5)

CTATCACC**GC**A**AGGGAT**AA
GAT**AGTGG**C**GT**TCCC**TATT**

The bases to which the protein makes contact are shown in bold-face. The protein interacts with both strands.

The eukaryotic homeodomain antennapedia

Homeodomains are highly conserved eukaryotic proteins, active in control of animal development. They regulate homeotic genes, i.e. genes that specify locations of body parts. Antennapedia is a *Drosophila* protein responsible for initiating leg development (see Figure 11.19). The earliest mutations found in antennapedia produced ectopic legs at the positions of antennae. Loss-of-function mutations produce antennae at the positions of legs.

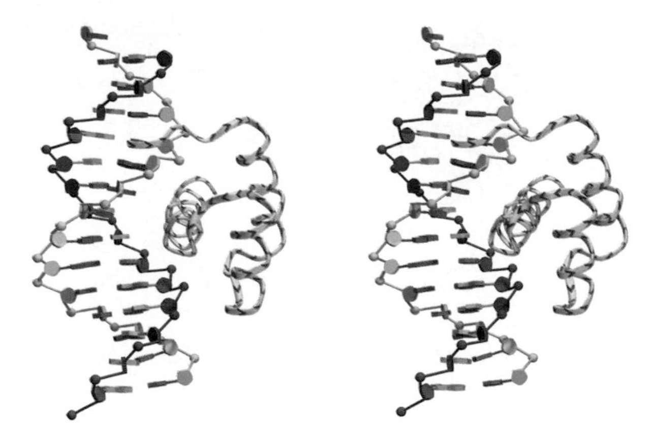

Figure 11.19 The homeodomain antennapedia–DNA complex [9ANT]. As with many DNA-binding proteins, an α-helix binds in the major groove of DNA. The structure of the antennapedia–DNA complex resembles, in some respects, prokaryotic helix-turn-helix proteins such as λ Cro. However, the tail that wraps around into the minor groove is N terminal to the helix-turn-helix motif in antennapedia, instead of C terminal as in λ Cro.

Leucine zippers as transcriptional regulators

Leucine zippers form another type of dimeric transcriptional regulator (see Figure 11.20). The Jun protein forms a homodimer consisting of an N-terminal domain with many positively charged sidechains that bind to DNA and a C-terminal leucine zipper domain involved in dimerization. Jun can dimerize not only with itself but with other related proteins, notably Fos. Different dimers have different DNA-sequence specificities and different affinities, affording subtle patterns of control.

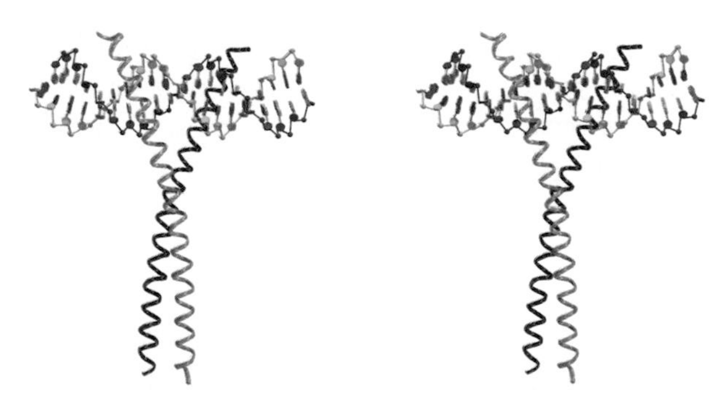

Figure 11.20 Jun dimer binding to DNA [1JNM]. The proteins grip the DNA as if they were picking it up with chopsticks. The α-helices bind in major grooves on opposite sides of the double helix. This structure shares with λ Cro, and many other DNA-binding proteins, the *symmetry* of the complex, which mimics the dyad symmetry of the DNA double helix. This requires, on the part of the protein, formation of symmetrical dimers, and on the part of the DNA, an approximately palindromic target sequence. For Jun dimers, the target sequence is ATGACGTCAT.

Zinc fingers

Zinc fingers are small modules found in eukaryotic transcription regulators. Each finger recognizes a triplet of bases in DNA. Tandem arrays of fingers recognize an extended region (see Figure 11.21). Understanding the relationship between the amino acid sequences of individual zinc fingers and the DNA sequences they bind would permit modular design of gene-specific repressors, by assembling a sequence of fingers.

The E. coli *Met repressor*

The Met repressor negatively regulates genes in the methionine biosynthesis pathway (see Figure 11.22).

The TATA box-binding protein

A TATA box is a sequence (consensus TATA[A/T]A[A/T]) upstream of the transcriptional start site of bacterial genes. Recognition of this sequence by the TATA box-binding protein (see Figure 11.23) initiates the formation of the basal transcription complex, a large multiprotein particle. This is an

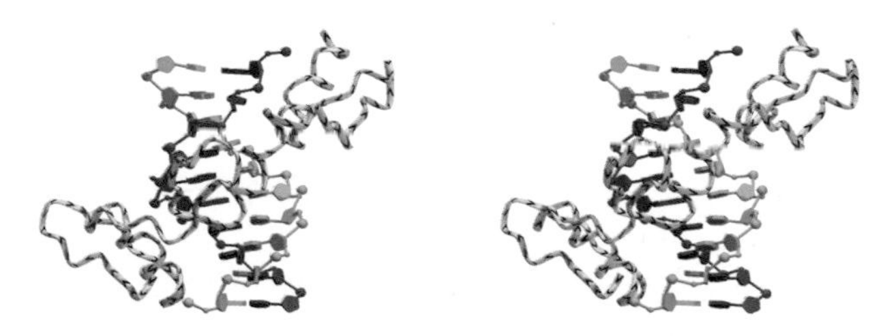

Figure 11.21 Zif268, a tandem three-finger structure binding the sequence GCGTGGGCG [1AAY]. Each finger interacts with three consecutive bases. In each finger, three positions along the α-helix, non-consecutive in the amino acid sequence, contain primary determinants of the DNA-sequence specificity.

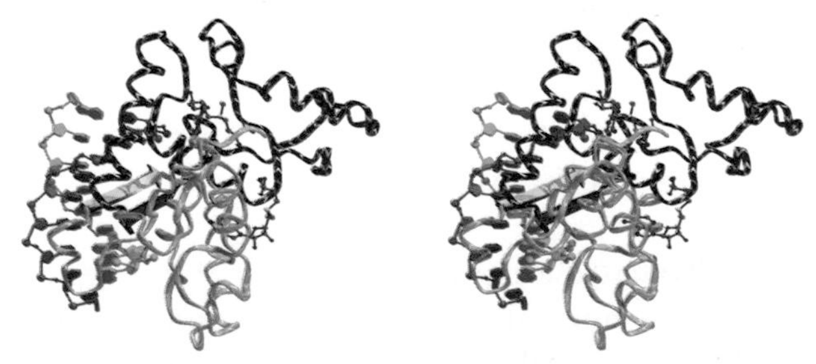

Figure 11.22 Like many other DNA-binding proteins, the Met repressor binds DNA as a symmetrical dimer [1CMA]. In the complex, each monomer contributes one strand of two-stranded β-sheet, which sits in the major groove, with sidechains making hydrogen bonds to bases. The co-repressor, *S*-adenosyl-methionine, is required for high-affinity binding.

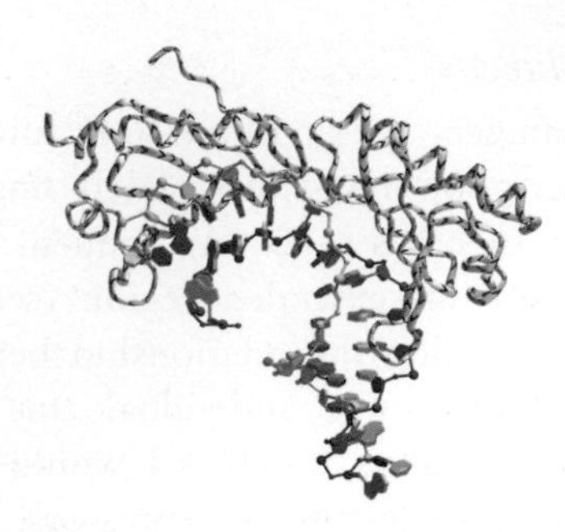

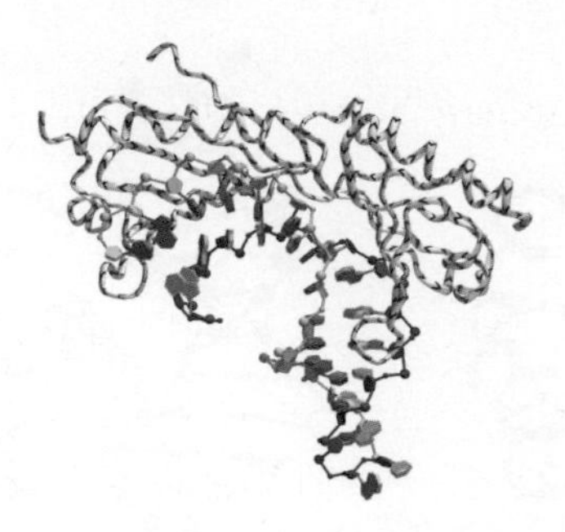

Figure 11.23 The TATA box-binding protein [1YTB]. The obvious feature of this complex is the very strong bending and unwinding induced in the DNA. A long curved β-sheet sits against an unusually flat surface on the DNA, the result of prying open of the minor groove. Phe sidechains intercalate between the bases.

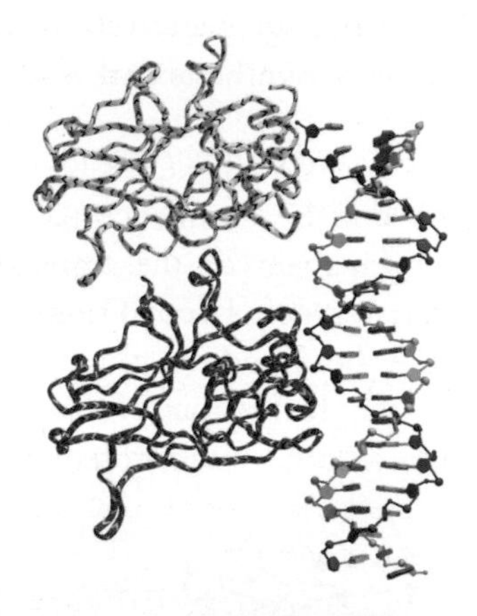

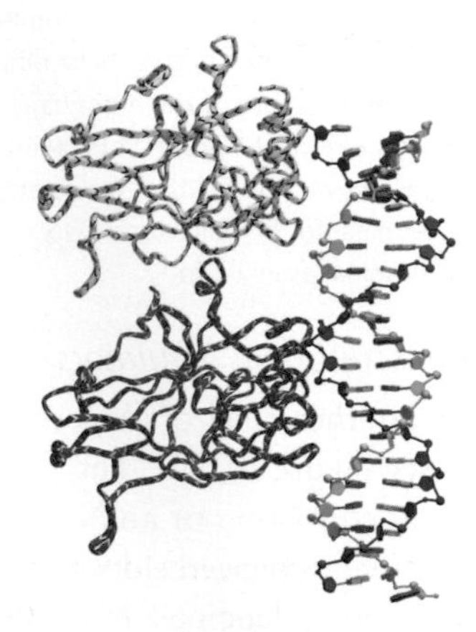

Figure 11.24 The structure of the DNA-binding subunit of p53 shows a double-β-sheet fold [1TSR]. A helix sits in the major groove and sidechains from loops connecting strands of the β-sheet insert into the minor groove.

example of initial binding of a protein to DNA followed by recruitment of other proteins to form an active complex.

p53

p53 is a transcriptional activator and a tumour suppressor (see Figure 11.24). It is of great clinical importance because mutations in the gene for p53 are very common in tumours.

p53 acts by surveilling genome integrity. Damage to DNA induces enhanced expression of p53, which stalls cell-cycle progression. This gives time for DNA repair; if repair is unsuccessful, the 'fail-safe' mechanism is apoptosis.

Gene regulation

Cells regulate the expression patterns of their genes. They sense internal cues to maintain metabolic stability, and external cues to respond to changes in the surroundings. The point of contact between genome and expression is the binding of RNA polymerase to promoter sequences, upstream of genes, to initiate transcription. This sensitive point is a juicy target for regulatory interactions (see Box 11.10).

The transcriptional regulatory network of *Escherichia coli*

Investigation of the mechanism of transcription regulation began with the work of F. Jacob and J. Monod on the lac operon in *E. coli*. The field has burgeoned, with comprehensive studies of the *coli* regulatory network, together with work on other organisms,

Vocabulary of gene regulation

Operator A control region associated with a gene.

Promoter A region upstream of a gene, the site of RNA polymerase binding to initiate transcription.

Repressor A DNA-binding protein that blocks transcription.

Operon A set of tandem genes in bacteria, usually catalysing consecutive steps in a metabolic pathway, under coordinated transcriptional control.

***cis*-Regulatory region** A segment of DNA that regulates expression of genes on the same DNA molecule. The *lac* repressor binding site is a *cis*-regulator of the adjacent protein-coding genes *lacZ*, *lacY*, and *lacA* (see p. 380).

Transcription start site The position in the gene that corresponds to the first residue in the mRNA.

Constitutive mutant A mutant defective in repression of a gene, which in consequence is expressed continously.

notably yeast. *E. coli* contains genes for 4398 proteins, 167 of which are recognized transcription factors. There are 2369 regulatory interactions, among the transcription factors and the genes they control (Figure 11.25).

- There are many fewer known regulatory interactions than genes. Many genes in *E. coli* are organized into operons, under coordinated control – one regulatory interaction controlling many genes. *E. coli* is estimated to contain ~2700 operons. Conclusion: more interactions remain to be discovered.

This network has been the subject of many investigations. There are questions about the *static* topology. Some of these address the local structure of the network, deriving the common types of small subgraphs, or the motifs. The fork, scatter pattern, and feed-forward loop are motifs in the regulatory networks of *E. coli* and other organisms (see Box 11.11). The network contains a large number of auto-regulatory connections. An auto-regulatory activator amplifies responses; an auto-regulatory repressor damps them out. In the *E. coli* regulatory network, links between different transcription factors are primarily activating, and auto-regulatory interactions are often repressive. A high density of repressive auto-regulatory interactions increases what might be thought of as the viscosity of the medium in which the network is active. The 'one-two punch', or feed-forward loop, motif, can also act to filter out random fluctuations, preventing the propagation of noise.

Combinations of the elementary motifs form modules within the network. These clusters of nodes are often dedicated to control of expression of genes with related physiological functions, such as a group of proteins responding to oxidative stress, or a group involved in aromatic amino acid biosynthesis. Such sets of genes need not be linked as a single operon.

Other analyses address the large-scale structure of the network. The distribution of degrees of the nodes; that is, the histogram of the number of edges meeting at a node, follows a power law:

$$\text{number of nodes with } k \text{ edges} \propto 10^{-\beta k}$$

with $\beta \approx 0.8$. The scale-free topology means that some nodes have many connections, and form the 'hubs' of the network.

Is there substantial feedback from downstream nodes? (A social analogy: is the network hierarchical or democratic? That is, does your boss listen to you, or just give orders? Ring Lardner's classic sentence, 'Shut up, he explained', emphasizes an absence, or dysfunction, of receptors for feedback signals from lower levels back to higher ones.) Although the *E. coli* transcriptional regulatory network contains many auto-regulatory interactions, it does not contain larger cycles; that is, paths in which gene A regulates gene B regulates gene C regulates . . . regulates gene A. Such cycles can lead to instabilities in the dynamics of a network.

Correlation of the topology of segments of the network with function suggests that short pathways, feed-forward loops, and repressive auto-regulatory interactions are involved in control of metabolic functions, such as a switch to alternative nutrients. This type of network topology is adapted to maintenance of homeostasis. Long hierarchical cascades, and activating auto-regulatory interactions, regulate developmental processes, such as biofilm formation, and flagellar development involved in mobility and chemotaxis.

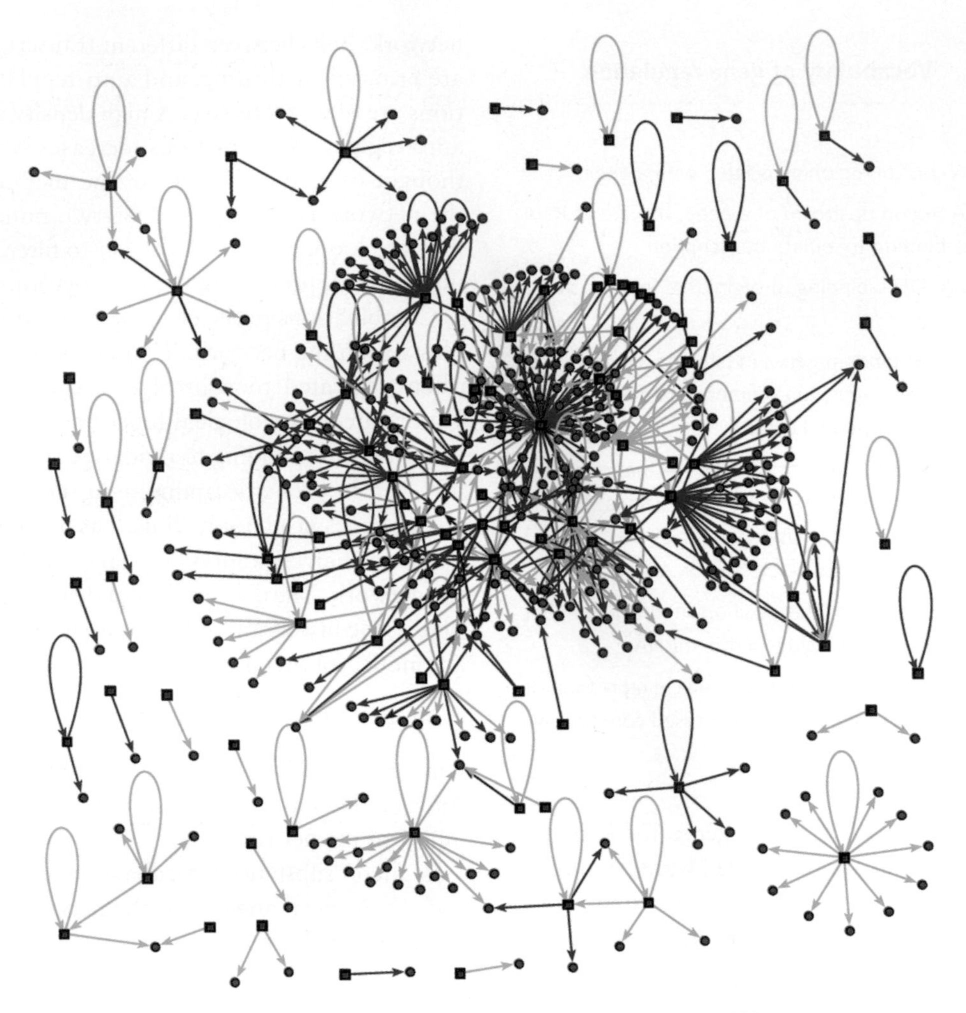

Figure 11.25 The *E. coli* transcriptional regulatory network represented as a directed graph. Colour-coding of nodes: transcription factors are shown as blue squares; regulated operons are shown as red circles. Colour-coding of links: activators, blue; repressors, green; indeterminate, brown.

From: Dobrin, R., Beg, Q.K., Barabàsi, A.L., & Oltvai, Z.N. (2004). Aggregation of topological motifs in the *Escherichia coli* transcriptional regulatory network. BMC Bioinformatics **5**, 10.)

BOX 11.11

Common motifs in biological control networks

Within the high complexity of typical regulatory networks, certain common patterns appear frequently. In the architecture of networks, these form building blocks which contribute to higher levels of organization. Shen-Orr, Milo, Mangan, & Alon* have described examples including: the *fork*, the *scatter*, and the *'one-two punch'* (a phrase from the boxing ring):

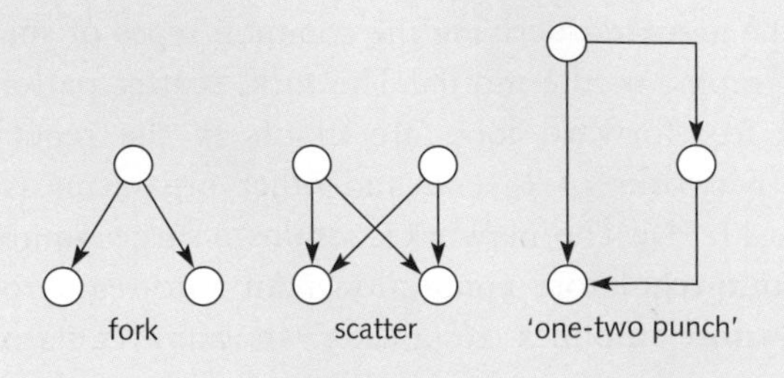

The *fork*, also called the single-input motif, transmits a single incoming signal to two outputs. Successive forks, or forks with higher branching degrees, are an effective way to activate large sets of genes from a single impulse. Generalizations of the binary fork include more downstream genes under common control (more tines to the fork), and auto-regulation of the control node. Forks can achieve general mobilization. Moreover, if the regulatory genes have different thresholds for activation, the dynamics of building up the signal can produce a temporal pattern of successive initiation of the expression of different genes.

The *scatter* configuration, also called the multiple input motif, can function as a logical 'or' operation: both downstream targets become active if *either* of the input impulses is active. Generalizations of the square scatter pattern shown may contain different numbers of nodes on both layers. Note that scatter patterns are superpositions of forks.

The *'one-two punch'*, also called the 'feed-forward loop', affects the output both directly through the vertical link; and indirectly and subsequently, through the intermediate link.

This motif can show interesting temporal behaviour if activation of the target requires simultaneous input from both direct and indirect paths (logical 'and'). Because build-up of the intermediate requires time, the direct signal will arrive before the indirect one. Therefore a short pulsed input to the complex will not activate the output – by the time the indirect signal builds up, the direct signal is no longer active. The system can thereby filter out transient stimuli in noisy inputs (Figure 11.26). Conversely, the active state of the system can shut down quickly upon withdrawal of the external trigger.

* Shen-Orr, S.S., Milo, R., Mangan, S., & Alon, U. (2002). Network motifs in the transcriptional regulation network of *Escherichia coli*. Nat. Genet. **31**, 64–68.

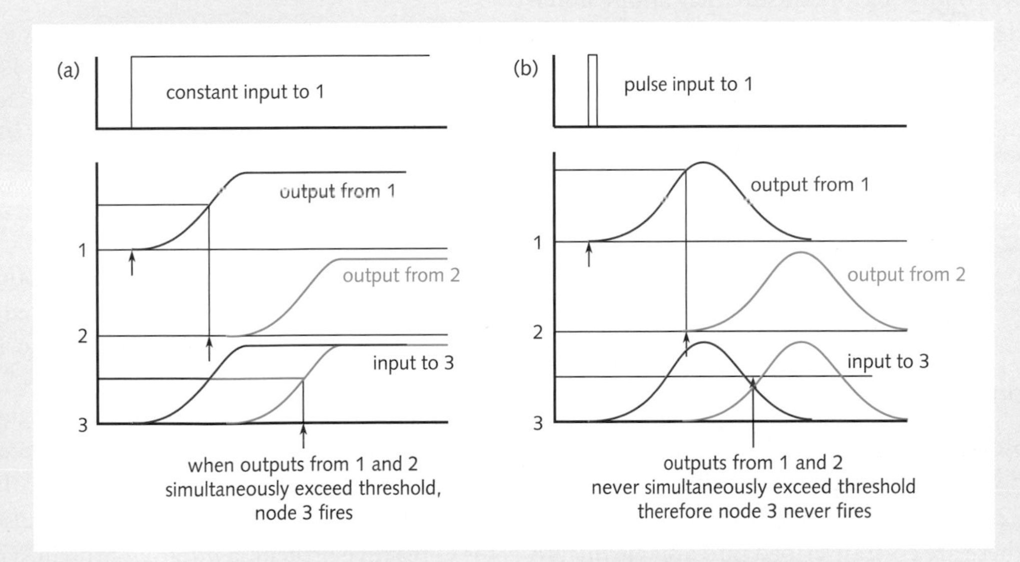

Figure 11.26 A 'one-two punch', or feed-forward loop, equipped with suitable AND logic at the downstream node, can filter out transient noise. (a) Constant input; (b) pulse input. The mechanism of signal transmission is the synthesis of a stimulatory molecule by an activated node. To avoid 'locking the signal on', this molecule must subsequently be removed. The effect described here depends on the time course of build-up and decay of the signal.

The dynamic properties of the network are also of interest. These include both the response of the network to changing conditions, as in the lac operon, and the comparison of regulatory networks in related organisms to understand how networks evolve.

Even a constant network can produce different outputs from different inputs. However, even within an organism, networks can change their structure in response to changes in conditions. This can affect even some of the hubs of the network, the points at which changes have the most far-reaching effects. This has been examined most closely in yeast (see p. 383).

Similarities and differences in the regulatory interactions in related organisms illuminate how their networks evolve. The evolutionary retention of transcription factors is smaller than that of target genes. Even transcription factors that serve as hubs are not more highly conserved. Different organisms are relatively free to explore different regulatory pathways, even to regulate orthologous genes. This may well be the 'other side of the coin' of the redundancy in the networks that provides robustness.

There is evidence that larger changes in regulatory networks reflect changes in lifestyle. Organisms with similar lifestyle – several species of soil bacteria, genuses *Bacillus*, *Corynebacterium*, and *Mycobacterium* – conserve regulatory interactions, as do intracellular parasites *Mycoplasma*, *Rickettsiae*, and *Chlamydiae*. Conversely, comparing organisms that adopt different lifestyles, individual transcription factors can be deleted.

- Regulatory networks are directed graphs. Some simple motifs, or common small subgraphs, form the lowest level of network structure. Networks can reprogram themselves, within an organism; and evolve, between species.

Regulation of the lactose operon in *E. coli*

The lactose operon of *E. coli* contains structural genes and regulatory regions (see Table 11.2 and Figure 11.27a). Three regions of the operon encode proteins; these are co-transcribed into a single mRNA and translated separately. The *lacI* gene (encoding lac repressor) lies upstream of the operator and is constitutively expressed.

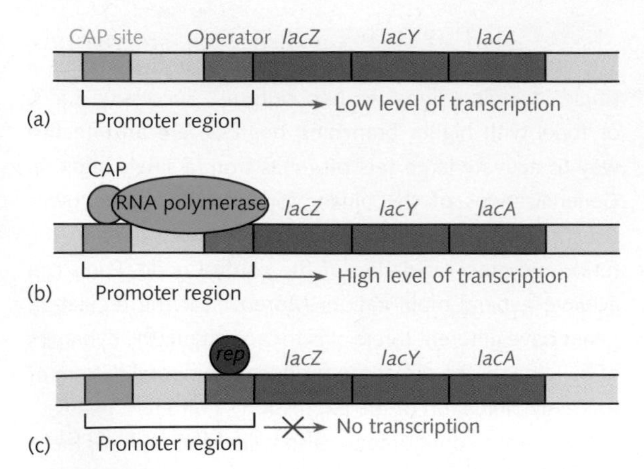

Figure 11.27 States of the lactose operon. (a) The promoter region contains regulatory sites upstream of the protein-encoding genes *lacZ*, *lacY*, and *lacA*. (b) Binding of CAP to its upstream site within the promoter enhances the binding affinity of RNA polymerase, turning transcription on. (c) Binding of repressor blocks binding of RNA polymerase, turning transcription off. Two additional subsidiary repressor binding sites are not shown.

- Figure 11.27(c) is a simplification: lac repressor actually binds to three sites. The cover of the 1 March 1996 issue of *Science* magazine shows a model of the lac repressor, binding to two sites on a ~100 bp section of DNA, plus the CAP protein. The DNA is bent into a loop.

The control regions of the lactose operon function as a switch, turning on and off transcription of the protein-encoding genes. Control is exerted through binding of CAP to the CAP site and lac repressor protein to the operator site. (The CAP–DNA complex is very similar to the structure shown in Figure 11.18. CAP stands for catabolite activator protein; CRP stands for cyclic-AMP receptor protein. These are the *same* protein. Unfortunately CRP *also* stands for C-reactive protein, a *different* protein synthesized in the liver and a useful marker in humans of inflammation.) Cyclic AMP (cAMP), produced in higher quantities if glucose is absent, and lactose

Table 11.2 Functions of the protein-encoding genes of the lactose operon

Gene	Enzyme	Function
lacZ	*β*-Galactosidase	Hydrolyses lactose → glucose + galactose; isomerizes lactose → allolactose
lacY	*β*-Galactoside permease	Pumps lactose into the cell
lacA	*β*-Galactoside transacetylase	Unknown, possibly detoxification?

analogues bind to CAP and lac repressor proteins, respectively, to control their binding affinities.

- Binding of lactose induces a conformational change in the repressor, from a tightly binding form to a weakly binding one. *Lac repressor will bind only in the absence of lactose.*
- The presence of glucose reduces the concentration of cyclic AMP, causing a conformational change in CAP to a weakly binding form. *CAP will bind only in the absence of glucose.*

The actual molecule that binds to repressor to reduce its affinity for its site on DNA is allolactose. Allolactose is an isomer of lactose, produced from lactose by β-galactosidase. Alternative lactose analogues also stimulate transcription. One that is useful in the laboratory is isopropylthiogalactoside (IPTG), for two reasons: (1) IPTG enters the cell even if the *lacY*-encoded transporter is dysfunctional or not expressed; and (2) IPTG is not metabolized; therefore its concentration stays constant during the course of an experiment.

The switch thereby responds to the type of sugar in medium.

- If both glucose and lactose are present, neither control protein binds (Figure 11.27a). RNA polymerase binds only weakly. Transcription occurs at a low basal level.
- If glucose is not present and lactose *is* present, the CAP–cAMP complex binds to the promoter. The binding of CAP–cAMP with RNA polymerase is cooperative. Interactions with CAP–cAMP increase the affinity of RNA polymerase, thereby stimulating transcription to approximately 40 times the basal level (see Figure 11.27b).
- If lactose is not present, the repressor binds to the operator site. This blocks RNA polymerase and turns off transcription (see Figure 11.27c).

In summary (– means 'absence of'; + means 'presence of'):

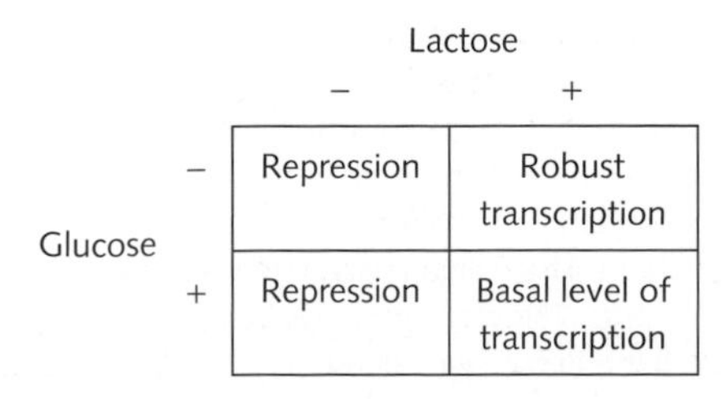

		Lactose –	Lactose +
Glucose	–	Repression	Robust transcription
Glucose	+	Repression	Basal level of transcription

The operator site is between the CAP site and the origin of transcription. As a result, *lactose absence 'trumps' glucose absence*. That is, in the absence of lactose, the binding of repressor stops transcription whether or not glucose is absent.

The effect is to express the proteins of the *lac* operon only if the medium contains only lactose. The bacteria are saying, in effect, 'I prefer to grow on glucose. If glucose is there, I don't want high expression levels of the genes for lactose transport and metabolism, *even if* lactose is present. Only if lactose is present and glucose is not present, express the genes that transport and cleave lactose.'

The lactose operon switch is an example of a 'fire-and-forget' mechanism. Once the mRNA is synthesized, what happens on the DNA does not affect it. However, the mRNA for the protein-coding genes of the *lac* operon (*lacZ*, *lacY*, and *lacA*) has a half-life of ~3 minutes. In the absence of continuous or repeated induction, synthesis of the lactose-metabolizing enzymes will cease within minutes. This resets the switch.

Logical diagram of the lac *operon*

Figure 11.28 represents the *lac* operon control logic as a network. This diagram is almost equivalent to the table showing the response to presence and absence of glucose (see Exercise 11.7).

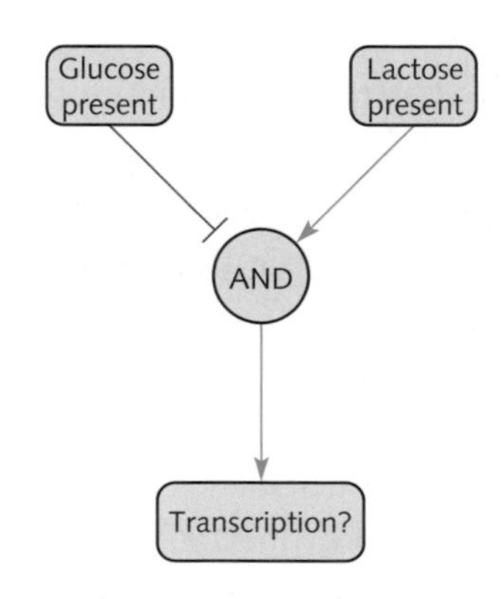

Figure 11.28 Logical diagram of *lac* operon. Green arrows show positive regulation. The red 'T' shows negative regulation. The circle containing AND will pass through a positive signal only if glucose is *not* present (red 'T') *and* lactose *is* present (green arrow). For many regulatory circuits, we know many of the inputs to a node but do not know the logic.

The genetic regulatory network of *Saccharomyces cerevisiae*

A recent study of transcription regulation in yeast treated a network containing 3459 genes, corresponding to approximately half of the known proteome of *S. cerevisiae*. The genes included 142 that encode transcription regulators and 3317 that encode target genes exclusive of transcription regulators. There are 7074 known regulatory interactions among these genes, including effects of regulators on one another and of regulators on non-regulatory targets.

Analysis of the overall network architecture revealed several features.

- The distribution of incoming connections to target genes has a mean value of 2.1 and is distributed exponentially. Most target genes receive direct input from about two transcriptional regulators. The probability that a gene is controlled by k transcription regulators, $k = 1, 2, \ldots,$ is proportional to $e^{-\alpha k}$, with $\alpha = 0.8$.
- The distribution of outgoing connections has a mean value of 49.8 and obeys a power law. The probability that a given transcriptional regulator controls k genes is proportional to $k^{-\beta}$, with $\beta = 0.6$. Power-law behaviour is common in networks and characterizes topologies in which a few nodes – the 'hubs' – have many connections and many nodes have few. In regulatory networks, hubs tend to be fairly far upstream, forming important foci of regulation with far-reaching control.
- The average number of intermediate nodes in a minimal path between a transcriptional regulator and a target gene is 4.7. The maximal number of intermediate nodes in a path between two nodes is 12.
- The clustering coefficient of a node is a measure of the degree of local connectivity within a network. If all neighbours of a node are connected to one another, the clustering coefficient of the node = 1. If no pair of neighbours of a node are connected to each other, the clustering coefficient of the node = 0. The mean clustering coefficient, averaged over all nodes, is a measure of the local density of the network. For the yeast transcriptional regulatory network, the mean clustering coefficient is 0.11.

Figure 11.29 is a cartoon-like sketch of a fragment of such a network indicating, rather loosely, some of its general features. Nodes are divided into **transcriptional regulators**, shown as circles, and **target genes**, shown as squares. Target genes are distinguished by having no output connections. There is extensive interregulation among the transcription factors, to a much higher density of interconnections than can intelligibly be shown in this diagram. Think of a seething broth of transcription factors, within the shaded area, sending out signals to target genes. The shaded area indicates only the *logical* clustering of the transcriptional regulators. There is no suggestion about physical localization; indeed, transcriptional regulators interact with DNA and almost never interact physically with the proteins whose expression they control.

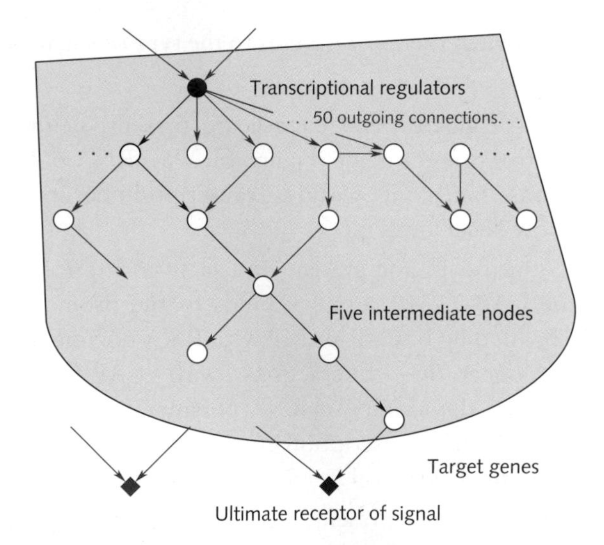

Figure 11.29 Simplified sketch illustrating some features of an 'average' segment of the pathways in the yeast interaction network. Transcriptional regulators appear as circles. Target genes appear as squares. A transcriptional regulator typically has direct influence over about 50 genes, indicated by multiple connections from the filled black circle to the circles on the line below it. Roughly one in ten of the neighbours of any node is connected to another neighbour, indicated by the horizontal arrow on the second row. The ultimate receptor of the signal lies at the end of a pathway typically containing about five intermediate nodes (shown in black). This ultimate target gene receives on average about two inputs. This diagram shows only a small fragment of a network that is in fact quite dense.

Each transcriptional regulator directly influences approximately 50 genes on average, although, as with other 'small-world' networks following power-law distributions of connectivities, the distribution is very skewed – some 'hubs' have very many output connections, but most nodes have very few. A few of the interregulatory connections between transcription factors are shown in red. In about 10% of cases, two neighbours of the same transcription factor interact with each other. A path from one regulator (filled black circle) to one ultimate receptor (filled black square), through five intermediate nodes, is shown in black. The intermediate nodes are other transcriptional regulators, connected both within the path drawn in black and off this path. Even the transcription factor used as the origin of the path receives input connections. Although it is possible to identify target genes from the absence of outgoing connections, it is more difficult to identify ultimate initiators of signal cascades.

The ultimate receptor is a target gene that receives regulatory input but itself has no output links. This target is expected to receive (on average) a second control input. The black target node receives input via a black arrow, along the selected path, and via a red arrow suggesting the second input. Of course the second input may arrive via a path that shares common nodes with the black path, including other routes from the filled black circle.

The dense forest of additional pathways from which this fragment is extracted is not shown. Some 'back-of-the-envelope' calculations indicate: (1) there are ~3500 nodes, each receiving an average of two input connections; (2) there are ~140 transcription factors, making an average of 50 output connections; (3) the number of input connections must equal the number of output connections, and indeed $3500 \times 2 = 140 \times 50 = 7000$.

Given the complexity, it is difficult to illustrate larger segments of the network in more detail than the simplified version appearing in Figure 11.29. Analysis of the structures of regulatory networks is an active current research topic. The motifs described in Box 11.11 are the 'secondary structures' of network architectures.

The high ratio of interactions to transcription regulators implies that we cannot expect to associate individual regulatory molecules with single, dedicated activities (as we can, for the most part, with metabolic enzymes). Instead, the activity of the network involves the coordinated activities of many individual regulatory molecules.

Adaptability of the yeast regulatory network

The yeast regulatory network achieves versatility and responsiveness by reconfiguring its activities. This is seen by comparing the changes in the activities of networks controlling yeast gene expression patterns in different physiological regimens of the organism: cell cycle, sporulation, diauxic shift (the change from anaerobic fermentative metabolism to aerobic respiration as O_2 levels increase), DNA damage, and stress response. Cell cycling and sporulation involve the unfolding of endogenous gene expression programmes; the others are responses to environmental changes.

Different states are characterized both by similarities and differences in gene expression patterns and by the components of the regulatory network that are active. There is considerable shift in expression of target genes. About a quarter of the target genes are specialized to individual physiological states. Of the total of about 3000 target genes, the expression levels of only about half do not show major changes in the different states. Of the 1906 that show altered expression levels in different states, almost half (803) are specialized to a single physiological state.

In contrast, different states show much more overlap in the usage of transcriptional regulators. For instance, for cell-cycle control, 280 target genes (8%) are differentially regulated by 70 (49%) of the transcription regulators. Clearly, there is a much greater degree of specialization in the target genes. In general, half of the transcription factors are active in at least three of the five physiological regimens. However, contrasting with the high overlap of usage of the transcriptional regulators (the nodes), the overlap of the activities within the network (the connections) is relatively low. Different components of the interaction network organize the different gene expression patterns in different states.

Whereas different physiological states are characterized by substitutions of different sets of synthesized proteins, the regulatory network uses much of the same structure but reconfigures the pattern of activity. Think of the transcription factors as 'hardware' and the connections as reprogrammable 'software'. The molecules do not change but the interactions do: in different states, many transcription regulators change most, or a substantial part, of their interactions. In particular, the set of transcription regulators that forms the hubs of the network – those with many outgoing nodes that form foci of control – are not a constant feature of the system. Some hubs are common to all states, but others step forward to take control in different physiological regimens. The result of the reconfiguration of activity is that over half of the regulatory interactions are *unique* to the different states.

The effect of the changes in the active interaction patterns is to alter the topological characteristics of the network in different states. For instance, under panic conditions – DNA damage and stress – the average number of genes under the control of individual transcriptional regulators increases; the average minimal path length between regulator and target decreases; and the clustering becomes less dense (i.e. there is less interregulation among transcription factors). This can be understood in terms of a need for fast and general mobilization – the equivalent of broadcasting 'Go! Go! Go!' over the radio. Normal circumstances – cell-cycle control, for instance – allow for a more dignified and precise regulatory state, which permits finer control over the temporal course of expression patterns. In cell-cycle control and sporulation, there is a much denser interregulation among transcription factors and longer minimal path lengths between transcriptional regulators and target genes.

Different physiological states also differ in their usage of the common motifs – fork, scatter, and 'one-two punch' (see Box 11.11). Forks are used more in conditions of stress, diauxic shift, and DNA damage. They are appropriate to the need for quick action. Requirements for a build-up of intermediates would delay the response. Conversely, the 'one-two punch' motif is more common in cell-cycle control. This is consistent with the need for a signal from one stage to be stabilized before the cell enters the next stage.

Much of evolution proceeds towards greater specialization. The human eye is a classic example. It is an intricate and fine-tuned structure, features that were once adduced as evidence *against* Darwin's theory. Many evolutionary pathways show a trade-off between specialized adaptation and generalized adaptability.

Regulatory networks are an exception. Evolution has produced structures that are both specialized *and* versatile. The reconfigurability of regulatory networks allows them to respond robustly to changes in conditions by creating many different structures specialized to the conditions that elicit them.

RECOMMENDED READING

- Several authors offer general overviews of the systems view in biology:

Cramer, F. & Loewus, D.I. (Translator) (1993). *Chaos and Order: The Complex Structure of Living Systems.* J. Wiley & Sons, New York.

Adami, C. (2002). What is complexity? BioEssays **24**, 1085–1094.

Wagner, A. (2005). *Robustness and Evolvability in Living Systems.* Princeton University Press, Princeton.

Alon, U. (2006). *An Introduction to Systems Biology*. Chapman & Hall, London.

Brenner, S, (2010). Sequences and consequences? Phil. Trans. Roy. Soc. Lond. B **365**, 207–212.

- Barabási and co-workers have studied general principles of networks. Biological networks have many properties in common with other networks:

Albert, R. & Barabási, A.-L. (2002). Statistical mechanics of complex networks. Rev. Mod. Phys. **74**, 47–97.

Barabási, A.-L. (2003). *Linked: How Everything Is Connected to Everything Else and What It Means*. Plume Books, New York.

- The following papers describe the structure, dynamics, and evolution of cellular signalling and regulatory networks:

  Ideker, T. (2004). A systems approach to discovering signaling and regulatory pathways – or, how to digest large interaction networks into relevant pieces. Adv. Exp. Med. Biol. **547**, 21–30.

  Babu, M.M., Luscombe, N.M., Aravind, L., Gerstein, M., & Teichmann, S.A. (2004). Structure and evolution of transcriptional regulatory networks. Curr. Opin. Struct. Biol. **14**, 283–291.

  Luscombe, N.M., Babu, M.M., Yu, H., Snyder, M., Teichmann, S.A., & Gerstein, M.B. (2004). Genomic analysis of regulatory network dynamics reveals large topological changes. Nature **431**, 308–312.

EXERCISES, PROBLEMS, AND WEBLEMS

Exercises

Exercise 11.1 In the undirected, unlabelled graph in Box 11.3 on page 344, (a) name two vertices such that if you add an edge between them at least one vertex has exactly four neighbours. (Note that two edges may cross without making a new vertex at their point of intersection.) (b) Name two vertices such that if you add an edge between them to the original graph, the graph becomes an (unrooted) tree. (c) Name two vertices (neither of them V_1) such that if you add an edge between them to the graph produced in (b), the resulting graph does not remain a tree. (d) Name two vertices such that if you add an edge between them to the original graph, there is exactly one path between V_1 and V_3, with no vertices repeated, and it has length 4. (e) Name two vertices such that if you add an edge between them to the original graph, there are alternative paths, of lengths 3 and 4, between V_1 and V_5, with no vertices repeated. (In determining the length of a path, you have to count the number of edges in the path. A path of length 2 between V_1 and V_5 contains one intermediate vertex.)

Exercise 11.2 Of the examples of graphs in Box 11.4, (a) which are directed graphs? (b) Which are labelled graphs? (c) In each example, what is the set of nodes? (d) In each example, what is the set of edges?

Exercise 11.3 What information is contained in Figure 11.3(b) that could not be recovered from the kind of data produced by the experiments shown in Figure 11.13?

Exercise 11.4 For which of the methods for determining interacting proteins (pp. 366ff) (a) must one of the proteins be purified; (b) must both of the proteins be purified?

Exercise 11.5 The binding site for λ Cro (p. 374) is an approximate palindrome, i.e. the two strands contain approximately the same sequence in reverse order. (a) On a copy of the binding site, indicate which six residues best fit the palindrome pattern. (b) A palindromic binding site can interact with a dimeric protein by presenting surfaces of similar structure to both protein subunits. How far apart are the six-residue regions identified in part (a)? How do you rationalize that distance in terms of features of the structure of DNA?

Exercise 11.6 From Figure 11.3, (a) what would be the effect of increased expression of *metJ* on the expression of *metC*? (b) What would be the effect of increased expression of *OxyP* on the expression of *metJ*? (c) What would be the effect of increased expression of *OxyP* on the expression of *metP*?

Exercise 11.7 Redraw Figure 11.28 with the top boxes containing glucose absent and lactose absent instead of glucose present and lactose present.

Exercise 11.8 Match mutant to phenotype:

Dysfunctional gene	Resulting phenotypes
1. *lacI*	(a) Operon expressed, no lactose uptake
2. Mutation in repressor binding site	(b) Operon expressed, no glucose or galactose produced
3. Mutation in RNA polymerase binding site	(c) No expression
4. *lacZ*	(d) Constitutive expression of operon because no repression is possible
5. *lacy*	

Exercise 11.9 In the London Underground: (a) What is the shortest path between Moorgate and Embankment stations? Note that, considered as a graph, the shortest path between two nodes is the path with the fewest intervening nodes, not the path that would take the minimal time or fewest interchanges. (b) What is the shortest cycle containing Kings Cross, Holborn, and Oxford Circus stations? (c) The clustering coefficient of a node in a graph is defined as follows. Suppose the node has k neighbours. Then the total possible connections between the neighbours is $k(k-1)/2$. The clustering coefficient is the observed number of neighbours divided by this maximum potential number of neighbours. If the neighbours of a station are the other stations that can be reached without passing through any intervening stations, what is the clustering coefficient of the Oxford Circus station? (If necessary, see http://www.bbc.co.uk/london/travel/downloads/tube_map.html)

Exercise 11.10 In the London Underground, (a) what is the maximum path length between any two stations? (That is, for which two stations does the shortest trip between them involve the maximum number of intervening stops?) (b) If the District Line were not active, what stations, if any, would be inaccessible by underground? (c) If the Jubilee Line were not active, what stations, if any, would be inaccessible by underground?

Exercise 11.11 On a photocopy of the three common network control motifs (Box 11.11, p. 378), (a) indicate which nodes are controlled by only *one* upstream node; (b) indicate which node exerts control over only *one* downstream node.

Exercise 11.12 On a photocopy of the simplified fragment of the yeast regulatory network (Figure 11.29) indicate examples of the network control motifs (a) star and (b) 'one-two punch'. (c) Add one arrow to create a scatter motif.

Exercise 11.13 In the overall yeast transcriptional regulatory network, the number of incoming connections to target genes follows an exponential distribution, i.e. the probability that a gene is controlled by k transcriptional regulators is proportional to $e^{-\alpha k}$, with $\alpha = 0.8$, $k = 1, 2, \ldots$. What is the ratio of the number of target genes receiving four input connections to the number receiving two input connections?

Problems

Problem 11.1 In one species, only enzyme A catalyses conversion $S \rightarrow P$, the rate-limiting step of a reaction pathway. What is the flux control coefficient of enzyme A? (b) In a related species, distinct but similar enzymes A and B both catalyse the $S \rightarrow P$ reaction. The kinetic characteristics of A and B are identical: $S \rightarrow P$ is still the rate-limiting step of the pathway. What is the flux control coefficient of A?

Problem 11.2 For dissociation of a complex involving a simple equilibrium:

$AB \rightleftharpoons A + B$, the equilibrium constant, $K_D = \dfrac{[A][B]}{[AB]}$, is equal to the ratio of

forward and reverse rate constants: $K_D = k_{off}/k_{on}$.

For avidin–biotin, $K_D = 10^{-15}$. Suppose k_{on} were as fast as the diffusion limit, ~ 10^{-9} M^{-1} s^{-1}. (a) What is the value of k_{off}? (b) What would be the half-life of the avidin–biotin complex? (c) Suppose k_{on} for avidin–biotin were 10^{-7} M^{-1} s^{-1}. What would be the half-life of the complex?

Problem 11.3 Write detailed structures for all of the metabolites that appear in Figure 11.3a.

Problem 11.4 The network of metabolic pathways must obey constraints of thermodynamics and physical-organic chemistry. Meléndez-Hevia and colleagues suggested the principle that metabolic pathways are optimized, subject to the constraints, for the minimum number of steps. The non-oxidative phase of the pentose phosphate pathway converts six five-carbon sugars to five six-carbon sugars:

6 ribulose-5-phosphate → 5 glucose-6-phosphate

A simplified model of a pathway for this conversion is a series of steps, each of which is either:

- transfer of a two-carbon unit from one sugar to another (a transketolase reaction); or
- transfer of a three-carbon unit from one sugar to another (a transaldolase or aldolase reaction).

Represent each sugar only by a number of carbon atoms. Starting with five five-carbon sugars, one possible initial step would be a transketolase step converting two five-carbon sugars to a three-carbon sugar and a seven-carbon sugar. Assume that all intermediates must have *at least* three carbon atoms.

Create a tableau with the following initial and final states (an initial transketolase (TK) step is also shown):

Step	Number of carbons in sugar molecules					
0	5 ↘	↙ 5	5	5	5	5
	TK					
1	3 ↙	↘ 7	5	5	5	5
		. . .				
N	6	6	6	6	6	0

Copy and fill in the tableau to find the shortest route from the top (step 0, six five-carbon sugars) to the bottom (five six-carbon sugars). Identify the intermediates created. Compare this with the observed metabolic pathway.

Problem 11.5 Choose 15 amino acids, by crossing off the ones most similar to others. Devise a doublet genetic code for these 15 residues, plus a stop signal, that is as close as possible to the actual triplet code.

Problem 11.6 (a) You create a strain of *E. coli* in which the order of promoter and operator in the *lac* operon are reversed. Will this strain express *lacZ*, *lacY*, and *lacA* in the absence of lactose and glucose? (b) You create a strain of *E. coli* by moving the operator from its normal position to a position between the *lacZ* and *lacY* genes. Will this strain express *lacZ*, *lacY*, and *lacA* in the absence of lactose and glucose? Will this strain express *lacZ*, *lacY*, and *lacA* in the presence of lactose and absence of glucose? (c) You add exogenous cyclic AMP to wild-type *E. coli*. Will *lacZ*, *lacY*, and *lacA* be expressed in the presence of glucose and lactose? (d) You create a strain of *E. coli* with a point mutation in *lacZ* that renders the enzyme completely dysfunctional. Will this strain express *lacY* in the presence of lactose? Will this strain express *lacY* in the presence of isopropylthiogalactoside (IPTG)?

Problem 11.7 Indicate how to connect a selection of the three common network control motifs so that a single input node can influence three output nodes.

Problem 11.8 What is the minimum number of 'yes-or-no' questions required to identify a specific letter of the upper-case alphabet: ABC . . . Z?

Weblems

Weblem 11.1 In the methionine biosynthetic pathway (see Figure 11.3), the product, methionine, inhibits an enzyme in the middle of the pathway, homoserine *O*-succinyltransferase. (a) The accumulation of which intermediate might this inhibition be expected to cause? (b) In what other pathways is this intermediate involved that might use it up?

Weblem 11.2 Figure 11.4 shows the amino acid biosynthesis leading from aspartate to methionine, threonine, and lysine. On a photocopy of this figure, write in the names of the omitted intermediates at the unlabelled positions between consecutive arrows.

Weblem 11.3 The genes encoded by *metC* and *malY* in *E. coli* convert cystathione to L-homocysteine. Each has another function in addition. What are these other functions?

Weblem 11.4 Compare the pathways for biosynthesis of chorismate in *E. coli*, *M. jannaschii*, and *Aeropyrum pernix*. What is the earliest common intermediate in these pathways? What are its precursors in the three species?

Weblem 11.5 Define the following terms: (a) interactome; (b) signalome. (c) This is more difficult: can you think of, and define, a reasonable 'ome' that has not yet been proposed?

Weblem 11.6 Compare the methionine biosynthesis pathway from asparate to methionine in *E. coli* and yeast. (a) Are there any differences in the series of intermediates? (b) Are the enzymes that catalyse similar transformations homologous? Show alignments of the amino acid sequences where possible. Use EcoCyc for *E. coli* (http://ecocyc.org) and the *Saccharomyces* Genome Database (http://www.yeastgenome.org) for yeast, searching in each case for 'methionine biosynthesis'.

Weblem 11.7 Draw Figure 11.27(a) to scale, using the known *E. coli* genome sequence as a source of the true sizes of the regions.

Weblem 11.8 Find the page in EcoCyc corresponding to Figure 11.3. Choose different levels of detail and list what types of information are presented at each level.

Weblem 11.9 An enzyme with a function related to peptidylglycine monooxygenase (EC 1.14.17.3), and linked to it in the ENZYME DB, is 1-aminocyclopropane-1-carboxylate oxidase. (EC 1.14.17.4). (a) What is the lowest common ancestor of the two reactions in the EC classification? (b) What is the lowest common ancestor of the two reactions in the Gene Ontology molecular function classification? (c) Are these two enzymes closely related in the Gene Ontology classification?

Weblem 11.10 What identifiers does Gene Ontology associate with *E. coli* asparate aminotransferase, in the molecular function category? Arrange them in a directed acyclic graph, indicating the parent–child relationships between these identifiers.

Weblem 11.11 According to EcoCyc, what reactions can orotidine 5′-monophosphate undergo? What enzymes catalyse these reactions? What genes encode these enzymes?

Weblem 11.12 Figure 11.5 shows the reductive carboxylate cycle, and the EC numbers of the enzymes that catalyse the individual steps. Find the corresponding information for the tricarboxylic acid cycle (or Krebs cycle), and for the glyoxylate cycle. Do an alignment of the metabolites participating in these cycles, display the EC numbers of the enzymes that correspond to different reactions in different cycles. Report what is common to pairs or to all three of these pathways, in terms of (a) metabolites, (b) links between metabolites, corresponding to reactions, (c) enzymes that catalyse the reactions.

EPILOGUE

The new century has already seen major achievements in genomics. Sequencing of the human genome is the jewel in the crown. And yet, the field is still in a preparative and anticipatory stage. We can be confident that:

- Methods of sequence determination will increase in power. There will be an explosion in the number of complete sequences of different human beings and of many other organisms.
- Tools for analysis will make progress. Better algorithms, making effective use of more data, will produce more reliable inferences. Modelling of structure and process will improve, towards the target goal of the simulation of the complete cell *in silicio*.
- We will achieve a more profound understanding of what life is and how it works.

We will also gain greater control over living systems. Many applications will emerge, in clinical, agricultural, and technological fields. Some of these are relatively simple extrapolations from what has already been achieved. Others seem more like the stuff of science fiction: a tightly coupled silicon–life interface and the *in vivo* deployment of nanoparticles sensing and interacting with our biochemical states. Nevertheless, they represent natural developments of the current state of the art.

Understanding the genome may ultimately release us from its constraints.

INDEX

C

D

R

Introduction to Genomics